THE MATHEMATICAL BASIS
OF THE ARTS

RESTFULNESS AND RESTLESSNESS IN A LANDSCAPE
As effected by a sine-curve and a complex curve.

THE MATHEMATICAL BASIS OF THE ARTS

BY

JOSEPH SCHILLINGER

Part One
SCIENCE AND ESTHETICS

Part Two
THEORY OF REGULARITY AND COORDINATION

Part Three
TECHNOLOGY OF ART PRODUCTION

PHILOSOPHICAL LIBRARY

NEW YORK

Copyright, 1948, by
FRANCES SCHILLINGER

All rights reserved
International Copyright Secured

PRINTED IN THE UNITED STATES OF AMERICA

JOSEPH SCHILLINGER possessed one of the brilliant analytical minds of our time. A composer and a world-famous teacher of musical composition, he was active also in the physical sciences and mathematics, and in the field of the visual arts. Born in Kharkov, Russia, in 1895, he died in New York in 1943, having become an American citizen in 1936. He held a series of important posts in his native country, including that of Dean of the faculty of composition at Kharkov Academy, and Professor at the State Institute of the History of Arts in Leningrad. From 1930 to 1936 he taught and lectured at various American schools, including The New School, New York University, and Teachers College, Columbia University. In addition to THE MATHEMATICAL BASIS OF THE ARTS, his works include *The Schillinger System of Musical Composition, Kaleidophone*, numerous essays and articles, and pure and industrial designs.

ACKNOWLEDGMENT

The Mathematical Basis of the Arts is Schillinger's masterwork. It represents 25 years of his discoveries and research. This book was completed and prepared for publication by Schillinger a year before his death in March 1943 at the age of 47.

I wish to express my appreciation to Arnold Shaw, executive director of the Schillinger Society, for his editorial contribution to *The Mathematical Basis of the Arts*. Mr. Shaw prepared the manuscript for the printers and supervised the production of the book.

Mrs. Joseph Schillinger

PREFACE

This work does not pretend to transform the reader into a proficient artist. Its goal is to disclose the mechanism of creatorship as it manifests itself in nature and in the arts. This system, which in a sense is itself a product of creation, *i.e.*, a work of art, opens new vistas long awaiting exploration.

As the range of readers will probably vary as widely as the respective fields covered by this theory, the author expects a correspondingly wide range of reactions to it. He has hopes, however, that this work will be a stimulus of high potential, which will lead the spirit of investigation into the most majestic of all playgrounds known.

Whereas one scientific theory overwhelms another only to be overwhelmed by new facts and new evidence, this system overwhelms the available facts and evidence. Hence its pragmatic validity.

Intentional art, as we know it, dates as far back as one hundred thousand years, a comparatively brief period in hypothetically established human history. The present system is newly born. The age of its oldest branches does not exceed a quarter of a century which, even at the pace of present development, is a negligible amount of time. Its youngest branches are only one decade old. It has been subjected to a very limited range of experimentation and verification. Yet evidence which has been accumulated with respect to the workability of this system and the quality of its products is so overwhelming that, if given a chance, it may well shatter the very foundation of the great myth of artistic creation as we have known it since the dawn of human history.

Let this system be put to a real test. Let it penetrate into all fields of knowledge, education and production for at least a generation, and then have judgment passed upon it. The risk is negligible and the dividends too great to neglect.

February 27, 1942.   Joseph Schillinger.

CONTENTS

PART ONE: SCIENCE AND ESTHETICS

CHAPTER 1. ART AND NATURE 3

CHAPTER 2. ART AND EVOLUTION 7

CHAPTER 3. MIMICRY, MAGIC AND ENGINEERING 11
    A. Development of Art 11
    B. Evolution of Rhythmic Patterns 11
    C. Biology of Sound 13
    D. Varieties of Esthetic Experience 14
    E. History of the Arts in Five Morphological Zones . . 17

CHAPTER 4. THE PHYSICAL BASIS OF BEAUTY 18

CHAPTER 5. NATURE OF ESTHETIC SYMBOLS 23
    A. Semantics 23
    B. Semantics of Melody 25
    C. Intentional Biomechanical Processes 28
    D. Definition of Melody 29

CHAPTER 6. CREATION AND CRITERIA OF ART 30
    A. Engineering vs. Spontaneous Creation 30
    B. Nature of Organic Art 32
    C. Creation vs. Imitation 34

CHAPTER 7. MATHEMATICS AND ART 38
    A. Uniformity and Primary Selective Systems . . . 38
    B. Harmonic Relations and Harmonic Coordination . . 41
    C. Other Techniques of Variation and Composition . . 44
    D. Pragmatic Validity of Theory of Regularity . . . 46

PART TWO: THEORY OF REGULARITY AND COORDINATION

CHAPTER 1. CONTINUUM 51
    A. Definition of an Art Product by the Method of Series . . 51
    B. Parametral Interpretation of a System 56
    C. The First Group of Art Forms 58
    D. Time (X_4) as a General Parameter 71
    E. The Second Group of Art Forms 74
    F. The Third Group of Art Forms 78
    G. Correspondences Between Art Forms 80

CONTENTS

Chapter 2.	CONTINUITY	85
	A. Series of Values	85
	B. Factorial-Fractional Continuity	92
	C. Elements of Factorial-Fractional Continuity	94
	D. Determinants	98
Chapter 3.	PERIODICITY	109
	A. Simultaneous Monomial Periodicities	112
	B. Polynomial Relations of Monomial Periodicities	123
	C. Polynomial Relations of Polynomial Periodic Series	129
	D. Practical Application in Art	140
	E. Synchronization of the Second Order	141
	F. Periodicity of Expansion and Contraction	146
	G. Progressive Symmetry	157
Chapter 4.	PERMUTATION	158
	A. Displacement	158
	B. General Permutation	162
	C. Permutations of the Higher Orders	168
	D. Mechanical Scheme for the Permutation of Four Elements	172
Chapter 5.	DISTRIBUTIVE INVOLUTION	173
	A. Powers	173
	B. General Treatment of Powers	179
Chapter 6.	BALANCE, UNSTABLE EQUILIBRIUM, AND CRYSTALLIZATION OF EVENT	184
	A. Formulae	186
	B. Unstable Equilibrium in Factorial Composition of Duration Groups	189
Chapter 7.	RATIO AND RATIONALIZATION	193
	A. Ratio—Rational—Relation—Relational	193
	B. Rationalization of the Second Order	193
	C. Rationalization of a Rectangle	195
	D. Ratios of the Rational Continuum	214
Chapter 8.	POSITIONAL ROTATION	215
	A. Dimensionality of Positional Rotation	215
Chapter 9.	SYMMETRY	219
	A. Symmetric Parallelisms	219
	B. Esthetic Evaluation on the Basis of Symmetry	223
	C. Rectangular Symmetry of Extensions in Serial Development	224
Chapter 10.	QUADRANT ROTATION	232
Chapter 11.	COORDINATE EXPANSION	244

CONTENTS

CHAPTER 12.	COMPOSITION OF DENSITY	254
	A. Technical Premise	255
	B. Composition of Density-Groups	256
	C. Permutation of the Sequent Density-Groups within the Compound Sequent Density-Group	260
	D. Phasic Rotation of Δ and $\Delta^{\rightarrow}$ through t and d	262
	E. Practical Application of $\Delta^{\rightarrow}$ to $\Sigma^{\rightarrow}$	266

PART THREE: TECHNOLOGY OF ART PRODUCTION

CHAPTER 1.	SELECTIVE SYSTEMS	273
	A. Primary and Secondary Selective Systems	273
	B. Temporal Scales	274
	C. Pitch Scales	277
	D. Scales of Linear Configurations and Areas	284
	(1) Periodicity of Dimensions	284
	(2) Periodicity of Angles	293
	(3) Rectilinear Segments Forming Angles in Alternating Directions	302
	(4) Monomial, Binomial and Trinomial Periodicity of Sector Radii	313
	(5) Periodicity of Radii and Angle Values	314
	(6) Periodicity of Arcs and Radii	324
	(7) $r_{4 \div 3}$ Rhythmic Groups in Linear Design	329
	(8) Planes	336
	(9) Closed Polygons	344
	E. Color Scales	346
CHAPTER 2.	PRODUCTION OF DESIGN	363
	A. Elements of Linear Design	363
	B. Rhythmic Design	365
	C. Design Analysis	375
	D. Three Compositions in Linear Design	391
	E. Problems in Design	397
CHAPTER 3.	PRODUCTION OF MUSIC	399
	A. Coordination of Temporal Structures	399
	B. Distribution of a Duration Group	401
	C. Synchronization of an Attack Group	402
	D. Distribution of a Synchronized Duration Group . . .	404
	E. Synchronization of an Instrumental Group	406
CHAPTER 4.	PRODUCTION OF KINETIC DESIGN	414
	A. Proportionate Distribution within Rectangular Areas . .	414
	B. Distributive Involution in Linear Design	418
	C. Application to Dimensions and Angles	423
	D. Positional Rotation Applied to Kinetic Design	423

CONTENTS

CHAPTER 5. PRODUCTION OF COMBINED ARTS 429
  A. The Time-Space Unit in Cinematic Design 429
  B. Correlation of Visual and Auditory Forms 432

APPENDICES

APPENDIX A. BASIC FORMS OF REGULARITY AND COORDINATION 447
 I. Binary and ternary synchronization of genetic factors . . 447
  (A) Binary synchronization 447
  (B) Binary synchronization with fractioning 458
  (C) Ternary synchronization 476
  (D) Generalization: synchronization of n generators . . 476
 II. Distributive involution groups 502
  A. Distributive square of binomials 502
  B. Distributive square of trinomials 520
  C. Generalization: distributive square of polynomials . . 542
  D. Distributive cube of binomials 563
  E. Distributive cube of trinomials 568
  F. Generalization: distributive cube of polynomials . . 586
 III. Groups of variable velocity 636

APPENDIX B. RELATIVE DIMENSIONS 643

APPENDIX C. NEW ART FORMS 663
 I. Double Equal Temperament 665
 II. Rhythmicon 665
 III. Solidrama 667

APPENDIX D. POETRY AND PROSE 669

APPENDIX E. PROJECTS 671
 I. Books 672
 II. Instruments 673

GLOSSARY (Compiled by Arnold Shaw) 675

INDEX 687

PART ONE

SCIENCE AND ESTHETICS

This Theory is Based on the Following Postulates:
1. *The fertility of a postulate.*
2. *Uniformity as the basic concept.*
3. *Fractioning of unity as the potential of evolution.*
4. *Unstable equilibrium as a genetic force.*
5. *The principle of interference as a factor of growth and evolution.*

CHAPTER 1

ART AND NATURE

IF art implies selectivity, skill and organization, ascertainable principles must underly it. Once such principles are discovered and formulated, works of art may be produced by scientific synthesis. There is a common misunderstanding about the freedom of an artist as it relates to self-expression. No artist is really free. He is subjected to the influences of his immediate surroundings in the manner of execution, and confined to the material media at his hand. If an artist were truly free, he would speak his own individual language. In reality, he speaks only the language of his immediate geographical and historical boundaries. There is no artist known who, being born in Paris, can express himself spontaneously in the medium of Chinese 4th century A.D., nor is there any composer, born and reared in Vienna, who possesses an inborn mastery of the Javanese gamelan.[1]

The key to real freedom and emancipation from local dependence is through scientific method. Authors, painters and composers have exercised their imaginations from time immemorial. And yet can any of their most daring dreams compare with what science offers us today? Man has always flown in his dreams; nevertheless, these never satisfied his urge for "real" flight. Since antiquity, a number of myths has persisted of man's attempts to fly by means of artificial wings. Such flights have always failed. Let this be a lesson to artists. We cannot liberate ourselves by imitating a bird. The real way to freedom lies in the discovery and mastery of the principles of flight. Creation directly from *principles*, and not through the *imitation* of appearances, is the real way to freedom for an artist. Originality is the product of knowledge, not guesswork. Scientific method in the arts provides an inconceivable number of ideas, technical ease, perfection, and, ultimately, a feeling of real freedom, satisfaction and accomplishment.

My life-long study, research, and accomplishments as a creative artist have been devoted to a search for facts pertaining to the arts. As a result of this work I have succeeded in evolving a *scientific theory of the arts*. The entire system emphasizes three main branches[2]:

1. The Semantics of Esthetic Expression
2. The Theory of Regularity and Coordination
3. The Technology of Art Production

---

[1] The gamelan (or gamelang) is a primitive type of orchestra, native to Java, and consisting of an instrument which resembles our xylophone, and several small buffalo hide drums, which are strummed rather than struck with the fingers. (Ed.)

[2] *The Theory of Regularity and Coordination* is Part II of the present work. *The Technology of Art Production* is Part III. Materials relating to *The Semantics of Esthetic Expression* will be found in Part I, particularly Chapter 5. The reader is also referred to *The Schillinger System of Musical Composition*, which offers basic ideas in Book XI, pp. 1410-1477. Although the last-mentioned pages deal with the "Semantic Basis of Music," they present fundamental aspects of the relationship between expression and geometrical forms, emotional patterns and spatial configurations. (Ed.)

The Semantics of Esthetic Expression deals with the relationship between form and sensation, and with the associational potential of form, thus establishing the meaning of esthetic perception. *The Theory of Regularity and Coordination* discloses the basic principles of creatorship. *The Technology of Art Production* embraces all details pertaining to the analysis and synthesis of works of art in individual and combined media. The second branch is more voluminous than the other two combined, and requires years of study.

This theory of art is not limited to the conventional forms but embraces all the possible forms that can be evolved in space and time, and perceived through *the organs of sensation*. In addition to individual art-forms, it discloses all the possible compound art-forms, and all forms of technique that make possible the transformation of individual art-forms into compound ones.

According to this theory, art is determined as a logical system in the Cartesian and Einsteinian manner[3], *i.e.*, as a system of correlated parameters (measuring lines). A work of art may be adequately expressed by means of graphs affording both analysis and synthesis. Each individual graph may express a special art component (such as pitch in sound, hue in color, etc.) in relation to time.

The laws of rhythm, formulated in this theory as general esthetic laws, are based on two fundamental processes:

(1) the generation of harmonic groups through interference[4]:
(2) the variation of harmonic groups through combinatory and involutionary techniques.

Certain conclusions drawn concerning the esthetic properties of art phenomena coincide with discoveries made concerning the harmonic structure of crystals (Goldschmidt, Whitlock) and the properties of tangent trajectories (Kasner). This can only mean that:

(1) either there are general laws of the empirical universe in which esthetic realities take their place among the physical; or
(2) our method of mathematical deduction, being limited by the laws of its own logic, cannot be divorced from the object analyzed.

In either case, we are bound to cancel the line of demarcation between esthetic and physical realities. The history of art may thereupon be described in the following form:

(1) Nature produces physical phenomena, which reveal an esthetic harmony to us; this harmony is the result of periodic and combinatory processes; esthetic actualities embody mathematical logic. This is the *pre-esthetic*, natural (physical, chemical, biological) period of art creation.

---

[3]The distinction implied here relates to the number of coordinates required to represent the different art forms. In Chapter 1, *Continuum*, of Part II, Schillinger classifies the individual art forms into eighteen groups depending on the sense organs they affect. Each art requires a system of special components (light, sound, mass, or surface, etc.) to represent it and from one to four general components (time = x_4, space = x_1, x_2, x_3). The Cartesian system, as distinguished from the Einsteinian, involves only two coordinates. (Ed.)

[4]The concept of interference is pivotal in Schillinger's approach to the arts. Arising from phenomena observed in all fields of wave motion, it refers to the coincidence of two waves which results in a new summary wave. This concept (and a graphic technique based on it) is developed in detail in Chapter 3, *Periodicity*, of Part II, and exemplified in Appendix A. (Ed.)

ART AND NATURE

(2) Man recreates esthetic realities by reproducing the appearance of the physical realities through his own body, or through a material at his command; this process of reproduction involves mathematical logic, regardless of whether the artist is conscious of it or not; imitating nature in a material medium, he expresses the laws of mathematical logic through his sensory experience; this is the *intuitive* period of art creation.

(3) Becoming more and more conscious in the course of his evolution, man begins to create directly from principles; with developments in the technique of handling material art media (special components) and the rhythm of the composition as a whole (general components: time, space), man is enabled to choose the desired product and allow the machine to do the rest; this is the *rational* and functional period of art creation.

Thus, the evolution of art falls into a closed system. An esthetic reality may be either a natural product, a product of human creative intuition, or a product of scientific synthesis, realized through computation by mathematical logic. In actuality, all three aspects coexist in perpetual interaction.

Every work of art conceived and executed by man is a modified (often merely reflected) counterpart of actuality. Music, for example, is a man-made illusion of actuality, and so is every art. Music is merely a mechanism simulating organic existence. Music makes one believe it is alive because it moves and acts like living matter. Even Aristotle had observed that "rhythms and melodies are movements as much as they are actions." The common belief that "music is emotional" has to be repudiated as a primeval animism, which still survives in the confused psyche of our contemporaries. This erroneous conception can be easily justified as "naive realism." Music appears emotional merely because it moves—since everything that moves associates itself with life and living. Actually, music is no more emotional than an automobile, locomotive or an airplane, which also move. Music is no more emotional than the Disney characters that make us laugh, but whose actual form of existence is not organic, but mechanical (a strip of pictures drawn on celluloid and projected on a screen).

Everything that moves is a mechanism, and the science of motion is mechanics. The art of making music consists in arranging the motion of sounds (pitch, volume, quality) in such a manner that they appear to be organic, alive. The science of making music thus becomes the mechanics of musical sounds. The technique of this science enables the art of music to serve its ultimate purpose: the conveyance of musical ideas to the listener .

Nature is the source and supplies the media and the instruments of the arts. The *sources* of the art of painting are the forms, the texture, and the coloring of rainbows, sunsets, birds' plumages, crystals, shells, plants, animal and human bodies. Minerals, plants and vegetables are the *media* (pigments) while the sense of vision is the *instrument* of that art.

Thunder, animal sounds, and the echo are as much the *sources* of music as all the inorganic and organic forms that provide the structural patterns for musical intonation and continuity. Lungs and vocal cords, reeds and animal skins, as well as electricity, are the *media* (sound-producing devices) of the art of music; and the sense of hearing is the *instrument* of the art.

Natural forms originated as a necessity: as instruments for efficient existence. Multiplication of forms and images, through optical and acoustical reflexion, and circumstantial mimicry, which provides aggressive and protective size, offensive and defensive shape and coloring, constitute the first steps in the evolution of art forms. Deduction of esthetic norms, combined with imitation and readjustment of appearances according to these norms, constitutes the succeeding step—intentional mimicry. The final step in the evolution of the arts is the scientific method of art production, whereby works of art are manufactured and distributed according to definite specifications. This final step becomes possible only after the laws of art have been disclosed.

Discovery of the laws of art has been an old dream of humanity. In the *Li-Ki*, or Memorial Rites of the Ancient Chinese, we read: "Music is intimately connected with the essential relations of beings. Thus, to know sounds, but not airs, is peculiar to birds and brute beasts; to know airs, but not music, is peculiar to the common herd; to the wise alone it is reserved to understand music. That is why sounds are studied to know airs, airs in order to know music, and music to know how to rule."

The science of art-making must be concerned with two fundamentals:

(1) the mechanics of pattern-making[5]; and
(2) the mechanics of reactions.[6]

A scientific theory of the arts must deal with the relationship that develops between works of art as they exist in their physical forms and emotional responses as they exist in their psycho-physiological form, *i.e.*, between the forms of excitors and the forms of reaction. As long as an art-form manifests itself through a physical medium, and is perceived through an organ of sensation, memory and associative orientation, it is a *measurable quantity*. Measurable quantities are subject to the laws of mathematics. Thus, analysis of esthetic form requires mathematical techniques, and the synthesis of forms (the realization of forms in an art medium) requires the *technique of engineering*.

There is no reason why music or painting or poetry cannot be designed and executed just as engines or bridges are. Today's technical progress offers ample evidence of the achievement possible through the method of engineering, *i.e.*, through the method of expedient economy and efficiency. And if this method has transformed the most daring dreams of yesterday into the actualities of today, it can be equally successful in the field of art.

---

[5]The mechanics of pattern-making are described in Parts II and III of the present work. (Ed.)
[6]The mechanics of reactions refers to the relationship between the forms of excitors and the forms of emotional response, or the semantics of expression. See footnote 2 above. (Ed.)

CHAPTER 2

ART AND EVOLUTION

WE usually think of art in terms of local dependence. Such an approach produces many "connoisseurs" and "experts." There is always somebody who knows "everything" about Sappho, Duerer, Shakespeare, Cezanne, or Surrealism. But as folk-wisdom suggests: "One cannot see the woods for the trees." What we want to know is art's fatal history, its *morphology*.

If we possessed some reliable information concerning the mechanics of fate, it might offer a definite clue to the solution of our problem. But as we have no such knowledge, we must be satisfied with the only reliable kind of information we can gather in any field—that achieved through application of scientific method. But scientific method furnishes only one kind of information on any object: the knowledge gathered from its behavior. What the object *is* ultimately, what its meaning is—we do not profess to know. But we may observe its actions and draw logical conclusions after such observation.

If we could trace the original parent-forms of what we know as the arts today, we might discover, through observation of these forms, evolutionary tendencies as they manifest themselves in the behavior of such forms. Then by further application of logic, we might reconstruct from such tendencies the morphological image of the whole. Art history is too young to let us see the goal; nevertheless, the evidence accumulated so far is sufficient to reveal some of its tendencies.

If we had developed a science of events, a sort of fatal technology or "eventology," then the destiny of the arts might be known to us. But today's theoretical sciences seem to be losing ground; a "theory of uncertainty" usurps the place of the theory of probability.

The pragmatic sciences, however, continue to function satisfactorily in the fields in which they have been evolved. For example, knowledge of topography permits us to locate a certain geographical point with any degree of precision required for practical purposes. But to locate an event in an evolutionary chain, which is either finite or infinite, is quite a different matter. The subject of our investigation is the *morphology of creatorship*, which should include all forms of inventiveness, *i.e.*, both the scientific and the artistic. This implies the necessity of a "general theory of regularity." But up to now we have not had such a theory; thus, we are unable to formulate the kind of regularity which controls the appearance of new ideas in our world.

Professor Dirac, a British mathematician, suggested during the summer of 1937, that there are new numbers constantly appearing in the universe. We know that the classical civilization of Europe did not possess the concept of a million or billion. Even within the 10,000 limit, their expressions were extremely cumbersome. Our laws of esthetics must necessarily be cumbersome for the present, as we have never had any general theory explaining esthetic theories as an evolutionary group. Yet we are able to deduce a scientific theory of art-making from the manifold of the art experiences of past cultures.

The span of the human race is an insignificant episode in world history. The span of the arts is only an insignificant episode in the history of the human race. The Age of Mammals, grass and land forests is about one-tenth of the entire period of life on our planet, starting with sea scorpions. Out of twenty million years of the Cainozoic age, the human race starting with the pithecanthropus erectus covers only about five hundred thousand years—that is, about one-fortieth of the whole period of existence of vertebrate mammals. According to recent discoveries (1936), some signs of intentional design have been discovered in the Himalayan region. These are supposed to go as far back as one hundred thousand years ago. In other words, the earliest art traced so far spans not more than one-fifth of the history of the human race. European art, if it is to be reckoned from the designs found in the caves of the early Heidelberg dwellers (Neanderthal Man), is about twenty-five thousand years old, or about one-fortieth of the history of the human race.

The human race is confined to five senses and associative orientation. Art forms are perceived through the five senses and stimulate associative impulses. The senses impose limitations on art materials. The materials of the tonal art, for example, are limited to low frequencies and low amplitudes. Amplitudes may be magnified but the possible range of frequencies depends entirely on further evolution of the sense of hearing. While the most developed organ of sensation in the human race today is sight, the sensation of hearing is very limited as to intensity and the range of frequencies. We do not hear anything beyond 18,000 vibrations per second. In dealing with higher frequencies, sound decreases in volume; instead of increasing in pitch, it fades out, remaining on a certain high pitch. Without electrical amplifiers, sound cannot be transmitted for any distance comparable with the distance covered by our vision. Ordinary human speech cannot be heard even a mile away. The loudest symphony orchestra playing outdoors and surrounded by silent areas for miles cannot be heard even for a distance of five miles. Sound wave frequencies, from sixteen to about five thousand, compose only an infinitesimal group within the range of all types of wave motion—those producing colors, heat waves and various forms of radiation.

Art appears at first as a necessity. Ornamental tendencies develop much later. Art will cease to exist when there is no need for it. The latest anthropological conclusions are that the mind develops to a much greater extent than the organs of sensation to serve the ends of progress and evolution. Animals have keener organs of sensation than ours because these are their tools for protection and defense. A human being may be warned by a telegram about a coming danger. An animal has to rely entirely on its senses.

Though in our empirical existence we deal with a world crowded with matter, the emptiness of the universe is beyond our imagination. Likewise, the quantity of "tonal matter" is very limited in the sensory continuum. Forms of energy transformed into matter are comparatively rare. Space saturated with matter is analogous to time saturated with events. We live in a world crowded with events. The space-time continuum of music consists of the alternation of silence and sounds. Soundless time empirically seems to be longer. Time saturated with sound variations produces the opposite effect. Concentration of musical matter in the time-space continuum is structural energy in sound. This means that the flow of time as perceived in a musical composition speeds up or slows down in direct relation to its structural constitution. A short musical composition containing many recurrences becomes unbearable. Our esthetic experience tells us that a short composition must sound longer in order to seem satisfactory and complete. A long composition must produce an illusion of being considerably shorter than its actual duration. These time sensations are common in our everyday experiences. The lack of activity and events about us makes time move at a slower pace than when we encounter a multitude of events. Moments of satisfaction, pleasure and ecstasy always seem too short. Moments of boredom always seem too long. Crowded events which assume simple periodic forms and do not require active participation on our part, lead us to boredom, that is, to time slowing down.

The mental growth of humanity, as revealed in scientific thinking, may be stated as a tendency to unite seemingly different categories into a complex unity into which previous concepts enter as component parts. The evolution of thought is a process of synthesizing concepts. Creation from principles should not be confused with imitation of appearances (mimicry).

It is time to admit that esthetic theories have failed in the analysis as well as synthesis of art. These have been unsuccessful both in interpreting the nature of art and in evolving a reliable method of composition. The artistic approach to art has proved to be inadequate in solving the problem of creative experience. An *adequate method* has to be found. But methodology is a scientific, not an artistic development. The evolution of method is the ultimate goal of science. Thus, science comes to the rescue of art. "Esthetic qualities" can be detected within art material, transformed into the geometrical relations of its components, and finally into corresponding number values. The entire problem of art emphasizes three phases:

(1) Scientific (graph) method of recording a work of art[1] through its components (in place of the present, inadequate systems of artistic notation), which permits us to measure, to analyze, to draw conclusions, and to deduce norms;
(2) Modification of a work of art through variation of its inherent geometrical properties in a corresponding graph record;
(3) Production of a synthetic work of art from a system of number values, transformed into geometrical relations, and, finally, into corresponding components of art material.

Thus, science establishes a precedent for the development of an art theory as a system satisfactory in any special case. Scientific method penetrates into another

---

[1] In music, for example, the horizontal axis represents note durations (each square may equal an eighth note, a sixteenth note, etc.) The vertical axis represents pitch, with each square equivalent to a semitone or a tone as the case may be. (Ed.)

realm of the unknown, and establishes premises for the analysis and synthesis of art as creative experience.

Scientific laws, such as the law of gravity, make prediction possible. Art, being an evolutionary group, must function through the laws of evolutionary groups. The differentiation of art forms corresponds to differentiation of the senses. Structural and associative pattern-making is universal. Art forms consist of structural as well as associative patterns. All evolutionary groups reveal the tendency of acceleration. The evolution of the human race as well as of our planet presents such evidence, and art is no exception to this law.

A structural evolutionary group may be expressed in the following concept series: impetus — motion — inertia — balance — stabilization — crystallization — deposition — disintegration (transformation). A pentacle in a starfish is a pattern crystallized for efficient existence. An abstract pentacle (geometrical pattern) becomes a source of new functional association, that is, it becomes the symbol of a fighting unit (the Red Army). It involves geological and biological as well as esthetic patterns. The appearance of new biological and esthetic patterns is necessitated by readjustment. Pattern-making has its general source in electro-chemical patterns of brain functioning.

According to Professor Barr, Yale anatomist, "Physiology becomes a branch of electrical engineering" (1936). Thus, the geometry of thought becomes the source of universal pattern-making. This bio-geometrical generator asserts certain tendencies, which in turn produce certain configurations and certain colors. Perhaps in the near future, we may learn that creative experiences are merely geometrical projections of the electro-chemical patterns of thought on various materials having sensory effects upon us.

The mysterious character of the prime number has always been a source of fascination to primitive cultures. This is probably why such numbers have been emphasized in the case of the Hindu (17), Javanese (5), and Siamese (7) musical cultures. Our musical culture deals with the number 12—and all kinds of irrelevant reasons and excuses have been found for this choice. The real justification for the value of 12 is its versatility with respect to division. It is the smallest number up to 60 which contains so many divisors. Versatility in division and other quantitative properties lead to greater combinatory variability. Theories a posteriori are very characteristic forms of art theories in general. Offering nothing in the analysis of the creative processes of art, such theories expose their futility in the contention that a genius is above theories, and that his creativity is free and does not conform to any laws or principles. This is the mythological period of esthetics. In 2000 B.C. lightning was a revelation of divine power and could not be explained in terms of human experience. Yet it was produced every hour on schedule at the New York World Fair of 1939, and was offered as a form of entertainment. There is less and less room for mystery and divinity so far as the manipulation of material elements is concerned.

CHAPTER 3

MIMICRY, MAGIC AND ENGINEERING[1]

A. Development of Art

INTEGRATION of esthetic experience assumes the following evolutionary cycle: mimicry (passive transformation), magic (active transformation), and engineering (scientific transformation). We shall illustrate this proposition in its application to a concrete art form.

The transformation of matter into energy and the transformation of a sensation into a concept find their analogues in the history of music. The development of musical instruments and the performance of music, as well as the development of the forms of musical composition and theories, follow a similar process of dematerialization; from the first intentional sound-production by means of the bodily organs, through the most elaborate material instruments (piano, organ), finally, to dematerialized electronic instruments. From reliance upon the organs (lungs, vocal cords, diaphragm, lips, fingers, arms, etc.) as the agents of performance, through utilization of electrical devices for the development of volume and tone-quality, finally, to elimination of the performer. From unintentional improvisation and imitation, through highly developed artistic creation, finally to scientific creation and engineering with automatic production of music and elimination of the composer. From spontaneous forms induced by biological pattern-making, through scholastic theories of rules and exceptions, finally to a scientific theory dealing with laws of intentional creation, and developed in accordance with general science: that is, from biological to mathematically logical patterns.

There are three fundamental periods in the history of musical instruments. First: a mammal or a man uses the organs of his body, vocal cords, palms, wings, etc. Second: man begins to utilize the objects of the surrounding world: cockleshells, horns, bamboo stems, etc. Through imitation of the appearances and materials, he tries to reproduce similar instruments. It takes centuries, if not millennia, to improve these crude forms offered by nature. Third: man discovers scientific methods of sound production. At the early stage of this period, he tries to utilize his physical knowledge of sound in order to improve existing instruments. At a later period he learns to produce sounds directly from oscillatory sources, such as electro-magnetic induction, interference in the electro-magnetic field, etc. At the same time he tries to increase the area covered by sound production. He develops ways and means of amplifying sound and of broadcasting.

B. Evolution of Rhythmic Patterns

All rhythmic patterns evolved by the human race, from prehistoric time until now, do not exceed the 12/12 series: that is, twelve is the largest genetic

---

[1]The material contained in this chapter was presented in part before meetings of the Music Teachers National Association in 1937. It is reprinted by permission from the M. T. N. A. *Volume of Proceedings* for 1937. (Ed.)

factor necessary to produce all the existing and known rhythmic patterns. Synchronized simple periodic groups of different frequencies generate the original unbalanced binomial, whose major genetic factor (generator), being the sum of the binomial, becomes the determinant of the series. The interference of the original binomial against its own converse generates a new resultant (which is a trinomial)[2].

Permutation of members of this group produces new derivative groups. These derivative groups fall into synchronization with their own permutations and produce the next interference group. The process follows in the same order until one of the interference groups produces uniformity. From this point on, the entire set of processes repeats all over again in the form of distributive squares. Any further distributive involution group may be assumed as a limit. The ultimate limit-group produces a hypothetical absolute uniformity. These compound processes migrate from one series to another, often not developing beyond the second power. The law of involution works in both coordinates, thus producing simultaneous and sequent groups.

Transition from one series of patterns to another takes as long a period as the growth and decline of a civilization. The evolution of chord structures took a very short period as compared to the evolution from unison to chord structures. It took about one hundred and fifty years to adopt a seventh-chord (tetrad) after the fifth-chord (triad) was known for centuries. It took about fifty years to adopt a ninth-chord (pentad) after the seventh-chord was assimilated. In the last few decades, the art of music has evolved more varied chord structures than in the entire past history of the human race.

While in the classical period of European history, the evolution of new musical systems did not occur more frequently than at intervals of a century, today so many systems are being evolved every decade that even the slightest attempt to apply them to the practice of musical composition would take a longer period than the entire period of the history of the human race. There is a decided rate of acceleration in all evolutionary processes. This applies to biological types as well as esthetic types. Perhaps it took more than a million years from the first sigh of the first mammal until the first intentional melody was devised. It took perhaps a few millennia until polyphony was devised, while in our day there are dozens of symphonies begun daily, each one trying to develop new experimental or theoretical approaches.

When existing tuning systems seem to lose their freshness, composers search for new intonations. Debussy was attracted by the Melanesian intonations while Ravel tried to adapt Madagascar songs. The alert system-makers of today offer quarter-tone and other devices. With all this immense raw material, a system of esthetic utilization becomes more and more urgent. It is more essential for the sake of esthetic efficiency to be able to manipulate in new ways two or three familiar elements than to get lost in a rich and variegated group of new elements that have not passed through our previous experience.

---

[2]The successive processes referred to in this paragraph and in the succeeding paragraphs are described in Part II, Chapter 3, *Periodicity*. (Ed.)

C. Biology of Sound

We have already referred to motion as the source of pattern-making. Muscular tension and release is one of the first sources of organic sound. Though commonly unknown and generally repudiated when brought into a discussion, this fundamental aspect of musical semantics had already been known to Aristotle who referred to rhythms and melodious sequences as movements. This is the first penetration into the true nature of musical language. Animal sound contains all the components of tonal art: intensity, frequency and duration.

The biological factors of sound are:
1. Reaction of an organism to sound as a signal of movement.
2. Connection between increase and decrease in intensity of sound, and analogous variations of intensity in the organism.

Movement itself is the first source of music: periodic vibrations occurring in nature produce sound—the material of music; organic movements (breathing, locomotion, expansion, contraction) produce the forms of music. The mechanical constitution of music varies with times and places, yet the patterns of it are familiar to us from our bio-mechanical experiences.

When we arrive at a conception of pattern-making as an experience general to all the perceptible world, musical phenomenon becomes merely a special case of esthetic phenomena in general. Its distinction from other esthetic phenomena depends not so much on the actual patterns as on the material in which these patterns are realized. Musical patterns do not necessarily signify the art of music. They may be created by a group of circumstances and not by the intentions of an individual or a collective artist. Thus, musical form may result from personal as well as impersonal expression.

The natural sources of music are in sounds as well as in the patterns of the organic and inorganic worlds. In the early history of mammals, sound probably was a spontaneous reflex of vocal cords, induced by fear and stimulated by the contraction pattern as a geometrical expression of fear. This sound became a signal of approaching danger. The process of crystallization itself was the result of repeated experiences through which the mammal learned of its efficiency. Evolution of the art of music from a signal has been substantiated by Karl Stumpf in *The Origin of Music*. A sound signal coordinates group reactions.

Here we have the origin of the organizing power of music. Efficiency (order, organization) results from two opposite processes: aggression (attack) and fear (defense). Thus, we acquire all the organizing forms of music: hunt, regimentation, emergency and labor signals. Hence, the deification of music as an organizing power. Music becomes a magical factor. By means of a sound signal, an animal tries to induce fear in another animal. This is the first source of the incantation of evil. If a sound signal can counteract the unfavorable and evil, it probably can attract the favorable and the good. Evocation of the favorable is the first religious function of music. After a time, primitive incantations are dissociated from their original magical connotations and disintegrate at the end of their evolutionary cycle into operatic, pseudo-mystical and nursery-rhyme forms.

If music has an influence upon the evil in the surrounding world, it may have the power to influence this evil in human bodies. Hence, the medicinal application of music through the course of many centuries. Music as a healing device penetrated, not only into such fields as psychotherapy, but into gynecology as well. Today forms of treatment by means of sound waves are being extended. Scientifically speaking, the difference between treatment by means of low frequency waves (sound waves) is only quantitatively different from treatment by means of waves of high frequency (x-rays).

D. Varieties of Esthetic Experience

Music as an idea-forming factor has been known since Plato, Aristotle and Aristoxenes. Plato in his *Politeia* discusses music as an ethical factor, and asserts that the purely emotional enjoyment of music is inherent in slaves as well as in animals. It was a part of the school curriculum at that time to know which musical scales stimulated virtues. Some of the scales were rejected because they had a bad influence upon the young generation. We have not progressed much since then. We meet people in our own society today who believe that certain patterns in musical scales have bad influences on our generation. They have in mind certain hybrids between the ecclesiastic and religious music of England, and the music of African cannibals. Apparently, this ethically injurious music is so alluring that it affects not only the "drifting" young generation, but some of the greatest composers of our time as well.

Contemplative music has its origin in the disintegration of labor processes. It is a form of movement by inertia or minute stimuli. Such are pastorals, barcarolles, and cradle songs. This is the music of satisfaction and of contemplation, that is, the lyrical form of ordinance. What is an obsession, caused by fear of unknown mysterious forces in a primitive man, assumes the form of obsession by forces that still contain a certain amount of mystery for the civilized man. Love is one of such forces. The active and passive forms of this obsession are nocturnes, love songs and serenades. Forms of dissatisfaction and unbalanced existence stimulate readjustment. Readjustment calls for organization and sometimes revolt. The expressions of dissatisfaction and revolt are revolutionary hymns and songs.

The evolution of ecclesiastic music into pure music assumed the following pattern: crystallized ecclesiastic dogmas influenced music patterns directly and indirectly, thus becoming esthetic dogmas. The admiration of divine harmony as a form of perfection resulted in admiration of musical harmony that would sound perfect to the human ear. Thus, the cult of concord was created. The evil of the primitive man assumed the form of dissonant chords for the civilized man. Music began to seek formal purity and became art for art's sake. From the bewitching concept through the glamorous, beautiful, charming, pretty, elegant, gallant, neat and orderly stages went the disintegration of musical patterns. Form became a crystallized scheme. Deposition and disintegration are the outcome of this evolutionary group. The cult of craftsmanship transforms into formalism and scholasticism, and leads to a dead end of musical theory and practice.

There has always been extensive speculation concerning the nature of music structures. Pythagoras attributed the meaning of music to the motion of celestial bodies. In the eighteenth century Saint-Martin compared the tones of a major triad with doubled root to the four elements. Schopenhauer, Novalis, Spencer and others tried to link music with architecture, poetry, and the processes of life itself.

There are many views on what music is supposed to represent. From its original medicinal connotations, music deviates into various influences in the field of psychology. Music often serves as a release of psychological obsessions. In other cases, music itself becomes an obsession. Frequently, musical abilities develop on account of other abilities. There are many musicians with subnormal mentality as well as people who are insane in the medical sense of this word but who possess extraordinary musical abilities and almost supernaturally retentive musical memories. In relation both to instruments and esthetic forms, musical trends are dependent upon sociological, economic and technical forms. These often determine the velocity make-up of the music of a corresponding era.

The educational value of music lies in the field of technical routines. In learning to play an instrument, an individual acquires the agility and the coordination of his muscles and respiratory technique. By writing and analyzing music, and studying intelligent music theories, an individual acquires similar agility and coordination of his mind. Rational musical education is more important than the immediate acquisition of one type of routine, which may be useless ten years later. The education of a professional musician must include all the technical training possible, combined with a thorough knowledge of sound as material, and a complete understanding of the general methods involved in all musical procedures. Musical instruments as well as musical forms go through their continuous evolution. It may happen that in the future neither finger agility nor sound-production will be necessary any longer.

It is the varieties of creative experience in music that makes the art of musical composition so intangible. Music may be composed in a rational as well as irrational way. The extreme of the latter is music appearing in a dream where the element of intention is zero. There is enough evidence among composers to substantiate this method of creation as not being uncommon. An intermediate form is a semi-rational intuitive process, and the extreme, a complete rationalization—the engineering of music.

With the adoption of an engineering technique, the entire approach to musical patterns becomes mathematical. Scientific analysis of musical composition reveals that all the processes involved in the creation of a musical composition may be represented by elementary mathematical procedures. For a number of centuries philosophers have suspected that there are unconscious mathematical procedures behind conscious musical intentions. Music becomes "the mathematics of the soul." The raw material of the mathematics of music begins with atomic structure and the life of living cells. It is quite simple to solve all the problems of musical creation with the mathematical equipment we possess at present. All musical procedures are only special cases of the general scheme of pattern-making. There is even an absolute identity among the series pertaining to the forms of organic growth, to crystal formations, to the ratios of curvature of the celestial trajectories and orbits, and to the forms of musical rhythm. Thus we come to the end of the cycle. Music is one of the phenomena of human experience. The integration of these experiences leads back to the fundamentals. We learn through music what we learn through astronomy and biology. We arrive at an idea. Music is one of the embodiments of the idea. In the remote future of human history through the continuous process of abstraction, this idea will emancipate itself from its functional associations in the way that a pentacle emancipated itself from a starfish or a sea urchin. This will be the logical end of music.

Before music disintegrates, it will acquire greater functional expediency. It will be manufactured and distributed in the way other industrial products are manufactured and distributed. Before music disintegrates, it will influence the allied arts and come into fusion with them. The compound art of primitive man in his ritual ceremonies develops into individual art forms, which later develop and acquire their independence. At the end of this evolutionary cycle, the relation between the allied arts increases anew and they begin at first to influence each other, and later to fuse with each other. A dance with musical accompaniment is one of the most trivial forms of such fusion. Not long ago it was visionary to admit the fusion of photography with speech and music, which is today commonplace entertainment.

The International Exposition in Paris (1937) presented the transformation of liquid masses, combined with a variation of projected color, and accompanied by music. This art of luminous fountains merged with music took place once more at the New York World Fair in 1939. Without overlooking the influence of musical forms upon the dance, we may note that music has influenced literary forms as well as painting. The patterns of musical composition take place in the new art of projected light (lumia). "The music of visible images" (abstract cinema), a comparatively recent development, calls for a greater precision in both design and music. The most recent and most successful of the new art forms is a new realism based on the fusion of the two arts: music and design. It is mechanical realism as we observe it in animated cartoons. These cartoons are the end of the cycle, beginning with ancient puppet plays. The art of the cinema has not yet reached its climax. On the contrary, it is too young to disintegrate in the near future; yet the amount of engineering technique employed in all phases of this art is incomparable with the amount of acoustical engineering that was necessary during the time of Bach or the amount of paint chemistry that was necessary in the time of Leonardo da Vinci. Television, being the ultimate achievement of the "engineering" of today, will undoubtedly stimulate further fusion of existing art-forms.

As physiology becomes a branch of electrical engineering in the study of brain functioning, esthetics becomes a branch of mathematics.

To sum up the evolutionary groups pertaining to art forms, we offer the following scheme of morphological zones. These zones may follow each other chronologically as well as overlap each other, and may differ in different localities.

E. History of the Arts in Five Morphological Zones

Zone One. Biological. Pre-Esthetic.

The struggle for existence. Defense reflexes. Tactile orientation. Adaptation to the medium. Automatic self-protection. Automatic self-destruction. Mimicry. Motor reflexes. Signaling.

Zone Two. Religious. Traditional-Esthetic.

Intentional mimicry. Reproduction. Performance. Magic. Ritual Art. Incantation. Religious art.

Zone Three. Emotional-Esthetic.

Emotion. Artistic expression of emotions. Self-expression as unconscious mimicry. Origination of an esthetic idea. Art for art's sake.

Zone Four. Rational-Esthetic.

Growth of esthetic ideas. Rationalizing. Rationalization. Experimenting. Novel art. Modernism. Experimental art.

Zone Five. Scientific. Post-Esthetic.

Analysis and synthesis of an art product. Scientific experiment. Art with a scientific goal. Scientifically functioning art. Manufacture, distribution and consumption of a perfect art product. Fusion of art materials and art forms. Disintegration of art. Abstraction and liberation of the idea.

CHAPTER 4

THE PHYSICAL BASIS OF BEAUTY

THE physical source of the arts as functional groups is the off-phase pair, or group, of sine-waves. One sine-wave is the limit of simplicity in action. Two sine-waves with a relatively negligible periodic difference produce beats or interference (pulse). This pulse is the life element, or the manifestation of life, of an esthetic entity. Thus, phasic differences, causing instability in wave motion, are the actual factor controlling esthetic varieties.

This proposition remains equally true whether it is applied to the human voice or a musical instrument. From an esthetic viewpoint, the quality of sound largely depends upon the form and frequency of this pulse, which in physical terms is the ratio produced by the difference between two component waves. When the frequency of this pulse is too low (below 5 cycles per sec.), the impression gained from tone-quality is that of insufficiency, of retarded life speed. When the pulse-frequency is normal (5 to 6 cycles per sec.), the impression gained from tone-quality is healthy existence, well-being. With pulse frequencies higher than normal (above 6 cycles per sec.), the impression gained from tone-quality is of accelerated, precipitated, tense existence.

The passive character of the first quality is due to the effect of unstable equilibrium, which is *not* sufficiently defined and seems to be below our biological rhythms. The normally active character of the second quality is due to the effect of unstable equilibrium, which seems to synchronize with the biological rhythms of a healthy body. The over-active character of the third quality is due to excessive instability, which makes the preservation of the equilibrium somewhat of an effort. It overstimulates our biological rhythms.

These three qualities can be defined respectively as sub-biological, biological and supra-biological. As the first quality can be properly compared with understimulation, the second, with normal stimulation, and the third, with overstimulation, the three qualities correspond to the psychological triad: subnormal-normal-supernormal. The first quality corresponds to depression, consciousness of weakness, melancholy, pessimism. The second quality corresponds to normal existence when well-being is not consciously noticeable. The third quality corresponds to joy, consciousness of vigor, heroic urge, overactivity and ecstasy.

As the second quality, axis of psychological equilibrium, is the persistent status quo, we are aware of its absence (not of its presence) only when we deviate from it. This sensation is comparable with the status quo of the horizon, which forms our visual axis. We become conscious of it only when it loses its horizontality. Consider the sight of familiar surroundings when viewed from an inclined surface, a boat, a float, a train, or a plane.

The psychological triad is in direct accord (and, perhaps, correspondence) with the frequencies stimulating visual and auditory perception. The low frequency of red affects us as understimulation; the middle frequencies of yellow, green and blue—as normal stimulation; and violet—as overstimulation. Compare these color frequencies with the sunset with its overabundance of red, the midday with a balanced spectrum and the forenoon with its ultra-violet predominance. In this picture, we get biological as well as photo-chemical evidence (reaction of the emulsion in color film).

The same is true of sound. Low pitches (low frequencies) produce a quiescent, nocturnal, sub-biological, understimulating effect. The middle range, particularly the one which corresponds to the range of the human voice (approximately between 64 and 1200 cycles) embraces the psychological range of normal stimulation. The high frequencies produce an effect of overstimulation, particularly when abundant with beats. Compare the shrill, ecstatic effects of high flutes.

Of course, frequency can be dissociated from intensity only for analytical purposes. It is the combination of intensity and frequency that forms the actual stimulus. In view of this, it is interesting to note that the alternation of two different intensities in rapid succession produces the beat or oscillating effect, and that the periodicity of intensity beats corresponds in frequency to that of phasic displacements.

We shall now examine our psychological triad in its correspondence to optical and acoustical mechanical frequencies, *i.e.*, speeds of projecting motion-picture films with their sound-tracks or playing phonograph records.

In projecting a motion-picture film at 8 intercepted images per sec., we obtain an insufficient degree of graduality of transition. The quality of motion appears to us as subnormal; it manifests itself in discontinuous movements. Sixteen intercepted images per second give satisfactory continuity of motion and suggest a nearly normal quality. Present-day 24 images per second produce a perfectly continuous motion of images if the photographed motion is of moderate speed. To get still more perfect normal effects from very fast motion, we must speed up both the photography and the projection. To judge which speed would appear as normal, we can take first the accelerated photography of 32, 64, 128 and more images per second and project them at 24 frames per second, thus obtaining "slow motion." If such slow motion gives perfect graduality of transition from one image to another (continuity of movement), let us say at 48 fr. per second, it would mean that normal projection of such a movement must also be at 48 fr. per second. Compare the cannon shots at 128 fr. per second which, being projected at 24 fr. per second, still do not produce the effect of continuity.

Empirically, however, we gain very little by photographing fast moving objects at excessive speeds and projecting them back at the same speed, as this realistic restoration, perfect in itself, does not correct the imperfection of our vision. Thus, we have two choices: either we see fast moving objects at a stretched time period, wherein we perceive all the details that we can perceive in continuity; or we see the actual image of a fast moving object realistically restored, something we cannot actually see.

Standard projection (16 or 24 fr. per second) of images taken in accelerated motion (at 32, 64, 128 and more fr. per second) produces an effect of the supernormal and even the supernatural. In the same way, the substandard projection, 8 or less fr. per sec., of normal motion (*i.e.*, motion which under standard projection and standard taking would appear continuous) produces a subnormal and even subnatural effect, due to the fact that the intercepted images do not form continuity.

The sub-biological speed, brought through lapse-shot cinematography to normal speed (projected at 16-24 fr. per sec.), is a source of particular fascination. In this connection, see the "growth of plants" filmed by various scientific organizations, which seem to present magical effects.

In sonic reproduction, the speed at which the sound is recorded requires a play-back at the same speed. As continuous images in sound are formed at the minimum of 16 cycles per sec., which is too low to transfer all components properly to the coating, and mechanically too difficult to achieve in sufficient uniformity, the standard 33.3 r.p.m. recording-reproducing speed is fully satisfactory. But we must remember that satisfactory results with this relatively low speed are a comparatively recent achievement. The commercial 78 r.p.m. speed is still used as a standard on most instruments.

The psychological triad corresponds to the recording and reproducing speeds in the following way: we obtain subnormal effects by playing at a speed lower than the recording speed. Under such condition the pulse of tone-quality, *i.e.*, the beat-frequency slows down more than twice, which is the ratio between 78 r.p.m. and 33.3 r.p.m. According to our previous analysis, this should result in sub-biological effects. And it does. The voice of tenor Enrico Caruso, under such conditions of reproduction, sounds like a cow (particularly when the attack[1] appears on "m" coupled with "oo" or "o" or "u.")

Normal effects are obtained, of course, by reproducing at the speed of recording. Supernormal effects are achieved by speeding up the mechanical frequencies in the reproducing apparatus. Thus, music recorded at 33.3 r.p.m. and played back at 78 r.p.m. displays supernatural agility and affects us as overstimulation.

Subnormal effects bordering on the sub-natural may be obtained by taking the original record made at 78 r.p.m., playing it at 33.3 r.p.m., re-recording it at 78 r.p.m. and playing it again at 33.3 r.p.m. In the same way, supernatural effects may be produced by reversing this operation. A record made at 78 r.p.m. must be played at 78 r.p.m., re-recorded at 33.3 r.p.m., and played again at 78 r.p.m. This procedure may be continued in both directions until we get no sound at all.

This zero of sound may be paralleled with the zero of motion that we obtain in a motion-picture by projecting just one individual frame upon the screen, or by standard projection of a very slow-moving object, photographed at an incredibly high speed. Under the latter condition, the object would appear sta-

---

[1] Schillinger uses "attack" to mean the production of sound, in this instance, by means of the vocal cords. (Ed.)

tionary, for the phasic changes that may be noticed by the eye would be dissociated by very low time intervals. Compare this with the stationary appearance of a starry sky.

It is important to realize that the ratio of audible frequency-limit equals 1000, whereas the ratio of the wave-length visible spectrum is $\frac{7}{4}$. It may be extended for convenience to the ratio 2, $i.e.$, $\frac{8}{4}$. The wave-length of extreme violet is .00040, and that of extreme red .00072. As infra-red has a wave-length of .001, such an extension of range as does not reach infra-red may be fully justified for empirical purposes of color composition. Thus, ranges do not correspond. Formation of perceptible kinetic visible or audible continuity, on the other hand, does have a common ground. The sequence of sound wave periods is 16 cycles per sec. minimum. Such a speed of impulses produces the effect of a continuous sound. The number of intercepted images in cinematic projection is set at 16 per sec. as a minimum producing continuous motion.

Esthetic pleasure grows with the increase of frequencies in both cases. However, sound and visible image have their own limits. Sixty-four to one hundred and twenty-eight frames per second are used in "slow motion" cinematography, which always delights audiences. The newly developed high speed cinematography (over a 1000 frames per second) transforms an ordinary phenomenon like a milk drop into a stirringly beautiful spectacle. In sound, it seems that the majority of our listeners prefer tones between one hundred and three hundred cycles per second (cello, male tenor, baritone and female contralto).

Since sense organs react to frequencies and intensities as such, and not to associative psychological forms and images, esthetic objects are capable of direct stimulation by number-harmonies and proportions present in the artistic media (sound, color, etc.)

Conditioned reflexes associated with pleasure and delight grow through repeated experiences. This is not to be confused with inherited, unconditioned reflexes. Thus, cultivation of positive reactions toward some kind of new art takes time and requires repeated experiences. Most people like either familiar music or music of a familiar kind.

It is only with the growth of refinement of perception in a certain field of art that an individual acquires an urge (desire, appetite) for the lesser known and more intriguing. Consider the delight which certain classes of professional musicians experience over the music of African cannibals.

The beauty of art material as a purely physical state may be described as follows:

1. beauty of the tone itself
2. beauty of the color itself
3. beauty of any component per se

Beauty of the composition as such would result from harmonic relations of harmonically developed components.

Under such conditions, any component may constitute beauty: spatial and temporal components, for example, may constitute beauty. This is where the harmony of numbers occurs. But a beautiful art work must have beautiful components in beautiful correlation. The resultant of synchronization of all components is equivalent to beauty. For instance: a melody which appears to be beautiful has a harmonic temporal flow (time rhythm), a harmonic relation of the sequent frequencies (pitch), a harmonic beat frequency combined with harmonic intensity (let us say 5 to 6 beats per second), and harmonic groups of intensity.

It is often true that the art material as such may overshadow other components as well as the composition as a whole. For instance, Caruso singing worthless and often stupid music may offer a vocal quality so harmonic that the listener's attention centers on it and he becomes unaware of other components of the artistic whole, including Caruso's own appearance.

Other examples are cumulus clouds, sunrise, rainbow, and an art form known as "lumia" and propagated by Thomas Wilfred.[2] The quality of celestial, luminous shapeless beauty is exactly in the same esthetic category as the images of the "clavilux." Some spectators attending the clavilux performances find a "cosmic touch" in the images created by this instrument, in spite of the great monotony of the temporal-spatial composition.

The beauty of the material per se, *i.e.*, the beauty of texture, differs only quantitatively, in terms of perceptible dimensions, from any other form of harmonic configuration. The components of texture are microscopically small as compared with the structural components of an art work. Though the degree of textural saturation varies, the aggregation of microscopic configurations (molecular structure of matter of the continuous periodic impulses producing sound) is so great that the configurations cannot be individually discriminated in sensory orientation and are perceived as a homogeneous qualitative whole instead.

[2]See art form No. 8 in Chapter 1, *Continuum* of Part II. (Ed.)

CHAPTER 5

NATURE OF ESTHETIC SYMBOLS

A. Semantics[1]

MUSIC in general—and melody in particular—has been considered, since time immemorial, a supernatural, magical medium. Many great philosophers in different civilizations have given their attention and directed their thoughts toward this elusive phenomenon. The more definitions of music you know, the more you wonder what music really is. It seems to fall into the category of life itself. It seems to have too many "x's."

People did not know much about lightning even ten thousand years ago, and ten millennia make only a one-hundredth in the range of *human* evolution. We tend to ascribe supernatural powers to any phenomenon we cannot explain. Today, we are surrounded by things more miraculous than any of the products of ancient imagination—and when you think of the achievements of modern technique, it seems to be incredible that a toy—as simple as melody—should still remain in the category of the irrational.

Following our method of analysis, however, we may assume that *any phenomenon can be interpreted and reconstructed*. To accomplish this, it is necessary to detect *all* the components and to determine the exact form of their correlation.

There are two sides to the problem of melody: one deals with the sound wave itself and its physical components and with physiological reactions to it. The other deals with the structure of melody as a whole, and esthetic reactions to it.

Further analysis will show this *dualism* is an illusion and is due to considerable quantitative differences. The shore-line of North America, for example, may be measured in astronomical, or in topographical, or in microscopic values. The difference between melody from a physical or musical standpoint is a *quantitative* difference. The differentials of the physical analysis become negligible values for purposes of musical (esthetic) analysis.

Melody is a complex phenomenon and may be analyzed from various standpoints. Physically, it can be measured and analyzed from an objective record, such as a sound track, a phonograph groove, an oscillogram or the like. Melody when recorded has the appearance of a curve. There are various families of curves, and the curves of one family have general characteristics. Melodic curve is a trajectory, *i.e.*, a path left by a moving body or a point. Variation of pitch in time continuity forms a melodic trajectory.

---

[1] This chapter is part of Book IV, *Theory of Melody*, of The Schillinger System of Musical Composition*, Copyright 1941 by Carl Fischer, Inc. Reprinted by permission. (Ed.)

Melody from a *physical* standpoint is a compound trajectory of frequency and intensity. Melody from a *musical* viewpoint is a compound trajectory of pitch, quality, and volume. The components of quality are timbre, attack, and vibrato.

Physically, pitch is an accelerated periodic attack. Physically, the difference between rhythm and melody is purely quantitative. Therefore, time-rhythm in a melody may have two forms.

1. Through periodicity of attacks of low frequency, which is unavoidable when the pitch-frequency is constant;
2. Through variation of frequencies, *i.e.*, through changes in pitch itself.

Frequency constitutes musical *pitch*. Any sound wave of a given frequency (constant or variable) generates its own frequency *subcomponents* (known as "partials" or "harmonics") resulting from purely physical causes. The latter are disturbances which convert a simple wave (known as a sine wave) into a complex one. The sound of a simple wave may be heard on specially made tuning forks and electronic musical instruments.

The intensity of a sound wave is one of the factors of disturbance, and the duration of intensity and its stability in time continuity are others. The latter are musical factors: depend on *form of attack* (or *accentuation*). Finally, the resultant of both components and all the subcomponents, *i.e.*, the interaction of all component frequencies and intensities in a sound wave, constitutes the musical component of *timbre and character* (quality) of sound.

The relative importance of musical components and subcomponents has already been measured, so to speak, by agreement among musicians and music lovers. The conclusion has been reached that two melodies are *identical* if their main components (time and pitch) are identical. For instance, a melody played on the piano, or sung, or played loud or soft, or with vibrato or without it, would be considered "the same" melody if rhythm (time) and intonation (pitch) are identical. The subcomponents and the sub-subcomponents pertain to execution, *i.e.*, to the performance of melody, not to its own structural actuality. The very neglect of subcomponents, on the one hand, relieves the composer of a certain amount of responsibility; on the other hand, it leads to loss in esthetic value of melody when the melody is wrongly executed by the performer. For then the performer has to supply the subcomponents without the benefit of any exact indication by the composer and therefore he acts at his own discretion, whether rightly or wrongly.

At this point we may adopt Helmholtz' definition of melody (which satisfies the musical aspect): *melody is a variation of pitch in time*.[2] Is *any* variation of pitch in time a melody? An attempt to answer this question leads into the *semantics of melody*.

---

[2] Hermann Ludwig Ferdinand von Helmholtz (1821-1894), the great German physicist and physiologist, sought to devise rules of musical science based on the physical nature of musical sounds. His most significant work is *On the Sensations of Tone as a Physiological Basis for the Theory of Music* published in 1863. (Ed.)

B. Semantics of Melody

The fundamental semantic requirements are that melody must "make sense," it must have (like words) associative power, *i.e.*, it must be able to convey an *idea* or *mood*, to "express something."

But these are also the requirements of language, and yet there is a distinct difference between *word* and *melody* as symbols of expression. The function of *words* is to express the *concept* of actuality, to find its verbal symbol. The function of melody is to express the *structural scheme* of actuality. Words have their origin in thought; melody has its origin in feeling, *i.e.*, originally in the reflexes. *Words generate concepts which may or may not stimulate feelings*. Melody, on the contrary, *stimulates feelings* (emotions, moods) as spontaneous reactions, *which may or may not generate concepts*. Melody expresses actuality *before* the concept is formed for that actuality. This is why, in listening to a melody, one is satisfied with its expression to such an extent that the quest for the concept, "What does it actually express," is never aroused. But, on the contrary, when a melody *does not convey sufficient associative power* (to stimulate reflexes, reactions or moods), then the listener looks for a verbal description of it, or, at least, for a title, a "label," a concept. Melody is *insufficient* whenever it calls for a verbal explanation. When a word does not convince through its own associative power, or in order to increase the latter, one resorts to intonation and gestures.

Words or melody may or may not be self-sufficient. Words that are not self-sufficient call for a specific form of intonation in order to acquire the necessary associative power. We may also state, reciprocally, that melody which is not self-sufficient as intonational form calls for word and often for a symbol in the form of a verbal concept. These two statements can be verified by simply studying the facts.

Here we arrive at the idea that although, in their developed forms, both word and melody are self-sufficient—in their early periods of formation they produce hybrid forms: an intonational form that calls for a concept—and a conceptual form that calls for intonation.

Here are a few of many references. According to the statements of George Herzog, Columbia anthropologist who made some pertinent recordings and demonstrated the phonograms, there are certain Central African tribes whose verbal language is just such a hybrid. A word of the same etymological constitution (spelling) has at least ten different forms of intonation, each attributing a different meaning to the word. In this case *intonation* is an *idiomatic factor*.

In other cases, as in some instances of Chinese music,[3] melody or even the single units of a scale become symbolic of a concept—*i.e.*, they assume the function of words.

The Stony Indians of Alberta, Canada, try in their songs to express the sound of a brook, the murmur of leaves, etc. Yet as a descriptive means it is not self-sufficient; it calls at least for a title. This is a case in which melody is a bad competitor of poetry.

---

[3] See Karl Stork, *History of Music*. (J. S.)

Out of many hypotheses as to the origin of music and word, I select the *reflexological* one.[4] Sound reflexes (of the vocal cords), before they crystallized into relatively distinguishable forms of word and melody, were spontaneous expressions of satisfaction or lack of satisfaction in an animal organism. Any cause of actual or potential disturbance that endangered an organism became a stimulus for the defensive reflex. This is probably the original form of the intonational *signal*. If such a form was at first an improvised reflex movement of the vocal cords expressing fear—a spontaneous reaction to danger—it may have crystallized later into the etymological form of the concept of "danger."

When an organism is on the verge of struggle for its own survival, it usually resorts to *intonational signaling* rather than to an etymological one. Even in our own time, a drowning man does not say: "I am drowning!" He generally shouts: "Help!"

The amount of semantic and acoustical elements in words or melodies varies greatly. There are all gradations from an exclamation to a polyconceptual polysyllabic word of the German language with the relative decrease of the acoustical (intonational) and the relative increase of the semantic element. In many undeveloped forms of speech, an outsider may in fact mistake such speech for melody.[5]

Melody always contains well-defined acoustical elements, although it may be alien to an ear trained to different systems of intonation. Melody offers also a scale of semantic gradations from imitative descriptive intonations, through symbolic abstractions, to the expression of mechanical forms.

Both imitative and symbolic functions of music tie it closely to verbal semantics. In this stage, melody is the language of a given community only. Tests show that even such commonplace moods as "gayety" and "sadness" cannot be expressed by means of melody that will mean the same thing to all nations. Melody is a *national* language or a language of a *given epoch* with regard to descriptive or symbolic qualities.

Arabian funeral music sounds anything but "sad" to us because of our association with major scales—which mean gayety, heroism, happiness and satisfaction to us. Gay Arabian dance-songs sound "sad" to us because of our association with harmonic minor scales, which mean exactly the opposite to us. It is similar with the forms of musical harmony. Through previous associations we react to major chords as we react to major scales. Yet we have the curious phenomenon of the Negro-American "blues," which is supposed to express depression, but which, nevertheless, has the richest scale of major chords.

All the controversies ascribing this or that semantic connotation (descriptive or symbolic) to music will vanish when we penetrate the real meaning of music, namely, *the expression of the forms of movement*. The objectification of this

---

[4] Here begins, in a partial form, Schillinger's exposition of his theory of the correspondences between music—melody, in particular—and the objective world of life. As such, his theory offers us the means whereby esthetic phenomena can be correlated in a scientific and materialistic way with the rest of human experience; in consequence, even this partial exposition is of the utmost philosophical importance. (Ed.)

[5] "Program music is a curious hybrid, that is, music posing as an unsatisfactory kind of poetry." —*Oxford History of Music*, Volume 6, Page 3. (J. S.)

meaning requires only one premise: *biomechanical, physiological experience, combined with a highly developed sensory system*. The requirement may be satisfied by any normal specimen of the higher animal forms.[6]

Though commonly unknown and generally repudiated when brought into a discussion, this fundamental form of musical semantics had already been known to Aristotle. Here is his definition: "Rhythms and melodious sequences are movements quite as much as they are actions." This is the first historical instance of penetration into the true nature of musical language.

The meaning of music evolves in terms of physico-physiological correspondences. These correspondences are *quantitative* and the quantities express form. This can be easily illustrated by the following example.

A sound of constant frequency and intensity and made up of a simple wave affects the eardrum and the hearing centers of the brain as an excitor of a simple pattern. Such a pattern may be projected by various means so that its structure becomes apparent to another more developed, and therefore more critical organ of sensation, that of sight. The complexity of reaction (*i.e.*, its form) is equivalent to the complexity of the form of the excitor. The number of components in a wave affects a corresponding number of the arches of the inner ear's Cortis organ, putting them into oscillatory motion. If a sine wave has one component, it will affect only that arch which reacts on the frequency corresponding to that transmitted through the air medium in the form of periodic compressions. When a wave of greater complexity affects the same organ, the reaction becomes more complicated.

It is a known fact that the ear can be trained. Therefore, the pattern of reaction is equivalent to the pattern of excitation with various degrees of approximation. All the components of sound work in similar patterns because these patterns are similar in all sensory experiences. Formation of the patterns is due to (1) *configuration* and (2) *periodicity*. The *configuration* may be simple or complicated in a mathematical sense, *i.e.*, its simplicity or complexity can be expressed in terms of components and their relations. This emphasizes both frequency and intensity in a sound wave, as well as the character of sound which is the resultant of the relations of the two components. *Periodicity* defines the form of recurrence and may be also of different degrees of complexity—for example, the periodicity of recurring monomials as compared to the periodicity of permutable groups.

Our physiological experience, combined with our awareness of that experience through our sensory and mental apparatus, makes it possible for us to understand the meaning of music in terms of "actions." Thus, *regularity* means *stability*, and *simplicity* means *relaxation*. Thus, the satisfied organism at rest is comparable to simple harmonic motion. The loss of stability is caused by powerful excitors affecting the very existence of the organism. Sex and danger are the excitors, and love and fear are the expressions of instability.

The awareness of such instability comes through variations in blood circu-

---

[6]Compare Plato's ideas on the meaning of music in his *Republic* and Ivan Pavlov's experiments with the pitch discrimination of dogs in his *Conditioned Reflexes*. (J. S.)

lation sensed through the heart-beat and variations in blood-pressure, resulting in respiratory movements. The whole existence of an organism is a variation of degrees of stability, fluctuating between certain extremes of restfulness and restlessness. The constitution of melody is equivalent to that of an organism. It is a variation of stability in frequency and intensity. Melody expresses those actions we know and feel through our very existence in terms of sound waves.

C. INTENTIONAL BIOMECHANICAL PROCESSES

We come now to a consideration of *intentional* biomechanical processes. Efficiency of action in relation to its goal is the foundation of evolution. The forms of action by which living organisms adapt themselves to the goal of survival in the existing medium may serve as a fundamental illustration. This efficiency comes about through "instinct" among the lower species, but through the conscious utilization of previous experiences leading to deliberate efficiency among the higher animals. Muscular tension and relaxation constitute the first instruments of such intentional action.

The mechanical constitution of melody varies with times and places, yet its patterns are familiar to us from our own biomechanical experiences.

The "contemplative" and the "dramatic" become two poles of our esthetic reactions. They grow out of the same biomechanical diads: restfulness-restlessness, and stability-instability.

Dramatic patterns themselves evolve out of two sources: the first is fear (defense—dispersed energy) and is caused by danger or aggression; it results in *contraction patterns*. The second is aggression (attack—concentrated energy) and is caused by an impulse or resistance; it results in *expansion patterns*. Confusion of patterns of compression with those of expansion (aberration of perception caused by instability) explains why the very same music sounds "passionate" to one listener but "weary" to another. This is a typical confusion observed by Professor Douglas Moore of Columbia in tests performed on students of non-musical departments at various universities, using Wagner's *Isolde's Love-Death* as material.

All the technical specifications for melodic pattern-making will be given later. The immediate question is: *how does it happen that the physiological patterns are identical with the esthetic patterns?* We can answer this question only hypothetically for we know very little about the technique of pattern formation at present. But as science progresses, we notice more and more correspondences in different fields. We find identical series in such seemingly remote fields as crystal formation, ratios of curvatures in the celestial trajectories, musical rhythms, design patterns, and, finally, in the very molecular structure of matter itself. Modern chemistry shows how by *geometrical* variation of mutual positions of the same group of electrons, entirely different substances are produced. Little as we know for the present about the electro-chemistry of brain-functioning, we may well suspect that all our pattern conception and pattern-making are merely the *geometrical* projection of electro-chemical processes, in the making, that occur in our brain. This geometrical projection is thought itself.

D. DEFINITION OF MELODY

The summary definition of *melody:*

(1) *Physiological definition:* Melody is an excitor existing in the form of a sound-wave which affects the organ of hearing. The latter being a receiver and a transmitter transfers it to the biomechanical pattern-making center of the brain.
(2) *Semantic definition:* Melody is an expression of biomechanical experiences in the sound medium.
(3) *Musical definition:* Melody is a variation of pitch in time, wherein pitch units follow a preselected scale of frequencies and express a relative stability of each individual unit.

The summary definition of *word:*

(1) *Physiological definition:* Word is an excitor existing in the form of a sound-wave which affects the organ of hearing. The latter, a receiver and a transmitter, transfers it to the concept-making center of the brain.
(2) *Semantic definition:* Word is an abstraction of biomechanical experiences in the sound medium. "Poetic image" is a variation of the original biomechanical abstraction.
(3) *Musical (tonal) definition:* Word is a variation of pitch in time, wherein pitch units express a relative instability of each individual unit and do not necessarily follow a preselected scale of frequencies.

It follows from these definitions: (1) that in symbolic notation (though different patterns are used)—printed letters or musical notes—both word and melody are identical; (2) a poem recited in a foreign or unknown language becomes an undeveloped form of music.

> *"Esthetic perception" is a tautology, as "aīsthētikos" means "perceptive." For this reason esthetic perception must be defined as a special form of selective perception.*
>
> *Selective (esthetic) perception is a capacity to discriminate relationships through senses and to associate such interrelatedness with the functionality of structure.*

CHAPTER 6

CREATION AND CRITERIA OF ART

A. Engineering vs. Spontaneous Creation

MATHEMATICAL *Basis of the Arts* is a scientific theory of art production. It classifies all the arts according to the organs of sensation through which they are perceived: *sight*, *hearing*, *touch*, *smell* and *taste*. It usually takes the lifetime of a genius to make a sizable contribution to any art, although some artists have attempted to paint, to compose music and to become mechanical inventors at the same time. And though a complete mastery of more than one art has been attributed to very few, the problem of scientifically coordinating several arts in one has never been accomplished by one individual.

The scientific theory of art production approaches this problem and solves it exactly by the same method as the problem of locomotion on the ground, in the water, and in the air is solved—that is, by *engineering*. The argument of spontaneous creation must be repudiated, particularly since works of art generally conceded to be among the greatest, have not been produced spontaneously. Quite on the contrary: the process of creation consumed an enormous amount of time and considerable mental and emotional effort.

The difficulty with spontaneous creation is due to the fact that an organic work of art is a combination of various components. For example, when a composer creates a theme, such a theme implies the co-existence of melody, rhythm, harmony, dynamics, phrasing, etc., all in one. A scientific approach first develops each component individually, then assembles them into a definite coordinated whole.

A building is not erected by magic. First comes the architect's idea. This idea is based in part on existing material forms and their properties. After the architect makes a blueprint and the contractor is called upon, only then does materialization of the idea begin. Excavation of the ground, draining, cementing, erection of steel girders, installation of the plumbing system, electric wiring, plastering, painting—all these phases of construction, following each other in a pre-determined manner, bring us finally to such achievements of contemporary engineering as the Empire State Building, the Golden Gate Bridge, and other important structures. A spontaneous creation in the field of architecture would probably result in nothing more complex than a log cabin. Compare the amount of engineering involved in an African canoe carved out of one log or trunk, and that necessary in the construction of a modern battleship.

Music, as well as the other arts, still relies upon cave-age methods of production. One may build spontaneously and without computation a simple hut, a cave, a tent, or a cabin; but spontaneous creation of a skyscraper would only result in disaster. Even inexpert engineering could not be relied upon to carry out such a task. When some Parisian company tried in 1928 to erect its first skyscraper following the American trend, the structure fell after the steel had been erected to the ninth floor. And that was in Paris, where the Eiffel Tower stands as a symbol of engineering genius. It is clear that one art form is complex enough to handle. Several arts cannot be integrated without the aid of engineering.

An art structure in its formation (the process of being created) may be expressed as a series of sub-structures in their sequent development and accumulation.

$$\left. \begin{array}{l} N \\ S \\ I \end{array} \right\} = S_I t_1 + (S_I \div S_{II}) t_2 + (S_I \div S_{II} \div S_{III}) t_3 + \ldots$$
$$\ldots + (S_I \div S_{II} \div S_{III} \div \ldots \div S_N) t_n$$

S = sub-structure
t_1, t_2, t_3 = consecutive moments in time

The difference between engineering and the artistic method of production is mainly that in the scientific method, each sub-structure or component is developed individually and correlated thereafter with the other individually developed components. In the artistic process all (or nearly all) components appear simultaneously as an *a priori* coordinated group. Any change in such a group, with the intent of perfecting one of the components, changes the balance-ratio of the entire group, thus necessitating laborious reconstruction of other components. The latter frequently changes the balance-ratio so completely that the final product only remotely resembles the originally intended structure. With the scientific process, on the contrary, the development of individual components may be carried out to the utmost perfection. At the same time their relationship with other components of the same structure may be constantly controlled and integrated.

Thus, in the scientific process, there are to be found all the consecutive points of the series through which individual components are correlated with other components. In the artistic process, in contrast, many terms of the series are missing. Moreover, groups of components are generated simultaneously, which lack logical and esthetic coherence with the final form of the product.

When the artist begins his process with an individual component (such as rhythmic pattern in music or linear configuration in design), the image of such a component frequently lacks clear definition. The process may be compared with seeing an off-focus image and gradually focusing it. Imagine, for example, a pentagon gradually transforming into a hexagon. Such transformation corresponds to a modulation from 180° to 120° uniform continuous motion.[1] All intermediate values between 180° and 120° do not produce closed polygons. Now imagine a diffused polygon (*i.e.*, one with an infinite number of sides) gradually approaching the 120° form and, finally, attaining it. This is the foundation from which a mathematical definition of the creative process may be deduced quantitatively. If the desired esthetic form is a regular hexagon, for example, the creative process can be defined as increase or decrease of angular values approaching the 120° limit under uniform periodic motion.

In the arts, the final product often does not reach its limit (focus) but remains in a partially diffused state. Definition of the degree of diffusion becomes an esthetic factor in such a case. In short, the degree of diffusion itself becomes an art component, as in the paintings of Seurat and Pissarro, and in music generally, where "un poco piu mosso" or "piano, crescendo, forte" only suggest variable limits.

B. Nature of Organic Art

The significance of art can be measured by its immediate appeal and the effect of naturalness. True art, which can be defined as *natural* and *organic* art, has a general appeal and does not require any explanations, just as birds' plumage, their singing, the murmur of a brook, leaves, mountains, glaciers, waterfalls, sunsets and sunrises do not require any explanation. The art "isms," despite lectures, commentaries and volumes of analysis, do not become a jot more appealing. Despite the propaganda, these "isms" cultivate little more than hypocrisy in the semi-literate and pseudo-cultured strata of the population.

What is the criterion of a natural and organic art? It possesses characteristics not present in the art "isms." These are coherence of structure which enables it to survive, and high associative (semantic) potential which results from such coherence. For example, a well-defined and economically expressed thought, or idea, has greater persuasive power than one which is vague and incoherently expressed.

Art is organic when its form can be traced back to its organic source, as the winding staircase can be traced back to antlers, horns and cockleshells. The ancient Egyptians and Greeks discovered the forms of organic art through the principle of "dynamic symmetry." This principle was based on a so-called summation series; in this case, the series in which every third number is the sum of the preceding two. This series, known to the Egyptians and Greeks, was brought to the attention of the American public in 1920 by Jay Hambidge in *Dynamic Symmetry*. Known through the ages, it was disclosed in Luca Pacioli's treatise, *De Divina Proporzione*. Leonardo da Vinci, Michelangelo and many others adhered to it in their creative work. Today, it is being applied to many things in everyday use, including book jackets and radio cabinets, as well as in still-life paintings, landscapes and portrait work by contemporary painters. Credit for

---

[1] The details of such transformation are to be found in Chapter 2, *Production of Design*, of Part III. (Ed.)

the mathematical formulation of this series goes to Fibonacci[2] (13th century) and the series itself bears his name.

It is the organic structural constitution of a pitch or color scheme that reveals itself as esthetic harmony to our ear, eye, or any other organ of sensation. If we had the olfactory organs of a dog, we would probably now possess an art of smell comparable with the art of music.

Experimental artists, the "ism" makers like Picasso, Kandinsky, Klee, not to mention hundreds of their followers, are not bringing us a step closer to the evaluation of organic art forms. Many of these "modern" artists merely impose upon us their unsound, insane, feverish, dreamy and distorted hallucinations of an actuality given to us by nature to be consumed and enjoyed, together with oxygen, the ultra-violet rays of the sun, the energy of the ocean, and the smell of the pine forests.

What gives us the right to call these highly imaginative works of art "insane"? The same criteria which would bring about a similar diagnosis, if we could submit the art products of these artists to the examination of an expert psychiatrist. But do we want to be driven to insanity? Is this the function of the arts? When I attended an epoch-making exhibit of Dada, Surrealism and Fantastic Art held at the New York Museum of Modern Art in 1938, I could not help thinking how unfair the Museum was to the poor souls held in various insane asylums, who often project equally as disturbing products of their imagination. When I reached the fourth floor, my call for justice was answered by that part of the exhibit, which was actually contributed by the inmates of insane asylums.

It is often believed that the greatness of a work of art lies in its persuasive realism. You look at a painting, and it just lives. And there are paintings that are true to life. But what is such realism save a mimetic reproduction of reflexes? The facial and figure expression associated with suffering appears on a canvas, and the critics praise it. Why? Look around, or go to the places where misery flourishes and you will see more than you care to. But isn't it ethically superior to suffer without outward expression? But then artists wouldn't have anything to paint. If painting were confined to slavish realistic mimicry of outward expressions, works like "Ivan the Terrible Killing His Son" by Repin would be among the greatest paintings, as in this composition blood stains are truly horrifying in their true-to-life quality. Yet to see real blood stains and witness a real scene of assassination is still much more horrifying. In the age of improved color cinema we can get not only a realistic record of such a scene, but we can make it still much more horrible by extending the torture in the film itself, *i.e.*, by the technique of "slow motion."

On the other hand, we hear and read that the art of Picasso is very important. Why? Apologists say because he has found a new way of expression. Is it really new? The idea of giving two heads to one rooster is very childish, particularly when you recall what the ancient Greeks did to a Hydra, and the Hindus to their

---

[2] Leonard of Pisa, who developed the additive series in 1202, was the son of Bonaccio (filius Bonacci), which resulted in his being known as Fibonacci, the name of the series. (Ed.)

own God Shiva. Of course, we can be referred to the technique by which this has been accomplished. But then it leads us back to kindergarten craft.

There should be one criterion in art and art appreciation: its sanity as revealed in its organic quality, that is, its life, and ultimately—growth. When artists try to be original either by robbing the kindergarten world of ideas or by projecting nightmares, it means that art as expressed by such artists has reached its dead end. The constant rejuvenation of art comes through the modernization of old and even ancient folk traditions; but one has to remember that these ancient art traditions were closely associated with natural forms and resulted in truly significant art, whereas the revised versions of it, usually with the prefix "neo," are merely degenerating traditions. There is no progress for art in the revision of old forms, that had their place, significance, and became monuments of history, as in the ancient Egyptian, Greek, Hindu, Javanese, Chinese, and other civilizations. Let us leave the foolish and meaningless productions of hybrids to the milliners, dressmakers, and other "creators of fashion."

C. Creation vs. Imitation

When artists, thirsty for the primeval source of art, imitate the art of extinct civilizations, they are on the wrong track. The true "elixir of life," which is progress itself, is never in imitation but in creation. Let us see how creation has revealed itself in the great civilizations of the past.

The difference between the creative and the imitative processes of art production lies mainly in the difference between projecting forms from a given set of principles, whether consciously or unconsciously used, into a given artistic medium, and merely imitating the appearance of such forms in that medium. In primitive ornamental Aztec design, the underlying principle arises from rectilinear, rectangular and diagonal elements. By means of these elements, Aztec artists accomplished the projection of human and animal forms. The ancient Egyptians and Greeks used, as an underlying structural principle, the equality of ratios which can be expressed by means of the summation series, and which the Greeks called symmetry.[3]

Greek sculptors did not try to make a figure of a deity through copying the appearance of some living model but by establishing a system of proportions *a priori*. Such regulative sets, as the set of proportionate relations in this case, or the rectilinear elements of the Aztecs, constitute limitations which make art what it is, *i.e.*, a system of symbols integrated in a harmonic whole. Thus, the Egyptians, Greeks, and Hindus established standards of beauty which were expressed through the symbols of harmonic relations.

The same is true also of more primitive civilizations. In some forms, esthetic expression derives from direct reproduction of the surrounding medium—as in the matching of colors typical of the surrounding sky, vegetation, or birds' plumage.

---

[3] The Greek concept of symmetry obviously had little in common with the modern conception of symmetry as the balanced distribution of elements around an axis. The Greeks conceived of symmetry as a correlation of proportions; hence the use of the word "dynamic" to distinguish it from the contemporary "static" concept. (Ed.)

Some of the media are directly supplied by the natural resources of the immediate vicinity, such as clay in Indian pottery. Nevertheless, the transformation of natural resources, as in the treatment of clay to render it black, or the invention of a simplified counterpart of the surrounding impressions (esthetic symbols), are in the domain of true creation.

Imitative art, on the other hand, has no underlying principles, except the reproduction of appearances as close to the appearance of the original as can be achieved by manual craftsmanship. In this respect, any scientific apparatus, either optical or acoustical, gives an infinitely more accurate reproduction of the visible or the audible image.

The same can be said about the acoustical perfection of present-day sound recording whether on coated discs, tape or film. The real difference between the art of painting and color photography, when "true to life" reproduction of the image is the goal, is purely quantitative. Painting requires the use of a brush, and no brush can have a point as small as the size of a grain, or a chemical molecule in photographic emulsion. One of the great artists of all time, Georges Seurat, reduced the optical elements of an image to the extreme limit possible with the use of a brush. He built his images with miniature spectral points. This art became known as "pointillism," "dotting" in literal translation. It evolved its images from the elements, which consisted of a material point of the spectrum. The final image was integrated by means of points of different luminosity. This approach, very close in idea to the principle of televised images, is far closer to nature than any artificial conception of structure and its elements. Breaking down an image into some system of elements, in such coordination that they reconstruct the image as an independent counterpart of the original, constitutes a true art.

However, neither brush technique nor photochemical synthesis in themselves constitute art in the esthetic sense. Nor can craftsmanship alone produce an original work of art. In order to attain the latter, the system of elements participating in the building of an image must be developed into a harmonic whole. Such a harmonic whole is the complex of harmonic groups of various orders. Proportions pertaining to space occupied by the created image; harmonic relations constituting the arrangement of elements in the image itself; correlation of the harmonic relations of the image with the harmonic relations of the space occupied by the image; components of color, such as hue, luminosity, saturation brought into harmonic relations with each other and correlated with the space and the image, implying both color harmony and quantitative spatial relations—all these taken together constitute an artistic whole that is harmonic. With this in view, color photography, which no doubt will be highly perfected as a technique in the near future, should not be considered an art inferior to painting. On the contrary, the elements of the latter are incomparably cruder, and therefore not capable of the refinement of expression possible with the microscopic elements.

In the artistic accomplishments of the future, where the problem of composition itself will be based on a refinement of expression worthy to be a counterpart of our complex psycho-physiological organism, the refinement of the medium will be a necessary asset. Artists will have to forego thinking in terms of imitation, and even in terms of creation based on the primitive systems of the great extinct civilizations of the past. They will be compelled to devise a creative system, which, in its refinement, exactitude and complexity, will be an adequate counterpart of the efficiency of present-day transmitting media, such as sound-recording, cinema, and television.

It is unfortunate that most artists and esthetes do not realize that the future of the visual arts lies not in the improvement of painting but in the development of kinetic visual forms, the crudest of which we succeeded in developing during the last thirty years. In spite of its youth as an art form, the motion picture already has become the dominant form of entertainment. Now, with the advent of color, it has already achieved outstanding works of art, which could not have been attained in any of the traditional art forms in a comparably short period.

Every art form we know or can imagine today was at one time or other a magical procedure. Whether in the form of symbolic movements, ritual dances, symbolic traces on sand, as in the Navajo sand paintings, or the sounding magic procedures (such as noisemaking and incantations), magic art always had a formula. In his primitive life, man learned that nature, as well as animals and human beings, could be acted upon by definite formulae. This notion has developed during late centuries into scientific formulae such as are used in chemistry or engineering. When we combine chemical elements in order to obtain a definite reaction, we use a formula which takes into account the molecular structure of the elements and their effects upon one another. This analysis discloses why some numbers and their relations have been believed to possess magical power. A stimulus of a certain kind, when applied, necessitates a certain kind of reaction, depending on the reacting entity and its response to a particular stimulus. In the past, magic formulae crystallized through numerous repeated experiences. Basically, there is no difference between the use of a magic formula and a scientific formula, except that the first is judged by the result of its use while the second, by all the functional relations of the interaction between the elements of the stimulus and the reaction.

There is a long evolutionary chain, virtually without demarcation lines, that has been established from physical and chemical reactions of so-called "inorganic" nature; through the biological, that is physico-physiological reactions of so-called "organic" nature; to reflexes and complex reflexes constituting instinct; and, ultimately, to responses known as associations. Depending on the state of development of the given substance, its reaction may be considered purely *biological*, as in the protective coloration of a chameleon; or *artistic*, as in the associative color groupings in the works of Kandinsky. They may range from the reflexes of muscular contraction of an animal overwhelmed by fear to the projections of diversified horror, "inspired" by uncoordinated associated processes, as in the dream projections of such experiences in the paintings of Dali.

One of the most remarkable consequences of primitive man's experiences, which placed him far above the animal world, was that effect is caused by the extensive repetition of identical or analogous processes. The first outstanding result of this knowledge was that fire results from rubbing wooden surfaces together. This information alone, through associative memory, led to the use of tools for carving, sewing, etc.

With the advent of science, it was discovered that the repetitious application of waves in motion may affect matter itself and be of constructive as well as destructive power. Physicists found that supersonic waves decomposed sugar. The "magic" of sympathetic vibration noted among the phenomena of wave motion is merely a different form of magic formulae. And if we look more analytically into the source of various stimuli producing both physical and psychological effects, we find that they spring in the end from the same source, *i.e.*, chemical reactions and electron arrangements in the molecules. In short, we discover that the fundamental aspect of all stimuli is one or another form of regularity of occurrence, *i.e.*, periodic motion.

It may be stated that periodic motion, usually known as "rhythm," and used as a term in the arts as well as in medicine and astronomy, is the foundation of all happenings in the world we know. This has been sensed from time immemorial. The character of rhythm, as its implications grew, has become more and more mysterious, as nobody has succeeded in analyzing it. In truth, no scientific study has gone beyond the elementary forms of regularity encountered in physics and chemistry.

The fact that artistic intuition frequently functions in primitive as well as in developed art, according to some mathematically conceivable regularity, emphasizes the point. If we recognize that this is so, it is clear that we must know more about the nature and forms of this regularity. The main branch of this book, *Theory of Regularity and Coordination*,[4] is devoted to the solution of this very problem. Since the solution has been found, questions pertaining to the origin and evolution and manifestation of the arts can be answered directly.

[4] See Part II and Appendix A.

CHAPTER 7

MATHEMATICS AND ART

A. Uniformity and Primary Selective Systems[1]

THE foundation of this system is the concept of *uniformity*. This concept may be evolved from or associated with various axiomatic propositions. The most basic of the latter are:

1. The system of count based on the so-called natural integers (1, 2, 3, . . .), where the addition of the constant minimal unit makes it appear fundamental;
2. The system of the measurement of space, where all minimal units forming the scale of measurement are equidistant and correspond to the set of natural integers;
3. The clock system of time measurement, based on the sequence of uniform time instants and corresponding to the set of natural integers.

These three forms of uniformity appear to be axiomatic in their respective fields. The first, in the field of reasoning, *i.e.*, logical association. The second, in the field of sight, *i.e.*, perceptive visual association. And the third, in the field of hearing, *i.e.*, perceptive auditory association as aroused by the ticking mechanism of a clock, a metronome, or of any mechanism producing audible attacks at uniform time intervals; also, in the field of vision, *i.e.*, where motion in space follows uniform intervals in time, and where each consecutive phase of such motion, or the spatial interval between the two symmetrically arranged positions, is immediately apparent to sight. Such is the case of a pendulum, where the two extreme positions are apparently dissociated by space and time intervals. These space and time intervals appear to our consciousness to be in axiomatic, *i.e.*, one-to-one, synchronized correspondence.

There are other correspondences which, while elementary, are not axiomatic. These take place when the relationship between the optical and acoustical images is more complex. When a wave of one component, *i.e.*, of simple harmonic motion, such as the tuning fork, is projected on the screen of an oscillograph by means of the cathode-ray tube, the visual image appears to be axiomatic, due to the recurrence of the phase pattern; but the sounding image, being continuous and of relatively high frequency does not disclose its periodicity to the sense of hearing, because the individual phases of oscillation are too fast for auditory discrimination. The usual synchronization-controls of an oscillograph

---

[1]This chapter represents the most complete and most compact statement available of the foundations and rationale of Schillinger's *"Theory of Regularity and Coordination."* In addition to summarizing the underlying ideas, it offers a comprehensive survey of the material presented in Part II of *The Mathematical Basis of the Arts*. (Ed.)

produce a fairly stationary image on the screen of even two phases, in which case the regularity can be observed immediately.

Finally we arrive at more complex forms of dependence in the field of uniformity. Such is the logarithmic dependence of ratios within a given limit-ratio. For instance, within the limit of $\frac{a}{b}$ ratio we can establish a scale of uniform n ratios. These correspond graphically, *i.e.*, geometrically, to equidistant symmetric points, which can be represented on a straight line, the extension of which would correspond in frequencies to the ratio of $\frac{a}{b}$. In such a case, the first point of the linear extension corresponds to b and the last point, to a. All the intermediate points of uniform symmetry become:

$$\sqrt[n]{\tfrac{a}{b}},\ \sqrt[n]{(\tfrac{a}{b})^2},\ \sqrt[n]{(\tfrac{a}{b})^3},\ \ldots\ \sqrt[n]{(\tfrac{a}{b})^n} = \frac{a}{b}$$

We shall consider such sets of uniform ratios as primary selective systems.[2] We shall also make a note that the logarithms of the uniform ratios, uniformly distributed, become sets of natural integers and are translatable into equidistant spatial extensions. Thus, we acquire a unified system where ratios correspond to space-time units.

The refinement of primary selective systems depends on the discriminatory capacities of perception. The fineness of unit in a selective system is in direct correspondence to the potential plasticity of expression. A drawing of a human face with many lines is potentially more plastic than a drawing of the same face with a few lines. In music, where primary selective systems are tuning-scales, greater expressiveness is made possible by a greater number of units. A tuning system of five units, like one of the Javanese scales, is less flexible as a medium for constructing musical images than some of the Hindu scales where the number of units reaches 22.

Space and time constitute a continuum which is expressible in terms of physico-mathematical dependence. The continuum of space and time, or the "space-time" continuum, corresponds to the "dense set" of number values, which includes all "real" numbers, *i.e.*, both rational and irrational. The geometrical, *i.e.*, spatial continuum, possesses the property of dense sets, meaning that the number of points in a straight-line segment (which is finite in itself) is infinite. It follows from the above that *primary selective systems are not dense sets*.

On the other hand, psychological perception and consciousness, under all conditions established as normal, form a psycho-physiological continuum, *i.e.*, the perception of space-time is continuous.

Space-time produces the constant background in which our rational orientation is capable of discriminating the relations of isolated and group-phases. These isolated and group-phases in their interaction produce *configurations* which may or may not contain any perceptible harmonic relations, *i.e.*, using a different terminology, "rhythm." Mathematically speaking, the continuum of space-time perception is an integral of many variables, constituting the manifold of impressions.

[2] See Chapter 1 of Part III. (Ed.)

As mentioned before, the density of a primary selective series implies a corresponding degree of refinement. However, any selective series or a set reaches a point of saturation, not in the mathematical sense, but in the sense of perception and rational orientation. Beyond this point, the refinement becomes empirically useless and meaningless. Mach's tuning scale of 720 units per octave may serve as an example of such a set, as such small intervals do not permit the auditory capacity, as we know it, to discriminate either configurations or ratios of the intervals, or any distinctly noticeable difference between the adjacent units of the scale.

For this reason, the density of set, even in view of ultimate refinement as a goal, is largely conditioned by the discriminatory capacity of the respective sense-organ, or the degree of coordination necessary between the respective sense-organs.

As sight is a more developed form of sensation and orientation, it allows the construction of primary selective systems that are denser sets than in the case of auditory orientation. It should be kept in mind, however, that any capacity can be considerably developed by training, particularly if we think of such capacity as pertaining to discrimination and not to the integral of perception. For example, a person can be trained to discriminate, and as a result of such discrimination, to enjoy finer relations in the configurations of pitches or hues; nevertheless, no lifetime training can augment the individual's perceptive range of audible frequencies or spectral wave-lengths. True, we do extend the spectral range to ultra-violet, but this is due to method and not to the growth of the range of perception.

Summing up this reasoning, we arrive at the following conclusion: discriminative orientation is capable of detecting uniformity in the form of perceptible, discontinuous phases, and configurations in the form of harmonic groups produced by the phases as units. This orientation is coexistent with the continuity of the integral of space-time perception and orientation.

There are two requirements to be added: first, that the configurations develop within the perceptive and discriminatory range; and secondly, that the phase-units, *i.e.*, scale units are great enough for the respective perceptive and discriminatory capacity.

The first requirement implies that perceptive range may be different from discriminatory range. For instance, the perception of pitch-frequencies as a range exceeds the capacity of pitch-discrimination. It is difficult to specify pitch-relations, even for a highly trained professional either in acoustics or music, beyond 5000 cycles per second. Yet the same individual can hear a sound produced by a frequency four times as great. This means that the discriminatory range may be two octaves shorter than the perceptive range.

The second requirement implies that no selective scale should be so dense a set as to approach the perceptible limit of continuity. One consideration must be added to the last point. In some arts the configurations are so fluent that the transitions between points representing the members of the selective set, *i.e.*, the fixed points, are continuous. What makes the fixed points stand out, is the fact that such points are *relatively more stable* than the intermediate points of the dense set, *i.e.*, of the continuity. The stability itself is expressed by means of intensity or duration-stress (accent or greater time-value). This is the characteristic trait in the execution of Chinese music. When the stability of fixed points is not sufficient, the configuration may merge with the continuum. This is what occurs to the sound image produced by a moderate, continuous self-overlapping surf. Thus, all primary selective scales in this system are based on the uniform distribution of the range of a component.

Space-time is considered a general component and is treated as *continuity*. All other components, which are perceived by specialized capacities, such as the visual, auditory, olfactory, gustatory and tactile, are considered special components and are treated as *discontinuity*.

B. Harmonic Relations and Harmonic Coordination

There are other more complex forms of regularity than uniformity. They are the outcome of combined forms of uniformity, obtained through superimposition of phases. From the physical viewpoint such forms represent periodicity of complex phases, *i.e.*, configurations or groups consisting of several simple components. Uniformity may thus be regarded merely as a special case of periodic regularity, which may be either simple or complex. Forms of regularity produced by the periodic recurrence of binomial or polynomial groups are subjected to detailed analysis in this system.[3]

The main theory of the origin of all configurations that are not axiomatic is based on harmonic relations and harmonic coordination, and explains in detail the source and the technique of composition. This branch makes possible the mastery of "rhythm" as the basis of composition, and establishes its own validity as the system of pattern-making and pattern-coordinating. It is called *Theory of Regularity and Coordination*, and in the not too distant future should have repercussions in all fields of scientific investigation embracing the *liberal*, the *fine* and the *technical arts*.

Application of this theory to music and visual arts has already produced far-reaching results. Facts demonstrate and prove beyond doubt that all efforts in the field of art-evolution, whether intentional or not, are the expressions of certain tendencies toward a certain goal. This theory can define the goal by detecting the tendencies as they reveal themselves in the course of evolution. It means that without undergoing the evolutionary stages of a certain final product, we can obtain such a product by direct scientific synthesis. This is true of an individual creation as it is of a style, a school, or of a whole culture.

These are the techniques embodied in the *Theory of Regularity and Coordination*, and these constitute the clue to the process of creation.

All non-uniform forms of regularity are the resultants of the interference of two or more uniform simple periodic waves of different frequencies, brought into synchronization. Such resultants, being the source of configurations, can be obtained either by direct computation or by graphs.[4]

When these resultants, the parent-shapes of all rhythms, are applied in

[3]See Appendix A.   [4]See Chapter 3. "Periodicity" of Part II. (Ed.)

direct sequence to any of the primary selective series, they, in turn, produce artistic scales, or sequent structures, which are the *secondary selective series*. These sets also vary in density, and, when they reach the point of saturation, become identical with the primary selective series. Thus, we can state that the secondary selective series is the result of rarefaction of a primary selective series, which is the limit, and, thus, the dense set of the secondary series.

Secondary selective systems produce sequences known in music as the rhythm of durations, of pitch, (*i.e.*, pitch-scales), of sequences of chord-progressions, of intensities, of qualities, of attacks; in fact, of any or all components. In linear design they are the source of sequences of linear, of plane or of solid motion, resulting either in static configurations like spirals or polygons, or in trajectories of various types, which include the participation of actual time in the process of performance. More concretely, the resultants of interference produce configurations from linear extensions, angles and arcs, which are the equivalents of scales. Thus, we can refer to scales of linear extension, scales of angles, scales of arcs,[5] etc.

The next stage of the *Theory of Regularity* deals with variations based on general or circular permutations. Both general and circular permutations produce compensatory scales to the original scales, which may be called derivative scales. The derivative scales together with their original, or parent-scale, constitute one family of secondary selective systems. However, in practical application, only such derivatives can be used as one family which result from permutations of the same subcomponents. For instance, pitch-scales constitute one family if the derivatives result from permutations either of the pitch-units or of the intervals between the latter, or of both such operations combined, but not from the combinations of several such sets resulting from different operations.

The originals and the derivatives, being brought into self-compensating sequence, at the same time produce self-compensating simultaneity. In many instances of artistic production, such simultaneity-continuity groups constitute a complete composition.

After a family of scales has been developed into a compound secondary selective system, new use may be made of the resultants of interference in the form of coefficients of recurrence, or coefficient groups. Such coefficient groups introduce a recurrence form into any original or derivative scale. The recurrence of scale-units, applied to any original or derivative scale, transforms the secondary selective scale into a master-pattern, or thematic motif, which in such a case becomes the original operand and the tertiary selective scale, set, or system.

Further permutation of the master-pattern results in derivative master-patterns, which, together with the original, produce a self-compensating simultaneity-continuity group.

The final compound set may constitute a complete composition or a complex theme from which a complete composition may be evolved by means of further permutation. This operation can be extended to any desired order and results in any desired complexity in the final configuration, which generally retains its homogeneous character.

[5] See Chapter 1. "Selective Systems" of Part III. (Ed.)

MATHEMATICS AND ART

The next basic technique, which follows the generation and the variation of the secondary, tertiary and family-sets, is the composition of harmonic contrasts.[6] Harmonic contrasts, evolved as counterparts to original sets of whatever origin, are based on distributive involution. Polynomials representing master-patterns of any origin, type and complexity can be proportionately, *i.e.*, harmonically, coordinated with any number of other components, which may belong to the same or to a different art form. Thus, not only two or more spatial configurations can be proportionately coordinated, but a spatial kinetic configuration may be coordinated even with a temporal configuration of sound, movement, lighting, etc. This technique is of particular value in coordinating either the different components of one art form, or the different art forms on the basis of harmonic contrasts.

Involution-groups derive from the same source as the interference groups, *i.e.*, from the uniform sets. The selection of binomials or of polynomials from a set is done according to the possible distributive form of the given set or series. The determinants of the series are associated with the set of natural integers, such as $\frac{2}{2}$ series, $\frac{3}{3}$ series, ... $\frac{n}{n}$ series. Some families are pure, *i.e.*, they belong to one series; and some are hybrid, *i.e.*, they are the composite of several determinants. Evidence shows that art works of superior quality are "purer" than those of inferior quality. This is true of traditional art as well as of individual artists. The greatest works of art have been created, not by the innovators or modernists, but by summarizers or synthesizers, who crystallized the experience of their predecessors to the highest degree of perfection. Such perfection is equivalent to consistency, and to correspondence between intentions and the forms in which they are projected. In this sense, consistency means consistency of the serial determinant, *i.e.*, strict adherence to a certain series: for instance, the adherence of honeycomb cells to hexagonal symmetry, *i.e.*, to $\frac{6}{6}$ series, of starfish to $\frac{5}{5}$ series, of the hereditary law of Mendel to $\frac{4}{4}$ series, and of the forms of growth to the first summation series, which consists of a group of determinants like $\frac{2}{2}, \frac{3}{3}, \frac{5}{5}, \frac{8}{8}$, etc. These are symmetric in the Hellenic sense, *i.e.*, they produce an equality (constancy!) of ratios.

But the same forms of series participate in design, in sculpture, in poetry, in dance, and in the other arts. In this sense, some forms of African and Asiatic music, dance, and ornamental design belong to $\frac{3}{3}, \frac{5}{5}$, and $\frac{8}{8}$ series. Other forms of European art belong to $\frac{3}{3}, \frac{4}{4}, \frac{6}{6}$, and $\frac{12}{12}$ series. The interference method, when applied to binomials of a certain series, generates the entire evolutionary group of that series. After reaching the point of saturation (binomials generate trinomials, trinomials generate quintinomials and so on), the resultant of the last order of that family is a dense set, *i.e.*, uniformity of the limit.

These family- or style-series, after exhausting themselves in the last interference, either undergo the above-described involutionary harmonic development, or hybridize themselves through intermarriage with other determinants. There are many hexa-octagonal hybrids, for instance, in both music and design, *i.e.*, configurations which are combined products of $\frac{6}{6}$ and $\frac{8}{8}$ series.

The method of interference, as applied to evolutionary series, is the basic

[6] See Chapter 5. "Production of Combined Arts" of Part III. (Ed.)

clue, possibly, to an understanding of the sciences and the arts. It is undoubtedly the most powerful analytic tool known, being at the same time the source of creation and of forecast. The latter is possible because once the tendency reveals itself, the succeeding evolutionary stages can be determined *a priori*.

Once all these resources of regularity and coordination are disclosed, the most ambitious undertakings in the field of art creation can be easily accomplished, as the series and the involution technique take care of both the factorial and the fractional sides. This means that form, growing as a whole, is at the same time coordinated with its own individual units and their configurations. Thus, the whole evolves itself on the basis of compensation and/or contrast.

C. Other Techniques of Variation and Composition

Further techniques evolve from other sets than those already described. Among these, the most prominent are the arithmetic progressions, geometric progressions, summation series, involution-series and various other types of progressive series, all of which may be used as the secondary selective systems, *i.e.*, configuration scales.[7] Such sets and series control the effects of variable velocities and of growth. The effects of positive and negative acceleration are the domain in which such scales provide the basic material of structure and coordination. Coordination is based on variability of phases, coefficients and direction (positive and negative acceleration). These scales, being coordinated with their own converses, produce resultants of interference of their own kind. These resultants, in turn, become scales, master-patterns, etc.

Another important phase of composition, pertaining to variability and extension, is quadrant rotation.[8] Optical and acoustical images can be subjected to this form of variation, which results in four equivalid variants of the original (including the original). The four variants can be further extended by means of the permutation technique, which, in turn, can be extended to any desired order. These variations do not change the inherent relations of the original configuration, but merely project it into a new geometrical position. Quadrant rotation is a technique which is both basic and natural.

Still another basic technique of variation and composition is coordinate expansion.[9] It may be performed either to the abscissa, thereby affecting the general component, or the ordinate expressing some special component. Coordinate expansion may be positive or negative (contraction); it may also assume various forms, such as arithmetic, geometric, logarithmic expansion or contraction, etc.

Different forms of coordination of the two axes of expansion, corresponding to the two coordinates, result in a variety of types of geometrical and optical projection. Control of proportions in homogeneous and heterogeneous space can be accomplished by this technique. As a resource of composition it is rather limited in the field of acoustical images, largely due to instrumental and perceptive limitations, but has an unlimited scope in the visual arts.

[7]See Chapter 2. "Continuity" of Part II. (Ed.)
[8]See Chapter 10. "Quadrant Rotation" of Part II. (Ed.)
[9]See Chapter 11. "Coordinate Expansion" of Part II. (Ed.)

In music, being applied to pitch-coordinate, geometrical expansion transforms music of one style of intonation into another, modernizing the original when used in its positive form. For example, expansions of Bach or Handel produce Debussy or Hindemith. Geometrical expansion of pitch affects the primary selective systems in such a way that their original points of symmetry become their own squares and cubes in the positive expansion, and their square or cube roots in the negative expansion. Thus, for instance, the contraction of a tuning system from the $\sqrt[12]{2}$ to the $\sqrt[24]{2}$ would permit the performance of semitonal music in the quarter-tone system. Positive expansions may be performed in the original system in which they produce configurations based on the rarefied set. In the visual arts, this technique produces forms of optical projection and aberration such as those for which El Greco and other artists of unusual perspectives are famous.

As sets and configurations grow through their own variability, resulting in different forms of saturation, the next important technique is that which pertains to density.[10] A density group must be considered a configuration of density or saturation. A dense group of density is the limit-group of saturation under specified conditions. A dense audible image requires the use of all parts of the score in a dense distribution of pitches. A dense visual image is saturated with configurations, which may overlap, and which are confined to the visual area, such as the canvas or the screen. In music, a monody is not a dense set, compared to harmonically accompanied polyphony. In painting, one flower occupying only a small portion of the total area is not a dense set, compared with a battle-scene completely crowding the picture.

The technique of composition of density-groups, their variation and coordination, controls the degree of saturation and its distribution throughout the whole. The variants of the original density configuration, *i.e.*, the master-pattern of density, in turn produce simultaneity-continuity density groups. The same configuration, of whatever origin, may coexist with its own variants of density, which often implies an evolutionary stage. Density groups, together with their variants as configurations, are subject to *positional rotation*. Such rotation may follow both coordinates and intercomposes their phases. This latter stage represents the ultimate degree of refinement in composition.

Homogeneity of character in the work of art defines the necessary quantity of master-patterns pertaining to various components. Such master-patterns constitute the thematic material of the composition, and may be coordinated into a whole by any of the basic techniques described above, or any combination of the latter.

One cannot fail in evolving a work of art according to this method of specification, selection and coordination. The specifications must be chosen in accordance with semantic requirements, which define the purpose of the production.

---

[10] See Chapter 12. "Composition of Density" of Part II. (Ed.)

D. Pragmatic Validity of Theory of Regularity

Seven points upon which the pragmatic validity of this theory rests:

(1) It establishes esthetic principles which remain true in any special instance.
(2) It provides a foundation for more efficient creation and a more objective criticism.
(3) It does not circumscribe the freedom of an individual artist, but merely releases him from vagueness by helping him to analyze and to realize his own creative tendencies. It gives him a universal knowledge of his material: the principles and the techniques of this system permit an infinite number of solutions, which satisfy any requirements set forth by art problems.
(4) It offers the student of this theory a manifold of techniques that enable him to handle individual and combined art-forms. Since these techniques are interchangeable and inter-related, designs and melodies may be plotted as graphs, and the student may dance or sing a design, or translate a melody into a drawing.
(5) It is scientifically valid in that it establishes the basic principles underlying creative processes and correlates esthetic reactions with generating excitors (*i.e.*, the works of art) of definite forms and defined variations.
(6) It stimulates art production and reduces the years of training, thereby making creation a process associated with pleasure, accomplishment and satisfaction.
(7) Its social significance lies in the fact that it leads toward unity and tolerance and away from disunity and intolerance through an understanding of the basic principles of inter-relatedness and functional interdependence.

This theory has been presented in part before various learned societies, including American Institute for the Study of Advanced Education, Mathematics Division of the American Institute of the City of New York, the Mathematicians Faculty Club of Columbia University, and the American Musicological Society. It has also been offered by the author in the form of courses and lectures at Teachers College of Columbia University (Departments of Mathematics, Fine Arts and Music), at New York University, and at the New School for Social Research.

Students of this theory included educators, architects, artists, designers, composers, and conductors. Some of them were celebrated artists when they came to study with me, and some attained prominence, partly as a result of their studies.

Some of the products of this theory, including my own compositions and the compositions of my students, have been presented in various forms in the field of symphonic, chamber and applied music (motion pictures, radio); in art exhibits, such as at the Architectural League of New York; and in science exhibits, such as that at the Mathematics Museum of Teachers College, Columbia University.

Most contemporary minds occupied with the thought of employing engineering to serve art, have restricted their research to telecasting, television, enlarging visible images, amplifying sound, expanding the range of visibility and audibility, etc. Mathematical analysis of the process of composition and production in art media and by means of electro-mechanical (acoustical, optical, etc.) synthesis has been for years the subject of my research. The theory formulated in this work is an objective system in the sense in which a system of algebra or geometry is objective. Through the application of principles and techniques evolved in this theory, works of art can be produced by computation and plotting, and therefore can also be realized mechanically in an art medium, *i.e.*, through automatic composition and performance.

PART TWO

THEORY OF REGULARITY AND COORDINATION

Each system is valid when functioning within its own strictly defined limits and its own operational conditions (laws).

We discover in the evolution of method that new processes call for new operational concepts, new terminology and new symbols.

Consecutive Selective Techniques

(1) NOTATION:
    a. universal system of mathematical notation for all art forms, structures and processes;
    b. graphs.

(2) CLASSIFICATION OF FORMULAE:
    a. series (selective)
    b. distributive and combinatory
    c. definitive (identities)
    d. quantitative, computative (equations)
    e. combined: computative-distributive (involution)

(3) SELECTIVE SERIES:
    a. primary
    b. secondary
        Components and configurations
            (structures, assemblages, forms)
        Continuity series (secondary selective series):
            closed and progressive forms of symmetry.

(4) DERIVATIVE HARMONIC GROUPS: the resultants of interference and their involutionary forms. Use of these groups as coefficient recurrence-groups. Evolutionary series (interference development).

(5) SYNCHRONIZATION (COORDINATION): primary, secondary, etc.

(6) KINETIC GEOMETRY (SPACE THEORY) [proportions, kinetic forms and images, diffusion forms, logarithmic variations of coordinates, quadrant rotation, cylindric rotation through abscissa or/and ordinate, projections: rectangular, spheric, etc.]

(7) PHYSICS: color, space, motion, frequency, intensity.

Applications of Mathematical Techniques

(1) ART FORMS (definition and classification).

(2) ART COMPONENTS (primary and secondary scales of components and their interrelation); (interrelation of components in an individual art form).

(3) ART PROCESSES (the techniques of production).

(4) CORRELATION OF ART FORMS.

(5) SAMPLES OF SYNTHETIC ART IN THE INDIVIDUAL AND COMBINED FORMS.

CHAPTER 1

CONTINUUM

A. DEFINITION OF AN ART PRODUCT BY THE METHOD OF SERIES

THE mental growth of humanity, as revealed in scientific thinking, may be stated as a tendency to fuse seemingly different categories into a complex unity, into which previous concepts enter as component parts. The evolution of thought is a process of synthesizing concepts.

Assuming this as a methodological premise, we can build complex concepts from concepts which previously seemed dissociated. Such concepts enter into a series as definitely located terms. Thus, a "record," an "impression," or a "stimulus" may be regarded as only one of a series of terms with respect to its place within different series.

The evolution of the concept of a spectrum provides an analogous case. Sanscrit literature shows that the three primary colors linguistically and otherwise evolved into the whole spectrum, and that this evolution may be determined through serial development from the primary colors.

Example: (1)

A simplified interpretation of color using standard terminology will easily show how the method of evolving series can be applied. Selecting from the infinite spectrum of hues a limit within a certain primary yellow and primary blue, we may observe the following growth of the series.

$$Y \ldots \ldots \ldots B$$
$$Y \ldots \ldots \ldots G \ldots \ldots \ldots B$$
$$Y \ldots \ldots \ldots YG \ldots \ldots G \ldots \ldots \ldots BG \ldots \ldots B \text{ (additive system)}$$

We find the same happening in musical pitch: a scale grows through inserting definite tones between tonal limits, whatever they may be: $(\frac{2}{1}, \frac{3}{2}, \frac{4}{3})$.

Example: (2)

$$\frac{2}{3} \ldots \ldots \ldots \frac{3}{2}$$
$$\frac{2}{3} \ldots \ldots \ldots \frac{2}{2} \ldots \ldots \ldots \frac{3}{2}$$
$$\frac{2}{3} \ldots \ldots \ldots \frac{2}{3} \ldots \ldots \ldots \frac{2}{2} \ldots \ldots \ldots \frac{3}{2} \ldots \ldots \ldots \frac{3}{2}$$

In a scale of increasing frequencies and applying the $\frac{2}{1}$ (octave) adjustment, the following development may be seen:

$$\frac{3}{4}\ldots\ldots\ldots \frac{8}{9}\ldots\ldots\ldots \frac{2}{2}\ldots\ldots\ldots \frac{9}{8}\ldots\ldots\ldots \frac{4}{3}$$

(The note equivalents of these ratios are: G, B♭, C, D, F)

As in this case, all the tones represented through definite ratios of frequency may be regarded simply as different terms of a series with respect to their places within the series.

It is not sufficient to examine a work of art in order to produce the phenomenon of an "art product." A number of conditions must be fulfilled in order to obtain an art product. A work of art is primarily an idea expressed through the means of art—its material. All art material has definite properties, *i.e.*, it is a complex of sound, a complex of light, a complex of mass, etc. Each art material consists of a number of components, such as pitch, intensity, quality of sound; hue, intensity, saturation of color; etc.[1]

An idea, therefore, can be expressed through correlation of the components of art material. When this condition is realized, we have the first requisite for the building of an art product.

Thus the first term is:—*an idea as realized through the correlated components of an art material.*

The value of such an idea is to produce a stimulus, sensory, motor, or mental. A sensory stimulus may or may not transform into a motor or mental form. A motor stimulus occurs, for example, when Charleston music stimulates muscular movements which result in an appropriate dance form, or when military music stimulates one human being to attack another. An illustration of a mental stimulus would be *Sposalizio* by Franz Liszt, which was stimulated by the *Sposalizio* of Raphael. The ancient Greeks did not believe that anything was achieved if a musical work did not produce mental stimulation.[2] Thus, *stimulus* is a term in a chain that results in the phenomenon of an art product. As soon as we have obtained these two terms ("idea" and "stimulus"), we can develop the rest of this complex concept by the method of series. "Idea" and "stimulus" are the two extreme terms of the series:

IDEA.........STIMULUS

There must be a *medium* between an idea and a stimulus—a purely physical medium like air or an electromagnetic field. Thus, we have a three term series:

IDEA.........MEDIUM.........STIMULUS

In order to exist in the medium, an idea should be generated physically. This gives us a *generator*. A "generator" in music can be an acoustical instrument; in the art of photography, an optical instrument. The "generator" will produce actual physical oscillations. To analyze these physical oscillations, a *transformer* is necessary. The human brain is such a "transformer."

---

[1] Art material is treated analytically as a system; its components, as parameters forming such a system.

[2] Plato, *Polyteia*.

The next stage in the development of this series is:

IDEA ..... generator ..... MEDIUM ..... transformer ..... STIMULUS

An idea can be expressed physically on a *record*. Such a "record" in music can be the sound track on a film, a record groove, a musical score in musical notation, a musical score in graphs, etc.

A "generator" requires a *transmitter* which will propagate the oscillations through a medium in a form adequate to the *record*. In music a human performer usually is the "transmitter." A mechanism may be substituted for the performer.[3] A "transformer" obtains its material supply from the *receiver*, which may be an organ of sensation such as the eye or ear. Finally, a *stimulus* can be obtained from a *transformer* only through the *impression* (made physiologically and electrochemically upon our brain cells, or psychologically by affecting our mood).

This completes the nine term series representing the complex concept of an Art Product:

IDEA ...... record ...... generator ...... transmitter ...... MEDIUM ...... receiver ...... transformer ...... impression ...... STIMULUS.

There can be two types of results in the development of an art product:

1. If the stimulus is sensory when the series is completed, the experience will fade out. This aspect of the series will be graphically a straight segment.
2. If the stimulus is mental or motor, the final term-stimulus will transform into a *new idea* (I_2) or into *action*. Then the evolution of the series will go through the same process, starting from the new point. This aspect of the series will be graphically an infinite cylinder built through consecutive coils,

Figure 1. *Chain relationship of idea and art product.*

where "I" will represent a new idea resulting from consuming and assimilating an art product.

[3] In an orchestra, a conductor is the transmitter and the orchestra is a generator. In the player-piano, a perforated paper roll is a record and a transmitter at the same time, while the piano is a generator. The generator is a passive mechanism; the transmitter is an active mechanism.

This chain process is logically similar to the continuation of a race through the reproduction of an individual. This is the only way to preserve the products of evolution. In some cases, for example in painting as compared to music, certain terms in the series will be missing because the record (canvas), the generator (canvas) and the transmitter (canvas) are identical. In the case of a compound art form (like sound cinema), we shall have parallel simultaneous series, absolutely alike or different.

An art idea can be transmitted from one individual to another if the sum total of their previous experiences in art is not too different. In order to produce an idea at the end of the series, the perceiving individual must be equipped with discriminating experience. This allows him to determine (or at least to have a general orientation in) the relations and correlations of parametral values. He must adapt himself to the absolute values of extensions, directions, positions and durations, in order to enjoy all the variations and modifications of an idea. He must be well-equipped particularly with discriminating experience in the field of special components.

Imagine an individual whose sense of pitch discrimination is such that he cannot recognize an increase or decrease in frequency of sound within $\frac{9}{8}$ ratio.[4] A melody by Chopin will be entirely distorted in his audible apprehension. Imagine a color-blind individual looking at a painting. The color relations will be beyond his visual apprehension. Finally, in order successfully to build an art product, concentrated and focused attention is necessary; otherwise, many elements will escape and the whole may be destroyed. In other words, the active cooperation of a consuming individual is absolutely necessary.

For the successful projection of an art idea:

1. all the terms of the series must be valuable or reliable in the adequacy of their functioning (the idea should have value, the record must express adequately the idea; the generator should be plastic, precise and reliable, etc.).

2. the consuming individual must be reliable in his focused attention and parametral discrimination.[5]

A change occurring in any term of a series might have an effect on some other terms of the same series.

If in a series forming an art product we take Rd (record), Rr (receiver) and In (impression):

$$Rd_1 — Rr_1 — In_1$$

---

[4] One whole tone.

[5] There is always a certain amount of the unknown in a work of art which we cannot discriminate until we go through several analogous experiences. If the amount of the unknown equals zero, we have perfect banality. If it is not too great, the majority enjoys it immensely. If it is too great no one enjoys it at first, for the simple reason that he does not understand the language in which it is brought to his attention. The ideal for a discriminating group of people lies between these extremes.

and if we change the Rr with the desire to preserve the same impression, we are forced to change the record:

$$Rd_2 — Rr_2 — In_1$$

Otherwise, the result will be:

$$Rd_1 — Rr_2 — In_2$$

In other words, if our perceptive apparatus changes, we cannot apprehend a work of art in the same way any longer. An authentic performance of Bach's *Fugues* on a harpsichord cannot impress us in the same manner and to the same degree that it impressed Bach's contemporaries. Here is an answer as to whether a transcription or an arrangement of old music is legitimate. It is not only legitimate but necessary if we do not wish to miss the enjoyment our ancestors experienced.

DEVELOPMENT
OF THE COMPLEX CONCEPT
OF AN "ART PRODUCT"
BY THE METHOD OF NORMAL SERIES

```
                    b....y
                b.....m.....y
           b.....h.....m.....s......y
   b.....e.....h.....j......m....p.....s......v......y
```

Idea..Stimulus
Idea..Medium..Stimulus
Idea..Generator..Medium..Transformer..Stimulus

Idea..   Stimulus..

   Record..   Impression..

     Generator..   Transformer..

       Transmitter..   Receiver..

              Medium..

Figure 2. Development of an art product by the method of normal series.

B. Parametral Interpretation of a System

In the first edition of his *General and Special Theory of Relativity*, Albert Einstein based his physical interpretation of the universe on Minkovsky's and Riemann's geometrical premises. Thus, he established for all space-time momenta, including their electro-magnetic behaviour, a hypothetical, geometrical system of parameters (measuring lines): X_1, X_2, X_3, X_4. It is neither essential for these parameters to have actual existence in this universe nor for them to correspond to our sensory perception. They are assumed in a purely mathematical (logical) sense as tools that enable us to explain what we know as the physical universe.[6]

There is no reason why analytical study of the world of art should methodologically differ from that of the physical universe. Our definition of an art product opens the door to such an interpretation.

What is art? We sit in a concert hall and listen to a piano recital. What stimulates us emotionally and mentally? The pianist touches the keys, which excite the strings; the strings move in transverse vibrations, which are sent through the air medium in longitudinal waves; the longitudinal waves affect our organ of hearing. This purely physical complex transforms into a physiological, *i.e.*, electro-chemical one, and, finally, into one of tone. The pianist learns which keys to strike, and in what progression and manner, from a highly symbolic and imperfect record called a musical score. He tries to present an adequate projection of this symbolic record, which contains the composer's idea as expressed in simultaneous and consecutive tone relations. Is this all? We believe it is, insofar as music is concerned, providing that the method of recording the composer's ideas is adequate and reliable.

One may ask: But what about the personality of the performer? his "magnetic" power? Then we may also ask: What about the personality of the alluring companion whom you brought to the recital and whose influence upon the resultant of your mood during the recital should be rated at 35% at least? All these factors must be taken into account since this is a problem of psychological resultants produced by the entire complex of stimuli acting upon us at the time. But these are not aspects of the art of music as such.

We can learn a great deal about moods stimulated by music if we learn enough about music itself. And this knowledge can be gained only from a reliable record. From a purely acoustical viewpoint, music can be analyzed from phonograph records, film sound tracks, oscillograms, etc. But this would provide little help in understanding a musical idea, and for the same reason that a microscope would be of little use in determining the contour of a continent. An oscillogram gives a splendid projection of a sound wave at a given moment, but with such detail and complexity that all the points of musical interest are

---

[6]During the year 1933 Einstein adopted the five parameter system (X_1, X_2, X_3, X_4, X_5) because the previous system failed to explain certain physical momenta, thus requiring the introduction of a new parameter.

too far removed from one another to mean anything musically. In both cases, the method is unsuited to the dimensions. From the physical point of view, it makes no difference whether chords change at one speed or another, while musically such variations in speed might completely transform the artistic meaning.

Why should one chord be followed by another and why should changes in speed be necessary? This cannot be answered by the physics of sound. The fact that the science of musical sound (not the science of musical composition), during the 5000 years of its existence, did not explain the mechanism of musical composition is sufficient evidence that acoustics is not adequate to provide such an explanation.

Herman von Helmholtz in the introductory chapter to his *Sensations of Tone* ("Plan of the Work") writes: "Questions relating to the equilibrium of the separate parts of a musical composition, to their development from one another, and their connections as one clearly intelligible whole, bear a close analogy to similar questions in architecture." Friedrich Schlegel said: "Gothic architecture is frozen music." By inverting this proposition we can say: music is fluent (or animated) architecture. But if gravitational tones are today to be explained geometrically, we can ascertain that music is a *fluent geometry*, *i.e.*, geometry using the time parameter. With a sufficient number of parameters, we can explain all the "laws" pertaining to a functioning system, whether it is a system of art or of the whole physical universe. When one ascertains that harmony or counterpoint can be explained, but not melody, it merely indicates that the necessary musical parameters have not yet been found, and that the interpretation provided for harmony and counterpoint may be correct only in certain cases.

The subtle points in music technically correspond to gravitational or electromagnetic phenomena (accumulation and discharge, attraction and repulsion), and once they acquire the proper geometric interpretation, all the "mysteries" of dramatic quality in music are revealed. This analysis can be made by means of the adequate record of an art idea. A comparative study of such records leads to the establishment of general laws of composition in art.[7]

A continuum is a system of unlimited parameters. In terms of measurement, parameters are the extensions. Any individual art form (music, sculpture, etc.) is a continuum, *i.e.*, a system of parameters representing the art components. Thus, we can speak of different art forms as different continua (continuum of sound-music; continuum of mass-sculpture, etc.) Every individual art continuum consists of two kinds of components:

1. general components (time, space);

2. special components (frequency, intensity, quality).

---

[7] The function of an interpreter of music, who is the "generator" in the whole complex of a "musical product," approaches zero when the "record" includes all the specifications of a "musical idea" and when his intentions are sincere; in other words, when he is trying to present an adequate projection of a "record." Thus, the elimination of a living performer is a natural result of the process of evolution. Interpretation will probably survive as a hobby.

An individual art continuum is related to a sensory continuum; the parameters of an art continuum correspond to the functioning of the organs of discrimination. If sound, as a musical continuum, consists of parameters of frequency (pitch), intensity (volume), quality (timbre and character), there must be corresponding discriminating units in our organ of hearing. These discriminating units produce corresponding reactions.[8] Thus, the cochlea in the inner ear, for instance, respond at definite rates of frequency with their microscopic stringlike units (Cortis arches) of varied length. The degree of air pressure of a sound wave, resulting from different amplitudes, affects the ear-drum likewise. Quality is a complex component resulting from frequency, intensity and wave-phases in their different relations. Discrimination of the quality of sound does not require new organs of discrimination.

The general categories of our mind, according to Kant, provide us with an orientation in space-time relations (general components), while our organs of sensation enable us to discriminate our sensory perceptions (special components).

Different individuals have differently developed organs of sensation. The discriminatory functioning of these organs can be developed greatly through training. These human abilities have usually been underestimated. The capacity in pitch discrimination of an average house dog, according to Pavlov's[9] experimentation in conditioned reflexes, is up to $\sqrt[240]{2}$, i.e., one fortieth of an equally tempered whole tone[10] in the twelve step equal temperament. A musically inclined individual, with fairly good pitch discrimination, notices the difference in about one one-hundredth of an equally tempered whole tone ($\sqrt[600]{2}$).[11] It does seem strange that some musicians think a quarter-tone too small an interval and too hard to discriminate.

C. The First Group of Art Forms (One System of Special Parameters)

We shall now discuss the different parameters of various art continua. We shall describe the individual art forms (art continua) that are possible, and the complex art forms (systems of continua) that can be deduced from them. Classifying art forms through the number of their general parameters, we obtain the following scheme:

The first group requiring one organ of sensation at a time and one system of special parameters.

---

[8] According to recent studies, these reactions are electro-chemical processes.

[9] Ivan Pavlov: *Conditioned Reflexes.*

[10] A whole tone in an equally tempered twelve-unit temperament = $\sqrt[6]{2}$.

[11] See Prof. Carl Seashore's materials on tests in pitch discrimination. I also had the opportunity of making tests by means of an electric organ constructed by Leon Theremin with variations in pitch up to $\frac{1}{300}$ of a whole tone in the twelve-unit equal temperament.

COMPLETE TABLE OF INDIVIDUAL ART FORMS

Sensation	General Component	System of Special Components		Title
Hearing [kinetic]	Time [1]	Sound	1.	The Art of Audible Sound
Touch	Time [1]	Mass	2.	The Art of Touchable Mass
Smell	Time [1]	Odor	3.	The Art of Smellable Odor
Taste	Time [1]	Flavor	4.	The Art of Tastable Flavor
Sight [static]	Space (X_1, X_2) [2]	Light	5.	The Art of Visible Light
Sight	Space (X_1, X_2) [2]	Pigment	6.	The Art of Visible Pigment
Sight	Space (X_1, X_2) [2]	Surface	7.	The Art of Texture of Visible Surface
Sight [kinetic]	Time [1], Space (X_1, X_2) [2]	Light	8.	Kinetic Art of Visible Light projected on a Plane Surface
		Pigment	9.	Kinetic Art of Visible Pigment transforming on a moving surface
		Surface	10.	Kinetic Art of Visible Texture transforming on a moving surface
Sight [static]	Space (X_1, X_2, X_3) [3]	Light	11.	Static Art of Visible Light placed inside of a 3-dimensional spatial form
		Pigment	12.	Static Art of Visible Pigment covering the surface of a 3-dimensional form
		Surface	13.	Static Art of Visible Texture of 3-dimensional forms
		Mass	14.	The Art of Static 3-dimensional visible mass
Sight [kinetic]	Time [1], Space (X_1, X_2, X_3) [3]	Light	15.	The Art of Visible Kinetic Light projected on a 3-dimensional or a 2-dimensional screen in motion
		Pigment	16.	The Art of Visible Kinetic Pigment covering 2- or 3-dimensional surfaces in motion
		Surface	17.	The Art of Visible Kinetic Texture of surface or volume
		Mass	18.	The Art of Visible Kinetic Mass

Figure 3. Complete table of individual art forms.

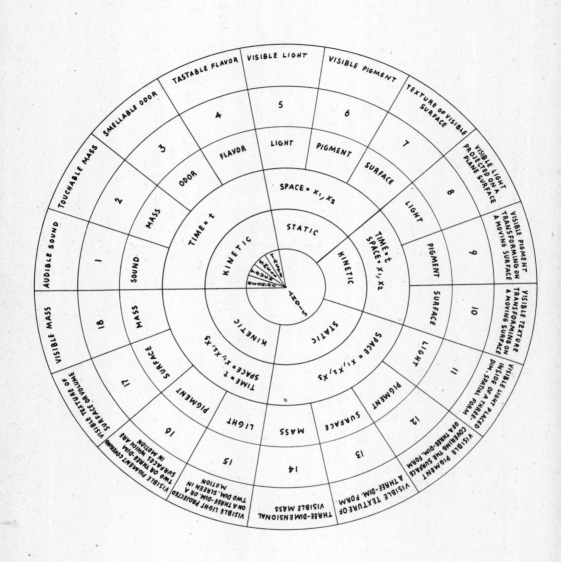

Figure 4. Diagram of art forms.

KINETIC ARTS (musical)

One general parameter—time.

Sensation	General Parameter	System (a complex) of special parameters	
Hearing	Time	Sound	(1)
Touch	Time	Mass[12]	(2)
Smell	Time	Odor	(3)
Taste	Time	Flavor	(4)

Figure 5. Kinetic arts with one general parameter.

STATIC ARTS (plastic)

Two general parameters—two dimensional space—area.

Sensation	General Parameters	System (a complex) of special parameters	
Sight	Coordinates X_1, X_2	Light[13]	(5)
Sight	Coordinates X_1, X_2	Pigment	(6)
Sight	Coordinates X_1, X_2	Surface[14]	(7)

Figure 6. Static arts with two general parameters.

---

[12](2) Marinetti proposed touch as an art form in his manifesto on *Tactilism* in 1920.

[13](5) Light as a source of illumination of a translucent (glass) or a transparent (dampened cloth) surface.

[14](7) Different visible textures of materials used as the material of art. Solid (wood, wire, glass, cork, rubber, etc.) and liquid; also photography and patterns used as elements of texture.

KINETIC ARTS (musical)

Three general parameters = time X_4 (1) + area X_1, X_2 (2).

Sensation	General Parameters	System (a complex) of special parameters	
Sight	X_1, X_2, X_4[15]	Light[16]	(8)
		Pigment[17]	(9)
		Surface[18]	(10)

Figure 7. Kinetic arts with three general parameters.

STATIC ARTS (plastic)

Three general parameters = volume (a system of planes).

Sensation	General Parameters	Special Parameters	
Sight	X_1, X_2, X_3	Light	(11)
		Pigment	(12)
		Surface	(13)
		Mass[19]	(14)

Figure 8. Static arts with three general parameters.

---

[15] $X_4 = t$.
[16] (8) Light source as such = mobile light.
[17] (9) Pigmented mobile liquid; pigmented images on a plane transforming in time, chemically, mechanically, or optically. *Archipentura* for instance.

[18] (10) Mobile textures of materials: transformations obtained chemically, mechanically or optically; geometrical transformations on a plane with or without perspective.
[19] (14) Solid mass; planimetric clusters.

KINETIC ARTS (musical)

Four general parameters = *volume* (X_1, X_2, X_3) + time (X_4).

Sensation	General Parameters	Special Parameters	
Sight	X_1, X_2, X_3, X_4	Light[20]	(15)
		Pigment[21]	(16)
		Surface[22]	(17)
		Mass[23]	(18)

Figure 9. Kinetic arts with four general parameters.

Considering the fact that our perception of the outer world is based on two categories of consciousness, *space* and *time* (corresponding to the four mathematical parameters, X_1, X_2, X_3, X_4), we may classify all forms of art into two groups—*static* and *kinetic*.

To the first group belong those arts defined as *forms crystallized* in space, which can be perceived by sight only. They do not change in time after the process of composition is completed. They can be observed from various stationary positions, each point of observation revealing a different form.

To the second group belong those arts that can be defined as a process of generation, transformation and degeneration of forms in time and continuity. They change in time after the process of composition is complete. This process can be perceived by any of the organs of sensation: sight, hearing, touch, smell or taste. All the art forms appealing to the different organs of sensation involve changeability in time, while sight may or may not require such a condition. Visual perception can be directed towards changeable as well as unchangeable forms. A *kinetic impression* may be obtained from *static optical forms* by a gradual displacement of the object, or of the point of observation (the position of the observer).

In art forms pertaining to touch, smell and taste, changeability in time can be produced in two ways: the excitor (*generator*) moves while the perceiving

---

[20](15) Mobile light.
[21](16) Pigmented mobile solids (solid masses; planimetric clusters), liquids, gases.
[22](17) Textures transforming in time chemically, mechanically or optically.
[23](18) Solid (solid masses; planimetric clusters) hard, soft, gummy, liquid, gaseous—transforming geometrically.

individual (smeller, toucher, taster) is in a stationary position; or the perceiving individual moves in space while the excitor remains stationary. For the hearer, only one form of apprehension is possible. He is stationary and the sound evolves before him in time.[24]

By eliminating our organs of sensation, an impression may approach zero. By excluding an organ of sensation that is not engaged, we concentrate better on the excitor appealing to other organs of sensation. Thus, many people close their eyes while listening to music.

Any system of special parameters (components of an individual art material) can be reduced to three parameters:

1. *Frequency* (or form)
2. *Intensity* (dimension or size)
3. *Quality* (texture or character)

Frequency and intensity are the conditioning parameters while quality is itself conditioned by the correlation of frequency and intensity, and is therefore a derivative product of both.[25]

A complete classification of all possible art continua through their general and special components (parameters) can now be achieved. We shall rely in part on the previous classification.

Art Form Number One (sense of hearing) has two aspects:

(a) *art of audible sound* (music)[26], with a system of special parameters—sound = frequency (pitch), intensity (volume = loudness), quality (timbre and character of sound);

(b) *art of the audible word* (poetry)[27], as an independent declamation or as part of a play with a system of special parameters—word (sound plus its semantic connotations) = frequency (pitch and inharmonic complexes), intensity (loudness), quality (1. sonorous character of the material; 2. sonorous character of the reciting voice), and the parameter of a plot built through a system of correlated poetic images as units.[28]

---

[24] The movement of a hearer or of the source of sound production varies pitch to an extent not to be taken into account in music, as we conceive it today.

[25] See D. C. Miller: *The Science of Musical Sounds* (1926).

[26] Music of sound (art of audible sound).

[27] Music of verbal images and sounding words (art of the audible word suggesting the relation of images).

[28] Many attempts have been made to establish poetry as a form free from the literary connotations of words (incantation).

CONTINUUM

Art Form Number Two (sense of touch):

Art of touchable mass with a system of special parameters—mass (in all its perceptible aspects) =

(a) quality of mass—texture[29], conductibility:
  1. due to the atomic structure.
  2. due to the speed of motion (movement of the mass).
(b) temperature:
  1. due to general conditions.
  2. due to the speed of motion.
(c) intensity:
  1. due to weight or pressure.
  2. due to electric conductibility (discharges).

Though this art form does not exist independently, we exercise our discrimination of tactile perception in every-day experience[30]: vibratory massage, sexual experience, the fondling of animals, a facial in a barber shop, diving, sailing, selecting fabrics, etc. The art of touch, now hypothetic, undoubtedly has a chance to play an important part in our lives in the near future when certain social standards are revised.[31]

Art Form Number Three (sense of smell):

Art of smellable odor with a system of special parameters—odor =

(a) quality (character of odor).
(b) intensity (of odor).
(c) density (quantity of odor in a given space).
(d) temperature (as perceived by the nose).
(e) humidity (of the air).

This is a hypothetic art form found in every-day experience.[32] Some applications have been made in commercial advertising in America.[33]

---

[29] Hard, gummy, soft, liquid.

[30] Look at advertisements: "a lovely skin invites romance," "the skin you love to touch," "that well-groomed look" (visual texture, but suggesting tactile impression [complex sensation]).

[31] There is an interesting problem in the art of touch; that of tactile, quantitative illusions obtained through displacement of organs of touch, fingers, doubling or tripling an object.

[32] Its application is as old as religion. Incense has been used in religious ceremonies and also in homes. Woolworth sells it!

[33] Perfumed paper, tobacconized paper, etc. "It is not the way you look, it is not the way you talk, it is the way you *smell*" (from a perfume advertisement).

Art Form Number Four (sense of taste):

Art of tastable flavor with a system of special parameters—flavor:
  (a) quality (character of flavor).
  (b) intensity (strong or weak flavor).
  (c) density (the degree of concentration).
  (d) temperature (of the material).
  (e) quantity (of the material taken at a time).
  (f) quality of resistance (texture of the flavor: liquid, soft, gummy, hard).

In order to experience works of art based on the sensation of taste, it is enough to taste or to chew without swallowing. This makes an essential difference between the art of taste and gastronomy. Such an art is potential in chewing (tobacco, betel, gum). Professional liquor tasters usually do not drink alcoholic beverages.

Temperature is a common component of the last three forms of art (Nos. *two, three* and *four*), which are at present hypothetical.

Art Form Number Five (sense of sight):

Art of visible light[34] with a system of special parameters—luminescent paint[35] =
  (a) frequency (hue).
  (b) intensity (luminosity).
  (c) quality (saturation).

This art form became a favorite several decades ago when it was used as a part of the setting of vaudeville shows. The luminescent effect is produced from especially prepared paints, illuminated by ultra-violet rays that make the paints visible in the dark. It might be considered as "luminescent painting" visible only in the dark. The color system is additive. This art form requires a light source in front of the painted surface.

Art Form Number Six (sense of sight):

Art of visible pigment with a system of special parameters—pigment[36] =
  (a) frequency (hue).
  (b) intensity (luminosity).
  (c) quality (saturation).

---

[34] Light as a system of parameters has been used and may be used in luminous screens (luminescent murals) with the source of illumination behind the screen.

[35] The source of illumination, ultra-violet rays, is an invariant and cannot be considered a parameter.

[36] Non-luminescent material (1) oil, water color, crayon, etc., in art done by hand with brush or without; (2) chemical pigmentation done by photographic chemicalia.

So far, two forms of this art exist:
1. Where the light source is in front of the surface (opaque surface) painting, etching, fresco, photography[37]; the color system is additive.
2. Where the light source is on both sides of the surface: (translucent surface) painting on glass as in lantern slides, curtains, etc.; the color system is subtractive.[38]

In this case the source of illumination (white light) is an invariant and as such cannot be considered a parameter.[39]

The essential difference between forms *five* and *six* is that while luminescent paint gives the effect of a light source, pigment, even though used on translucent material, does not.

Art Form Number Seven (sense of sight):

Art of the texture of visible surface with a system of parameters — texture:
(a) frequency of recurring elements and their dimensions[40] (density of structure).
(b) intensity[41] (extension into the third dimension).
(c) quality[42] (molecular structure of matter, absorbing-reflecting reaction on light).

This art form was presented as a definite movement at the end of the first decade of the twentieth century by the "futurists." It has failed, probably because of poor ideas and poor realization. Fabrics, fragments of photographs, newspapers, household tools and what-not were used as materials. Worthwhile attempts were later made to apply it to window displays, commercial advertising and interior decorating. Some successful experiments were made in book covers, using different types, and photomontage as elements of texture.[43]

Strictly speaking, the difference between design as an art material and texture of visible surface is purely quantitative: microscopic configurations form texture.

There are many hybrid forms where the effects of visible color and the texture of visible surface are combined: "pointillism" in painting; illumination of Niagara Falls by powerful light sources, where falling water forms a most impressive texture and provides a unique screen offering unusual opportunity for luminous color effects.

---

[37] In this classification an optico-chemical process substitutes for pigment and is realized through a mechanical device (camera).
[38] A static color composition projected on a screen is also in this group.
[39] When the source of light is variable, such an art becomes kinetic.
[40] Patterns of design as elements of texture.
[41] The bas-relief quality, producing light-shade effects.
[42] Solid (from "hard" to "soft"), liquid, gaseous.
[43] A honeycomb is an illustration of homogeneous texture with small but distinctly noticeable elements (patterns).

Art Form Number Eight (sense of sight):

Kinetic art of visible light projected on a plane surface. (Art form number *five* in its kinetic state, functioning in time.)

There are two distinctly different forms of the kinetic art of light. The first deals with the light source as such (Rimington, Wilfred, Klein) and is known as the art of color-music, producing a visible reality.

The second uses the light source as a medium for projecting successive photographic records representing consecutive phases of a moving actuality. It reproduces an actuality formerly existing as an event. This event is first analytically photographed and then synthetically reproduced in motion. It is the art of the cinema, a typical 20th-century product. It has a number of possibilities:

(a) It can record and reproduce an actual event (action and acting).
(b) It can record an actual event and reproduce it at an entirely new speed (much slower or much faster).[44]
(c) It can produce an effect of actuality by recording a number of drawings that represent consecutive phases of motion. This has been developed in "trailers," "cartoons" and "kinetic abstractions," and is based on "frame by frame" (single shot) technique.
(d) Optical distortions and multiplications of a recorded actuality (system of mirrors).
(e) The multidimensional effects obtained by montage intercomposition of continuity and of frame: interpenetrating translucent actualities. This method is an essential supplement to the three-dimensional effect (depth), obtained through motion on a two-dimensional surface (screen).

All these possibilities provide a splendid opportunity for the development of an independent art of kinetic images, so-called "abstract cinema," until proper optical instruments using light source as such (not a record, not a film) are devised.[45]

Art Form Number Nine (sense of sight):

Kinetic art of visible pigment transforming on a moving surface. (Art form number *six* functioning in time).

This art form is based on the movement of pigmented surfaces consisting of very small partial areas. The surfaces exist on a flexible material and revolve on large cylinders of which only a small part can be seen. Thus, the coincidence of different phases of the designs on both cylinders produces a design transforming in time. A device of this nature, the *Archipentura* was presented to New York audiences in 1931 by the sculptor Archipenko. This form offers many possibilities through the correlation of different speeds and directions of motion (mobile mosaics).

---

[44] See "growth of plants," "slow motion" and "accelerated motion" shown in various films.

[45] Even when such instruments are devised, these compositions might be distributed by means of film reprints. Compare phonograph records.

Art Form Number Ten (sense of sight):

Kinetic art of visible texture transforming on a moving surface. (Art form number *seven* functioning in time.)

This art form is similar to number *seven* in its special parameters and to number *nine* in its general parameters. It is a hypothetical art form, although, in reproduced form, it can be found in certain motion pictures.[46]

Art Form Number Eleven (sense of sight):

Static art of visible light placed inside a three-dimensional spatial form. (Art form number *seven* with one additional general parameter X_3).

To produce the spatial form, we may use frosted glass, marble, or any other translucent material through which the shape of the source of illumination is not clearly visible. The resulting light may be white or colored. Many modern lamps belong to this art form.[47] In conventional terms it might be called "sculpture illuminated from within." Comparatively little of real artistic value has as yet been produced.

Any form of illumination of an interior belongs to this group. Many examples were to be found at the Institute of Light at the Grand Central Palace in New York. In this case the interior itself formed a three-dimensional shape, with the spectator inside.

Art Form Number Twelve (sense of sight):

Static art of visible pigment covering the surfaces of a three-dimensional form. (Art form number *nine* with one additional general parameter X_3).

This art form has been known as "painted sculpture" for thousands of years and is almost as old as sculpture itself. Many experiments in furniture making, interior decoration, and window displays, are directed toward perfecting this art form. In such cases, texture is often more important than color.[48]

Art Form Number Thirteen (sense of sight):

Static art of visible texture of three-dimensional forms. (Art form number *seven* with one additional parameter X_3).

In this case, combinations of two-dimensional surfaces produce three-dimensional formations:

1. Planimetric clusters.
2. Surfaces, as parts of solids.
3. Mixture of both: planimetric clusters intercomposed with solids, the exterior of which is seen.

---

[46] The beginning of the *Fall of the House of Usher* by Watson and Webber. Of course, in this case, the illusion of moving textures is devised optically. Ralph Steiner's *Surf and Seaweed* and H_2O.

[47] A most effectively lighted tower is the top of the *Titania Palast* in Berlin.

[48] Also inlay woodwork.

Such an art form necessarily borrows its geometric aspect (spatial structure) from sculpture, and its recurring patterns (texture as such), from design. It may be described as "abstract" sculpture with emphasis on the material used. Its highest form of development has been achieved in modern stage settings, furniture, window displays, etc.—in other words, in applied forms.

Art Form Number Fourteen (sense of sight):

Static art of three-dimensional visible mass involving solid matter that can retain its original form. This art can be described as a literary or "abstract" sculpture, with emphasis on the structural expressiveness of the spatial form. Examples are innumerable from the neolithic era to Brancusi. It consists of planimetric systems and solids.[49]

The last four art forms described become kinetic with the addition of the fourth general parameter—time (X_4).

Art Form Number Fifteen (sense of sight):

Kinetic art of light projected on a three-dimensional or on a two-dimensional screen in motion. The light source and the screen are subjected to general parameters.

In art forms of this group, the light source changes in time continuity with respect to its special parameters. Screens of various quality[50] and form can be used. They may be static or kinetic.[51] The following forms can be adopted:

1. "Quasi-two-dimensional" screens (the third dimension approaches zero: "minimum thickness"); both sides can be used. They are planimetric surfaces.
2. Three-dimensional spheric surfaces: both concave and convex sides can be used.
3. Systems of screens involving planimetric surfaces, spheric surfaces, or both (screen clusters).
4. Actual solids, showing different phases in motion, used as screens. Human body in motion.
5. Multi-screen effects through systems of mirror reflections, and using 1, 2, 3 and 4 forms of screen.

Various applications of this art form can be found in the theatre and kinetic forms of advertisement. There are innumerable instrumental possibilities in modern engineering technique for its further evolution.

---

[49]Bas-relief belongs here, although it is a hybrid form between sculpture and visible texture.

[50]Solid, liquid, gaseous (smoke screens). Niagara Falls illuminated; "Les Fontaines Lumineuses" Paris International Exposition, 1937.

[51]Moving only, or moving and transforming.

Art Form Number Sixteen (sense of sight):

Kinetic art of visible pigment covering two or three-dimensional surfaces that are in motion.

The surfaces can be classified into the same five forms as in number *fifteen*. The kinetic process in a pigment can be stimulated chemically or optically.[52] Optical transformation has been used in many vaudeville shows: for example, changing the color of the dancers' dresses in rhythm with the music of the dance.

Art Form Number Seventeen (sense of sight):

Kinetic art of visible texture of surface or volume.

Similar in every respect to form number *sixteen* in its general parameter and based on form number *thirteen* in its special parameter. It would be a three-dimensional kinetic *Archipentura* in a way, but with emphasis on texture. Many industrial processes in metallurgy, in textiles, etc., suggest very powerful impressions of this kind. For the geometric classification of surfaces see form number *fifteen*. With the advancement of physico-chemical knowledge, the possibility of full range variation in visible texture may be realized.[53]

Art Form Number Eighteen (sense of sight):

Kinetic art of visible mass.

This form deals with planimetric systems and volumes that move and transform in time continuity with respect to one another. This art form can be described as a kinetic abstract sculpture. It is a geometric art par excellence.[54] The dance belongs to this art form.

D. Time (X_4) as a General Parameter

The foregoing concludes the description of special parameters in the general classification of art forms dealing with one organ of sensation at a time and with one system of special parameters.

As to general parameters, they may be classified as follows:

Time, $t = X_4$, is duration, an unavoidable condition underlying any real, conditioned, or imaginary existence. It is a psycho-physical category and a mathematical parameter. As a psycho-physical category, it has only one direction: from the past, through the present, to the future. We cannot make it run otherwise. As a mathematical parameter it is convertible. If any event has been recorded, it can be reproduced in all forms of changeability in time, thus adopting the property of any other single parameter.

---

[52]There is also a mechanical possibility that can be realized through moving very small portions of the surfaces. It would amount to something like three-dimensional *Archipentura*.

[53]The control over transformations of matter, its solid, liquid and gaseous states. At present, preliminary efforts in this direction can be realized through the cinema.

[54]Solidrive, an instrument for the realization of Solidrama, designed by the author; a working model of Solidrive, 45″ in diameter, is at the Schillinger Studio.

The record of an event, in its special parameter of co-existing points along the parametral extension, may be represented thus:

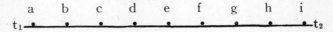

We may regard any point along this extension as the *present*. Then, any point to the left will be the *past* and any point to the right will be the *future*, providing that the direction of our movement is from a to i.[55] By inverting the direction of movement, or by inverting the position of this parameter by means of a new parameter, the former future may become the past, and the former past may become the future. This may be illustrated in actual[56] time by running the record (a film or phonograph disc) backwards. By freezing portions of time continuity in the form of recorded events, one can arrange them in new ways, which amount to a complete transformation of the continuity of these events.[57] This process has been used in motion picture montage, one of the most important forms of cinema technique. By means of this process, besides the forward-backward progression of events, all forms of continuity are obtained through permutation of portions of time.

$$a \longrightarrow b, b \longrightarrow a$$
$$b \longrightarrow c, c \longrightarrow b$$
$$c \longrightarrow d, d \longrightarrow c, \text{etc.},$$

can be used consecutively or simultaneously, thus building an entirely new temporal actuality. This can be performed with any art form realized in any system of special parameters: sound, odor, flavor, light, etc.

The relationship of psycho-physical time, and time as a general parameter in an art continuum, may be easily illustrated through the following scheme: If $t_1 - t_2$ is a portion of time in an art continuum, and $T_m - T_n$ a portion of the general time category, then $t_1 - t_2$ can be superimposed at any point on $T_m - T_n$. Let us assume t_a, t_b, t_c . . . as intermediate points on the parametral extension $T_m - T_n$:

$$T_m \underset{\bullet \quad \bullet \quad \bullet \quad \bullet \quad \bullet \quad \bullet}{\overset{t_a \quad t_b \quad t_c \quad t_d \quad t_e \quad t_f}{\rule{6cm}{0.4pt}}} T_n$$

We can superimpose $t_1 - t_2$ on any point t_a, t_b, t_c . . . For instance, we can make t coincide with t_c:

$$T_m \underset{\bullet \quad \bullet \quad \bullet \quad \bullet \quad \bullet \quad \bullet}{\overset{t_a \quad t_b \quad t_1 \quad t_2 \quad t_e \quad t_f}{\rule{6cm}{0.4pt}}} T_n$$

---

[55]Or $t_1 \longrightarrow t_2$.
[56]Psycho-physical.

[57]While they are being reproduced in motion.

CONTINUUM

This means that any given art product can be performed at any time and that it will preserve the same time relations within its own time limits (t_1—t_2) and regardless of events formerly recorded. Any work of art can be reproduced at any time and in any place, provided that there are technical facilities.

The three parameters (X_1, X_2, X_3) of space in the world of Euclidean geometry represent the rectangular coordinates (length, width, depth). In the curvilinear space of immense dimensions, (Lobachevsky and Riemann), this Euclidean space is infinitesimal. And, within this infinitesimal space, all the Euclidean premises remain true.[58] In other words, there is nothing wrong with our interpretation of space in terms of the three rectangular coordinates. We accept this statement as a premise for the definition of spatial relations in visible art forms. .

We can produce any geometric formation with any visible art material (light, paint, clay) by superimposing this sensuously perceptible material upon the imaginary geometric extension. Thus, a straight line may be generated by moving an ink point along the geometric extension of one of the parameters (X_1, X_2, X_3).

In the previous classification of art forms, a visual art of one dimension was not considered. In linear design, only one kind of design exists, a straight line of any desirable length. This is true so long as time does not enter as the second component. When time enters, the kinetic form of art gives only two very limited possibilities:

(1) a point moving forward-backward at different velocities, and
(2) a straight line changing its dimension in time.

A line moving on a plane (behavior of X_1 with respect to X_2) provides all the forms of planimetric linear design, and also different forms of optical illusion: perspective (suggestion of X_3) and the distortion of angles and dimensions.

The components of linear design on a *plane*, due to X_1, are:

1. dimension
2. direction

The component of linear design on a *plane*, due to X_2 is:
3. angle

Through the combination of constant and variable dimensions, directions and angles, an infinite variety of linear design patterns can be produced.[59]

A line moving in space (behavior of X_1, with respect to X_2 and X_3) provides all the forms of stereometric linear design.[60]

The components of a linear design in *space* due to X_1, are:

1. dimension
2. direction

---

[58] Einstein's general theory of relativity is related to his special theory of relativity as these two postulates of space; one is a special case of the other.

[59] See Part III, Chapter 3, Design.

[60] Practical realization is possible through wire or waxed thread. The third dimension (2r) is negligible on account of its small value.

The component of a linear design in space due to X_2 and X_3 is the relation of the values of angles on the two coordinate planes.

In kinetic art forms, any linear design becomes a trajectory, *i.e.*, the trace left by a material point moving through X_1, or through X_1X_2, or through $X_1X_2X_3$.

A line moving perpendicularly to its own direction forms a plane. A plane can move with respect to X_1 and X_2, thus producing a planimetric design on a plane. A plane can move with respect to X_1, X_2 and X_3, thus producing a planimetric design in space. The components of X_1, X_2 and X_3, previously described, are used together with one additional component of *dimension* for X_2.

The kinetic aspect of a planimetric design is a planimetric trajectory (when it moves with respect to X_1 or X_2) or a stereometric trajectory (when it moves with respect to X_3).

A plane moving perpendicularly to X_1 and X_2 forms a solid. A solid can move with respect to X_1X_2, or X_1X_3, or X_2X_3, thus producing a stereometric design in space limited by two parallel planes.[61] A solid can move with respect to $X_1X_2X_3$, thus producing a stereometric design in space with no boundaries.[62]

The kinetic aspect of a stereometric design is a stereometric trajectory (when it moves with respect to X_1X_2, X_1X_3, or X_2X_3), or a hyper-stereometric[63] trajectory (when it moves with respect to $X_1X_2X_3$).

Simultaneous combinations of spatial components here described produce simultaneous spatial systems. The properties of such systems are:

(1) motion
(2) transformation
(3) inter-penetration[64]

These are to be found in different designs as well as in relation to one another.

E. Second Group of Art Forms (More Than One Organ of Sensation and More Than One System of Special Parameters)

In the second group of this classification, art forms previously classified under numbers one, two, three and four do not yield any combinations since they require more than one organ of sensation at a time.

Classifying combinations which are complex homogeneous art forms by the number of systems representing special parameters, we obtain:

1. two systems:

   light and pigment   (5 and 6)
   light and visible texture (5 and 7)
   pigment and visible texture (6 and 7)

2. three systems:

   light, pigment and visible texture   (5, 6 and 7)

---

[61]Conditioned three-dimensional space.
[62]Unconditioned three-dimensional space.
[63]Some geometric schools (Hinton) consider such a trajectory as four-dimensional solid. We believe it is a misconception because the $X_1X_2X_3$ does not move with respect to spatial X_4, perpendicular to X_1, X_2 and X_3.
[64]At present they can be realized through multi-exposures on film.

CONTINUUM

A complex homogeneous form may consist of combinations by two or three elements, which, in this case, are light, pigment, visible texture.

In the same way, in the kinetic group, we obtain:
$$8 + 9;\ 8 + 10^{65} \text{ and}$$
$$8 + 9 + 10.$$

In the next group:
$$11 + 12;\ 11 + 13;\ 11 + 14$$
$$12 + 13;\ 12 + 14$$
$$13 + 14 \text{ and}$$
$11 + 12 + 13;\ 11 + 12 + 14;\ 11 + 13 + 14;\ 12 + 13 + 14$ and $11 + 12 + 13 + 14$, and, in the last group:
$$15 + 16;\ 15 + 17;\ 15 + 18$$
$$16 + 17;\ 16 + 18 \text{ and}$$
$$15 + 16 + 17;\ 15 + 16 + 18;\ 15 + 17 + 18;\ 16 + 17 + 18$$
and
$$15 + 16 + 17 + 18.$$

All the primary forms of each group 5 — 18 can combine with one another by 2 to 18:

By Two:

$5 + 8;\ 5 + 9;\ 5 + 10;\ 5 + 11;\ 5 + 12;\ 5 + 13;\ 5 + 14;\ 5 + 15;$
$\quad\quad 5 + 16;\ 5 + 17;\ 5 + 18.$

$6 + 8;\ 6 + 9;\ 6 + 10;\ 6 + 11;\ 6 + 12;\ 6 + 13;\ 6 + 14;\ 6 + 15;$
$\quad\quad 6 + 16;\ 6 + 17;\ 6 + 18.$

$7 + 8;\ 7 + 9;\ 7 + 10;\ 7 + 11;\ 7 + 12;\ 7 + 13;\ 7 + 14;\ 7 + 15;$
$\quad\quad 7 + 16;\ 7 + 17;\ 7 + 18.$

$8 + 11;\ 8 + 12;\ 8 + 13;\ 8 + 14;\ 8 + 15;\ 8 + 16;\ 8 + 17;\ 8 + 18.$

$9 + 11;\ 9 + 12;\ 9 + 13;\ 9 + 14;\ 9 + 15;\ 9 + 16;\ 9 + 17;\ 9 + 18.$

$10 + 11;\ 10 + 12;\ 10 + 13;\ 10 + 14;\ 10 + 15;\ 10 + 16;\ 10 + 17;\ 10 + 18.$

$11 + 15;\ 11 + 16;\ 11 + 17;\ 11 + 18.$

$12 + 15;\ 12 + 16;\ 12 + 17;\ 12 + 18.$

$13 + 15;\ 13 + 16;\ 13 + 17;\ 13 + 18.$

$14 + 15;\ 14 + 16;\ 14 + 17;\ 14 + 18.$

By Three:

$5 + 8 + 9;\ 5 + 8 + 10;\ 5 + 8 + 11;\ 5 + 8 + 12;\ 5 + 8 + 13;$
$5 + 8 + 14;\ 5 + 8 + 15;\ 5 + 8 + 16;\ 5 + 8 + 17;\ 5 + 8 + 18.$
$5 + 9 + 10;\ 5 + 9 + 11;\ 5 + 9 + 12;\ 5 + 9 + 13;\ 5 + 9 + 14;$
$5 + 9 + 15;\ 5 + 9 + 16;\ 5 + 9 + 17;\ 5 + 9 + 18.$
$5 + 10 + 11;\ 5 + 10 + 12;\ 5 + 10 + 13;\ 5 + 10 + 14;\ 5 + 10 + 15;$
$\quad\quad 5 + 10 + 16;\ 5 + 10 + 17;\ 5 + 10 + 18.$

[65] These numbers, suggesting various combined arts, refer to the art forms described above (Ed.)

5 + 11 + 12; 5 + 11 + 13; 5 + 11 + 14; 5 + 11 + 15; 5 + 11 + 16;
5 + 11 + 17; 5 + 11 + 18.
5 + 12 + 13; 5 + 12 + 14; 5 + 12 + 15; 5 + 12 + 16; 5 + 12 + 17;
5 + 12 + 18.
5 + 13 + 14; 5 + 13 + 15; 5 + 13 + 16; 5 + 13 + 17; 5 + 13 + 18.
5 + 14 + 15; 5 + 14 + 16; 5 + 14 + 17; 5 + 14 + 18.
5 + 15 + 16; 5 + 15 + 17; 5 + 15 + 18.
5 + 16 + 17; 5 + 16 + 18.
5 + 17 + 18.

Combining 6 with the rest by Three[66]

(6 + 8) + 9, + 10, + 11, + 12, + 13, + 14, + 15, + 16, + 17, + 18.
(6 + 9) + 10, + 11, + 12, + 13, + 14, + 15, + 16, + 17, + 18.
(6 + 10) + 11, + 12, + 13, + 14, + 15, + 16, + 17, + 18.
(6 + 11) + 12, + 13, + 14, + 15, + 16, + 17, + 18.
(6 + 12) + 13, + 14, + 15, + 16, + 17, + 18.
(6 + 13) + 14, + 15, + 16, + 17, + 18.
(6 + 14) + 15, + 16, + 17, + 18.
(6 + 15) + 16, + 17, + 18.
(6 + 16) + 17, + 18.
  6 + 17 + 18.

Combining 7 with the rest by Three:

(7 + 8) + 9, + 10, + 11, + 12, + 13, + 14, + 15, + 16, + 17, + 18.
(7 + 9) + 10, + 11, + 12, + 13, + 14, + 15, + 16, + 17, + 18.
(7 + 10) + 11, + 12, + 13, + 14, + 15, + 16, + 17, + 18.
(7 + 11) + 12, + 13, + 14, + 15, + 16, + 17, + 18.
(7 + 12) + 13, + 14, + 15, + 16, + 17, + 18.
(7 + 13) + 14, + 15, + 16, + 17, + 18.
(7 + 14) + 15, + 16, + 17, + 18.
(7 + 15) + 16, + 17, + 18.
(7 + 16) + 17, + 18.
  7 + 17 + 18.

Combining 8 with the rest by Three:

(8 + 11) + 12, + 13, + 14, + 15, + 16, + 17, + 18.
(8 + 12) + 13, + 14, + 15, + 16, + 17, + 18.
(8 + 13) + 14, + 15, + 16, + 17, + 18.
(8 + 14) + 15, + 16, + 17, + 18.
(8 + 15) + 16, + 17, + 18.
(8 + 16) + 17, + 18.
  8 + 17 + 18.

[66]These numbers, suggesting various combined arts, refer to the art forms described above (Ed.)

CONTINUUM

Combining 9 with the rest by Three:

(9 + 11) + 12, + 13, + 14, + 15, + 16, + 17, + 18.
(9 + 12) + 13, + 14, + 15, + 16, + 17, + 18.
(9 + 13) + 14, + 15, + 16, + 17, + 18.
(9 + 14) + 15, + 16, + 17, + 18.
(9 + 15) + 16, + 17, + 18.
(9 + 16) + 17, + 18.
   9 + 17 + 18.

Combining 10 with the rest by Three:[68]

(10 + 11) + 12, + 13, + 14, + 15, + 16, + 17, + 18.
(10 + 12) + 13, + 14, + 15, + 16, + 17, + 18.
(10 + 13) + 14, + 15, + 16, + 17, + 18.
(10 + 14) + 15, + 16, + 17, + 18.
(10 + 15) + 16, + 17, + 18.
(10 + 16) + 17, + 18.
 10 + 17 + 18.

Combining 11 with the rest by Three:

11 + 15 + 16; 11 + 15 + 17; 11 + 15 + 18.
11 + 16 + 17; 11 + 16 + 18.
11 + 17 + 18.

Combining 12 with the rest by Three:

12 + 15 + 16; 12 + 15 + 17; 12 + 15 + 18.
12 + 16 + 17; 12 + 16 + 18.
12 + 17 + 18.

Combining 13 with the rest by Three:

13 + 15 + 16; 13 + 15 + 17; 13 + 15 + 18.
13 + 16 + 17; 13 + 16 + 18.
13 + 17 + 18.

Combining 14 with the rest by Three:

14 + 15 + 16; 14 + 15 + 17; 14 + 15 + 18.
14 + 16 + 17; 14 + 16 + 18.
14 + 17 + 18.

Combinations by four and more elements can be obtained by the process described below.

[68]These numbers, suggesting various combined arts, refer to the art forms described above (Ed.)

F. Third Group of Art Forms (More Than One Organ of Sensation and More Than One System of Parameters.)

Art forms numbers one, two, three and four combined with one another or with any of the remaining (5—18) forms belong to this group.

Classifying combinations, which are complex heterogeneous art forms by the number of organs of sensations required, we obtain:

1. **Two Organs of Sensation:**

Sound and texture	(1 and 2)
Sound and odor	(1 and 3)
Sound and flavor	(1 and 4)
Texture and odor	(2 and 3)
Texture and flavor	(2 and 4)
Odor and flavor	(3 and 4)

2. **Three Organs of Sensation:**

Sound, texture and odor	(1 + 2 + 3)
Sound, texture and flavor	(1 + 2 + 4)
Sound, odor and flavor	(1 + 3 + 4)
Texture, odor and flavor	(2 + 3 + 4)

3. **Four Organs of Sensation:**

Sound, texture, odor and flavor	(1 + 2 + 3 + 4)

A complete heterogeneous form may consist of combinations by two, three and four elements, which in this case are sound, texture, odor and flavor.

The rest of the complex heterogeneous art forms is obtainable from the simple form 5—18 combined with forms 1, 2, 3 and 4 by two. To combine 3, 4, 5—18, use complex homogeneous forms and combine them with forms 1, 2, 3 and 4.

For general orientation in finding any desirable combination by any number of elements, the chart on page 79 is supplied. To find any combination of the forms represented by consecutive series, or partly consecutive series, use one direction. If the numbers cease to be consecutive, move perpendicularly to the previous direction, until the desired number occurs. Always move *down* or from *left* to *right*. The proper method for discovering combinations is by underlining the path at the bottom of a number for the horizontal progression, and at the left side of the number for vertical progression.

CONTINUUM

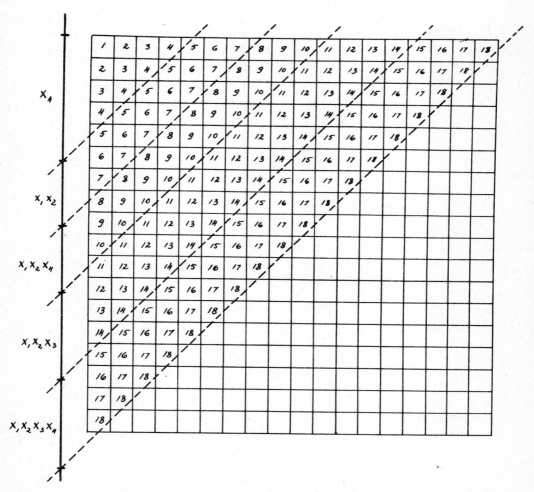

The diagonal lines indicate borders between the groups classified through the number of general parameters.

X_4 = time X_3 = depth X_2 = width X_1 = length

Figure 10. Chart for the combination of complex art forms.

G. Correspondences Between Art Forms

We shall conclude our classification of art forms with an examination of the correspondences among different art forms.

Two kinds of correspondences are possible:

(1) absolute correspondence of *values* in terms of measurement.
(2) absolute correspondence of *relations* in terms of measurement.

Both of these kinds may be either subjective or objective, personal or impersonal.

1. Subjective Correspondences of Value

A great many tests have been performed and a number of books written on the *subjective* correspondences between color and sound, odor and sound, and even flavor and sound. Some individuals, with so-called "color audition," insist that the pitch of a certain tone is green, another, red, etc. Similar associations have been obtained in relation to musical chords and even tonalities. Thus, some people believe that F sharp major is purple. With reference to tone quality, they consider the oboe tone quality a dark orange and the trumpet, a bright yellow. In Huysmans' novel, *A Rebours*, the hero, DesEsseintes, classifies his liquors by associating them with the tone qualities of different musical instruments, and labels them "flute," "clarinet," etc. There are also subjective associations with temperature. Some people consider c minor cold, and d minor warm, while E major is supposed to suggest the temperature of a flame.

All these associations are different with different individuals and therefore cannot be obligatory for anyone else. Subjective correspondences of values reveal no definite system of relations with physical values (intensity, frequency) forming an excitor.

2. Objective Correspondences of Values

Objective correspondences of physical frequencies or intensities, adopted by many authors of color audition theories, establish correlations between the lowest audible pitch (frequency of 16 periods per sec.) and the shortest visible wavelength (.00072 mm.)[69]. Then the authors assume that the highest audible pitch must correspond to the longest visible wave-length. In some cases, the correspondence is built on an attempt to correlate the 2:1 ratio of sound wave frequencies (an "octave" in music) with the whole visible spectrum, thus associating all the hues with one octave in pitch. In order to derive other color values corresponding to other octaves of pitch, intensification (increase of luminosity) of the same progression of hues is applied.

These systems of absolute correspondences of values in frequencies and intensities are arbitrary and pseudo-scientific. Firstly, the approximate ratio of pitch relations is 1:1000 while that of the visible spectrum equals approximately 7:4. Secondly, the elementary ratio of pitch relations is 2:1 while the

---

[69] Robert W. Wood, *Physical Optics*.

ratios of primary colors are very different. The approximate ratio of blue and green equals 9:10; green and red[70] equals 10:13; blue and yellow equals 3:4; yellow and red equals 7:8. Thirdly, no natural association has been observed which would apply to everyone. Thus, the correspondence of an extremely low red with the tone "c" of 16 vibrations per second cannot be enforced as a law.

3. General Correspondences of Relations

According to the geometric concept of extension, one can insert as many points as are desired between two given points, "a" and "b." In terms of rectilinear measurement, this amounts to dividing length into a number of uniform units (linear values). In terms of circular measurement, these uniform units may be represented through angular values (degrees, π). In order to secure a uniform scale of units between two given limits, it is necessary to determine the ratio of the limits in terms of frequencies. If a rectilinear or a circular extension is the graph of a special parameter, one can determine the scale of units in actual frequencies, such as pitch or hue.

Algebraically it may be expressed as follows:

$$u = \sqrt[n]{\frac{a}{b}}$$

where "u" is a unit in terms of frequencies (oscillations); $\frac{a}{b}$, a ratio expressing the limits in terms of frequencies; exponent "n"—the number of units required in a scale. This formula has general meaning and may be applied to any parameter of art material.[71]

One of the well-known applications of this formula is in the system of the equally tempered scale of tonal pitch where the unit called the semitone is:

$$\mu = \sqrt[12]{2}$$

The numeral 2 indicates the 2:1 ratio of vibrations (256:128, for instance), and the number 12, the number of such units between one and two.

On a logarithmic rectilinear graph of pitch values, this system of units will consist of twelve equidistant points between logarithm 1 and logarithm 2, if the base of the logarithm equals $\sqrt[12]{2}$. On a circular graph serving the same purpose, there would be 12 equidistant points on the arc between its extreme points.

---

[70] This is the approximate ratio of complementary colors.

[71] The scale of units $\sqrt[n]{\frac{a}{b}}$, can be graphed as the scale of logarithms from ba to a, to the base $\sqrt[n]{\frac{a}{b}}$; thus $\frac{a}{b} = b$, $\sqrt[n]{\frac{a}{b}}$, $\sqrt[n]{(\frac{a}{b})^2}$, $\sqrt[n]{(\frac{a}{b})^3}$ ..... $\sqrt[n]{(\frac{a}{b})^n}$ $(= a)$. (Ed.)

The intermediate points can be found by measuring the segments of the arc through corresponding angles.

$$A = 360° \qquad \smile\mu = \frac{360°}{12} = 30°$$

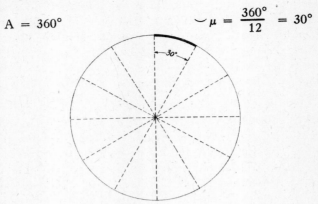

Figure 11. *The twelve divisions represent the equal temperament tuning system.*

$$A = 120° \qquad \smile\mu = \frac{120°}{12} = 10°$$

$$ob = 120°$$

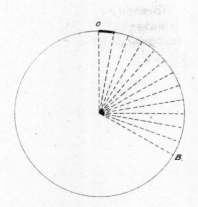

Figure 12. *Tuning system as represented in 120° of a circular graph.*

Thus, the relations of one parameter may be made to correspond with the relations of another. In a musical composition, transposition made from one tonal base to another changes the general quality only, without affecting the pitch-time-intensity relations. It is considered the same composition appearing in a different form of instrumental sonority. The essential components are pitch, time and intensity in their relations, perceived as different degrees of similarity and contrast. In the same sense, the relations of hue, intensity and saturation in color admit transposition from one color base to another. And in the same sense, these relations are perceived as degrees of similarity and contrast.

With such a premise, a given portion of one parameter, split into "n" units, will adequately correspond to any chosen portion of another parameter split into the same number of units. Thus, for example, once we establish a correspondence between the equally-tempered twelve-step scale of pitch units and the full range of visible hues arranged as a uniform progression of twelve color-units, we can transcribe a musical melody into a color progression. The absolute value of a tonal base and the absolute value of a color base, with respect to the tonal base, is not essential. The correspondence of relations in two systems is essential. The tone "c" as a tonal base might be used with any color base, green, yellow, red or other. But as long as "c" moves to "e flat," for example, the selected color base must change adequately, *i.e.*, on 90° of the spectral circle.[72] The color base being yellow, for example, will change to orange (↻), or to green (↺).

We shall determine correspondences as being normal through the probability of their recurrence and through their pragmatic physical adequacy.[73] Such correspondences are predominant in art. The improbable correspondences serve the purpose of distortion and exaggeration.

All correspondences realized through associations, "normal" as well as "abnormal," are due to conditioned physiological reflexes caused by the physical excitors affecting an organ of sensation. All the problems of artistic taste must be studied from this angle. "Beauty" is a psychological complex and a derivative of physiological reflexes. It is a form of satisfaction induced by a relation or a system of relations of simultaneous or consecutive excitations. In order to produce an effect of "beauty," the percentage of excitations previously experienced must be quite high. Insofar as the normal musical taste of the trained majority is concerned, for instance, the amount of familiar excitations should be quite high, about 85% or more.

One of the necessary moments in artistic enjoyment is a standard deviation from the nearest simple relation. For example, if a progression of two uniform durations becomes too obvious, human intuition finds a value of standard devia-

---

[72] "c" to "e♭" represents one-quarter of an octave, or one-quarter of 360° = 90°. This can be performed either clockwise or counterclockwise. It is also possible to establish and use a system of secondary relations such as: $P_xP_y = n:mn$. All the correspondences may also be used in oblique and inverse relations. (Ed.)

[73] An example of normal correspondence: A *giant* (dimension) must be *heavy* (gravity), speak with a *bass voice*, (dimensions of the vocal cords) and possess *red* cheeks (coloration as a result of good blood circulation). By changing some of these components, one can easily illustrate an abnormal correspondence.

84   THEORY OF REGULARITY AND COORDINATION

tion. For example, instead of performing ♩♩ exactly (as $\frac{4}{8} + \frac{4}{8}$) American "crooners" will sing ♩♪♩. (which is $\frac{5}{8} + \frac{3}{8}$), or ♩.♪♩ (which is $\frac{3}{8} + \frac{5}{8}$). One eighth, in this case, will be a standard deviation, an "artistic infinitesimal."

From a mathematical viewpoint, beauty may be expressed as a *differential variation of a rational term* (a relation, or a system of relationships), where the rational term and the differential derive from one homogeneous harmonic series. The rational term is usually a commonly known idiom, and the differential ("artistic differential") comes as a further refinement.

Esthetic satisfaction comes mainly from the sensation of being off balance, but in an obvious relation to balance. Here a mechanical experience becomes an artistic one through the discrimination of the "artistic differential" by our senses. Then the joy of discriminating simple relations in a sensory form possessing a standard deviation is psychologically similar to solving problems or riddles. The element of the unknown stimulates curiosity, and the process of associating it with the known produces a feeling of satisfaction. Here lies the success of one work of art and the failure of another. This explains why a jig-saw puzzle hobby may become an epidemic.[74]

---

[74]See "Periodicity of Expansion and Contraction" in Chapter 3, *Periodicity*.

CHAPTER 2

CONTINUITY

CERTAIN points in a continuum sometimes acquire special significance. This happens either because they form a class, or because they happen to belong to a particular class. By a particular class, we mean one that is simple or axiomatic, such as the class of natural integers; or, one that involves some harmonic relationship, such as a class that develops according to a constant ratio: a geometric progression, a summation series, etc.

Continuity is a finite portion of continuum. It is formed by the progression of extensions along any coordinate or parameter. The terms of such progression are expressed through the relations of their numerical values. The manner of transition (gradual or sudden, through rational numbers or differentials) from the antecedent to the following term in a continuity is dependent upon the relation of the values determining these terms, and upon the value of the differential between the terms. The relative values are determined by the terms of periodicity of different forms and orders.

A. SERIES OF VALUES

A continuity consists of coordinate and parametral components. Parametral components determine coordinate components in sensory forms. Parametral extensions are the positive values. Coordinate extensions as such are the negative values.

Both positive and negative values produce a series based on one or another form of regularity. The type of series upon which a certain continuity is based determines the potential forms of development and growth of such continuity.

1. Natural Series.

Integer Series—A natural integer series is an infinite progression of all the integer numbers between one and any integer value approaching infinity. $\sum_{1}^{n} = 1, 2, 3, \ldots\ldots n.$ There can be finite portions of a natural integer series between two given limits.

Thus, a series from 1 to 5 will be:

$$\sum_{1}^{5} = 1, 2, 3, 4, 5 \text{ or generally,}$$

$$\sum_{a}^{m} = a, b, c, \ldots\ldots m$$

THEORY OF REGULARITY AND COORDINATION

Factorial Continuity.

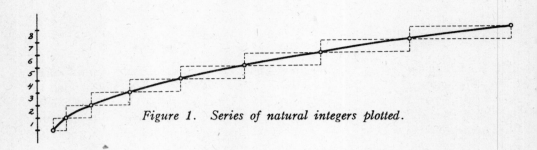

Figure 1. *Series of natural integers plotted.*

If we plot a curve through points that are the terms of a natural integer series, its curvature will approach zero and will reach zero at the infinity point of the series. Therefore, the degree of curvature is in inverse proportion to the number value of a plotted point reached by the curve. The degree of curvature will be greater at the point $p = 2$ than at the point $p = 7$.

Fractional Series—A natural fractional series is an infinite progression of all the rational fractions with the numerator one and denominators with any integer value.

$$\sum_{\frac{1}{2}}^{\frac{1}{n}} = \frac{1}{2}, \frac{1}{3}, \frac{1}{4}, \ldots \ldots \frac{1}{n}$$

where n approaches infinity.

Thus a series from $\frac{1}{2}$ to $\frac{1}{5}$ will be:

$$\sum_{\frac{1}{2}}^{\frac{1}{5}} = \frac{1}{2}, \frac{1}{3}, \frac{1}{4}, \frac{1}{5}, \text{ or generally,}$$

$$\sum_{\frac{1}{a}}^{\frac{1}{m}} = \frac{1}{a}, \frac{1}{b}, \frac{1}{c}, \ldots \ldots \frac{1}{m}$$

If we plot a curve through the points which are the terms of a natural fractional series, the degree of curvature of such a curve will approach zero, and will reach zero at the point $\frac{1}{\infty}$ of the series. The degree of curvature will increase in its first few terms and decrease to zero with each succeeding term.

CONTINUITY

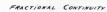

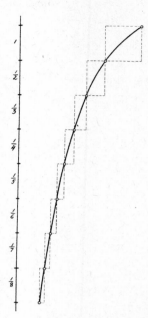

Figure 2. Natural fraction series plotted.

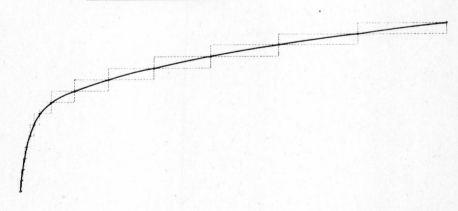

Figure 3. Natural integer and fraction series plotted together.

2. Other Series.

Various other series of integral or fractional values can be formed, though these series are not "natural."

(1) Arithmetical Progression Series. A series whose values increase in an arithmetical progression, such as:

$$\sum\nolimits_{x}^{x+mc} {}_{+c} = x, x + c, (x + c) + c, (x + 2c) + c, \ldots + (x + mc)$$

If $x = 3$ and $c = 2$, then:

$$\sum\nolimits_{3}^{3+2m} {}_{+2} = 3, 3 + 2, (3 + 2) + 2, 3 + (2 \times 2) + 2, \ldots$$

$$\ldots + 3 + 2m = 3, 5, 7, 9, \ldots 3 + 2m$$

The same values may be used as denominators for the fractional series.

$$\sum\nolimits_{\frac{1}{x}}^{\frac{1}{x+mc}} {}_{+c} = \frac{1}{x}, \frac{1}{x+c}, \frac{1}{(x+c)+c}, \frac{1}{(x+2c)+c}, \cdots \frac{1}{(x+mc)}$$

$$\sum\nolimits_{\frac{1}{3}}^{\frac{1}{3+2m}} {}_{+2} = \frac{1}{3}, \frac{1}{5}, \frac{1}{7}, \frac{1}{9}, \cdots \frac{1}{3+2m}$$

(2) Geometrical Progression Series. A series whose values increase in a geometrical progression, such as:

$$\sum\nolimits_{x}^{c^m x} {}_{c} = x, cx, c^2 x, \ldots c^m x$$

If $x = 3$ and $c = 2$, then

$$\sum\nolimits_{3}^{2^m x} {}_{2} = 3, 3 \times 2, 3 \times 2^2, \ldots 3 \times 2^m = 3, 6, 12, \ldots 3 \times 2^m$$

The same values may be used as denominators for the fractional series.

$$\sum_{\frac{1}{x}}^{\frac{1}{c^m x}} c = \frac{1}{x}, \frac{1}{cx}, \frac{1}{c^2 x}, \ldots \frac{1}{c^m x}$$

$$\sum_{\frac{1}{3}}^{\frac{1}{3x^{2m}}} 2 = \frac{1}{3}, \frac{1}{6}, \frac{1}{12}, \ldots \frac{1}{3x^{2m}}$$

When x = c, then:

$$\sum_{x}^{x^m} c=x = x, x^2, x^3, \ldots x^m$$

This last is the most frequent form of musical rhythm.

(3) Summation Series ("Fibonacci Series"). A series in which each succeeding term equals the sum of the two preceding terms.

$$\sum S = x, y, x+y, x+2y, 2x+3y \ldots\ldots 3x+5y, 5x+8y, \ldots$$

Example:

$$\sum_{1}^{55} S = 1, 2, 3, 5, 8, 13, 21, 34, 55$$

This series is the foundation of the theory of Dynamic Symmetry evolved by Jay Hambidge and his group. Hambidge found that this series had been applied in ancient Egyptian and ancient Greek art. This series was also observed by Professor Church of Oxford as the mechanical basis of the growth of certain plants.

Here are various summation series:

1, 2, 3, 5, 8, 13, 21
1, 3, 4, 7, 11, 18
1, 4, 5, 9, 14, 23
1, 5, 6, 11, 17, 28
1, 6, 7, 13, 20, 33, ...
1, 7, 8, 15, 23, 38, ...
1, 8, 9, 17, 26, 43, ...
1, 9, 10, 19, 29, 48, ...
1, 10, 11, 21, 32, 53, ...

* * * *

2, 4, 6, 10, 16, 26, ...
2, 5, 7, 12, 19, 31, ...
2, 6, 8, 14, 22, 36, ...
2, 7, 9, 16, 25, 41, ...
  *   *   *   *
3, 6, 9, 15, 24, 39, ...
3, 7, 10, 17, 27, 44, ...
3, 8, 11, 19, 30, 49, ...
3, 9, 12, 21, 33, ...

The inverted fractional form is:

$$\sum\nolimits_S = \frac{1}{x}, \frac{1}{y}, \frac{1}{x+y}, \frac{1}{x+2y}, \frac{1}{2x+3y}, \ldots$$

$$\sum\nolimits_S {}_{\frac{1}{2}}^{\frac{1}{13}} = \frac{1}{2}, \frac{1}{3}, \frac{1}{5}, \frac{1}{8}, \frac{1}{13}$$

(4) Series of Natural Differences. A series in which the difference between preceding and succeeding terms grows in the natural integer series.

$$\sum\nolimits_D = x, x+1, (x+1)+2, (x+3)+4, \ldots$$

$$\sum\nolimits_D {}_1^{37} = 1, 2, 4, 7, 11, 16, 22, 29, 37, \ldots$$

(5) Series of Prime Numbers. A series in which the whole progression consists of consecutively increasing prime numbers.

If $x, x_2, x_3 \ldots$ are the prime numbers arranged in increasing values, we may write:

$$\sum\nolimits_{x=pn} {}_{x_1}^{x_n} = x_1, x_2, x_3, \ldots x_n$$

$$\sum\nolimits_{pn} {}_1^{23} = 1, 2, 3, 5, 7, 11, 13, 17, 19, 23$$

3. Rational Ratios of the Natural Series Fractional Continuity.

$$R_{\frac{1}{2}}^{\frac{1}{3}} = \frac{2}{1}; \quad R_{\frac{1}{3}}^{\frac{1}{4}} = \frac{5}{3}; \quad R_{\frac{1}{4}}^{\frac{1}{5}} = \frac{3}{2}; \quad R_{\frac{1}{5}}^{\frac{1}{6}} = \frac{7}{5}$$

$$R_{\frac{1}{6}}^{\frac{1}{7}} = \frac{4}{3}; \quad R_{\frac{1}{7}}^{\frac{1}{8}} = \frac{9}{7}; \quad R_{\frac{1}{8}}^{\frac{1}{9}} = \frac{5}{4}; \quad R_{\frac{1}{9}}^{\frac{1}{10}} = \frac{11}{9}$$

$$d\left(\frac{\frac{1}{3} \longleftrightarrow \frac{1}{2}}{\frac{1}{4} \longleftrightarrow \frac{1}{3}}\right) = 2$$

$$d\left(\frac{\frac{1}{4} \longleftrightarrow \frac{1}{3}}{\frac{1}{5} \longleftrightarrow \frac{1}{4}}\right) = \frac{5}{3}$$

$$d\left(\frac{\frac{1}{5} \longleftrightarrow \frac{1}{4}}{\frac{1}{6} \longleftrightarrow \frac{1}{5}}\right) = \frac{3}{2}$$

$$d\left(\frac{\frac{1}{6} \longleftrightarrow \frac{1}{5}}{\frac{1}{7} \longleftrightarrow \frac{1}{6}}\right) = \frac{7}{5}$$

$$^d2n\left(\frac{\frac{1}{m-1} \longleftrightarrow \frac{1}{m-2}}{\frac{1}{m} \longleftrightarrow \frac{1}{m-1}}\right) = \frac{k}{k-1}$$

$$^d2n+1\left(\frac{\frac{1}{m-1} \longleftrightarrow \frac{1}{m-2}}{\frac{1}{m} \longleftrightarrow \frac{1}{m-1}}\right) = \frac{k_1}{k_1-2}, \text{ where}$$

^d2n is any even number place.

$^d2n+1$ is any odd number place.

$$\frac{\frac{1}{m-1} \longleftrightarrow \frac{1}{m-2}}{\frac{1}{m} \longleftrightarrow \frac{1}{m-1}}$$ is the relation of distances of two consecutive terms of fractional continuity.

k is any numerator of the ratio taking an odd place in the series.
k_1 is any numerator of the ratio taking an even place in the series.

. . . .

Thus, the consecutive division of a rectilinear segment produces a complex alternating series of fractional values.

Terms in the odd places form:

$$\sum\nolimits_{+1}{}_{\frac{1}{2}}^{\frac{1}{m}} = \frac{2}{1}, \frac{3}{2}, \frac{4}{3}, \cdots \frac{m}{m-1}$$

Terms in the even places form:

$$\sum\nolimits_{+2}{}_{\frac{1}{3}}^{\frac{1}{n}} = \frac{5}{3}, \frac{7}{5}, \frac{9}{7}, \cdots \frac{n}{n-2}$$

Then the whole series takes the following form:

$$\sum\nolimits_{+1,+2}{}_{\frac{1}{2}}^{\frac{1}{n}} = \frac{2}{1}, \frac{5}{3}, \frac{3}{2}, \frac{7}{5}, \frac{4}{3}, \frac{9}{7}, \cdots \frac{m}{m-1}, \frac{n}{n-2}$$

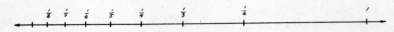

Figure 4. *Natural series fractional continuity.*

B. Factorial-Fractional Continuity

May be represented through normal series with any desirable number of places by filling out the intervals between the already existing terms of a series.

The zero series has two terms:

$$\sum\nolimits_0 = 0 \ldots\ldots\ldots \infty$$

CONTINUITY

The first series has three terms:

$$\sum\nolimits_{1} = 0 \ldots \ldots \frac{x}{x} \ldots \infty$$

The second series has five terms:

$$\sum\nolimits_{2} = 0 \ldots \ldots \frac{1}{x} \ldots \frac{x}{x} \ldots x \ldots \infty$$

The third series has nine terms:

$$\sum\nolimits_{3} = 0 \ldots \ldots \left(\frac{1}{x}\right)^n \ldots \frac{1^{n>\text{\textcircled{m}}>1}}{x} \ldots \frac{1}{x} \ldots \frac{1}{x} \ldots x \ldots x^{n>\text{\textcircled{m}}>1} \ldots$$
$$\ldots x^n \ldots \infty$$

Any value, integral or fractional, may be attributed to x.

Example:

A five term series, when x = 5

$$\sum\nolimits_{2\,x=5} = 0 \ldots \frac{1}{5} \ldots \frac{5}{5} \ldots 5 \ldots \infty$$

If the limits are given, for instance, between $\frac{1^n}{x}$ and x^n, then:

$$\sum\nolimits_{\left(\frac{1}{x}\right)^n}^{x^n}{}_{2\,x=5} = \left(\frac{1}{5}\right)^n \ldots \frac{1}{5} \ldots \frac{5}{5} \ldots 5 \ldots 5^n$$

If n = 3, then:

$$\sum\nolimits_{\left(\frac{1}{x}\right)^3}^{x^3}{}_{2\,x=5} = \left(\frac{1}{5}\right)^3 \ldots \frac{1}{5} \ldots \frac{1}{5} \ldots 5 \ldots 5^3$$

The middle term of the first series, if the latter is a normal one, is unity; the value attributed to its numerator and denominator will determine the succeeding development of the series, provided that the series is a homogeneous one. Thus, $\frac{x}{x}$ is the determinant of a series.

The values on the left side of a series determine the terms of *fractional continuity*, which will be designated by "f". The values on the right side of a series determine the terms of *factorial continuity*, which will be designated by "F".

In a normal series the product of a term of F by a term of f, if they are equidistant from the middle term, is unity.

$$F_n f_n = 1$$

Therefore, the relation of equidistant terms of Factorial and Fractional continuity can be expressed through

$$F_n = \frac{1}{f_n} \quad \text{and}$$

$$f_n = \frac{1}{F_n}$$

If $F_n = x^3$, then: $f_n = \frac{1}{x^3}$, and vice versa.

$$F_n = 1 : \frac{1}{x^3} = x^3$$

When $x = 3$

$$f_n = \frac{1}{F_n} = \frac{1}{x^3} = \frac{1}{9},$$

then the three term series will be:

$$\sum_{\left(\frac{1}{x}\right)^3}^{x^3} 1^{x=3} = \frac{1}{9} \ldots \frac{9}{9} \ldots 9$$

C. Elements of Factorial-Fractional Continuity Expressed through Series

1. Monomials.

$$\sum_0^\infty 1^x = 0 \ldots \frac{x}{x} \ldots \infty$$

$$\sum_0^\infty 2^x = 0 \ldots \frac{1}{x} \ldots \frac{x}{x} \ldots x \ldots \infty$$

$$\sum_0^\infty 3^x = 0 \left(\frac{1}{x}\right)^n \ldots \left(\frac{1}{x}\right)^2 \ldots \frac{1}{x} \ldots \frac{x}{x} \ldots x \ldots x^2 \ldots x^n \ldots \infty$$

CONTINUITY

2. Binomials.

$$\sum_0^\infty 1^{x+y} = 0 \ldots \frac{x+y}{x+y} \ldots \infty$$

$$\sum_0^\infty 2^{x+y} = 0 \ldots \frac{1}{x+y} \ldots \frac{x+y}{x+y} \ldots x+y \ldots \infty$$

$$\sum_0^\infty 3^{x+y} = 0 \ldots \left(\frac{1}{x+y}\right)^n \ldots \left(\frac{1}{x+y}\right)^2 \ldots \frac{1}{x+y} \ldots \frac{x+y}{x+y} \ldots$$

$$\ldots x+y \ldots (x+y)^2 \ldots (x+y)^n \ldots \infty$$

3. Polynomials.

$$\sum_0^\infty 1^{a+b+c+\ldots+m} = 0 \ldots \frac{a+b+c+\ldots+m}{a+b+c+\ldots+m} \ldots \infty$$

$$\sum_0^\infty 2^{a+b+c+\ldots+m} = 0 \ldots \frac{1}{a+b+c+\ldots+m} \ldots \frac{a+b+c+\ldots+m}{a+b+c+\ldots+m} \ldots$$

$$\ldots (a+b+c+\ldots+m) \ldots \infty$$

$$\sum_0^\infty 3^{a+b+c+\ldots+m} = 0 \ldots \left(\frac{1}{a+b+c+\ldots+m}\right)^n \ldots$$

$$\ldots \left(\frac{1}{a+b+c+\ldots+m}\right)^2 \ldots \frac{1}{a+b+c+\ldots+m} \ldots \frac{a+b+c+\ldots+m}{a+b+c+\ldots+m} \ldots$$

$$\ldots (a+b+c+\ldots+m) \ldots (a+b+c+\ldots+m)^2 \ldots$$

$$\ldots (a+b+c+\ldots+m)^n \ldots \infty$$

Formulae for $\sum_{(\frac{1}{x})^n}^{x^n}$, $\sum_{(\frac{1}{x+y})^n}^{(x+y)^n}$, and $\sum_{(\frac{1}{a+b+c+\ldots+m})^n}^{(a+b+c+\ldots+m)^n}$

$$\sum_{\left(\frac{1}{x}\right)^n}^{x^n} = \left(\frac{1}{x}\right)^n \ldots \left(\frac{1}{x}\right)^2 \ldots \frac{1}{x} \cdot \frac{x}{x} \ldots x \ldots x^2 \ldots x^n$$

$$\sum_{\left(\frac{1}{x+y}\right)^n}^{(x+y)^n} = \left(\frac{1}{x+y}\right)^n \ldots \left(\frac{1}{x+y}\right)^2 \ldots \frac{1}{x+y} \cdot \frac{x+y}{x+y} \ldots x+y \ldots (x+y)^2 \ldots (x+y)^n$$

$$\sum_{\left(\frac{1}{a+b+c+\ldots+m}\right)^n}^{(a+b+c+\ldots+m)^n} = \left(\frac{1}{a+b+c+\ldots+m}\right)^n \ldots \left(\frac{1}{a+b+c+\ldots+m}\right)^2 \ldots$$

$$\ldots \frac{1}{a+b+c+\ldots+m} \cdot \frac{a+b+c+\ldots+m}{a+b+c+\ldots+m} \ldots$$

$$\ldots (a+b+c+\ldots+m) \ldots (a+b+c+\ldots+m)^2 \ldots (a+b+c+\ldots+m)^n \ldots$$

4. Binomial Series. — $x = a + b$.

$$\sum_{\left(\frac{1}{x}\right)^n}^{x^n}{}_{x=a+b} = \left(\frac{a}{a+b} + \frac{b}{a+b}\right)^n \ldots \left(\frac{a}{a+b} + \frac{b}{a+b}\right)^2 \ldots \frac{a}{a+b} + \frac{b}{a+b} \cdot \frac{a+b}{a+b} \ldots$$

$$\ldots a+b \ldots (a+b)^2 \ldots (a+b)^n$$

5. Polynomial Series. — $x = a+b+c+\ldots+m$.

$$\sum_{\left(\frac{1}{x}\right)^n}^{x^n}{}_{x=a+b+c+\ldots+m} = \left(\frac{a}{a+b+c+\ldots+m} + \frac{b}{a+b+c+\ldots+m} + \right.$$

$$\left. + \frac{c}{a+b+c+\ldots+m} + \ldots \frac{m}{a+b+c+\ldots+m}\right)^n \ldots$$

$$\ldots \left(\frac{a}{a+b+c+\ldots+m} + \frac{b}{a+b+c+\ldots+m} + \frac{c}{a+b+c+\ldots+m} + \ldots\right.$$

$$\left. \ldots + \frac{m}{a+b+c+\ldots+m}\right)^2 \ldots \left(\frac{a}{a+b+c+\ldots+m} + \frac{b}{a+b+c+\ldots+m} + \right.$$

$$\left. + \frac{c}{a+b+c+\ldots+m} + \ldots + \frac{m}{a+b+c+\ldots+m}\right) \cdot \frac{a+b+c+\ldots+m}{a+b+c+\ldots+m} \ldots$$

$$\ldots (a+b+c+\ldots+m) \ldots (a+b+c+\ldots+m)^2 \ldots (a+b+c+\ldots+m)^n$$

CONTINUITY

6. Series Consisting of Positive and Negative Values.

$$\sum_{-\infty}^{+\infty} n = -\infty \ldots -x^n \ldots -x^2 \ldots -x \ldots -\frac{x}{x} \ldots -\frac{1}{x} \ldots \left(-\frac{1}{x}\right)^2 \ldots \left(-\frac{1}{x}\right)^n \ldots$$

$$\ldots 0 \ldots \left(\frac{1}{x}\right)^n \ldots \left(\frac{1}{x}\right)^2 \ldots \frac{1}{x} \ldots \frac{x}{x} \ldots x \ldots \ldots x^2 \ldots x^n \ldots \infty$$

In a factorial-fractional continuity of rational values, the exponents are in inverse proportion to their bases.

Averages observed in music of the civilized world:

Bases	Exponents
$\frac{1}{2}$	4
2	3
$\frac{1}{3}$	2
3	1
$\frac{1}{5}$	1
$\frac{1}{7}$	1

Values for Factorial and Fractional Continuity

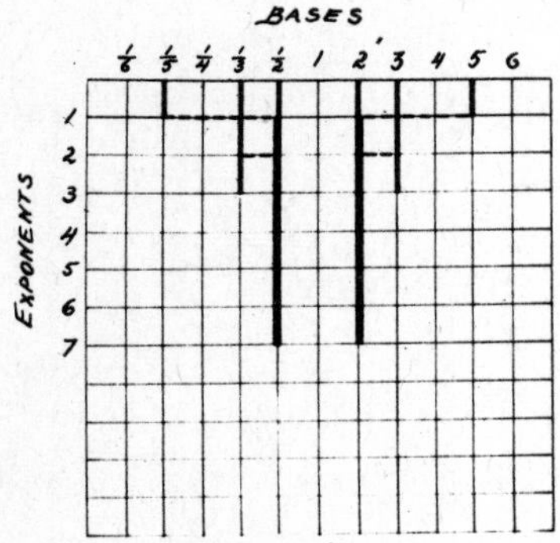

Figure 5. Averages in music of the civilized world.

THEORY OF REGULARITY AND COORDINATION

7. Series of Averages Observed in Music of the Civilized World.[1]

$$\cdots \left(\frac{1}{2}\right)^4 \cdots \left(\frac{1}{2}\right)^3 \cdots \left(\frac{1}{2}\right)^2 \cdots \cdots \frac{t}{t} \cdots \frac{1}{2} \cdots \frac{2}{2} \cdots 2 \cdots 2^2 \cdots 2^3 \cdots$$

$$\cdots \left(\frac{1}{2}\right)^2 \times \left(\frac{1}{3}\right)^2 \cdots \frac{1}{2} \times \left(\frac{1}{3}\right)^2 \cdots \left(\frac{1}{3}\right)^2 \cdots \cdots \frac{1}{3} \cdots \frac{3}{3} \cdots 2 \cdots 2^2 \cdots 2^3 \cdots$$

$$\cdots \cdots \left(\frac{1}{2}\right)^2 \times \frac{1}{5} \cdots \frac{1}{2} \times \frac{1}{5} \cdots \frac{1}{5} \cdots \frac{5}{5} \cdots 2 \cdots 2^2 \cdots 2^3 \cdots$$

$$\cdots \cdots \cdots \cdots \frac{1}{2} \times \frac{1}{7} \cdots \frac{1}{7} \cdots \frac{7}{7} \cdots 2 \cdots 2^2 \cdots 2^3 \cdots$$

$$\cdots \cdots \cdots \cdots \frac{1}{2} \times \frac{1}{9} \cdots \frac{1}{9} \cdots \frac{9}{9} \cdots 2 \cdots 2^2 \cdots 2^3 \cdots$$

$$\cdots \cdots \cdots \cdots \frac{1}{2} \times \frac{1}{11} \cdots \frac{1}{11} \cdots \frac{11}{11} \cdots 2 \cdots 2^2$$

8. Some of the Series Observed in Folklore.

$$\cdots \left(\frac{1}{3}\right)^3 \cdots \left(\frac{1}{3}\right)^2 \cdots \frac{1}{3} \cdots \frac{3}{3} \cdots 3 \cdots 3^2 \cdots \quad \text{(American Indians)}[2]$$

$$\cdots \frac{1}{2} \times \frac{1}{5} \cdots \frac{1}{5} \cdots \frac{5}{5} \cdots 5 \cdots 2 \times 5 \cdots \quad \text{(Great Russians)}$$

D. Determinants

The value of any term in a factorial-fractional series with one determinant can be expressed through:

$$F_{n°} = x^n$$

$$f_{n°} = \left(\frac{1}{x}\right)^n$$

$F_{n°}$ represents a factorial unit of the nth place in the series, counting from the initial unit $\frac{x}{x}$, the determinant of the series; $f_{n°}$ represents a fractional unit of the nth place in the series, counting to the left from the initial unit $\frac{x}{x}$.

---

[1] The fractions to the left of $\frac{t}{t}$ represent rhythms within measures while the numbers to the right represent the rhythms of the measures themselves. To Schillinger, musical form is an integrated unity of both these quantities, the measure-rhythm being structurally related to the phrase-rhythm. (Ed.)

[2] Helen Roberts, *Forms of Primitive Music* (American Library of Musicology, 1932), page 39.

CONTINUITY

The consecutive terms of a series, following to the right from the initial term $\frac{x}{x}$, produce units of factorial periodicity in their increasing orders. The consecutive terms of a series, following to the left from the initial term $\frac{x}{x}$, produce units of fractional periodicity in their increasing orders. Thus, the value n expresses:

(1) the order of a unit in a series;
(2) its place in a series;
(3) the power index of a base x.

In order to find the value of x at a given order n, it is sufficient to raise x to the nth power.

$$F_{n°} = x_{n°} = x^n, \text{ because:}$$

x	...	x^2	...	x^3	...	x^n	[powers]
1°		2°		3°		n°	[orders]

Likewise, $f_{n°} = \left(\frac{1}{x}\right)^{n°} = \left(\frac{1}{x}\right)^n$, for:

$\left(\frac{1}{x}\right)^n$	...	$\left(\frac{1}{x}\right)^3$	...	$\left(\frac{1}{x}\right)^2$	...	$\frac{1}{x}$	[powers]
n°		3°		2°		1°	[orders]

Example:

$\frac{x}{x} = \frac{2}{2}$. Find the value of x and of $\frac{1}{x}$ at their fourth order.

$$F_{4°} = 2^4 = 16$$
$$f_{4°} = \left(\frac{1}{2}\right)^4 = \frac{1}{16}$$

The determinant of a series in its factorial-fractional form may be expressed through the following identity:

$$F_{1°} f_{1°} = T = \frac{x}{x} = x\left(\frac{1}{x}\right) = \frac{t_1}{x} + \frac{t_2}{x} + \ldots + \frac{t_x}{x}$$

and the general form of a factorial-fractional unit of the nth order:

$$F_{n°} f_{n°} = T^n = \left(\frac{x}{x}\right)^n = \frac{t_1}{x^n} + \frac{t_2}{x^n} + \ldots + \frac{t_x}{x^n}$$

If the determinant of a series is m, and $x \neq m$, then the orders grow through the powers of m.

$$F_{1°} f_{1°} = T = m\left(\frac{x}{x}\right) = \left[mx\left(\frac{1}{x}\right)\right] = m\left(\frac{t_1}{x} + \frac{t_2}{x} + \ldots + \frac{t_x}{x}\right) \text{ and}$$

$$F_{n°} f_{n°} = T_n = m^n\left(\frac{x}{x}\right) = m^n\left(\frac{t_1}{x} + \frac{t_2}{x} + \ldots + \frac{t_x}{x}\right)$$

Here, m is the coefficient of growth of the term x.

Example:

$x = 3; n = 4$

$$T = \frac{3}{3} = 3\left(\frac{1}{3}\right) = \frac{t_1}{3} + \frac{t_2}{3} + \frac{t_3}{3}$$

$$T_{4°} = T^4 = \left(\frac{3}{3}\right)^4 = 3^4\left(\frac{1}{3^4}\right) = 81\left(\frac{1}{81}\right) = \frac{t_1}{81} + \frac{t_2}{81} + \ldots + \frac{t_{81}}{81}$$

$x = 3; m = 2; n = 4$

$$T = 2\left(\frac{3}{3}\right) = 2\left[3\left(\frac{1}{3}\right)\right] = 2\left(\frac{t_1}{3} + \frac{t_2}{3} + \frac{t_3}{3}\right)$$

$$T_{4°} = 2^4\left(\frac{3}{3}\right) = 16\left[3\left(\frac{1}{3}\right)\right] = 16\left(\frac{t_1}{3} + \frac{t_2}{3} + \frac{t_3}{3}\right)$$

When it is desirable to plan a continuity in advance, with respect to the number of units and their values at a given order n, the following formulae should be used:

1. Series With One Determinant.

(a) $m = x$

$$T_{n°} = T^n = x(x^{n-1}), x^2(x^{n-2}), x^3(x^{n-3}), \ldots x^{n-3}(x^3), x^{n-2}(x^2), x^{n-1}(x).$$

The continuity T at the nth order equals T to the nth power, equals x units of the value x^{n-1}, x^2 units of the value x^{n-2}, etc.

Examples:

$x = 2; n = 5; T^5 = x^5 = 2^5 = 32$

$T_{5°} = T^5 = 2(2^4), 2^2(2^3), 2^3(2^2), 2^4(2) = 2(16), 4(8), 8(4), 16(2).$

$x = 3; n = 5; T^5 = x^5 = 3^5 = 243$

$T_{5°} = T^5 = 3(3^4), 3^2(3^3), 3^3(3^2), 3^4(3) = 3(81), 9(27), 27(9), 81(3).$

$x = 4; n = 5; T^5 = x^5 = 4^5 = 1024$

$T_{5°} = T^5 = 4(4^4), 4^2(4^3), 4^3(4^2), 4^4(4) = 4(256), 16(64), 64(16), 256(4).$

$x = 5; n = 4; T^4 = x^4 = 5^4 = 625$

$T_{4°} = T^4 = 5(5^3), 5^2(5^2), 5^3(5) = 5(125), 25(25), 125(5).$

$x = 7; n = 4; T^4 = x^4 = 7^4 = 2401$

$T_{4°} = T^4 = 7(7^3), 7^2(7^2), 7^3(7) = 7(343), 49(49), 343(7).$

CONTINUITY

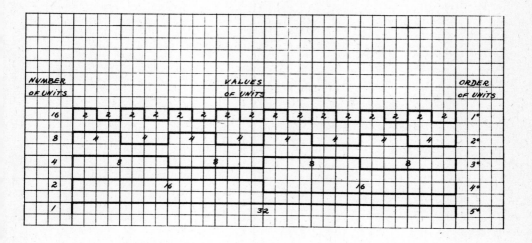

Figure 6. Series with one determinant.

(b) $m \neq x$

$T_{n°} = T^n = mx(m^{n-1}x), m^2x(m^{n-2}x), m^3x(m^{n-3}x), \ldots$

$\ldots m^{n-3}x(m^3x), m^{n-2}x(m^2x), m^{n-1}x(mx).$

The continuity T at the nth order equals T to the nth power, equals mx units of the value $m^{n-1}x$, m^2x units of the value $m^{n-2}x$, etc.

Examples:

$x = 3; m = 2; n = 5; T^5 = m^5x^2 = 2^5 \times 3^2 = 288$

$T_{5°} = T^5 = 2 \times 3(2^4 \times 3), 2^2 \times 3(2^3 \times 3), 2^3 \times 3(2^2 \times 3), 2^4 \times 3(2 \times 3) = 6(48), 12(24), 24(12), 48(6).$

$x = 2; m = 3; n = 4; T^4 = m^4x^2 = 3^4 \times 2^2 = 324$

$T_{4°} = T^4 = 3 \times 2(3^3 \times 2), 3^2 \times 2(3^2 \times 2), 3^3 \times 2(3 \times 2) = 6(54), 18(18), 54(6).$

$x = 5; m = 2; n = 4; T^4 = m^4x^2 = 2^4 \times 5^2 = 400$

$T_{4°} = T^4 = 2 \times 5(2^3 \times 5), 2^2 \times 5(2^2 \times 5), 2^3 \times 5(2 \times 5) = 10(40), 20(20), 40(10).$

2. Series With Two Determinants.

x and y are the two periodically alternating determinants of the series.

(a) The order of T is $2n + 1$.

$T_{2n+1} = x(x^n y^n), xy(x^n y^{n-1}), x^2 y(x^{n-1} y^{n-1}), x^2 y^2 (x^{n-1} y^{n-2}), \ldots$
$\ldots x^{n-1} y^{n-2}(x^2 y^2), x^{n-1} y^{n-1}(x^2 y), x^n y^{n-1}(xy), x^n y^n(x).$

Example:

$x = 2; y = 3; 2n + 1 = 7$

$T_7 = x(x^3 y^3), xy(x^3 y^3), x^2 y(x^2 y^2), x^2 y^2(x^2 y), x^3 y^2(xy), x^3 y^3(x) = 2(2^3 \times 3^3), 2 \times 3(2^3 \times 3^2), 2^2 \times 3(2^2 \times 3^2), 2^2 \times 3^2(2^2 \times 3), 2^3 \times 3^2(2 \times 3), 2^3 \times 3^3(2) = 2(216), 6(72), 12(36), 36(12), 72(6), 216(2).$

(b) The order of T is 2n.

$T_{2n} = x(x^n y^{n-1}), xy(x^{n-1} y^{n-1}), x^2 y(x^{n-1} y^{n-2}), \ldots \ldots x^{n-1} y^{n-2}(x^2 y),$
$x^{n-1} y^{n-1}(xy), x^n y^{n-1}(x).$

Example:

$x = 2; y = 3; 2n = 8$

$T_8 = x(x^4 y^3), xy(x^3 y^3), x^2 y(x^3 y^2), x^2 y^2(x^2 y^2), x^3 y^2 (x^2 y), x^3 y^3 (xy), x^4 y^3(x) = 2(2^4 \times 3^3), 2 \times 3(2^3 \times 3^3), 2^2 \times 3(2^3 \times 3^2), 2^2 \times 3^2(2^2 \times 3^2), 2^3 \times 3^2(2^2 \times 3), 2^3 \times 3^3(2 \times 3), 2^4 \times 3^3(2) = 2(432), 6(216), 12(72), 36(36), 72(12), 216(6), 432(2).$

When m is the determinant of a series and x and y are periodically alternating coefficients of growth, m enters as a factor in both co-factors.

$T_{2n+1} = mx(mx^n y^n), mxy(mx^n y^{n-1}), \ldots$
$T_{2n} = mx(mx^n y^{n-1}), mxy(mx^{n-1} y^{n-1}), \ldots$

Example:

$x = 2; y = 3; m = 5; 2n+1 = 7$

$T_7 = 10(1080), 30(360), 60(180), 360(30), 1080(10).$

$x = 2; y = 3; m = 4; 2n = 6$

$T_6 = 8(288), 24(144), 48(48), 144(24), 288(8).$

3. Series With Three Determinants.

x, y and z are the three periodically alternating determinants of the series.

(a) The order of T is $3n+1$

$T_{3n+1} = x(x^n y^n z^n),\ xy(x^n y^n z^{n-1}),\ xyz(x^n y^{n-1} z^{n-1}),\ x^2yz(x^{n-1}y^{n-1}z^{n-1}),$
$x^2y^2z(x^{n-1}y^{n-1}z^{n-2}),\ x^2y^2z^2(x^{n-1}y^{n-2}z^{n-2}),\ \ldots\ x^{n-1}y^{n-2}z^{n-2}(x^2y^2z^2),$
$x^{n-1}y^{n-1}z^{n-2}(x^2y^2z),\ x^{n-1}y^{n-1}z^{n-1}(x^2yz),\ x^n y^{n-1} z^{n-1}(xyz),\ x^n y^n z^{n-1}(xy),$
$x^n y^n z^n(x).$

Example:

$x = 2;\ y = 3;\ z = 5;\ 3n+1 = 7$

$T_7 = 2(2^2 \times 3^2 \times 5^2),\ 2\times 3(2^2 \times 3^2 \times 5),\ 2\times 3 \times 5(2^2 \times 3 \times 5),\ 2^2 \times 3 \times 5(2\times 3\times 5),$
$2^2 \times 3^2 \times 5(2\times 3),\ 2^2 \times 3^2 \times 5^2 (2) = 2(900),\ 6(180),\ 30(60),\ 60(30),\ 180(6),\ 900(2).$

(b) The order of T is 3n.

$T_{3n} = x(x^n y^n z^{n-1}),\ xy(x^n y^{n-1} z^{n-1}),\ xyz(x^{n-1}y^{n-1}z^{n-1}),\ x^2yz(x^{n-1}y^{n-1}z^{n-2}),$
$x^2y^2z(x^{n-1}y^{n-2}z^{n-2}),\ x^2y^2z^2(x^{n-2}y^{n-2}z^{n-2}),\ \ldots\ x^{n-2}y^{n-2}z^{n-2}(x^2y^2z^2),$
$x^{n-1}y^{n-2}z^{n-2}(x^2y^2z),\ x^{n-1}y^{n-1}z^{n-2}(x^2yz),\ x^{n-1}y^{n-1}z^{n-1}(xyz),\ x^n y^{n-1}z^{n-1}(xy),$
$x^n y^n z^{n-1}(x).$

Example:

$x = 2;\ y = 3;\ z = 5;\ 3n = 6$

$T_6 = 2(2^2 \times 3^2 \times 5),\ 2\times 3(2^2 \times 3 \times 5),\ 2\times 3 \times 5(2\times 3\times 5),$
$2^2 \times 3 \times 5(2\times 3),\ 2^2 \times 3^2 \times 5(2) = 2(180),\ 6(60),\ 30(30),\ 60(6),\ 180(2).$

As in previous cases x, y and z may become coefficients of growth; then m as a determinant of the series becomes a constant complementary factor of both variable co-factors of the series.

4. Generalization.

Let $x_1, x_2, x_3, \ldots x_{k-2}, x_{k-1}, x_k$ be the periodically alternating coefficients of growth and m the determinant of a series.

(a) The order of T is $kn+1$

$$T_{kn+1} = mx_1 \left[m \begin{pmatrix} k \\ x \\ 1 \end{pmatrix}^n \right]$$

$$T_{kn} = mx_1 x_2 \left\{ m \left[\begin{pmatrix} k-1 \\ x \\ 1 \end{pmatrix}^n x_k^{n-1} \right] \right\}.$$

$$T_{kn-1} = mx_1x_2x_3 \left\{ m\left[\begin{pmatrix} k-2 \\ x \\ 1 \end{pmatrix}^n x_{k-1}^{n-1} \; x_k^{n-1} \right] \right\}$$

. . . .

$$T_{[k(n-1)]+1} = m \begin{matrix} k \\ x \\ 1 \end{matrix} \left[m \begin{pmatrix} k \\ x \\ 1 \end{pmatrix}^{n-1} \right]$$

. . . .

$$T_{[k(n-2)]+1} = m \begin{pmatrix} k \\ x \\ 1 \end{pmatrix}^2 \left[m \begin{pmatrix} k \\ x \\ 1 \end{pmatrix}^{n-2} \right]$$

. . . .

$$T_{[k(n-3)]+1} = m \begin{pmatrix} k \\ x \\ 1 \end{pmatrix}^3 \left[m \begin{pmatrix} k \\ x \\ 1 \end{pmatrix}^{n-3} \right]$$

. . . .

$$\left. \begin{array}{l} T_{k\left(\frac{n}{2}\right)+1} \quad * \\ \\ T_{k\left(\frac{n}{2}\right)} \quad ** \end{array} \right| = m \begin{pmatrix} k \\ x \\ 1 \end{pmatrix}^{\frac{n}{2}} \left\{ m \left[x_1^{\frac{n}{2}+1} \begin{pmatrix} k \\ x \\ 1 \end{pmatrix}^{\frac{n}{2}} \right] \right\}$$

$$\left. \begin{array}{l} T_{k\left(\frac{n}{2}\right)+1} \quad * \\ \\ T_{k\left(\frac{n}{2}\right)+2} \quad ** \end{array} \right| = \left\{ m \left[x_1^{\frac{n}{2}+1} \begin{pmatrix} k \\ x \\ 1 \end{pmatrix}^{\frac{n}{2}} \right] \right\} m \begin{pmatrix} k \\ x \\ 1 \end{pmatrix}^{\frac{n}{2}}$$

. . . .

$$T_{3k} = \left[m \begin{pmatrix} k \\ x \\ 1 \end{pmatrix}^{n-3} \right] m \begin{pmatrix} k \\ x \\ 1 \end{pmatrix}^3$$

. . . .

*Odd number.
**Even number.

CONTINUITY

$$T_{2k} = \left[m \begin{pmatrix} k \\ x \\ 1 \end{pmatrix}^{n-2} \right] m \begin{pmatrix} k \\ x \\ 1 \end{pmatrix}^2$$

....

$$T_k = \left[m \begin{pmatrix} k \\ x \\ 1 \end{pmatrix}^{n-1} \right] m \begin{pmatrix} k \\ x \\ 1 \end{pmatrix}$$

....

$$T_3 = \left\{ m \left[\begin{pmatrix} k-2 \\ x \\ 1 \end{pmatrix}^n x_{k-1}^{n-1} x_k^{n-1} \right] \right\} m x_1 x_2 x_3$$

$$T_2 = \left\{ m \left[\begin{pmatrix} k-1 \\ x \\ 1 \end{pmatrix}^n x_k^{n-1} \right] \right\} m x_1 x_2$$

$$T_1 = \left[m \begin{pmatrix} k \\ x \\ 1 \end{pmatrix}^n \right] m x_1$$

(b) The order of T is kn.

$$T_{kn} = m x_1 \left\{ m \left[\begin{pmatrix} k-1 \\ x \\ 1 \end{pmatrix}^n x_k^{n-1} \right] \right\}$$

$$T_{kn-1} = m x_1 x_2 \left\{ m \left[\begin{pmatrix} k-2 \\ x \\ 1 \end{pmatrix}^n x_{k-1}^{n-1} x_k^{n-1} \right] \right\}$$

$$T_{kn-2} = m x_1 x_2 x_3 \left\{ m \left[\begin{pmatrix} k-3 \\ x \\ 1 \end{pmatrix}^n x_{k-2}^{n-1} x_{k-1}^{n-1} x_k^{n-1} \right] \right\}$$

....

$$T_{k(n-1)} = m \begin{pmatrix} k \\ x \\ 1 \end{pmatrix} \left[m \begin{pmatrix} k \\ x \\ 1 \end{pmatrix}^{n-1} \right]$$

....

$$T_{k(n-2)} = m \begin{pmatrix} k \\ x \\ 1 \end{pmatrix}^2 \left[m \begin{pmatrix} k \\ x \\ 1 \end{pmatrix}^{n-2} \right]$$

....

$$T_{k(n-3)} = m \begin{pmatrix} k \\ x \\ 1 \end{pmatrix}^3 \left[m \begin{pmatrix} k \\ x \\ 1 \end{pmatrix}^{n-2} \right]$$

....
....

$$T_{k(\frac{n}{2})} = m \begin{pmatrix} k \\ x \\ 1 \end{pmatrix}^{\frac{n}{2}} \left[m \begin{pmatrix} k \\ x \\ 1 \end{pmatrix}^{\frac{n}{2}} \right]$$

$$T_{k(\frac{n}{2})+1} = \left[m \begin{pmatrix} k \\ x \\ 1 \end{pmatrix}^{\frac{n}{2}} \right] m \begin{pmatrix} k \\ x \\ 1 \end{pmatrix}^{\frac{n}{2}}$$

....
....

$$T_{3k} = \left[m \begin{pmatrix} k \\ x \\ 1 \end{pmatrix}^{n-3} \right] m \begin{pmatrix} k \\ x \\ 1 \end{pmatrix}^3$$

....

$$T_{2k} = \left[m \begin{pmatrix} k \\ x \\ 1 \end{pmatrix}^{n-2} \right] m \begin{pmatrix} k \\ x \\ 1 \end{pmatrix}^2$$

....

$$T_k = \left[m \begin{pmatrix} k \\ x \\ 1 \end{pmatrix}^{n-1} \right] m \begin{pmatrix} k \\ x \\ 1 \end{pmatrix}$$

....

$$T_3 = \left\{ m \left[\begin{pmatrix} k-3 \\ x \\ 1 \end{pmatrix}^n x_{k-2}^{n-1} x_{k-1}^{n-1} x_k^{n-1} \right] \right\} m x_1 x_2 x_3$$

....

$$T_2' = \left\{ m \left[\begin{pmatrix} k-2 \\ x \\ 1 \end{pmatrix}^n x_{k-1}^{n-1} x_k^{n-1} \right] \right\} m x_1 x_2$$

$$T_1 = \left\{ m \left[\begin{pmatrix} k-1 \\ x \\ 1 \end{pmatrix}^n x_k^{n-1} \right] \right\} m x_1$$

CONTINUITY

The limit indication in these formulae determines the products of all coefficients of growth within the given limits.

Thus, $x_1 = x_1$

$x_1 x_2 = x_1 \times x_2$

$\overset{k}{\underset{1}{x}} = x_1 \times x_2 \times x_3 \times \ldots \ldots x_k$

Figure 7. Rhythm is the law of regularity. A Schillinger design.

CHAPTER 3

PERIODICITY[1]

RHYTHM is the Law of Regularity in a factorial-fractional continuity. It is the resultant of the synchronized component periodicities, their progression, modification and powers.

The simplest form of periodicity may be observed in longitudinal sine waves, where the wave forms have definite regularity with respect to their frequency and amplitude. Certain sound waves, moving through space, have such periodicity.

A pendulum in its transverse oscillations[2], adjusted to leave a trace, will produce a wave form similar to the simplest wave. Such a wave motion expresses uniformity in frequency, amplitude and phase, and is known in mechanics as "simple harmonic motion".

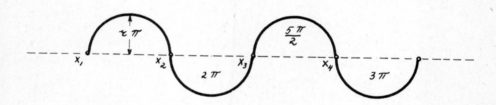

Figure 1. *Transverse oscillations of a pendulum.*

The distance $x_1 - x_2$ forms a *phase*, and the distance $x_1 - x_3$ forms a *period*.

Periodicity is a continuity of periods, or a continuity of phases. For the sake of simplicity and graphic considerations we shall use the *periodicity in phases* as units of measurement in further exposition. In order to establish values of periodicities, we may neglect the amplitude of a phase (r). The graph of simple harmonic motion, representing a progression of values along any parameter, may be drawn in the following form.

---

[1]The reader is referred to Appendix A *Basic Forms of Regularity and Coordination* for a detailed elaboration of the ideas and formulae presented in this chapter. See pages 445-639. (Ed.)

[2]If a sheet of paper is placed under a pendulum and moved in a direction perpendicular to its line of motion, the pendulum will describe a sine curve. (Ed.)

Figure 2. Simple harmonic motion graphed (sine curve).

The horizontal segments represent a progression of phases and the vertical segments define the limits between the adjacent phases.

Single phases in a rectangular graph will be considered as standard units of measurement.

We can define the periodicity representing "simple harmonic motion" as *monomial periodicity*. The general form of monomial periodicity may be expressed through consecutive terms: $t_1, t_2, t_3 \ldots$, which correspond to phases on the graph; T is a definite periodic portion of continuity along any parametric extension:

$$T = t_1 + t_2 + t_3 + t_4 + \ldots t_n \tag{1}$$

The relative values of consecutive terms (which may correspond to absolute values in any standard of measurement) may be expressed through a, b, c, ... The expression for a definite portion of monomial periodic continuity will be:

$$T_a = at_1 + at_2 + at_3 + \ldots$$
$$T_b = bt_1 + bt_2 + bt_3 + \ldots$$
$$T_c = ct_1 + ct_2 + ct_3 + \ldots$$
$$\ldots\ldots\ldots\ldots\ldots\ldots$$
$$T_m = mt_1 + mt_2 + mt_3 + \ldots$$

By defining the limits "a" and "n" a more generalized form may be obtained.

$$\overset{n}{\underset{a}{T_m}} = mt_1 + mt_2 + mt_3 + \ldots + mt_n \tag{2}$$

"m" may be equal to one inch, one second, one degree of an angle, one gram, etc. These values, although arbitrarily chosen, will represent nevertheless an absolute system of relations between the different periodicities.

Assuming that a = 2 and b = 3, then:

$$T_a = 2t_1 + 2t_2 + 2t_3 + \ldots$$
$$T_b = 3t_1 + 3t_2 + 3t_3 + \ldots$$

$t_a : t_b = 2 : 3$, *i. e.*, the term of periodicity "a" is related to the term of periodicity "b" as 2 to 3. From this progression t_a can be expressed through t_b:

PERIODICITY

and t_b can be expressed through t_a:

$$t_b = \frac{3t_a}{2}$$

or, generally, if $a = m$ and $b = n$,

$$t_a : t_b = m : n, \text{ then}$$

$$t_a = \frac{mt_b}{n}; \quad t_b = \frac{nt_a}{m}$$

A relation of two monomial periodicities through their absolute values may be represented graphically. If $a = 1$ and $b = 2$ in terms of $\frac{1''}{4}$, then:

$$T_a = t_1 + t_2 + t_3 \ldots = \frac{1''}{4} + \frac{1''}{4} + \frac{1''}{4} + \ldots$$

$$T_b = 2t_1 + 2t_2 + 2t_3 \ldots = \frac{1''}{2} + \frac{1''}{2} + \frac{1''}{2} + \ldots$$

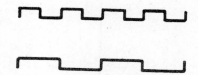

Figure 3. *Relation of two monomial periodicities.*

This is the ratio of $1:2$, *i. e.*, $t_a : t_b = 1:2$. Thus, $t_a = \frac{t_b}{2}$ or $t_b = 2t_a$.

In terms of relative values ($a = 1$, $b = 2$) these periodicities may also be expressed

$$T_a = 1 + 1 + 1 + 1 + \ldots$$
$$T_b = 2 + 2 + 2 + 2 + \ldots$$

Assuming $a = 1 = \frac{1''}{2}$, the graph will be enlarged accordingly.

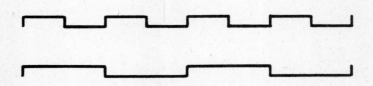

Figure 4. Relative values of a and b increased.

Though the value of "a" has changed, the relation of "a" to "b" remains the same.

A. SIMULTANEOUS MONOMIAL PERIODICITIES. (Synchronization)

1. *Binomial relations of monomial periodicities.*
(Synchronization of two periodicities)

The whole range of monomial periodicities, with respect to their relative values and progression, may be represented through the series of natural fractions previously described. We may write:

$$\frac{t}{t}$$

$$\frac{t_1}{2} + \frac{t_2}{2}$$

$$\frac{t_1}{3} + \frac{t_2}{3} + \frac{t_3}{3}$$

$$\frac{t_1}{4} + \frac{t_2}{4} + \frac{t_3}{4} + \frac{t_4}{4}$$

$$\ldots\ldots\ldots\ldots\ldots$$

$$\frac{t_1}{n} + \frac{t_2}{n} + \frac{t_3}{n} + \ldots + \frac{t_n}{n}$$

$$\ldots\ldots\ldots\ldots\ldots$$

$$\frac{t_1}{\infty} + \frac{t_2}{\infty} + \frac{t_3}{\infty} + \ldots + \frac{t_n}{\infty} + \ldots + \frac{t_\infty}{\infty}$$

The series from $\frac{1}{1}$ to $\frac{1}{\infty}$ bt T consists of an infinite number of series, where

$$T_a = t, \ T_b = 2t, \ T_c = 3t, \ T_d = 4t,$$
$$T_n = nt \text{ and } T_\infty = \infty t.$$

PERIODICITY

The *derivative series* resulting from the synchronization of T_a and T_b, where a and b are fractions with any desirable denominator and a numerator of unity, are obtainable through the interference of synchronized T_a and T_b. This process will be realized as follows:

In order to express t_a and t_b in the same unities, it is necessary to find the common denominator of a and b.

But, $T_a = at$ and
$T_b = bt$,   then

$T_a = a(t_1 + t_2 + t_3 + \ldots\ldots)$
$T_b = b(t_1 + t_2 + t_3 + \ldots\ldots)$,   then

$t_a = \dfrac{1}{a}$ and

$t_b = \dfrac{1}{b}$

The common denominator is ab.
Therefore, b is the complementary factor of a.

Assuming $a = 2$ and $b = 3$, we obtain

$$T_2 = \frac{t_1}{2} + \frac{t_2}{2}$$

$$T_3 = \frac{t_1}{3} + \frac{t_2}{3} + \frac{t_3}{3}$$

The common denominator is $2 \times 3 = 6$.

Therefore:
$$T_2 = 2\left(\frac{3t}{6}\right) = \frac{3t_1}{6} + \frac{3t_2}{6}$$

$$T_3 = 3\left(\frac{2t}{6}\right) = \frac{2t_1}{6} + \frac{2t_2}{6} + \frac{2t_3}{6}$$

Now t_a and t_b are equal: $t_{a:b} = \dfrac{1}{6}$ We may graph:

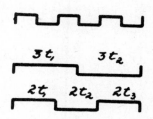

Figure 5. Synchronization of two monomial periodicities.

We now have two synchronized monomial periodicities, T_2 and T_3. The derivative periodicity $T_{2:3}$ resulting from the interference of phases, can be obtained graphically by dropping perpendiculars from the beginning of each phase of T_a and T_b.

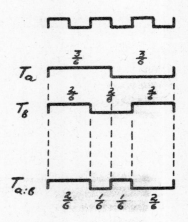

Figure 6. *Derivative periodicity of* $T_{2:3}$.

Reading from the graph, we derive values for $T_{a:b}$.

$$T_{a:b} = \frac{2t_1}{6} + \frac{t_2}{6} + \frac{t_3}{6} + \frac{2t_4}{6}$$

Omitting the t's but preserving the same order of progression and writing the numerators only, we discover the following rhythmic series:

$$T_{2:3} = 2 + 1 + 1 + 2$$

Any values may be selected for a and b. If $\frac{a}{b}$ is a reducible fraction it should be reduced first; otherwise the resulting rhythmic series will repeat itself. For instance, $\frac{4}{6} = \frac{2}{3}$, $\frac{6}{12} = \frac{1}{2}$, etc.

We shall now assume a = 3 and b = 4. Then:

$$T_3 = \frac{t_1}{3} + \frac{t_2}{3} + \frac{t_3}{3}$$

$$T_4 = \frac{t_1}{4} + \frac{t_2}{4} + \frac{t_3}{4} + \frac{t_4}{4}$$

The common denominator is $3 \times 4 = 12$.

Therefore,

$$T_3 = 3\left(\frac{4t}{12}\right) = \frac{4t_1}{12} + \frac{4t_2}{12} + \frac{4t_3}{12}$$

$$T_4 = 4\left(\frac{3t}{12}\right) = \frac{3t_1}{12} + \frac{3t_2}{12} + \frac{3t_3}{12} + \frac{3t_4}{12}$$

$t_{a:b} = \frac{1}{12}$, and we may graph:

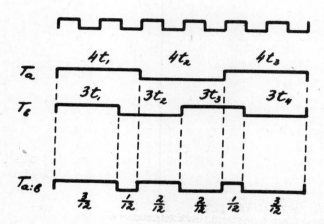

Figure 7. Synchronization and derivative periodicity of $T_{3:4}$.

The rhythmic series

$$T_{3:4} = \frac{3t_1}{12} + \frac{t_2}{12} + \frac{2t_3}{12} + \frac{2t_4}{12} + \frac{t_5}{12} + \frac{3t_6}{12}$$

Or, using the numerators only:

$$T_{3:4} = 3 + 1 + 2 + 2 + 1 + 3$$

Assuming various values for a and b, we get the following results:

$T_{2:5} = 2 + 2 + 1 + 1 + 2 + 2$
$T_{3:5} = 3 + 2 + 1 + 3 + 1 + 2 + 3$
$T_{4:5} = 4 + 1 + 3 + 2 + 2 + 3 + 1 + 4$
$T_{5:6} = 5 + 1 + 4 + 2 + 3 + 3 + 2 + 4 + 1 + 5$
$T_{2:7} = 2 + 2 + 2 + 1 + 1 + 2 + 2 + 2$
$T_{3:7} = 3 + 3 + 1 + 2 + 3 + 2 + 1 + 3 + 3$
$T_{4:7} = 4 + 3 + 1 + 4 + 2 + 2 + 4 + 1 + 3 + 4$
$T_{5:7} = 5 + 2 + 3 + 4 + 1 + 5 + 1 + 4 + 3 + 2 + 5$
$T_{6:7} = 6 + 1 + 5 + 2 + 4 + 3 + 3 + 4 + 2 + 5 + 1 + 6$
$T_{3:8} = 3 + 3 + 2 + 1 + 3 + 3 + 1 + 2 + 3 + 3$
$T_{5:8} = 5 + 3 + 2 + 5 + 1 + 4 + 4 + 1 + 5 + 2 + 3 + 5$
$T_{7:8} = 7 + 1 + 6 + 2 + 5 + 3 + 4 + 4 + 3 + 5 + 2 + 6 + 1 + 7$

Rhythmic series resulting from the interference of two monomial periodic groups a and b may be obtained by graphs or by computation. We shall consider such a series as the resultant of interference, and we shall designate it as $r(a \div b)$.

The total number of terms in $r(a \div b)$ is:

$$N_{a:b} = a + b - 1 \qquad (1)$$

The resultant always starts and ends with the term b, which also occurs in the center of r, when

$$a + b - 1 = 2n + 1, \; i.\, e.,$$

when $N_{a:b}$ has an odd number of terms. We shall consider the b-terms, appearing at the beginning and at the end of r, as end-terms.

The number of b-terms in the entire $r(a \div b)$ is:

$$N_b = a - b + 1 \qquad (2)$$

We shall consider $\frac{a}{b} = m$, where m is the integral part of the quotient. The b-term appears at each end m times. Thus if $m = 1$ it appears once at the beginning and once at the end of r; if $m > 1$, b-term, appearing m times at each end of r, produces $G_{b'}$, i. e., the end groups of b. The total number of b-groups in the entire r is:

$$NG_b = b \qquad (3)$$

The number of b-terms, appearing in the end-groups of r, is:

$$N_{b'}(G_{b'}) = 2m, \qquad (4)$$

as it is m at each end. Hence, the number of remaining b-terms is:

$$N_{b''} = N_b - 2m = a - b + 1 - 2m \qquad (5)$$

It follows from the above, that the number of remaining groups of b-terms is:

$$NG_{b''} = b - 2 \qquad (6)$$

Hence the number of b''-terms, appearing in each $G_{b''}$ in succession, is:

$$N_{b''}(G_{b''}) = \frac{a - b + 1 - 2m}{b - 2}, \qquad (7)$$

when $G_b = 2n$, and

$$N_{b''}(G_{b''}) = \frac{a - b + 1 - 2m}{b - 3}, \qquad (7a)$$

when $G_b = 2n + 1$.

The number of groups of x-terms appearing between the b-groups is:

$$NG_x = a - b, \text{ when } m = 1, \qquad (8)$$

and $\quad NG_x = b - 1$, when $m > 1$. $\qquad (8a)$

The number of x-terms in each Gx is:

$$N_x(G_x) = \frac{2b - 2}{a - b}, \text{ when } m = 1, \qquad (9)$$

and $\quad N_x(G_x) = \dfrac{2b - 2}{b - 1}$, when $m > 1$. $\qquad (9a)$

When such a division is impossible without a remainder, two kinds of groups develop between the b-groups: G_x and G_y. The number of all groups of G_x and G_y remains the same, as if they all were G_x, *i.e.*,

$$NG_{xy} = NG_x, i.e.: (8) \text{ and } (8a).$$

The distribution of G_x and G_y can be obtained as follows. Through the method of normal series, we can determine the possible places for the groups x and y. Every resultant is a group with bifold symmetry. The geometric center of the group is a point of symmetry. Thus, the second half of the series is the inversion of the first half of it. Therefore, if the number of b-terms does not exceed 3 when $a < 2b$, or 5 when $a > 2b$, the two groups being on symmetrical places can be x only.

$$b \ldots x \ldots b \ldots x \ldots b$$
or $\quad b + b \ldots x \ldots b \ldots x \ldots b + b$

With 4b when $a < 2b$ and 6b when $a > 2b$, there is only one solution for the places of groups between the b's.

$$b \ldots x \ldots b \ldots y \ldots b \ldots x \ldots b$$
or $\quad b + b \ldots x \ldots b \ldots y \ldots b \ldots x \ldots b + b$

With 5b when $a < 2b$ and 7b when $a > 2b$, one of the b's coincides with the point of symmetry. Thus, we obtain x and y on each side of the series.

$$b \ldots x \ldots b \ldots y \ldots b \ldots y \ldots b \ldots x \ldots b$$
$$b + b \ldots x \ldots b \ldots y \ldots b \ldots y \ldots b \ldots x \ldots b + b$$

Some other examples:

b appears 6 times —

$$b\ldots x\ldots b\ldots y\ldots b\ldots x\ldots b\ldots y\ldots b\ldots x\ldots b$$
$$b\ldots x\ldots b\ldots x\ldots b\ldots y\ldots b\ldots x\ldots b\ldots x\ldots b$$

b appears 7 times —
```
b...x...b...y...b...x...b...x...b...y...b...x...b
b...x...b...y...b...y...b...y...b...y...b...x...b
b...x...b...x...b...y...b...y...b...x...b...x...b
```

As 2 has always two identical conversely arranged halves, it is necessary to find individual terms only for $\frac{r}{2}$. The values of terms x and y in the groups of x and y are controlled by two basic conditions:

(1) the values are represented by continuously alternating binomials: $a - b$ and $b - a$;

(2) the coefficients of these binomials express the following regularity:

   (a) cases, where $a - b \not> 2$:
      coefficients of a: 1, 1, 2, 2, 3, 3, ... m, m;
      coefficients of b: 1, 1, 2, 2, 3, 3, ... m, m;

   (b) cases, where $a - b > 2$:
      coefficients of a: 1, 1, 2, 2, 3, 3, ... m, m;
      coefficients of b: 1, 2, 3, 4, 5, 6, ... m.

Examples:

(1) $r(7 \div 6) = b + (a - b) + (2b - a) + (2a - 2b) + (3b - 2a) + \ldots$

(2) $r(9 \div 5) = b + (a - b) + (2b - a) + (3b - 2b) + (2a - 3b) +$
            $+ (4b - 2a) + \ldots$
   where $3b - 2b = b$.

2. Synchronization of two monomial periodicities with consecutive displacements of periodicity b with regard to the integral phases of periodicity a.

Through this process the resulting periodicity becomes fractioned on both sides of the point of symmetry. When $b = n$, n single units will appear in succession in the center of the resulting periodicity. In order to obtain such a result, it is necessary to use b as a factor for the number of terms in both a and b periodicities.

Thus, $T_a = \frac{t_1}{a} + \frac{t_2}{a} + \ldots$ b times

or $T_a = b\left(\frac{t}{a}\right)$

This periodicity being in synchronization with periodicity b will appear only once.

Also, $T_b = \frac{t_1}{b} + \frac{t_2}{b} + \ldots$ b times

or $T_b = b\left(\frac{t}{b}\right)$

PERIODICITY

Periodicity T_b in its consecutive displacements towards the phases of periodicity T_a in synchronization will appear $a - b + 1$ times.

$$NT_a = 1$$
$$NT_b = a - b + 1$$

Now we can represent the whole scheme of such synchronization:

$$\left[\begin{array}{l} T_a \quad \dfrac{t_1}{a} + \dfrac{t_2}{a} + \dfrac{t_3}{a} + \dfrac{t_4}{a} + \ldots \\[6pt] T_{a_1} \quad \dfrac{t_1}{b} + \dfrac{t_2}{b} + \dfrac{t_3}{b} + \dfrac{t_4}{b} \quad \ldots \\[6pt] T_{a_2} \qquad\qquad \dfrac{t_1}{b} + \dfrac{t_2}{b} + \dfrac{t_3}{b} + \dfrac{t_4}{b} + \ldots \\[6pt] T_{a_3} \qquad\qquad\qquad\qquad \dfrac{t_1}{b} + \dfrac{t_2}{b} + \dfrac{t_3}{b} + \dfrac{t_4}{b} + \ldots \end{array}\right.$$

. . . .

Assuming values for a and b

$$a = 3 \; ; b = 4,$$ we can represent this process graphically.

$$NT_b = 4 - 3 + 1 = 2$$

As in previous cases of synchronization we find the common denominator first

$$\frac{t}{a} = \frac{1}{3} \; ; \frac{t}{b} = \frac{1}{4} \qquad t_{a \div b} = \frac{1}{12}$$

The complementary factor for T_a is b and for b the factor is a. Therefore, the unit of periodicity a in synchronization is

$$T_a = \frac{b}{ab} = \frac{4}{12}$$

$$T_b = \frac{a}{ab} = \frac{3}{12}$$

b being a factor for both periodicities in synchronization will give

$$T_a = \frac{t_1 b}{ab} + \frac{t_2 b}{ab} + \frac{t_3 b}{ab} + \frac{t_4 b}{ab}$$

$$T_b = \frac{t_1 a}{ab} + \frac{t_2 a}{ab} + \frac{t_3 a}{ab} + \frac{t_4 a}{ab}$$

or

$$T_3 = 4\left(\frac{4}{12}\right) \text{ and } T_4 = 4\left(\frac{3}{12}\right)$$

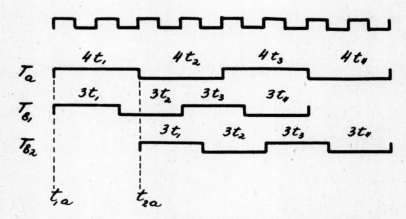

Figure 8. *Consecutive displacement of periodicity b.*

Through interference, we may obtain the desired resulting series with fractioning around the point of symmetry.

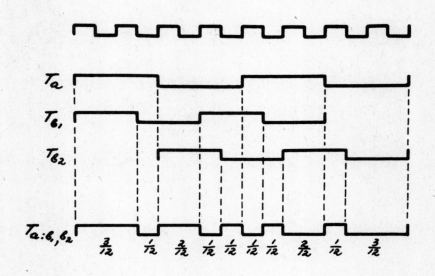

Figure 9. *Derivative periodicity.*

PERIODICITY

$$T_{3:(2 \times 4)} = \frac{3t_1}{12} + \frac{t_2}{12} + \frac{2t_3}{12} + \frac{t_4}{12} + \frac{t_5}{12} + \frac{t_6}{12} + \frac{t_7}{12} +$$

$$+ \frac{2t_8}{12} + \frac{t_9}{12} + \frac{3t_{10}}{12}, \text{ or using numerators only,}$$

$$T_{3:(2 \times 4)} = 3 + 1 + 2 + 1 + 1 + 1 + 1 + 2 + 1 + 3$$

Let us now assume that $a = 2$ and $b = 5$. Then,

$$t_a = \frac{1}{2}; \quad t_b = \frac{1}{5}, \text{ therefore } t_{a:b} = \frac{1}{10}$$

In synchronization, their values will be:

$$T_a = \frac{b}{ab} = \frac{5}{10}; \quad T_b = \frac{a}{ab} = \frac{2}{10}$$

b being a factor for both periodicities in synchronization will give:

$$T_a = \frac{t_1 b}{ab} + \frac{t_2 b}{ab} + \frac{t_3 b}{ab} + \frac{t_4 b}{ab} + \frac{t_5 b}{ab}$$

$$T_b = \frac{t_1 a}{ab} + \frac{t_2 a}{ab} + \frac{t_3 a}{ab} + \frac{t_4 a}{ab} + \frac{t_5 a}{ab}$$

or

$$T_2 = 5\left(\frac{5}{10}\right) \text{ and } T_5 = 5\left(\frac{2}{10}\right)$$

$$NT_b = 5 - 2 + 1 = 4$$

Now we can synchronize T_a with T_b

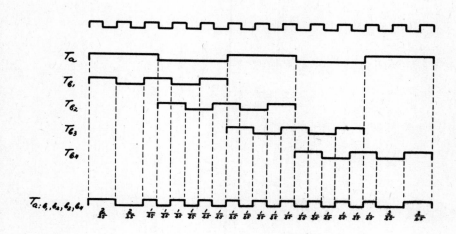

Figure 10. Synchronization and derivative periodicity of $T_{2:5}$ with consecutive displacement of b.

$$T_{2:(4\times5)} = \frac{2t_1}{25} + \frac{2t_2}{25} + \frac{t_3}{25} + \frac{t_4}{25} + \frac{t_5}{25} + \frac{t_6}{25} + \frac{t_7}{25} + \frac{t_8}{25} + \frac{t_9}{25} + \frac{t_{10}}{25} + \frac{t_{11}}{25} +$$

$$+ \frac{t_{12}}{25} + \frac{t_{13}}{25} + \frac{t_{14}}{25} + \frac{t_{15}}{25} + \frac{t_{16}}{25} + \frac{t_{17}}{25} + \frac{t_{18}}{25} + \frac{t_{19}}{25} + \frac{2t_{20}}{25} + \frac{2t_{21}}{25}$$

or, using numerators only

$T_{2:(4\times5)} = 2 + 2 + 1 + 1 + 1 + 1 + 1 + 1 + 1 + 1 + 1 + 1 + 1 + 1 + 1 + 1 + 1 + 1 + 1 + 2 + 2$

Assuming various values for a and b we get the following results:

$T_{2:(2\times3)} = 2 + 1 + 1 + 1 + 1$

$T_{3:(2\times4)} = 3 + 1 + 2 + 1 + 1 + 1 + 1 + 2 + 1 + 3$

$T_{3:(3\times5)} = 3 + 2 + 1 + 2 + 1 + 1 + 1 + 1 + 1 + 1 + 1 + 1 + 1 + 2 + 1 + 2 + 3$

$T_{4:(2\times5)} = 4 + 1 + 3 + 1 + 1 + 2 + 1 + 2 + 1 + 1 + 3 + 1 + 4$

$T_{4:(4\times7)} = 4 + 3 + 1 + 3 + 1 + 2 + 1 + 1 + 2 + 1 + 1 + 1 + 1 + 1 + 1 + 1 + 1 + 1 + 1 + 1 + 1 + 1 + 2 + 1 + 1 + 2 + 1 + 3 + 1 + 3 + 4$

$T_{5:(2\times6)} = 5 + 1 + 4 + 1 + 1 + 3 + 1 + 2 + 2 + 1 + 3 + 1 + 1 + 1$

$T_{5:(3\times7)} = 5 + 2 + 3 + 2 + 2 + 1 + 2 + 2 + 1 + 1 + 1 + 2 + 1 + 2 + 1 + 1 + 1 + 2 + 2 + 1 + 2 + 2 + 3 + 2 + 5$

After synchronization of both periodicities, **a** being a major term and **b** being a minor term, we may obtain the whole number of terms in the resulting periodicity through the following formula:

$$NT = a(a - b + 1) + (b - 1) = a^2 - ab + a + b - 1 \qquad (1)$$

3. Balance.

Formulae:

(1) When $a \geqslant 2b$, $a \geqslant 3b$, $a \geqslant 4b$, ... $a \geqslant mb$

$$\boxed{B_{a>b} = r_{a \div b} + r_{a \div b} + a(a - b)}$$

(2) When $a > 2b$, $a > 3b$, $a > 4b$, ... $a > mb$

$$B_{a>2b} = r_{a \div b} + 2r_{a \div b} + (a^2 - 2ab)$$
$$B_{a>3b} = r_{a \div b} + 3r_{a \div b} + (a^2 - 3ab)$$
$$B_{a>4b} = r_{a \div b} + 4r_{a \div b} + (a^2 - 4ab)$$

$$\boxed{B_{a>mb} = r_{a \div b} + mr_{a \div b} + (a^2 - mab)}$$

. . . GENERAL FORMULA . . .

$$\boxed{B_{a>mb} = r_{a \div b} + mr_{a \div b} + (a^2 - mab)}$$

PERIODICITY

4. Application

$5 \div 2$

(1) $B_{5 \div 2} = \underline{r_{5 \div 2}} + r_{5 \div 2} + 5(5 - 2) = 25t + 10t + 15t$
(2) $B_{5 \div 2} = \underline{r_{5 \div 2}} + 2r_{5 \div 2} + (25 - 20) = 25t + 10t + 10t + 5t$

$7 \div 3$

(1) $B_{7 \div 3} = \underline{r_{7 \div 3}} + r_{7 \div 3} + 7(7 - 3) = 49t + 21t + 28t$
(2) $B_{7 \div 3} = \underline{r_{7 \div 3}} + 2r_{7 \div 3} + (49 - 42) = 49t + 21t + 21t + 7t$

$8 \div 3$

(1) $B_{8 \div 3} = \underline{r_{8 \div 3}} + r_{8 \div 3} + 8(8 - 3) = 64t + 24t + 40t$
(2) $B_{8 \div 3} = \underline{r_{8 \div 3}} + 2r_{8 \div 3} + (64 - 48) = 64t + 24t + 24t + 16t$

$9 \div 4$

(1) $B_{9 \div 4} = \underline{r_{9 \div 4}} + r_{9 \div 4} + 9(9 - 4) = 81t + 36t + 45t$
(2) $B_{9 \div 4} = \underline{r_{9 \div 4}} + 2r_{9 \div 4} + (81 - 72) = 81t + 36t + 36t + 9t$

$7 \div 2$

(1) $B_{7 \div 2} = \underline{r_{7 \div 2}} + r_{7 \div 2} + 7(7 - 2) = 49t + 14t + 35t$
(2) $B_{7 \div 2} = \underline{r_{7 \div 2}} + 3r_{7 \div 2} + (49 - 42) = 49t + 14t + 14t + 14t + 7t$

$9 \div 2$

(1) $B_{9 \div 2} = \underline{r_{9 \div 2}} + r_{9 \div 2} + 9(9 - 2) = 81t + 18t + 63t$
(2) $B_{9 \div 2} = \underline{r_{9 \div 2}} + 4r_{9 \div 2} + (81 - 72) = 81t + 18t + 18t + 18t + 18t + 9t$

B. Polynomial Relations of Monomial Periodicities.

(Synchronization of several periodicities).

As in the two previous cases, the resultant rhythmic series derives from the interference of monomial periodicities, synchronized through a common denominator. If we have three monomial periodicities T_a, T_b and T_c, they should be synchronized through $T_{a:b:c}$ where each unit in synchronization is $\dfrac{t}{abc}$.

The complementary factor for a is bc.
The complementary factor for b is ac.
The complementary factor for c is ab.

Each unit of the synchronized continuity will take the following form:

$$T_a = a\left(\frac{bct}{abc}\right) = \frac{bct_1}{abc} + \frac{bct_2}{abc} + \ldots \text{ [a times]}$$

$$T_b = b\left(\frac{act}{abc}\right) = \frac{act_1}{abc} + \frac{act_2}{abc} + \ldots \text{ [b times]}$$

$$T_c = c\left(\frac{abt}{abc}\right) = \frac{abt_1}{abc} + \frac{abt_2}{abc} + \ldots \text{ [c times]}$$

Assuming $a = 2$, $b = 3$ and $c = 5$, we obtain

$$T_2 = \frac{t_1}{2} + \frac{t_2}{2}$$

$$T_3 = \frac{t_1}{3} + \frac{t_2}{3} + \frac{t_3}{3}$$

$$T_5 = \frac{t_1}{5} + \frac{t_2}{5} + \frac{t_3}{5} + \frac{t_4}{5} + \frac{t_5}{5}$$

The common denominator is $2 \times 3 \times 5 = 30$

Therefore $T_2 = 2\left(\frac{15t}{30}\right) = \frac{15t_1}{30} + \frac{15t_2}{30}$

$$T_3 = 3\left(\frac{10t}{30}\right) = \frac{10t_1}{30} + \frac{10t_2}{30} + \frac{10t_3}{30}$$

$$T_5 = 5\left(\frac{6t}{30}\right) = \frac{6t_1}{30} + \frac{6t_2}{30} + \frac{6t_3}{30} + \frac{6t_4}{30} + \frac{6t_5}{30}$$

Thus, $t_{2:3:5} = \frac{1}{30}$

Now we may graph:

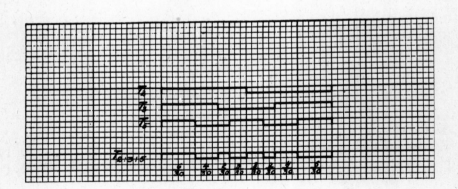

Figure 11. *Synchronization of three monomial periodicities.*

The resulting rhythmic series:

$$T_{2:3:5} = \frac{6t_1}{30} + \frac{4t_2}{30} + \frac{2t_3}{30} + \frac{3t_4}{30} + \frac{3t_5}{30} + \frac{2t_6}{30} + \frac{4t_7}{30} + \frac{6t_8}{30}$$

or using numerators only:

$$T_{2:3:5} = 6 + 4 + 2 + 3 + 3 + 2 + 4 + 6$$

Another example of three synchronized monomial periodicities.

$T_{3:4:5}$

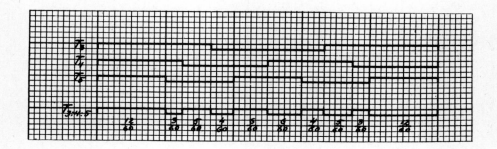

Figure 12. Synchronization of $T_{3:4:5}$.

The resulting rhythmic series

$$T_{3:4:5} = \frac{12t_1}{60} + \frac{3t_2}{60} + \frac{5t_3}{60} + \frac{4t_4}{60} + \frac{6t_5}{60} + \frac{6t_6}{60} +$$

$$+ \frac{4t_7}{60} + \frac{5t_8}{60} + \frac{3t_9}{60} + \frac{12t_{10}}{60}$$

Or, using numerators only:

$$T_{3:4:5} = 12 + 3 + 5 + 4 + 6 + 6 + 4 + 5 + 3 + 12$$

Likewise:

$$T_{3:4:7} = 12 + 9 + 3 + 4 + 8 + 6 + 6 + 8 + 4 + 3 + 9 + 12$$

In some cases of the synchronization of several monomial periodicities, the component periodicities neutralize each other through interference, because of their numerical relations. For instance, in the case

$$T_{2:3:7} = 7 + 7 + 7 + 7 + 7 + 7,$$

results in a monomial periodicity.

In order to obtain a resulting rhythmic series from four or more monomial periodicities synchronized, it is necessary to follow the procedure previously described.

Here is the process for synchronizing monomial periodicities in more generalized form:

(1) Select the number of component monomial periodicities to be synchronized. For instance, suppose we take n periodicities.

(2) Assume certain values (in integer terms) for all the component periodicities. For instance,
$$n = a, b, c, \ldots m$$

(3) The values a, b, c, ... will not contain a common divisor. Find the common denominator for the periodicities to be synchronized. It is a product of a by b, by c, etc. If T_a represents the periodicity which will appear a times in synchronization, an analogous interpretation will be given to
T_b, T_c, ... etc.

Then the common denominator or the unit of periodicities T_a, T_b, T_c, ...T_m in synchronization can be expressed through
$$T_{a:b:c:\ldots:m} = abc\ldots m \text{ or } t = \frac{1}{abc\ldots m}$$

(4) Find the complementary factor for each component periodicity in synchronization. The complementary factor equals the product of the remaining terms. Thus, for

T_a the complementary factor equals
$$bc\ldots m \text{ or } \frac{abc\ldots m}{a}$$
T_b the complementary factor equals
$$ac\ldots m \text{ or } \frac{abc\ldots m}{b}$$
T_c the complementary factor equals
$$ab\ldots m \text{ or } \frac{abc\ldots m}{c}$$
....
T_m the complementary factor equals
$$abc\ldots e \text{ or } \frac{abc\ldots m}{e}$$

(5) Now the component periodicities can be expressed in common units and synchronized
$$T_a = a\left(\frac{bcd\ldots mt}{abcd\ldots m}\right)$$
$$T_b = b\left(\frac{acd\ldots mt}{abcd\ldots m}\right)$$
$$T_c = c\left(\frac{abd\ldots mt}{abcd\ldots m}\right)$$
....
$$T_m = m\left(\frac{abcd\ldots et}{abcd\ldots m}\right)$$

PERIODICITY

Let us assume $n = 5$, where $a = 2, b = 3, c = 5, d = 7, e = 11$
Then, $T_{2:3:5:7:11} = 2 \times 3 \times 5 \times 7 \times 11 = 2310$

$$t = \frac{1}{2310}$$

The complementary factor for

$$T_2 = \frac{2310}{2} = 1155$$

$$T_3 = \frac{2310}{3} = 770$$

$$T_5 = \frac{2310}{5} = 462$$

$$T_7 = \frac{2310}{7} = 330$$

$$T_{11} = \frac{2310}{11} = 210$$

Now we may synchronize:

$$T_2 = 2\left(\frac{1155t}{2310}\right) = \frac{1155t_1}{2310} + \frac{1155t_2}{2310}$$

$$T_3 = 3\left(\frac{770t}{2310}\right) = \frac{770t_1}{2310} + \frac{770t_2}{2310} + \frac{770t_3}{2310}$$

$$T_5 = 5\left(\frac{462t}{2310}\right) = \frac{462t_1}{2310} + \frac{462t_2}{2310} + \frac{462t_3}{2310} + \frac{462t_4}{2310} + \frac{462t_5}{2310}$$

$$T_7 = 7\left(\frac{330t}{2310}\right) = \frac{330t_1}{2310} + \frac{330t_2}{2310} + \ldots + \frac{330t_7}{2310}$$

$$T_{11} = 11\left(\frac{210t}{2310}\right) = \frac{210t_1}{2310} + \frac{210t_2}{2310} + \ldots + \frac{210t_{11}}{2310}$$

The general number of terms in any resultant rhythmic series, produced through synchronization of n monomial periodicities, equals the sum of all the component periodicities, minus $n - 1$.

$$GNT_{a:b:c:\ldots:m} = (a + b + c + \ldots m) - (n - 1) =$$
$$a + b + c + \ldots + m - n + 1$$

In order to produce more extended rhythmic series, where the number of terms will be considerably greater, while their value is considerably smaller, it is necessary to synchronize the component monomial periodicities through the common product, instead of the common denominator. This can be performed with three or more component periodicities.

If T_a, T_b and T_c are the component periodicities, their common product will be $T_{a:b:c} = abc$ and $t = \frac{1}{abc}$.

Then, $T_a = bc \, (at) = at_1 + at_2 + \ldots$ [bc times]
$T_b = ac \, (bt) = bt_1 + bt_2 + \ldots$ [ac times]
$T_c = ab \, (ct) = ct_1 + ct_2 + \ldots$ [ab times]

Assuming $a = 5, b = 3, c = 2$, we get

$$T_{5:3:2} = 30 \quad \text{and} \quad t = \frac{1}{30}$$

$T_5 = 6(5t) = 5t_1 + 5t_2 + 5t_3 + 5t_4 + 5t_5 + 5t_6$
$T_3 = 10(3t) = 3t_1 + 3t_2 + 3t_3 + \ldots + 3t_{10}$
$T_2 = 15(2t) = 2t_1 + 2t_2 + 2t_3 + \ldots + 2t_{15}$

Now we can graph

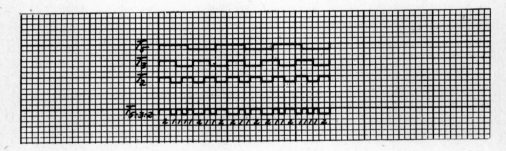

Figure 13. Synchronization of $T_{5:3:2}$.

Thus, $T_{5:3:2} = 2t_1 + t_2 + t_3 + t_4 + t_5 + 2t_6 + t_7 + t_8 + 2t_9 + 2t_{10} +$
$+ t_{11} + t_{12} + 2t_{13} + 2t_{14} + t_{15} + t_{16} + 2t_{17} + t_{18} + t_{19} +$
$+ t_{20} + t_{21} + 2t_{22}$

Or, simply, $T_{5:3:2} = 2 + 1 + 1 + 1 + 1 + 2 + 1 + 1 + 2 + 2 + 1 + 1 +$
$+ 2 + 2 + 1 + 1 + 2 + 1 + 1 + 1 + 1 + 2$

Here are some other rhythmic series obtained through the common product of three periodicities:

$T_{5:4:3} = 3 + 1 + 1 + 1 + 2 + 1 + 1 + 2 + 3 + 1 + 2 + 2 + 1 + 3 +$
$+ 1 + 2 + 1 + 2 + 2 + 1 + 2 + 1 + 3 + 1 + 2 + 2 + 1 + 3 +$
$+ 2 + 1 + 1 + 2 + 1 + 1 + 1 + 3$

$T_{7:3:2} = 2 + 1 + 1 + 2 + 1 + 1 + 1 + 1 + 2 + 2 + 1 + 1 + 2 + 2 +$
$+ 1 + 1 + 2 + 2 + 1 + 1 + 2 + 2 + 1 + 1 + 1 + 1 + 2 + 1 +$
$+ 1 + 2$

$T_{7:5:2} = 2+2+1+1+1+1+2+2+2+1+1+2+2+1+$
$+1+2+1+1+2+2+2+2+1+1+2+2+2+2+$
$+1+1+2+1+1+2+2+1+1+2+2+2+1+1+$
$+1+1+2+2$

$T_{7:4:3} = 3+1+2+1+1+1+3+2+1+1+2+2+1+3+$
$+3+1+2+2+1+2+1+3+1+2+2+2+1+3+$
$+1+2+1+2+2+1+3+3+1+2+2+1+1+2+$
$+3+1+1+1+2+1+3$

$T_{7:5:4} = 4+1+2+1+2+2+2+1+1+4+1+3+1+3+$
$+2+2+3+1+4+2+2+1+3+1+1+2+3+1+$
$+4+3+1+1+3+2+2+3+1+1+3+4+1+3+$
$+2+1+1+3+1+2+2+4+1+3+2+2+3+1+$
$+3+1+4+1+1+2+2+2+1+2+1+4$

$T_{7:5:3} = 3+2+1+1+2+1+2+2+1+3+2+1+3+1+$
$+2+1+2+3+2+1+3+1+2+3+3+1+1+1+$
$+3+1+1+1+3+3+2+1+3+1+2+3+2+1+$
$+2+1+3+1+2+3+1+2+2+1+2+1+1+2+$
$+3$

C. Polynomial Relations of Polynomial Periodic Series.
(Synchronization of various periodic series).

Once the method of synchronizing through common denominator or product has been established, all other cases of polynomial relations of various periodic series can be classified, and the resulting series deduced.

1. *Binomial relations of binomial periodic series.*

(1) If T_{a+b} and $T'_{a'+b'}$ are equal in the sum of their binomial as well as in their corresponding terms (a = a, b = b), the resulting periodic series will be equal to any of the two component series T. In this case, the results valuable for art purposes may be obtained by applying various coefficients to T and T' and by synchronizing the latter.

For example, if we assume

$$T_{a+b:a'+b'=m:n},$$

then, $\quad T = m(a+b) = (a+b)t_1 + (a+b)t_2 + \ldots + (a+b)t_m$

$\quad T' = n(a'+b') = (a'+b')t_1 + (a'+b')t_2 + \ldots + (a'+b')t_n$

Now we can synchronize.

$$T_m = m\left(\frac{na+nb}{mn}\right)t = \left(\frac{na+nb}{mn}\right)t_1 + \left(\frac{na+nb}{mn}\right)t_2 + \ldots +$$
$$+ \left(\frac{na+nb}{mn}\right)t_m$$

$$T_n = n\left(\frac{ma+mb}{mn}\right)t = \left(\frac{ma+mb}{mn}\right)t_1 + \left(\frac{ma+mb}{mn}\right)t_2 + \ldots +$$

$$+ \left(\frac{ma+mb}{mn}\right)t_n$$

Let us assume that $a = 2$, $b = 1$, and $m = 4$, $n = 3$.

Then, $T_4 = \dfrac{(3\times 2)+(3\times 1)}{12}t_1 + \dfrac{(3\times 2)+(3\times 1)}{12}t_2 + \dfrac{(3\times 2)+(3\times 1)}{12}t_3 +$

$+ \dfrac{(3\times 2)+(3\times 1)}{12}t_4 = \dfrac{6+3}{12}t_1 + \dfrac{6+3}{12}t_2 + \dfrac{6+3}{12}t_3 + \dfrac{6+3}{12}t_4$

$T_3 = \dfrac{(4\times 2)+(4\times 1)}{12}t_1 + \dfrac{(4\times 2)+(4\times 1)}{12}t_2 + \dfrac{(4\times 2)+(4\times 1)}{12}t_3 =$

$= \dfrac{8\times 4}{12}t_1 + \dfrac{8\times 4}{12}t_2 + \dfrac{8\times 4}{12}t_3$

or, graphically
    scale
    $t_a = 4$
    $t_b = 3$

The whole scheme of synchronization will take

$T_{m:n(a+b)} = mn(a+b)$
$T_{4:3(2\times 1)} = 12 \times 3 = 36$

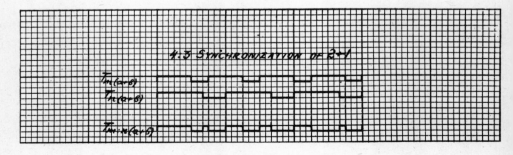

Figure 14. 4:3 synchronization of 2 + 1.

PERIODICITY

The resulting series is

$$T_{4:3(2\times 1)} = \frac{6t_1}{36} + \frac{2t_2}{36} + \frac{t_3}{36} + \frac{3t_4}{36} + \frac{3t_5}{36} + \frac{3t_6}{36} + \frac{2t_7}{36} + \frac{4t_8}{36} + \frac{3t_9}{36} +$$

$$+ \frac{5t_{10}}{36} + \frac{t_{11}}{36} + \frac{3t_{12}}{36}$$

or, $6+2+1+3+3+3+2+4+3+5+1+3$

(2) $a \neq a'$ and $b \neq b'$, but, $a + b = a' + b' = s$

In this case the whole scheme of synchronization can be performed directly from s in $1 : 1 = s : s$ synchronization.

Let us assume that $a = 5, b = 3$
$\qquad\qquad\qquad\quad a' = 6, b' = 2$
Then

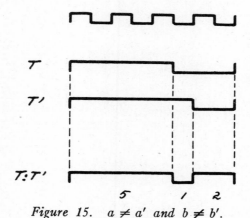

Figure 15. $a \neq a'$ and $b \neq b'$.

$$T_{(5+3):(6+2)} = \frac{5t_1}{8} + \frac{t_2}{8} + \frac{2t_3}{8} \text{ or } 5 + 1 + 2$$

A special case when $a = b'$ and $b = a'$ is typical in art. It yields a synchronization of a binomial and its converse.

$a = 3, b = 1$
$a' = 1, b' = 3$

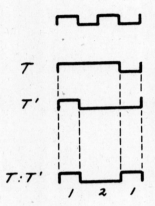

Figure 16. $a = b'$ and $b = a'$.

$$T_{(3+1):(1+3)} = \frac{t_1}{4} + \frac{2t_2}{4} + \frac{t_3}{4} \text{ or } 1 + 2 + 1$$

Synchronization may also be performed through the coefficients independent of the value of **s** or its component terms. This will be realized through a common product.

For example, $\begin{matrix} 4 + 1 = 5 \\ 1 + 4 = 5 \end{matrix}$ $s = 5, m = 3, n = 2$

$$T_{a+b} = 3\left[\frac{2(4+1)}{6}\right]t = \frac{8+2}{6}t_1 + \frac{8+2}{6}t_2 + \frac{8+2}{6}t_3$$

$$T_{a'+b'} = 2\left[\frac{3(1+4)}{6}\right]t = \frac{3+12}{6}t_1 + \frac{3+12}{6}t_2$$

but $T_{m:n(s)} = mns$

$T_{3:2(s)} = 3 \times 2 \times 5 = 30$

$t = \dfrac{1}{30}$

or graphically:

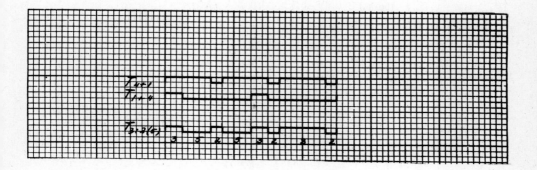

Figure 17. $S_{a+b} = S'_{a'+b'}$.

$$T_{3:2(5)} = \frac{3t_1}{30} + \frac{5t_2}{30} + \frac{2t_3}{30} + \frac{5t_4}{30} + \frac{3t_5}{30} + \frac{2t_6}{30} + \frac{8t_7}{30} + \frac{2t_8}{30}$$

or, $3+5+2+5+3+2+8+2$

(3) $S_{a+b} \neq S'_{a'+b'}$

When the sum of one binomial does not equal the sum of another, S and S' act as complementary factors in synchronization.

$$T_s = S'\left(\frac{St}{SS'}\right) = \frac{St_1}{SS'} + \frac{St_2}{SS'} + \ldots + \frac{St_{s'}}{SS'}$$

$$T_{s'} = S\left(\frac{S't}{SS'}\right) = \frac{S't_1}{SS'} + \frac{S't_2}{SS'} + \ldots + \frac{S't_s}{SS'}$$

Let us assume that $S = 3 + 2 = 5$

and $S' = 4 + 3 = 7$

$$T_{5=3+2} = 7\left(\frac{5t}{35}\right) = \frac{5t_1}{35} + \frac{5t_2}{35} + \ldots + \frac{5t_7}{35}$$

$$T_{7=4+3} = 5\left(\frac{7t}{35}\right) = \frac{7t_1}{35} + \frac{7t_2}{35} + \ldots + \frac{7t_5}{35}$$

134 THEORY OF REGULARITY AND COORDINATION

Now we may graph

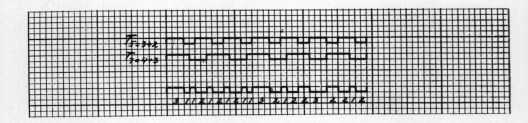

Figure 18. $S_{a+b} \neq S'_{a'+b'}$.

$$T_{5(3+2):7(4+3)} = \frac{3t_1}{35} + \frac{t_2}{35} + \frac{t_3}{35} + \frac{2t_4}{35} + \frac{t_5}{35} + \frac{2t_6}{35} + \frac{t_7}{35} + \frac{2t_8}{35} + \frac{t_9}{35} +$$

$$+ \frac{t_{10}}{35} + \frac{3t_{11}}{35} + \frac{2t_{12}}{35} + \frac{t_{13}}{35} + \frac{2t_{14}}{35} + \frac{2t_{15}}{35} + \frac{3t_{16}}{35} + \frac{2t_{17}}{35} + \frac{2t_{18}}{35} + \frac{t_{19}}{35} + \frac{2t_{20}}{35}$$

or,

$$3+1+1+2+1+2+1+2+1+1+3+2+1+2+2+3+2+2+1+2$$

2. Synchronization of a Motive.

$$2+1+1$$

in $1 \div 2 \div 3 \div 4$

Common product = 12   12 (2 + 1 + 1)   1

6 (2 + 1 + 1)   2

$1 = ♪$ 4 (2 + 1 + 1)   3

3 (2 + 1 + 1)   4

PERIODICITY

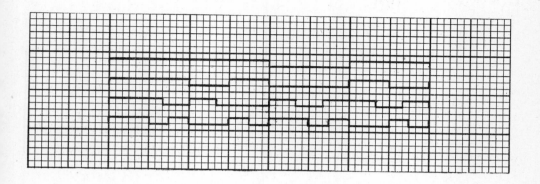

Figure 19. Synchronization of a musical motive 2 + 1 + 1.

3. *Polynomial Relations of Binomial Periodic Series.*

This case involves the synchronization of several binomial periodic series. Let us assume that

$$A = a_1 + a_2$$
$$B = b_1 + b_2$$
$$C = c_1 + c_2$$
$$\dots\dots\dots\dots$$
$$M = m_1 + m_2,$$

where $A, B, C, \dots M$ are the sums of the corresponding binomials $a_1 + a_2$, $b_1 + b_2$, $c_1 + c_2$, $\dots m_1 + m_2$. The sums of the component binomial periodicities enter as factors of the common product $ABC\dots M$. The complementary factor for each component periodicity equals the general product divided by a given periodicity. The complementary factor for

A equals $BC\dots M$
B " $AC\dots M$
C " $AB\dots M$
$\dots\dots\dots\dots$
M " $ABC\dots$

Now the component binomial periodicities can be expressed in common units and synchronized.

$$T_A = a_1 + a_2 = BC\dots M \left(\frac{At}{ABC\dots M}\right) = \frac{(a_1 + a_2)t_1}{ABC\dots M} + \frac{(a_1 + a_2)t_2}{ABC\dots M} + \dots$$
$$\dots + \frac{(a_1 + a_2)t_{BC\dots M}}{ABC\dots M}$$

$$T_B = b_1 + b_2 = AC\dots M \left(\frac{Bt}{ABC\dots M}\right) = \frac{(b_1 + b_2)t_1}{ABC\dots M} + \frac{(b_1 + b_2)t_2}{ABC\dots M} + \dots$$
$$\dots + \frac{(b_1 + b_2)t_{AC\dots M}}{ABC\dots M}$$

$$T_C = c_1 + c_2 = AB\dots M \left(\frac{Ct}{ABC\dots M}\right) = \frac{(c_1 + c_2)t_1}{ABC\dots M} + \frac{(c_1 + c_2)t_2}{ABC\dots M} + \dots$$
$$\dots + \frac{(c_1 + c_2)t_{AB\dots M}}{ABC\dots M}$$

$$\dots\dots\dots\dots\dots\dots\dots\dots\dots\dots\dots\dots\dots\dots$$

$$T_M = m_1 + m_2 = ABC\dots \left(\frac{Mt}{ABC\dots M}\right) = \frac{(m_1+m_2)t_1}{ABC\dots M} + \frac{(m_1+m_2)t_2}{ABC\dots M} + \dots$$
$$\dots + \frac{(m_1+m_2)t_{ABC\dots}}{ABC\dots M}$$

PERIODICITY

Example:

Let us assume that $3 = 2 + 1, 5 = 3 + 2, 7 = 4 + 3, 8 = 5 + 3$

$$T_{3:5:7:8} = 3 \times 5 \times 7 \times 8 = 840$$

$$t = \frac{1}{840}$$

The complementary factors are:

$$t_3 = \frac{3}{840} = 280$$

$$t_5 = \frac{5}{840} = 168$$

$$t_7 = \frac{7}{840} = 120$$

$$t_8 = \frac{8}{840} = 105$$

Now we can write:

$$T_3 = 280\left(\frac{3t}{840}\right) = \frac{(2+1)t_1}{840} + \frac{(2+1)t_2}{840} + \ldots + \frac{(2+1)t_{280}}{840}$$

$$T_5 = 168\left(\frac{5t}{840}\right) = \frac{(3+2)t_1}{840} + \frac{(3+2)t_2}{840} + \ldots + \frac{(3+2)t_{168}}{840}$$

$$T_7 = 120\left(\frac{7t}{840}\right) = \frac{(4+3)t_1}{840} + \frac{(4+3)t_2}{840} + \ldots + \frac{(4+3)t_{120}}{840}$$

$$T_8 = 105\left(\frac{8t}{840}\right) = \frac{(5+3)t_1}{840} + \frac{(5+3)t_2}{840} + \ldots + \frac{(5+3)t_{105}}{840}$$

The graph may be made according to the method previously used.

4. Binomial relations of polynomial periodic series.

Let us assume two polynomial periodic series:

$$S = a + b + c + d \quad \text{and}$$
$$S' = a' + b' + c' + d' + e' \quad \text{where}$$

(1) all of the terms are respectively equal—$a = a'$, $b = b'$, $c = c'$, $d = d'$, while the number of terms in S and S' may vary;

(2) some of the terms are respectively equal—$a = a'$, or $d = d'$, or irrespectively equal $a = c'$, or $d = b'$;

(3) none of the terms of the two series is equal. As in case 2 above,

$$T_{S:S'} = S \times S' = SS'; \quad t = \frac{1}{SS'}$$

$$T_S = S'\left(\frac{St}{SS'}\right) = \frac{(a+b+c+d)t_1}{SS'} + \frac{(a+b+c+d)t_2}{SS'} + \ldots$$

$$\ldots + \frac{(a+b+c+d)t_{s'}}{SS'}$$

$$T_{S'} = S\left(\frac{S't}{SS'}\right) = \frac{(a+b+c+d+e)t_1}{SS'} + \frac{(a+b+c+d+e)t_2}{SS'} + \ldots$$

$$\ldots + \frac{(a+b+c+d+e)t_S}{SS'}$$

In the case of the two polynomial periodic series, the number of terms in each series may be the same or different.

$$S = m, S' = m \quad \text{or} \quad S = m, S' = n$$

Example:

$$S = 5 = 2 + 2 + 1 \qquad \text{(3 terms)}$$

$$S' = 7 = 3 + 2 + 1 + 1 \qquad \text{(4 terms)}$$

$$T_{S:S'} = 5 \times 7 = 35; \quad t = \frac{1}{35}$$

$$T_5 = 7\left(\frac{5t}{35}\right) = \frac{(2+2+1)t_1}{35} + \frac{(2+2+1)t_2}{35} + \ldots + \frac{(2+2+1)t_7}{35}$$

$$T_7 = 5\left(\frac{7t}{35}\right) = \frac{(3+2+1+1)t_1}{35} + \frac{(3+2+1+1)t_2}{35} + \ldots$$

$$\ldots + \frac{(3+2+1+1)t_5}{35}$$

5. Polynomial relations of polynomial periodic series.

As in Case 3 above, some of the terms of one periodic series may or may not be equal to the terms of another series. The process of synchronization develops as in Case 2. Let us assume m polynomial periodic series with a variable number of terms.

PERIODICITY

$A = a_1 + a_2$ \hfill (2 terms)
$B = b_1 + b_2 + b_3 + b_4 + b_5$ \hfill (5 terms)
$C = c_1 + c_2 + c_3$ \hfill (3 terms)
....
$M = m_1 + m_2 + m_3$ \hfill (3 terms)

$$T_{A:B:C:\ldots M} = ABC\ldots M; \quad t = \frac{1}{ABC\ldots M}$$

Now we can synchronize.

$$T_A = BC\ldots M\left(\frac{At}{ABC\ldots M}\right) = \frac{(a_1+a_2)t_1}{ABC\ldots M} + \frac{(a_1+a_2)t_2}{ABC\ldots M} + \ldots$$
$$\ldots + \frac{(a_1+a_2)t_{BC\ldots M}}{ABC\ldots M}$$

$$T_B = AC\ldots M\left(\frac{Bt}{ABC\ldots M}\right) = \frac{(b_1+b_2+b_3+b_4+b_5)t_1}{ABC\ldots M} +$$
$$+ \frac{(b_1+b_2+b_3+b_4+b_5)t_2}{ABC\ldots M} + \ldots + \frac{(b_1+b_2+b_3+b_4+b_5)t_{AC\ldots M}}{ABC\ldots M}$$

$$T_C = AB\ldots M\left(\frac{Ct}{ABC\ldots M}\right) = \frac{(c_1+c_2+c_3)t_1}{ABC\ldots M} + \frac{(c_1+c_2+c_3)t_2}{ABC\ldots M} + \ldots$$
$$\ldots + \frac{(c_1+c_2+c_3)t_{AB\ldots M}}{ABC\ldots M}$$

....

$$T_M = ABC\ldots\left(\frac{Mt}{ABC\ldots M}\right) = \frac{(m_1+m_2+m_3)t_1}{ABC\ldots M} +$$
$$+ \frac{(m_1+m_2+m_3)t_2}{ABC\ldots M} + \ldots + \frac{(m_1+m_2+m_3)t_{ABC\ldots}}{ABC\ldots M}$$

Example:

$A = 4 = 2 + 1 + 1$
$B = 5 = 3 + 2$
$C = 7 = 3 + 1 + 2 + 1$

$$T_{4:5:7} = 4 \times 5 \times 7 = 140; \quad t = \frac{1}{140}$$

$$T_{4=2+1+1} = 35\left(\frac{4t}{140}\right) = \frac{(2+1+1)t_1}{140} + \frac{(2+1+1)t_2}{140} + \ldots$$
$$\ldots + \frac{(2+1+1)t_{35}}{140}$$

$$T_{5-3+2} = 28\left(\frac{5t}{140}\right) = \frac{(3+2)t_1}{140} + \frac{(3+2)t_2}{140} + \ldots + \frac{(3+2)t_{28}}{140}$$

$$T_{7-3+1+2+1} = 20\left(\frac{7t}{140}\right) = \frac{(3+1+2+1)t_1}{140} + \frac{(3+1+2+1)t_2}{140} + \ldots$$

$$\ldots + \frac{(3+1+2+1)t_{20}}{140}$$

D. Practical Application in Art.

For practical application in art, any binomial or polynomial periodic series may be chosen from the rhythmic series resulting through interference. All the rhythmic series obtained in such a way have two fundamental characteristics:

(1) inverted symmetry

(2) maximum of balance at the point of symmetry and minimum of balance at the starting and ending terms.

In the series with fractioning around the point of symmetry, the balance is produced by a group of uniform units. In the series with an odd number of terms, the most balanced binomial usually falls at the beginning and at the end of the series. For instance:

$$T_{5:3} = \underbrace{3+2}+1+3+1+\underbrace{2+3}$$

In this case $3 + 2$ is the most balanced binomial with $S = 5$.

For expressive, contrasting, dramatic effects it is more appropriate to use the unbalanced binomials. For calm, inexpressive, persistent, monotonous or contemplative effects, it is more appropriate to use the balanced binomial or polynomial groups. Using both these types of groups, unbalanced and balanced, produces an effect of relaxation and the relief of tension. The progression of balanced and unbalanced groups may also be subjected to interference, as two elements U (unbalanced) and B (balanced).

All resulting rhythmic series may be used in their entirety, or in halves (from the beginning to the middle term or to the point of symmetry), or in selected segments. A rhythmic series (or a part of it) once selected becomes a *thematic entity*, and often the foundation not only of one individual composition, but of the whole style of an epoch or a nation. We shall analyze this more thoroughly in the succeeding chapters, and refer now to a few typical cases.

European music of the 17th to 19th centuries operates mainly on binomials, $2 + 1$ and $3 + 1$, and a trinomial, $2 + 1 + 1$, with permutations. In its original idiom, Russian folk music employs $3 + 2$ and $2 + 3$, $2 + 3 + 2$ and $3 + 2 + 3$. Today's American jazz has a Charleston foundation, probably imported from the

Caribbean, where it is most common in folk music (compare the dance songs of Puerto Rico). Its binomial is $5 + 3$ and $3 + 5$; its trinomial is further fractioning of the same binomial, $3 + 2 + 3$, with permutations.[3]

To fit any particular style, one may select the corresponding balanced or unbalanced groups from the appropriate rhythmic series. For instance, if one desires to operate on Charleston rhythms, in other words, on a binomial $S = 8$, we can produce this binomial from the whole series of possible binomials within the sum 8:

$$7 + 1, 6 + 2, 5 + 3, 4 + 4, 3 + 5, 2 + 6, 1 + 7.$$

The most balanced binomial in this case is $5 + 3$ or $3 + 5$ ($4 + 4$ is excluded as reducible to $1 + 1$). By selecting these binomials from the appropriate rhythmic series, we gain because we find all the variation of a selected idiom and in a properly arranged progression, ready to use. If $S = 8$ and $a + b = 5 + 3$, all the series resulting from the interference of 8:5, 8:3, 8:5:3 will satisfy the style.

In this respect the general formula of style ratios may be expressed:

$$S = a + b + c + \ldots + m$$
$$S:a; \; S:b; \; S:c; \ldots S:m$$
$$S:a:b; \; S:a:c; \ldots S:a:m$$
$$S:b:c; \; S:b:d; \ldots S:b:m$$
$$S:a:b:c; \; S:a:b:d; \ldots S:a:b:m$$
$$\ldots\ldots\ldots\ldots\ldots\ldots\ldots\ldots\ldots\ldots\ldots\ldots$$
$$S:a:b:c:\ldots\ldots:m$$

Here S, a, b, c, ...m are the *style determinants*.

E. SYNCHRONIZATION OF THE SECOND ORDER.

A polynomial rhythmic series can be synchronized with itself or with any other rhythmic series under various coefficients, and as many times as desired.

1. Synchronization of a rhythmic series with itself through various coefficients: m, n, o, p, q, ...

Suppose we have a rhythmic series

$$T_{x:y} = a + b + c + c + b + a$$

Let us synchronize it through m, n, p.

Then

$$T_{m:n:p(x:y)} = mnp \, (x:y); \qquad t = \frac{1}{mnp \, (x:y)}$$

---

[3] These quantities refer mainly to rhythmic durations. (Ed.)

$$T_{m(x:y)} = np\left[\frac{m(x:y)t}{mnp}\right] = np\left[\frac{m(a+b+c+c+b+a)t}{mnp}\right] =$$

$$= np\left[\frac{(ma+mb+mc+mc+mb+ma)t}{mnp}\right] =$$

$$= \frac{(ma+mb+mc+mc+mb+ma)t_1}{mnp} +$$

$$+ \frac{(ma+mb+mc+mc+mb+ma)t_2}{mnp} + \ldots$$

$$\ldots + \frac{(ma+mb+mc+mc+mb+ma)t_{np}}{mnp}$$

$$T_{n(x:y)} = mp\left[\frac{n(x:y)t}{mnp}\right] = mp\left[\frac{n(a+b+c+c+b+a)t}{mnp}\right] =$$

$$= mp\left[\frac{(na+nb+nc+nc+nb+na)t}{mnp}\right] =$$

$$= \frac{(na+nb+nc+nc+nb+na)t_1}{mnp} +$$

$$+ \frac{(na+nb+nc+nc+nb+na)t_2}{mnp} + \ldots$$

$$\ldots + \frac{(na+nb+nc+nc+nb+na)t_{mp}}{mnp}$$

$$T_{p(x:y)} = mn\left[\frac{p(x:y)t}{mnp}\right] = mn\left[\frac{p(a+b+c+c+b+a)t}{mnp}\right] =$$

$$= mn\left[\frac{(pa+pb+pc+pc+pb+pa)t}{mnp}\right] =$$

$$= \frac{(pa+pb+pc+pc+pb+pa)t_1}{mnp} +$$

$$+ \frac{(pa+pb+pc+pc+pb+pa)t_2}{mnp} + \ldots$$

$$\ldots + \frac{(pa+pb+pc+pc+pb+pa)t_{mn}}{mnp}$$

Example:

$T_{3:2} = 2 + 1 + 1 + 2$ synchronized with itself through $2:3:5$

$T_{2:3:5(3:2)} = 2 \times 3 \times 5\,(3:2) = 30\,(3:2)$

$$T_{2(3:2)} = 15\left[\frac{2(3:2)t}{30}\right] = 15\left[\frac{2(2+1+1+2)t}{30}\right] = 15\left[\frac{(4+2+2+4)t}{30}\right] =$$

$$= \frac{(4+2+2+4)t_1}{30} + \frac{(4+2+2+4)t_2}{30} + \ldots + \frac{(4+2+2+4)t_{15}}{30}$$

$$T_{3(3:2)} = 10\left[\frac{3(3:2)t}{30}\right] = 10\left[\frac{3(2+1+1+2)t}{30}\right] = 10\left[\frac{(6+3+3+6)t_1}{30}\right] +$$

$$+ \frac{(6+3+3+6)t_2}{30} + \ldots + \frac{(6+3+3+6)t_{10}}{30}$$

$$T_{5(3:2)} = 6\left[\frac{5(3:2)t}{30}\right] = 6\left[\frac{5(2+1+1+2)t}{30}\right] = 6\left[\frac{(10+5+5+10)t}{30}\right] =$$

$$= \frac{(10+5+5+10)t_1}{30} + \frac{(10+5+5+10)t_2}{30} + \ldots + \frac{(10+5+5+10)t_6}{30}$$

In graphic representation, the resulting rhythmic series of the second order through interference would take the following form.

Figure 20. Derivative periodicities of second order.

$$T_{2:3:5(3:2)} = (4+2+2+1+1+2+3+1+2+2+4+3+1+2) +$$
$$+ (2+4+4+2+2+1+3+2+2+2+2+4) +$$
$$+ (3+1+2+2+2+2+3+1+2+2+1+3+4+2) +$$
$$+ (2+4+3+1+2+2+1+3+2+2+2+2+1+3) +$$
$$+ (4+2+2+2+2+3+1+2+2+4+4+2) +$$
$$+ (2+1+3+4+2+2+1+3+2+1+1+2+2+4).$$

2. *Synchronization of different rhythmic series* through various coefficients.

(a) The sum of the terms of one rhythmic series *equals* the sum of the terms of another series.

Let us assume that
$$T' = 3 + 1 + 2 + 2 + 1 + 3 = 12$$
$$T'' = 4 + 2 + 2 + 4 = 12$$

In this case both series are ready for synchronization under any coefficient. We shall assume now
$$T' : T'' = 3 : 2$$

The complementary factor for T' is 2 and for T'' is 3.

Thus, $T'_2 = 3[2(3 + 1 + 2 + 2 + 1 + 3)]$

$T''_3 = 2[3(4 + 2 + 2 + 4)]$

$T'_2 = 3(6 + 2 + 4 + 4 + 2 + 6)t$

$T''_3 = 2(12 + 6 + 6 + 12)t$

$T'_2 = (6 + 2 + 4 + 4 + 2 + 6)t_1 + (6 + 2 + 4 + 4 + 2 + 6)t_2 +$
$\quad\quad + (6 + 2 + 4 + 4 + 2 + 6)t_3$

$T''_3 = (12 + 6 + 6 + 12)t_1 + (12 + 6 + 6 + 12)t_2$

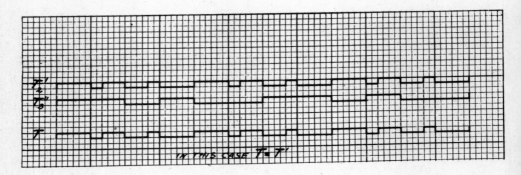

Figure 21. Synchronization of rhythmic series through different coefficients.

PERIODICITY

(b) The sum of the terms of one rhythmic series *does not* equal the sum of the terms of another series.

This case involves the double process of synchronization. In the first synchronization, we equalize both series through complementary factors of the common product. In the second synchronization, we select coefficients for the equalized series.

Let us take $T_{4:3} = 3+1+2+2+1+3 = 12$

and $\quad T_{4:(2\times 3)} = 3+1+2+1+1+1+1+2+1+3 = 16$

$$\frac{T_{4:3}}{T_{4:(2\times 3)}} = \frac{12}{16} = \frac{3}{4}$$

The first synchronization gives

$$T_{4:3} = 4(3+1+2+2+1+3) = 12+4+8+8+4+12 = 48$$

$$T_{4:(2\times 3)} = 3(3+1+2+1+1+1+1+2+1+3) =$$
$$= 9+3+6+3+3+3+3+6+3+9 = 48$$

Now, after both series are equalized, the second synchronization through selected coefficients can be performed. Let us designate the first series T' and the second T'', and let us use the second series twice. We shall illustrate this synchronization through coefficients 3, 4 and 5. Then,

$$T' : T'' : T'' = 3 : 4 : 5$$

Now we may write:

$$T_3 = 20(3T')$$
$$T_4 = 15(4T'')$$
$$T_5 = 12(5T'')$$

Substituting the values for T' and T'' previously obtained, we get:

$T_3 = 20[3(12+4+8+8+4+12)t] = 20(36+12+24+24+12+36)t =$
$= (36+12+24+24+12+36)t_1 + (36+12+24+24+12+36)t_2 + \ldots$
$\ldots + (36+12+24+24+12+36)t_{20}$

$T_4 = 15[4(9+3+6+3+3+3+3+6+3+9)t] = 15(36+12+24+12+$
$+12+12+12+24+12+36)t = (36+12+24+12+12+12+12+24+$
$+12+36)t_1 + (36+12+24+12+12+12+12+24+12+36)t_2 + \ldots$
$\ldots + (36+12+24+12+12+12+12+24+12+36)t_{15}$

$$T_5 = 12[5(9+3+6+3+3+3+3+6+3+9)t] = 12(45+15+30+$$
$$+15+15+15+15+30+15+45)t = (45+15+30+15+15+15+15+$$
$$+30+15+45)t_1 + (45+15+30+15+15+15+15+30+15+$$
$$+45)t_2 + \ldots + (45+15+30+15+15+15+15+30+15+45)t_{12}$$

The sum of each series after the second synchronization equals $48 \times 60 = 2880$. To be represented on a 12×12 per square inch graph, using $\frac{1}{12}$ of an inch as a unit, this graph would require 20′.

Although such results might seem impractical at first for artistic application, they actually are of great assistance in spatial design, particularly in covering large areas, such as murals, for instance. In music their practicability will depend on the speed with which they are mechanically performed. For example, if we wanted to have the shortest unit $\frac{1}{2880}$ run in $\frac{1}{12}$ of a second (a considerable musical speed), the 20′ of a graphed roll would take 4 minutes of performance. This is very practical, considering that at this rate the roll would move at a speed of 1 mile in 5 hours, 52 minutes.

F. PERIODICITY OF EXPANSION AND CONTRACTION.

The periodicity of expansion and contraction determines the constant forms of periodic variability. In *general parameters* it amounts to continuous increase or decrease of spatial dimension, or of velocity in temporal dimension. In *special parameters* it means continuous increase or decrease of frequency of sound waves, which amounts to sliding pitch; continuous increase or decrease in amplitude, which amounts to crescendo and diminuendo. In spectral hues, it means continuous transition from one part of the spectrum to another (continuous increase and decrease in wave-length); continuous change of intensity of a light source; etc.

Speeding up and slowing down time values, whether in long or in short portions of continuity, is inherent in any temporal or spatial-temporal art form. A comparative analysis of folk art provides a large variety of illustrations. In many a folk dance of Hungary and other countries, this variability of speed becomes one of the most substantial thematic components.

Our perception of the visible world implies a form of spatial contraction, namely, *optical perspective*. All equidistant intervals of space, in the direction perpendicular to our field of vision, seem to contract as distance grows. After about 1150 feet, equidistant intervals lose all distinction, and everything beyond this distance appears "flat".

In observing "slow motion" vibratory phenomena, such as sound waves, we find that, though all the possible frequencies within the audible range form a mathematical and a psycho-physiological continuum, we recognize only those that present simpler relations as harmonic ones. The series of overtones called

PERIODICITY

harmonic or partial series produce a natural fraction series with regard to their wave-length contraction and a natural integer series for the corresponding frequencies. If we designate "f" as frequency and "w" as wave-length, the corresponding relations will take the form of the following series:

$$\text{When } f = 1, 2, 3, 4, \ldots\ldots n$$

$$w = 1, \frac{1}{2}, \frac{1}{3}, \frac{1}{4}, \ldots\ldots \frac{1}{n}$$

The relationship of frequency to the corresponding wave-length is in square interdependence:

$$\frac{f}{w} = \frac{n}{1:n} = n^2 \text{ or } \frac{w}{f} = \frac{1:n}{n} = \frac{1}{n^2}.$$

The decrease of wave-length in such a series produces a psycho-physiological effect of contraction. As the difference between wave-lengths decreases mathematically

$$\frac{1}{1} - \frac{1}{2} = \frac{1}{2}; \quad \frac{1}{2} - \frac{1}{3} = \frac{1}{6}; \quad \frac{1}{3} - \frac{1}{4} = \frac{1}{12}; \quad \frac{1}{4} - \frac{1}{5} = \frac{1}{20}; \quad \frac{1}{5} - \frac{1}{6} = \frac{1}{30}; \text{ etc.}$$

the difference in pitch of the corresponding tones seems to approach zero, as we hear it.[4]

Thus, the natural series provide us with the full range of rhythmic possibilities. Rhythms based on constant velocities are either continuous repetitions of the terms of a monomial periodicity, or of several monomial or polynomial periodicities synchronized. Rhythms based on variable velocities are progressions of single terms belonging to different periodicities.

Proceeding from $\frac{1}{2}$ to $\frac{1}{n}$, we can produce the full scale of consecutive *contraction* within the limits $\frac{1}{2}$ and $\frac{1}{n}$. Proceeding from $\frac{1}{n}$ to $\frac{1}{2}$, we can produce the full scale of consecutive *expansion* within the same limits. By way of generalizing, we may make the following statement: *In a series* $\sum_{\frac{1}{a}}^{\frac{1}{n}} = \frac{1}{a}, \frac{1}{b}, \frac{1}{c}, \ldots \frac{1}{n}$ *where the values of denominators constantly increase from a to n, the degree of expansion or contraction increases as it approaches a and decreases as it approaches n*. It depends upon the area of the series selected for operations. For example, the degree of expansion or contraction will be greater in the area between $\frac{1}{2} \longleftrightarrow \frac{1}{7}$ than in the area between $\frac{1}{7} \longleftrightarrow \frac{1}{12}$.

---

[4] This is true only with frequencies not exceeding about 5000 vib. per second. Further increase of frequencies (above 18,000 per second), we apprehend as decrease of intensity instead of increase in pitch.

In the first case, by using all the intermediate terms of the series, we get $\frac{1}{2}+\frac{1}{3}+\frac{1}{4}+\frac{1}{5}+\frac{1}{6}+\frac{1}{7}$ where the differences are $\frac{1}{6}+\frac{1}{12}+\frac{1}{20}+\frac{1}{30}+\frac{1}{42}$. The initial difference in velocity $\left(\frac{1}{6}\right)$ existing between the first two terms grows into $\frac{1}{42}$ between the last two terms, *i.e.*, it increases 7 times $(42:6=7)$. In the second case we get $\frac{1}{7}+\frac{1}{8}+\frac{1}{9}+\frac{1}{10}+\frac{1}{11}+\frac{1}{12}$ where the differences are $\frac{1}{56}+\frac{1}{72}+\frac{1}{90}+\frac{1}{110}+\frac{1}{132}$. The initial difference between the corresponding terms grows from $\frac{1}{56}$ to $\frac{1}{132}$, *i.e.*, it increases $2\frac{5}{14}$ times $\left(132:56 = 2\frac{5}{14}\right)$.

To be applied in art, all these progressions must be expressed in the same terms, *i.e.*, through a common denominator. Thus, if we wish to see the expansion-contraction $\sum_{\frac{1}{2}}^{\frac{1}{6}}$ as a rhythmic reality, we must write $\frac{1}{2}+\frac{1}{3}+\frac{1}{4}+\frac{1}{5}+\frac{1}{6}=\frac{30}{60}+\frac{20}{60}+\frac{15}{60}+\frac{12}{60}+\frac{10}{60}$ where $T = 60t$, $t = \frac{1}{60}$ and $\sum_{\frac{1}{2}}^{\frac{1}{6}} = 87t$.

Here is the graph of this rhythm:

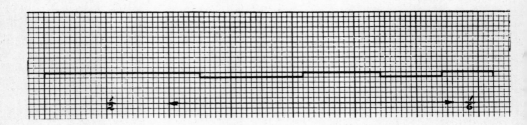

Figure 22. *Expansion—contraction* $\sum_{\frac{1}{2}}^{\frac{1}{6}}$.

Although the natural expansion-contraction series follows the consecutive terms of the natural fraction series, taken within selected limits, this is only a token form of such regularity. Besides selecting the limits for Σ, one can attribute various other forms of regularity to the progression of its terms, using rhythmic or other series previously described. Expansion-contractions, being

PERIODICITY

combined with each other into units of a higher order, become themselves subject to regularity, and may be treated as elements of the latter. The simplest and most practical way of evolving a *rhythmic series of deviation* (various forms of expansion and contraction) is to produce them from a constant unit added to a unit of growth. A term (t) in a rhythmic series of deviation equals a constant unit (t′) plus a unit of growth (τ):

$$t = t' + \tau$$

τ is subjected to regularity and varies according to the series to which it belongs. When the series is a monomial periodic series, τ is constant; otherwise it varies. The simplest and one of the most typical relations of t′ to τ occurs when $t = \dfrac{t'}{t'}$ and $\tau_1 = \dfrac{1}{t'}$. The general expression for a rhythmic series of deviation is:

$$Tt':\tau = t + (t' + \tau_1) + (t' + \tau_2) + (t' + \tau_3) + \ldots$$

where τ may be positive (expansion) or negative (contraction).

1. *Series of deviation based on various forms of growth for τ.*

(a) The numerator of τ grows through the natural integer series.

$$\tau_1 = \frac{1}{t'}$$

$$\sum_{t}^{2t} = t + \left(t + \frac{1}{t'}\right) + \left(t' + \frac{2}{t'}\right) + \left(t' + \frac{3}{t'}\right) + \ldots + \left(t' + \frac{2t-1}{t'}\right) + 2t$$

Example:

$$t = \frac{8}{8} \qquad \tau_1 = \frac{1}{8}$$

$$\sum_{1}^{2} = 1 + \left(1 + \frac{1}{8}\right) + \left(1 + \frac{2}{8}\right) + \left(1 + \frac{3}{8}\right) + \ldots + \left(1 + \frac{7}{8}\right) + 2 =$$

$$= \frac{8}{8} + \frac{9}{8} + \frac{10}{8} + \frac{11}{8} + \frac{12}{8} + \frac{13}{8} + \frac{14}{8} + \frac{15}{8} + \frac{16}{8}$$

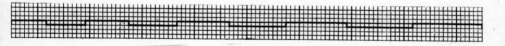

Figure 23. Series of deviation based on growth of τ through natural integer series.

150 THEORY OF REGULARITY AND COORDINATION

(b) The numerator of τ grows through an arithmetical progression.

$$\tau_1 = \frac{1}{t'}$$

$$\sum_{t}^{2t} = t + \left(t' + \frac{1}{t'}\right) + \left(t' + \frac{3}{t'}\right) + \left(t' + \frac{5}{t'}\right) + \ldots + \left(t' + \frac{2t-2}{t'}\right) + 2t$$

Example:

$$t = \frac{5}{5} \quad \tau_1 = \frac{1}{5}$$

$$\sum_{1}^{2} = 1 + \left(1 + \frac{1}{5}\right) + \left(1 + \frac{3}{5}\right) + 2 = \frac{5}{5} + \frac{6}{5} + \frac{8}{5} + \frac{10}{5}$$

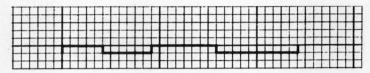

Figure 24. Numerator τ grows through arithmetical progression.

(c) The numerator τ grows through a geometrical progression.

$$\tau_1 = \frac{1}{t'}$$

$$\sum_{t} = t + \left(t' + \frac{1}{t'}\right) + \left(t' + \frac{2}{t'}\right) + \left(t' + \frac{4}{t'}\right) + \ldots$$

or, generally:

$$\sum_{t} = t + \left(t' + \frac{m}{t'}\right) + \left(t' + \frac{2m}{t'}\right) + \left(t' + \frac{2^2 m}{t'}\right) + \ldots$$

Examples:

$$t = \frac{8}{8} \quad \tau_1 = \frac{3}{8}$$

$$\sum_{1}^{4} = 1 + \left(1 + \frac{3}{8}\right) + \left(1 + \frac{6}{8}\right) + \left(1 + \frac{12}{8}\right) + \left(1 + \frac{24}{8}\right) =$$

$$= \frac{8}{8} + \frac{11}{8} + \frac{14}{8} + \frac{20}{8} + \frac{32}{8}$$

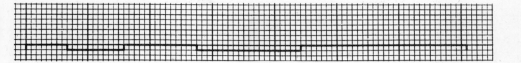

Figure 25. Numerator τ grows through geometrical progression.

(d) *The numerator τ grows through a summation series.*

$$\tau_1 = \frac{1}{t'}$$

$$\sum_t = t + \left(t' + \frac{1}{t'}\right) + \left(t' + \frac{2}{t'}\right) + \left(t' + \frac{4}{t'}\right) + \left(t' + \frac{7}{t'}\right) + \left(t' + \frac{12}{t'}\right) +$$

$$+ \left(t' + \frac{20}{t'}\right) + \ldots$$

Example:

$$t = \frac{7}{7} \quad \tau_1 = \frac{1}{7}$$

$$\sum_1^2 = 1 + \left(1 + \frac{1}{7}\right) + \left(1 + \frac{2}{7}\right) + \left(1 + \frac{4}{7}\right) + \left(1 + \frac{7}{7}\right) =$$

$$= \frac{7}{7} + \frac{8}{7} + \frac{9}{7} + \frac{11}{7} + \frac{14}{7}$$

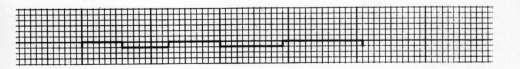

Figure 26. Numerator τ grows through a summation series.

(e) The numerator τ grows through the series of natural differences.

$$\tau_1 = \frac{1}{t'}$$

$$\sum_t = t + \left(t' + \frac{1}{t'}\right) + \left(t' + \frac{2}{t'}\right) + \left(t' + \frac{4}{t'}\right) + \left(t' + \frac{8}{t'}\right) + \left(t' + \frac{15}{t'}\right) +$$

$$+ \left(t' + \frac{26}{t'}\right) + \dots$$

Example:

$$t = \frac{9}{9} \quad \tau_1 = \frac{2}{9}$$

$$\sum_1^4 = 1 + \left(1 + \frac{2}{9}\right) + \left(1 + \frac{3}{9}\right) + \left(1 + \frac{5}{9}\right) + \left(1 + \frac{9}{9}\right) + \left(1 + \frac{16}{9}\right) +$$

$$+ \left(1 + \frac{27}{9}\right) = \frac{9}{9} + \frac{11}{9} + \frac{12}{9} + \frac{14}{9} + \frac{18}{9} + \frac{25}{9} + \frac{36}{9}$$

Figure 27. Numerator τ grows through the series of natural differences.

(f) The numerator τ grows through the series of prime numbers.

$$\tau_1 = \frac{1}{t'}$$

$$\sum_t = t + \left(t' + \frac{1}{t'}\right) + \left(t' + \frac{2}{t'}\right) + \left(t' + \frac{3}{t'}\right) + \left(t' + \frac{5}{t'}\right) + \left(t' + \frac{7}{t'}\right) +$$

$$+ \left(t' + \frac{11}{t'}\right) + \left(t' + \frac{13}{t'}\right) + \left(t' + \frac{17}{t'}\right) + \dots$$

PERIODICITY

Example:

$$t = \frac{5}{5} \quad \tau_1 = \frac{1}{5}$$

$$\sum_{1}^{\frac{47}{5}} = 1 + \left(1 + \frac{1}{5}\right) + \left(1 + \frac{2}{5}\right) + \left(1 + \frac{3}{5}\right) + \left(1 + \frac{6}{5}\right) + \left(1 + \frac{11}{5}\right) +$$

$$+ \left(1 + \frac{18}{5}\right) + \left(1 + \frac{29}{5}\right) + \left(1 + \frac{42}{5}\right) + \ldots =$$

$$= \frac{5}{5} + \frac{6}{5} + \frac{7}{5} + \frac{8}{5} + \frac{11}{5} + \frac{16}{5} + \frac{23}{5} + \frac{34}{5} + \frac{47}{5} + \ldots$$

Figure 28. Numerator τ grows through the series of prime numbers.

All the cases described here, when read backwards, represent contraction. Any consecutive contraction may be obtained through $\Sigma = t + \left(t' - \frac{m}{t'}\right) + \ldots$ where m may grow through any preselected series.

2. Forms of standard deviation in a binomial or a polynomial.

This form is known to musicians as *rubato*. We have already discussed standard deviation as an art determinant[5]. Mathematically it is a special case of τ, when $\tau = \frac{1}{t'}$. If a balanced binomial is thrown out of balance by a unit of standard deviation, we obtain:

balanced binomial $B_2 = a + a$

unbalanced binomial $U_2 = (a + \tau) + (a + \tau)$

or, through permutation, $U_2 = (a - \tau) + (a - \tau)$

In other words, in order to throw a balanced binomial out of balance, it is neces-

---

[5]See Chapter 1, Section G. "Correspondence Between Art Forms." (Ed.)

sary to add the unit of standard deviation to one of the terms of a binomial and to subtract it from another. If $T = 2t = t_1 + t_2$, the unbalanced form will be $T = 2t = \left(t_1 + \dfrac{1}{2t}\right) + \left(t_2 + \dfrac{2}{2t}\right)$. For example, if $T = 10$, $t = \dfrac{10}{2} = 5, \dfrac{1}{2t} = \dfrac{1}{10}$; then $\dfrac{10}{10} = \dfrac{5}{10} + \dfrac{5}{10}$ in balance and $\dfrac{10}{10} = \dfrac{5+1}{10} + \dfrac{5-1}{10}$, or $\dfrac{6}{10} + \dfrac{4}{10}$ out of balance.

In order to bring a polynomial out of balance, it is necessary to bring one of its binomials out of balance. Further variations can be obtained through consecutive displacement of the unbalanced binomial. If we have $T = mt = t_1 + t_2 + t_3 + \ldots + t_m$, (where all the t's are equal) and if we select any two neighboring terms as a binomial, (for instance, $t_2 + t_3$), then the unbalanced form of the binomial will be $(t_2 + \tau) + (t_3 - \tau)$, and the whole polynomial will be

$$T_m \begin{cases} t_1 + (t_2 + \tau) + (t_3 - \tau) + t_4 + \ldots + t_m \\ t_1 + t_2 + (t_3 + \tau) + (t_4 + \tau) + t_5 \ldots + t_m \\ \ldots \ldots \ldots \ldots \ldots \ldots \ldots \ldots \ldots \ldots \ldots \ldots \ldots \\ t_1 + t_2 + t_3 + \ldots + (t_{m-1} + \tau) + (t_m - \tau) \\ (t_1 + \tau) + (t_2 - \tau) + t_3 + t_4 + \ldots + t_m \end{cases}$$

The opposite process would bring the binomial back to balance.

Example:

$$T_4 = \frac{8}{8} = 4\left(\frac{8}{8}\right); \quad t = \frac{2}{8}; \quad \tau = \frac{1}{8}$$

$$T_4 = \frac{2t_1}{8} + \frac{2t_2}{8} + \frac{2t_3}{8} + \frac{2t_4}{8}$$

Bringing the first binomial out of balance, we get $T_4 = \left(\dfrac{2}{8} + \dfrac{1}{8}\right) + \dfrac{2}{8} + \dfrac{2}{8} + \dfrac{3t_1}{8} + \dfrac{t_2}{8} + \dfrac{2t_3}{8} + \dfrac{2t_4}{8}$, or $\dfrac{2t_1}{8} + \dfrac{3t_2}{8} + \dfrac{t_3}{8} + \dfrac{2t_4}{8}$, or $\dfrac{2t_1}{8} + \dfrac{2t_2}{8} + \dfrac{3t_3}{8} + \dfrac{t_4}{8}$

Or, by interior inversion of the binomial:

$$T_4 = \frac{t_1}{8} + \frac{3t_2}{8} + \frac{2t_3}{8} + \frac{2t_4}{8}, \text{ or } \frac{2t_1}{8} + \frac{t_2}{8} + \frac{3t_3}{8} + \frac{2t_4}{8}, \text{ or } \frac{2t_1}{8} + \frac{2t_2}{8} + \frac{t_3}{8} + \frac{3t_4}{8}$$

3. Other forms of deviation in a binomial or a polynomial.

Any unit t may grow through any of the series previously described. This process brings more intense expansion-contraction than through the growth of a unit of deviation added to t. Allowing t to grow through the natural integer series, we obtain:

$$\sum_t = t, 2t, 3t, 4t, \ldots$$

PERIODICITY

Allowing t to grow through the summation series, we obtain:

$$\sum_t = t,\ 2t,\ 3t,\ 5t,\ 8t,\ 13t,\ \ldots$$

Thus, we may produce various forms of factorial continuity previously described. Besides this procedure, there is a possibility of varying one of the terms of a binomial or a polynomial, while the rest of the terms remain constant or vary in a different manner.

Here are a few of the possibilities:

Binomials

(a) The first term of a binomial remains constant, while the second grows through any series.

Example:

1 + 2 is a given binomial. The second term grows through the natural series.

(1 + 2) + (1 + 3) + (1 + 4) + (1 + 5) + (1 + 6) + ...

(b) The combined form = forward-backward progression of the previous series.

Example:

(1 + 2) + (1 + 3) + (1 + 4) + ... + (4 + 1) + (3 + 1) + (2 + 1)

(c) Both terms of a binomial grow through the same or different series.

Example:

(1 + 2) + (2 + 3) + (3 + 4) + ... or

(1 + 2) + (2 + 4) + (3 + 8) + (4 + 16) + ...

(d) Forward-backward progression of the same series.

Example:

(1 + 2) + (2 + 3) + (3 + 4) + ... + (4 + 3) + (3 + 2) + (2 + 1)

or

(1 + 2) + (2 + 4) + (3 + 8) + ... + (8 + 3) + (4 + 2) + (2 + 1)

(e) Forms of interrupted expansion-contraction obtained from the previous series through permutations.

Examples:

Theme: $(1 + 2) + (3 + 4) + (5 + 6) + (7 + 8) + \ldots$
Var. I: $(1 + 2) + (2 + 1) + (3 + 4) + (4 + 3) + \ldots$
Var. II: $(1 + 2) + (5 + 6) + (3 + 4) + (7 + 8) + \ldots$
Var. III: $(1 + 2) + (7 + 8) + (3 + 4) + (5 + 6) + \ldots$
Var. IV: $(1 + 2) + (2 + 1) + (3 + 4) + (5 + 6) + (6 + 5) +$
          $+ (7 + 8) + \ldots$
Var. V: $(1 + 2) + (3 + 4) + (2 + 1) + (4 + 3) + \ldots$
Var. VI: $(1 + 2) + (4 + 3) + (5 + 6) + (8 + 7) + \ldots$
Var. VII: $(1 + 3) + (2 + 4) + (3 + 5) + (4 + 6) + (5 + 7) + \ldots$
Var. VIII:$(1 + 3) + (3 + 2) + (2 + 4) + (4 + 3) + (3 + 5) +$
          $+ (5 + 4) + \ldots$

Trinomials

(a) Growth of the first term.

$(1 + 2 + 3) + (2 + 2 + 3) + (3 + 2 + 3) + \ldots$
$(1 + 2 + 3) + (3 + 2 + 3) + (5 + 2 + 3) + \ldots$

(b) Growth of the middle term.

$(1 + 2 + 3) + (1 + 3 + 3) + (1 + 4 + 3) + \ldots$
$(1 + 2 + 3) + (1 + 4 + 3) + (1 + 8 + 3) + \ldots$

(c) Growth of the last term.

$(1 + 2 + 3) + (1 + 2 + 4) + (1 + 2 + 5) + \ldots$
$(1 + 3 + 1) + (1 + 3 + 3) + (1 + 3 + 5) + (1 + 3 + 7) + \ldots$

(d) Growth of the first two terms.

$(1 + 2 + 3) + (2 + 3 + 3) + (3 + 4 + 3) + \ldots$
$(1 + 2 + 3) + (2 + 4 + 3) + (4 + 8 + 3) + \ldots$
$(1 + 2 + 3) + (2 + 3 + 3) + (4 + 4 + 3) + (8 + 5 + 3) + \ldots$

(e) Growth of the last two terms.

$(1 + 2 + 2) + (1 + 3 + 4) + (1 + 4 + 8) + (1 + 5 + 16) + \ldots$
$(1 + 2 + 3) + (1 + 3 + 4) + (1 + 5 + 5) + (1 + 8 + 6) + \ldots$
$(3 + 2 + 3) + (3 + 4 + 5) + (3 + 5 + 6) + (3 + 6 + 7) + \ldots$

(f) One term expands, one contracts, one remains constant.

$(1 + 2 + 3) + (2 + 1 + 3)$ | $(2 + 3 + 4) + (1 + 4 + 4)$
$(1 + 2 + 3) + (2 + 2 + 2)$ | $(2 + 3 + 4) + (1 + 3 + 5)$
$(1 + 2 + 3) + (1 + 3 + 2)$ | $(2 + 3 + 4) + (2 + 2 + 5)$

PERIODICITY

There are many other possibilities in expansion-contraction techniques, such as juxtaposition of various coefficients and powers. They all form rhythmic series of the second order and are valuable in composition.[6]

G. PROGRESSIVE SYMMETRY

1. Forms of Harmonic Continuity.[7]

(a) One Subject: A.

$$A$$

(b) Two Subjects: A, B.

$$A + (A + B) + A$$

(c) Three Subjects: A, B, C.

$$A + (A + B) + (A + B + C) + (B + C) + C$$

(d) Four Subjects: A, B, C, D.

$$A + (A + B + C) + (A + B + C + D) + (B + C + D) + D$$

$$A + (A + B) + (A + B + C + D) + (C + D) + D$$

(e) Five Subjects: A, B, C, D, E.

$$A + (A + B) + (A + B + C) + (A + B + C + D + E) + (C + D + E) +$$
$$+ (D + E) + E$$

---

[6] These series of the second order are treated in greater detail in Chapter 4, Section 6. (Ed.)

[7] These abstract forms are more readily understood if one thinks of them in relation to music.

Each letter may be regarded as representing either a theme or a fragment of a theme. Thus, each of the alternatives may suggest the development of themes in a symphony. (Ed.)

CHAPTER 4

PERMUTATION

(Compensatory Coordination and Continuity)

A. Displacement

ANY group of elements representing different values of the same component may be expressed as a polynomial $P = a+b+c+ \ldots + m$. Such a group can be varied by means of *permutation*, thus producing a modified version of the same style.

Mechanical simplicity of variation determines the kinship of various groups. In the closest relation to the original are groups obtained through *displacement*.[1] The latter may be regarded as a consecutive displacement of terms following one direction. This can be performed clockwise (forward) or counterclockwise (backward).

Original polynomial: $\quad P = (a + b + c + \ldots + m)$

Derivative polynomials obtained through displacement:

$$P_1 = b + c + \ldots + m + a$$
$$P_2 = c + \ldots + m + a + b$$
$$P_3 = \ldots + m + a + b + c$$
$$------------$$
$$P_m = m + a + b + c + \ldots$$

The number of displacements equals the number of terms in a polynomial minus one.

$$N_d = N_t - 1$$

A polynomial of 2 terms gives one variation through displacement; a polynomial of 3 terms gives two variations; a polynomial of n terms gives $n-1$ variations.

Displacement in a Rhythmic Group.

$$P_{4:3} = 3 + 1 + 2 + 2 + 1 + 3$$
$$P_1 \;\; = 1 + 2 + 2 + 1 + 3 + 3$$
$$P_2 \;\; = 2 + 2 + 1 + 3 + 3 + 1$$
$$P_3 \;\; = 2 + 1 + 3 + 3 + 1 + 2$$
$$P_4 \;\; = 1 + 3 + 3 + 1 + 2 + 2$$
$$P_5 \;\; = 3 + 3 + 1 + 2 + 2 + 1$$

---
[1] Displacement, a special case of permutation, is also known as *circular permutation*.

Graphically these displacements appear as follows:

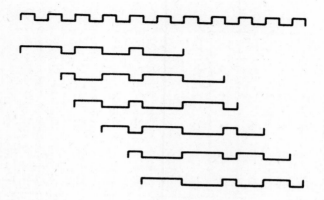

Figure 1. Graphic representation of displacement.

Rearranged in simultaneity these displacements take the following form:

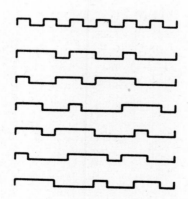

Figure 2. Displacements arranged in simultaneity.

THEORY OF REGULARITY AND COORDINATION

Displacement is the basic process for evolving simultaneity and sequence in composition. Using the same polynomial, $P_{4:3}$, we may evolve the following sequences of simultaneity through displacement.

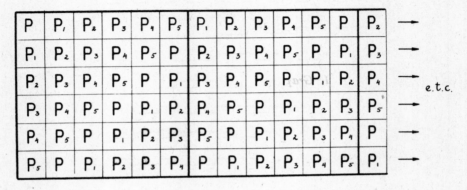

Figure 3. Sequences of simultaneity based on displacements.

Graphically this may be represented as follows:

$$P_{4:3},$$

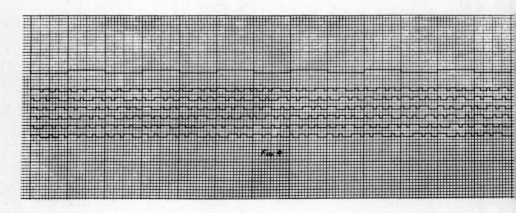

Figure 4. Graphic representation of material in preceding Figure.

PERMUTATION

The corresponding artistic expression for the *sequence of simultaneity* is counterpoint. The method of displacement provides an ultimate form of rhythmic counterpoint. Here is a typical example of rhythmic counterpoint in music:

Representing this in figures, we obtain the following:

$$2 + 1 + 1$$
$$1 + 2 + 1$$
$$1 + 1 + 2$$

Graphically this sequence may be represented as follows:

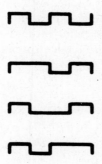

Figure 5. Rhythmic counterpoint in graph form.

The first term displaces itself forward. Generalizing the principle of displacement for the composition of sequences in simultaneity, we may state that the number of simultaneous terms equals the number of terms in a given polynomial; and the number of polynomial groups in the sequence of the first order equals the same number (the number of terms in a given polynomial).

B. General Permutation

Permutations in general are more numerous, mechanically more complex, and artistically a little less obvious than displacements. Nevertheless, all the permutations of one polynomial produce modifications of one style or one national artistic language. For example, if a certain group of numbers specifies the pitch scale characteristic of a certain nation, all the derivative scales obtained through permutation will be recognized as pertaining to the musical language of the same nation. This device permits us to evolve a great many new variations on styles already established and seemingly exhausted. The problem of creating a new style likewise becomes an easy procedure which can be accomplished in a much more reliable way than by mere feeling.

The number of permutations possible with a given polynomial $P = a + b + c + \ldots + m$ equals the product of the integer numbers from one to the number of terms of the polynomial. When P has n terms, the number of permutations equals the product 1 by 2, by 3, by 4, ... by n.

$$N_p = 1 \times 2 \times 3 \times \ldots \times n.$$

A polynomial consisting of two terms has two permutations (including the original).

$$1 \times 2 = 2$$

A polynomial consisting of three terms has six permutations (including the original).

$$1 \times 2 \times 3 = 6$$

A polynomial consisting of 12 terms has over half a billion permutations.

There are two ways of evolving the sequences of permutations. We shall call the first one—*mechanical permutation*, and the second—*logical permutation*. In the case of 2 elements the mechanical permutations and the logical permutations are identical:

ab   ba

In order to obtain mechanical permutations of 3 elements it is necessary to take each of the permutations of 2 elements, add the third element and displace it for each from left to right.

ab<u>c</u>   ba<u>c</u>
a<u>c</u>b   b<u>c</u>a
<u>c</u>ab   <u>c</u>ba

PERMUTATION

Logical permutations may be obtained from the first group abc by exchanging the positions of b and c; then starting a group with b and exchanging the positions of a and c; and, finally, starting a group with c and exchanging the positions of a and b.

$$\underline{a}bc \qquad \underline{b}ac \qquad \underline{c}ab$$
$$\underline{a}cb \qquad \underline{b}ca \qquad \underline{c}ba$$

The displacements abc, bca and cab are symmetrically located in both of the above groups.

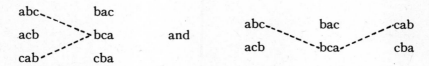

The general method of producing a table of permutations for n elements in a mechanical sequence requires the addition of the nth element to each group of the table of permutations for n − 1 elements, and the consecutive displacement of the added element from right to left. For example, if abcde is one of the group to which the new element f is added, then f moves consecutively backwards without affecting the sequence of the remaining terms.

$$abcde\underline{f}$$
$$abcd\underline{f}e$$
$$abc\underline{f}de$$
$$ab\underline{f}cde$$
$$a\underline{f}bcde$$
$$\underline{f}abcde$$

This holds true for every group. The mechanical sequence of permutations produces an inverted symmetry from its center, a property already observed in rhythmic groups resulting from the interference of uniform periodicities.[2]

Examples of inverted symmetry obtained through mechanical permutation.

Example 1. *Color.*

a = yellow, b = blue, c = red

Y — B — R — Y — R — B — R — Y — B — B — Y — R — B — R — Y — R — B — Y

[2] See Chapter 3, *Periodicity*. (Ed.)

THEORY OF REGULARITY AND COORDINATION

Example 2. *Sound.*

1. Table of permutations in mechanical sequence.

2 Elements 1 × 2 = 2 permutations
ab ba

3 Elements 1 × 2 × 3 = 6 permutations
abc bac
acb bca
cab cba

4 Elements 1 × 2 × 3 × 4 = 24 permutations
abcd bacd
abdc badc
adbc bdac
dabc dbac

acbd bcad
acdb bcda
adcb bdca
dacb dbca

cabd cbad
cadb cbda
cdab cdba
dcab dcba

5 Elements 1 × 2 × 3 × 4 × 5 = 120 permutations
abcde bacde
abced baced
abecd baecd
aebcd beacd
eabcd ebacd

PERMUTATION

abdce	badce
abdec	badec
abedc	baedc
aebdc	beadc
eabdc	ebadc
adbce	bdace
adbec	bdaec
adebc	bdeac
aedbc	bedac
eadbc	ebdac
dabce	dbace
dabec	dbaec
daebc	dbeac
deabc	debac
edabc	edbac
acbde	bcade
acbed	bcaed
acebd	bcead
aecbd	becad
eacbd	ebcad
acdbe	bcdae
acdeb	bcdea
acedb	bceda
aecdb	becda
eacdb	ebcda
adcbe	bdcae
adceb	bdcea
adecb	bdeca
aedcb	bedca
eadcb	ebdca
dacbe	dbcae
daceb	dbcea
daecb	dbeca
deacb	debca
edacb	edbca
cabde	cbade
cabed	cbaed
caebd	cbead
ceabd	cebad
ecabd	ecbad

cadbe	cbdae
cadeb	cbdea
caedb	cbeda
ceadb	cebda
ecadb	ecbda
cdabe	cdbae
cdaeb	cdbea
cdeab	cdeba
cedab	cedba
ecdab	ecdba
dcabe	dcbae
dcaeb	dcbea
dceab	dceba
decab	decba
edcab	edcba

2. Table of permutations where some of the elements are identical (mechanical sequence).

These sequences are evolved through the previous table. All the coinciding groups are eliminated in their corresponding places.

3 Elements

2 identical elements.   3 permutations (= displacements).

aab
aba
baa

4 Elements

3 identical elements.   4 permutations (= displacements).

aaab
aaba
abaa
baaa

2 identical pairs.   6 permutations.

aabb	baba
abab	abba
baab	bbaa

2 identical elements.   12 permutations.

aabc	aacb	acab	caab
abac	abca	acba	caba
baac	baca	bcaa	cbaa

PERMUTATION

5 Elements

4 identical elements. 5 permutations (= displacements).

        aaaab
        aaaba
        aabaa
        abaaa
        baaaa

3 identical + 2 identical. 10 permutations.

aabba	aabab	aaabb
ababa	abaab	
baaba	baaab	
babaa		
abbaa		
bbaaa		

2 identical + 2 identical + 1. 30 permutations.

aabcb	aacbb	acabb	caabb
abacb	abcab	acbab	cabab
baacb	bacab	bcaab	cbaab
aabbc			
ababc			
baabc			
	abcba	acbba	cabba
abbac	abbca		
			cbaba
babac	babca	bacba	bcaba
		bcbaa	cbbaa
bbaac	bbaca	bbcaa	

2 identical + 3. 60 permutations.

aabcd	aacbd	acabd	caabd
abacd	abcad	acbad	cabad
baacd	bacad	bcaad	cbaad
aabdc	aacdb	acadb	caadb
abadc	abcda	acbda	cabda
baadc	bacda	bcada	cbada
aadbc	aadcb	acdab	cadab
abdac	abdca	acdba	cadba
badac	badca	bcdaa	cbdaa
adabc	adacb	adcab	cdaab
adbac	adbca	adcba	cdaba
bdaac	bdaca	bdcaa	cdbaa

daabc	daacb	dacab	dcaab
dabac	dabca	dacba	dcaba
bdaac	dbaca	dbcaa	dcbaa

C. Permutations of the Higher Orders.

A group (*combination*) of two or more elements may be regarded as an element of the succeeding higher order. Thus, permutations can be performed ad infinitum with any number of elements. This method is especially useful in evolving an extended continuity from a limited number of elements.

1. Binomial elements. Binomial permutations.

Assuming that a_1 and b_1 are elements of the first order, we may evolve a continuity through permutation.

$$(a_1 + b_1) + (b_1 + a_1)$$

Now let these two binomials be elements of *the second order*.

Then, $\quad a_1 + b_1 = a_2 \quad$ and
$$b_1 + a_1 = b_2$$

Permutation of elements of the second order will produce the following continuity:

$$(a_2 + b_2) + (b_2 + a_2) = [(a_1 + b_1) + (b_1 + a_1)] + [(b_1 + a_1) + (a_1 + b_1)]$$

Now, $\quad a_2 + b_2 = a_3$
$$b_2 + a_2 = b_3$$

Then, $(a_3 + b_3) + (b_3 + a_3) = [(a_2 + b_2) + (b_2 + a_2)] + [(b_2 + a_2) + (a_2 + b_2)] = \{[(a_1 + b_1) + (b_1 + a_1)] + [(b_1 + a_1) + (a_1 + b_1)]\} + \{[(b_1 + a_1) + (a_1 + b_1)] + [(a_1 + b_1) + (b_1 + a_1)]\}$

Or, in general, $\quad a_n = a_{n-1} + b_{n-1}$
$$b_n = b_{n-1} + a_{n-1}$$

Then, $(a_n + b_n) + (b_n + a_n) = [(a_{n-1} + b_{n-1}) + (b_{n-1} + a_{n-1})] + [(b_{n-1} + a_{n-1}) + (a_{n-1} + b_{n-1})]$

PERMUTATION

Example:

$$a = 2; \quad b = 1$$

$$(a_1 + b_1) + (b_1 + a_1) = (2 + 1) + (1 + 2)$$

$$a_2 = a_1 + b_1 = 2 + 1$$

$$b_2 = b_1 + a_1 = 1 + 2$$

$$(a_2 + b_2) + (b_2 + a_2) = [(2 + 1) + (1 + 2)] + [(1 + 2) + (2 + 1)]$$

$$a_3 = a_2 + b_2 = (2 + 1) + (1 + 2)$$

$$b_3 = b_2 + a_2 = (1 + 2) + (2 + 1)$$

$$(a_3 + b_3) + (b_3 + a_3) = \{[(2 + 1) + (1 + 2)] + [(1 + 2) + (2 + 1)]\} +$$
$$+ \{[(1 + 2) + (2 + 1)] + [(2 + 1) + (1 + 2)]\}.$$

A two-element combination may be chosen from any group. A trinomial in its six variations may be assumed an element of the second order. Thus,

$$a_1 + b_1 + c_1 = a_2 \qquad b_1 + a_1 + c_1 = d_2$$
$$a_1 + c_1 + b_1 = b_2 \qquad b_1 + c_1 + a_1 = e_2$$
$$c_1 + a_1 + b_1 = c_2 \qquad c_1 + b_1 + a_1 = f_2$$

The following 30 binomial combinations are possible from these 6 elements:

$a_2 + b_2$	$b_2 + a_2$	$c_2 + a_2$
$a_2 + c_2$	$b_2 + c_2$	$c_2 + b_2$
$a_2 + d_2$	$b_2 + d_2$	$c_2 + d_2$
$a_2 + e_2$	$b_2 + e_2$	$c_2 + e_2$
$a_2 + f_2$	$b_2 + f_2$	$c_2 + f_2$
$d_2 + a_2$	$e_2 + a_2$	$f_2 + a_2$
$d_2 + b_2$	$e_2 + b_2$	$f_2 + b_2$
$d_2 + c_2$	$e_2 + c_2$	$f_2 + c_2$
$d_2 + e_2$	$e_2 + d_2$	$f_2 + d_2$
$d_2 + f_2$	$e_2 + f_2$	$f_2 + e_2$

By continuous inversion of these binomials in their consecutive orders, we obtain the *binomial growth of trinomials*.

Let us develop $a_2 + f_2$ where

$$a_2 = a_1 + b_1 + c_1 \quad \text{and}$$
$$f_2 = c_1 + b_1 + a_1$$

Thus, $\quad a_3 = a_2 + f_2 \quad$ and
$$b_3 = f_2 + a_2$$

Substituting a_2 and f_2 for their trinomial expressions, we obtain:

$$a_3 + b_3 = (a_2 + f_2) + (f_2 + a_2) = (a_1 + b_1 + c_1) +$$
$$+ (c_1 + b_1 + a_1) + (c_1 + b_1 + a_1) + (a_1 + b_1 + c_1).$$

Likewise, $\quad a_4 = a_3 + b_3 \quad$ and
$$b_4 = b_3 + a_3$$

Thus,

$$a_4 + b_4 = (a_3 + b_3) + (b_3 + a_3) = [(a_2 + f_2) + (f_2 + a_2)] +$$
$$+ [(f_2 + a_2) + (a_2 + f_2)] = \{[(a_1 + b_1 + c_1) + (c_1 + b_1 + a_1)] +$$
$$+ [(c_1 + b_1 + a_1) + (a_1 + b_1 + c_1)]\} + \{[(c_1 + b_1 + a_1) +$$
$$+ (a_1 + b_1 + c_1)] + [(a_1 + b_1 + c_1) + (c_1 + b_1 + a_1)]\}.$$

Assuming $\quad a_1 = 3, b_1 = 2, c_1 = 1,$

we obtain
$$a_2 = 3 + 2 + 1$$
$$f_2 = 1 + 2 + 3$$
$$a_3 = a_2 + f_2 = (3 + 2 + 1) + (1 + 2 + 3)$$
$$b_3 = f_2 + a_2 = (1 + 2 + 3) + (3 + 2 + 1)$$
$$a_3 + b_3 = [(3 + 2 + 1) + (1 + 2 + 3)] + [(1 + 2 + 3) + (3 + 2 + 1)]$$
$$a_4 = a_3 + b_3$$
$$b_4 = b_3 + a_3$$
$$a_4 + b_4 = \{[(3 + 2 + 1) + (1 + 2 + 3)] + [(1 + 2 + 3) + (3 + 2 + 1)]\} +$$
$$+ \{[(1 + 2 + 3) + (3 + 2 + 1)] + [(3 + 2 + 1) + (1 + 2 + 3)]\}.$$

In general, any polynomial $(a + b + c + \ldots + m)$ may become an element of the succeeding order and may be combined with other polynomials. It is not necessary that the polynomials P′ and P″, etc., be derivatives of P. They may belong either to the same or to a different series. Thus, any trinomial and its variations produce 6 elements of the following order. These 6 elements may combine by 2, by 3, by 4, by 5 and by 6.

PERMUTATION

Three elements of the first order combined by two, produce, through binomial permutations, 12 terms.

$$a_1 + b_1 + c_1 = a_2$$
$$a_1 + c_1 + b_1 = b_2$$

a_2 has three terms.

$a_2 + b_2 + b_2 + a_2$ yields 12 terms.

Three elements of the first order combined by three, produce, through trinomial permutations, 54 terms.

$$a_1 + b_1 + c_1 = a_2$$
$$a_1 + c_1 + b_1 = b_2$$
$$c_1 + a_1 + b_1 = c_2$$

a_2 has three terms

b_2 ” ” ”

c_2 ” ” ”

$a_2 + b_2 + c_2$ yields 9 terms.

$$6 (3 \times 3) = 54$$

The increase in the number of combined elements causes rapid increase in the number of permutations. Thus, by combining trinomials of the first order by 6, we obtain six elements of the second order. They produce $1 \times 2 \times 3 \times 4 \times 5 \times 6 = 720$ permutations. Each element is a trinomial in this case. This yields in the second order

$$3 \times 720 = 2160 \text{ terms.}$$

In general, any polynomial of the first order, $a_1 + b_1 + c_1 + \ldots + m_1$ produces through permutations $a_2, b_2, c_2 \ldots m_2$. These being combined by 2, by 3, by 4,... by n, yield respectively, 4, 6, 24, ... or $1 \times 2 \times 3 \times \ldots \times n$ term groups. The latter become elements of the third order. Expansion of the number of elements in their consecutive higher orders grows, through permutation, very rapidly into an enormous number of terms. For example, 3 elements of the first order produce 6 elements of the second order, 720 elements of the third order, etc.

The artistic value of this method lies in the production of simpler groups through their higher orders. Such continuity achieves the effect of variety in thematic unity.

D. Mechanical Scheme for the Permutation of Four Elements

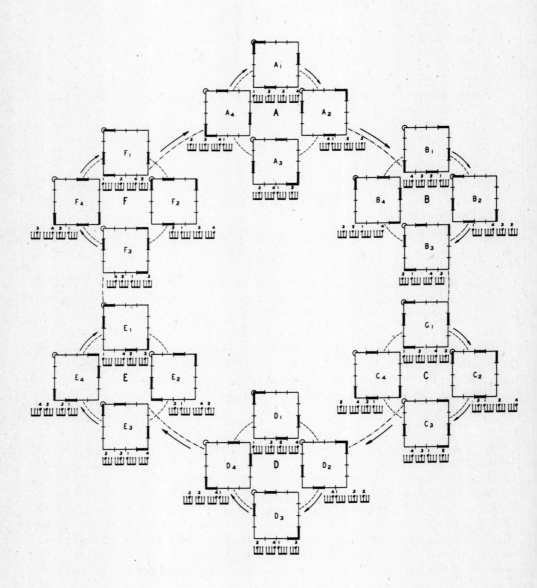

Figure 6. Mechanical scheme for permuting four elements.

> *"From shelf to crossbar to cage-top and back again the gibbon swarmed, breaking down rhythm and timing into their cube roots and building them up again, without so much as a hairsbreadth of misjudgment."*
>
> "With the Greatest of Ease,"
> by Paul Annixter.

CHAPTER 5

DISTRIBUTIVE INVOLUTION

(Coordination and Continuity of Harmonic Contrasts)

A. Powers

ALGEBRAIC powers are the basis of factorial-fractional continuity. They build rhythm within a unit (a bar in music, an area in design) and organize the whole continuity rhythmically.

The power formulae found in standard algebra do not express the *distributive properties* of a resulting polynomial, which are most important in the construction of an art form.[1] They may be used although they are not quite satisfactory because of

(1) the limited number of terms in the resulting polynomials;

(2) the extremely unbalanced values of the terms; and

(3) the lack of distinct representation of qualities, values and relationships in the resulting terms and groups.

The following formulae establish full correlation between the terms and the groups of a polynomial according to the initial ratio. Raising to power should be performed through consecutive multiplication of all the terms of a preceding power by each term of the given polynomial. Thus, the factoring coefficients are related to each other as the terms of the initial polynomial.

Example:

$$(a + b)^2 = (a^2 + ab) + (ab + b^2).$$

Here a^2 is related to ab, and ab is related to b^2 as a is related to b. $a^2 + ab$ is related to $ab + b^2$ as a to b.

---
[1] The distributive properties are treated in combinatory analysis.

$$\frac{a}{b} = \begin{vmatrix} \dfrac{a^2}{ab} \\ \dfrac{ab}{b^2} \end{vmatrix} \quad \dfrac{a^2 + ab}{ab + b^2}$$

$$a = 2; \quad b = 1$$

$$\frac{2}{1} = \begin{vmatrix} \dfrac{4}{2} \\ \dfrac{2}{1} \end{vmatrix} \quad \dfrac{4 + 2}{2 + 1}$$

(1) *The First Power*

Factorial monomial periodicity with one determinant, x.

$$F_x = xt_1 + xt_2 + xt_3 + \ldots + xt_x$$

Factorial binomial periodicity with two determinants, x and y.

$$F_{x+y} = (xt_1 + yt_2) + (xt_3 + yt_4) + (xt_5 + yt_6) + \ldots + (xt_{x+y-1} + xt_{x+y}).$$

Factorial trinomial periodicity with three determinants, x, y and z.

$$F_{x+y+z} = (xt_1 + yt_2 + zt_3) + (xt_4 + yt_5 + zt_6) + (xt_7 + yt_8 + zt_9) + \ldots$$
$$\ldots + (xt_{x+y+z-2} + yt_{x+y+z-1} + zt_{x+y+z}).$$

Factorial polynomial periodicity with n determinants, a, b, c, ... m, where $a + b + c + \ldots + m = k$.

$$F_{a+b+c+\ldots+m} = (at_1 + bt_2 + ct_3 + \ldots + mt_n) + (at_{n+1} + bt_{n+2} + ct_{n+3} + \ldots$$
$$\ldots + mt_{2n}) + (at_{2n+1} + bt_{2n+2} + ct_{2n+3} + \ldots + mt_{3n}) + (at_{3n+1} +$$
$$+ bt_{3n+2} + ct_{3n+3} + \ldots + mt_{4n}) + \ldots + (at_{k-n} + bt_{k-n+1} + ct_{k-n+2} + \ldots$$
$$\ldots + mt_k).$$

Fractional monomial periodicity with one determinant, $\dfrac{1}{x}$.

$$f_{\frac{1}{x}} = \frac{t_1}{x} + \frac{t_2}{x} + \frac{t_3}{x} + \ldots + \frac{t_x}{x}$$

Fractional binomial periodicity with two determinants, $\dfrac{x}{x+y}$ and $\dfrac{y}{x+y}$.

$$f_{\frac{x}{x+y}+\frac{y}{x+y}} = \left(\frac{xt_1}{x+y} + \frac{yt_2}{x+y}\right) + \left(\frac{xt_3}{x+y} + \frac{xt_4}{x+y}\right) + \ldots + \left(\frac{xt_{x+y-1}}{x+y} + \frac{xt_{x+y}}{x+y}\right).$$

DISTRIBUTIVE INVOLUTION

Fractional trinomial periodicity with three determinants,

$$\frac{x}{x+y+z}, \frac{y}{x+y+z} \text{ and } \frac{z}{x+y+z}.$$

$$f_{\frac{x}{x+y+z}+\frac{y}{x+y+z}+\frac{z}{x+y+z}} = \left(\frac{xt_1}{x+y+z} + \frac{yt_2}{x+y+z} + \frac{zt_3}{x+y+z}\right) + \left(\frac{xt_4}{x+y+z} + \frac{yt_5}{x+y+z} + \frac{zt_6}{x+y+z}\right) + \ldots + \left(\frac{xt_{x+y+z-2}}{x+y+z} + \frac{yt_{x+y+z-1}}{x+y+z} + \frac{zt_{x+y+z}}{x+y+z}\right).$$

Fractional polynomial periodicity with m determinants,

$$\frac{a}{a+b+c+\ldots+m}, \frac{b}{a+b+c+\ldots+m}, \frac{c}{a+b+c+\ldots+m}, \ldots \frac{m}{a+b+c+\ldots+m}$$

$$f_{\frac{a}{a+b+c+\ldots+m}}^{\frac{m}{a+b+c+\ldots+m}} = \left(\frac{at_1}{a+b+c+\ldots+m} + \frac{bt_2}{a+b+c+\ldots+m} + \frac{ct_3}{a+b+c+\ldots+m} + \ldots + \frac{mt_n}{a+b+c+\ldots+m}\right) + \left(\frac{at_{n+1}}{a+b+c+\ldots+m} + \frac{bt_{n+2}}{a+b+c+\ldots+m} + \frac{ct_{n+3}}{a+b+c+\ldots+m} + \ldots + \frac{mt_{2n}}{a+b+c+\ldots+m}\right) + \left(\frac{at_{2n+1}}{a+b+c+\ldots+m} + \frac{bt_{2n+2}}{a+b+c+\ldots+m} + \frac{ct_{2n+3}}{a+b+c+\ldots+m} + \frac{mt_{3n}}{a+b+c+\ldots+m}\right) + \ldots + \left(\frac{at_{k-n}}{a+b+c+\ldots+m} + \frac{bt_{k-n+1}}{a+b+c+\ldots+m} + \frac{ct_{k-n+2}}{a+b+c+\ldots+m} + \ldots + \frac{mt_k}{a+b+c+\ldots+m}\right)$$

Here n is the number of determinants a, b, c, ... m and $a + b + c + \ldots + m = k$.

In the following formulae, the expressions with n determinants will be simplified as follows:

$$\overset{k}{\underset{1}{F}} = (at_1 + bt_2 + \ldots + m_n) + \ldots + (at_{k-n} + bt_{k-n+1} + \ldots + mt_k), \text{ and}$$

$$\overset{k}{\underset{1}{f}} = \frac{at_1}{k} + \frac{bt_2}{k} + \ldots + \frac{mt_n}{k} + \ldots + \frac{at_{k-n}}{k} + \frac{bt_{k-n+1}}{k} + \ldots + \frac{mt_k}{k}.$$

(2) *The Second Power*

$$F_{x^2} = x^2 t_1 + x^2 t_2 + x^2 t_3 + \ldots + x^2 t_{x^2}.$$

$$F_{(x+y)^2} = (x^2 t_1 + xy t_2) + (xy t_3 + y^2 t_4).$$

$$F_{(x+y+z)^2} = (x^2 t_1 + xy t_2 + xz t_3) + (xy t_4 + y^2 t_5 + yz t_6) + (xz t_7 + yz t_8 + z^2 t_9).$$

$$F_{(a+b+c+\ldots+m)^2} = (a^2 t_1 + ab t_2 + ac t_3 + \ldots + am t_n) + (ab t_{n+1} + b^2 t_{n+2} + bc t_{n+3} + \ldots + bm t_{2n}) + \ldots + (am t_{n^2-n} + bm t_{n^2-n+1} + cm t_{n^2-n+2} + \ldots + m^2 t_{n^2}).$$

General Formula of the Second Power of a Polynomial, Preserving the Initial Distribution of its Terms.

$$(a + b + c + \ldots + m)^2 = a(a + b + c + \ldots + m) + b(a + b + c + \ldots + m) + c(a + b + c + \ldots + m) + \ldots + m(a + b + c + \ldots + m)$$
$$= (a^2 + ab + ac + \ldots + am) + (ab + b^2 + bc + \ldots + bm) + (ac + bc + c^2 + \ldots + cm) + \ldots + (am + bm + cm + \ldots + m^2).$$

$$f_{(\frac{1}{x})^2} = \frac{t_1}{x^2} + \frac{t_2}{x^2} + \frac{t_3}{x^2} + \ldots + \frac{t_{x^2}}{x^2}$$

$$f_{(\frac{x}{x+y}+\frac{y}{x+y})^2} = \left[\frac{x^2 t_1}{(x+y)^2} + \frac{xy t_2}{(x+y)^2}\right] + \left[\frac{xy t_3}{(x+y)^2} + \frac{y^2 t_4}{(x+y)^2}\right]$$

$$f_{(\frac{x}{x+y+z}+\frac{y}{x+y+z}+\frac{z}{x+y+z})^2} = \left[\frac{x^2 t_1}{(x+y+z)^2} + \frac{xy t_2}{(x+y+z)^2} + \frac{xz t_3}{(x+y+z)^2}\right] +$$
$$+ \left[\frac{xy t_4}{(x+y+z)^2} + \frac{y^2 t_5}{(x+y+z)^2} + \frac{yz t_6}{(x+y+z)^2}\right] + \left[\frac{xz t_7}{(x+y+z)^2} +\right.$$
$$\left. + \frac{yz t_8}{(x+y+z)^2} + \frac{z^2 t_9}{(x+y+z)^2}\right]$$

$$f_{(\frac{a}{k}+\frac{b}{k}+\frac{c}{k}+\ldots+\frac{m}{k})^2} = \left(\frac{a^2 t_1}{k^2} + \frac{ab t_2}{k^2} + \frac{ac t_3}{k^2} + \ldots + \frac{am t_n}{k^2}\right) + \left(\frac{ab t_{n+1}}{k^2} + \frac{b^2 t_{n+2}}{k^2} +\right.$$
$$\left. + \frac{bc t_{n+3}}{k^2} + \ldots + \frac{bm t_{2n}}{k^2}\right) + \left(\frac{ac t_{2n+1}}{k^2} + \frac{bc t_{2n+2}}{k^2} + \frac{c^2 t_{2n+3}}{k^2} + \ldots\right.$$
$$\left. \ldots + \frac{cm t_{3n}}{k^2}\right) + \left(\frac{am t_{n^2-n}}{k^2} + \frac{bm t_{n^2-n+1}}{k^2} + \frac{cm t_{n^2-n+2}}{k^2} + \ldots + \frac{m^2 t_{n^2}}{k^2}\right)$$

DISTRIBUTIVE INVOLUTION 177

(3) *The Third Power*

$$F_{x^3} = x^3 t_1 + x^3 t_2 + x^3 t_3 + \ldots + x^3 t_{x^3}$$

$$F_{(x+y)^3} = [(x^3 t_1 + x^2 y t_2) + (x^2 y t_3 + x y^2 t_4)] + [(x^2 y t_5 + x y^2 t_6) + (x y^2 t_7 + y^3 t_8)]$$

$$F_{(x+y+z)^3} = [(x^3 t_1 + x^2 y t_2 + x^2 z t_3) + (x^2 y t_4 + x y^2 t_5 + x y z t_6) + (x^2 z t_7 +$$
$$+ x y z t_8 + x z^2 t_9)] + [(x^2 y t_{10} + x y^2 t_{11} + x y z t_{12}) + (x y^2 t_{13} + y^3 t_{14} +$$
$$+ y^2 z t_{15}) + (x y z t_{16} + y^2 z t_{17} + y z^2 t_{18})] + [(x^2 z t_{19} + x y z t_{20} + x z^2 t_{21}) +$$
$$+ (x y z t_{22} + y^2 z t_{23} + y z^2 t_{24}) + (x z^2 t_{25} + y z^2 t_{26} + z^3 t_{27})]$$

$$F_{(a+b+c+\ldots+m)^3} = [(a^3 t_1 + a^2 b t_2 + a^2 c t_3 + \ldots + a^2 m t_n) + (a^2 b t_{n+1} +$$
$$+ a^2 b^2 t_{n+2} + a b c t_{n+3} + \ldots + a b m t_{2n}) + (a^2 c t_{2n+1} + a b c t_{2n+2} +$$
$$+ a c^2 t_{2n+3} + \ldots + a c t m_{3n}) + \ldots + (a^2 m t_{3n+1} + a b m t_{3n+2} + a c m t_{3n+3} + \ldots$$
$$\ldots + a^2 m^2 t_{4n})] + [(a^2 b t_{4n+1} + a b^2 t_{4n+2} + a b c t_{4n+3} + \ldots + a b m t_{5n}) +$$
$$+ (a b^2 t_{5n+1} + b^3 t_{5n+2} + b^2 c t_{5n+3} + \ldots + b^2 m t_{6n}) + (a b c t_{6n+1} +$$
$$+ b^2 c t_{6n+2} + b^2 c^2 t_{6n+3} + b c m t_{7n}) + \ldots + (a b m t_{7n+1} + b^2 m t_{7n+2} +$$
$$+ b c m t_{7n+3} + \ldots + b m^2 t_{8n})] + [(a^2 c t_{8n+1} + a b c t_{8n+2} + a c^2 t_{8n+3} + \ldots$$
$$\ldots + a c m t_{9n}) + (a b c t_{9n+1} + b^2 c t_{9n+2} + b c^2 t_{9n+3} + \ldots + b c m t_{10n}) +$$
$$+ (a c^2 t_{10n+1} + b c^2 t_{10n+2} + c^3 t_{10n+3} + \ldots + c^2 m t_{11n}) + \ldots + (a c m t_{11n+1} +$$
$$+ b c m t_{11n+2} + c^2 m t_{11n+3} + \ldots + c m^2 t_{12n})] + \ldots + [(a^2 m t_{n^2-4n} +$$
$$+ a b m t_{n^2-4n+1} + a c m t_{n^2-4n+2} + \ldots + a m^2 t_{n^2-3n-1}) + (a b m t_{n^2-3n} +$$
$$+ b^2 m t_{n^2-3n+1} + b c m t_{n^2-3n+2} + \ldots + b m^2 t_{n^2-2n-1}) + (a c m t_{n^2-2n} +$$
$$+ b c m t_{n^2-2n+1} + c^2 m t_{n^2-2n+2} + \ldots + c m^2 t_{n^2-n-1}) + \ldots + (a m^2 t_{n^2-n} +$$
$$+ b m^2 t_{n^2-n+1} + c m^2 t_{n^2-n+2} + \ldots + m^3 t_{n^2})]$$

General Formula of the Third Power of a Polynomial, Preserving the Initial Distribution of its Terms.

$$(a + b + c + \ldots + m)^3 = [(a^3 + a^2 b + a^2 c + \ldots + a^2 m) + (a^2 b +$$
$$+ a^2 b^2 + a b c + \ldots + a b m) + (a^2 c + a b c + a c^2 + \ldots + a c m) + \ldots$$
$$\ldots + (a^2 m + a b m + a c m + \ldots + a^2 m^2)] + [(a^2 b + a b^2 + a b c + \ldots$$
$$\ldots + a b m) + (a b^2 + b^3 + b^2 c + \ldots + b^2 m) + (a b c + b^2 c + b^2 c^2 + \ldots$$
$$\ldots + b c m) + \ldots + (a b m + b^2 m + b c m + \ldots + b m^2)] + [(a^2 c + a b c +$$

$+ ac^2 + \ldots + acm) + (abc + b^2c + bc^2 + \ldots + bcm) + (ac^2 + bc^2 +$
$+ c^3 + \ldots + c^2m) + \ldots + (acm + bcm + c^2m + \ldots + cm^2)] + \ldots$
$\ldots + [(a^2m + abm + acm + \ldots + am^2) + (abm + b^2m + bcm + \ldots$
$\ldots + bm^2) + (acm + bcm + c^2m + \ldots + cm^2) + \ldots + (am^2 +$
$+ bm^2 + cm^2 + \ldots + m^3)]$

$$f_{(\frac{1}{x})^3} = \frac{t_1}{x^3} + \frac{t_2}{x^3} + \frac{t_3}{x^3} + \ldots + \frac{t_x}{x^3}$$

$$f_{(\frac{x}{x+y}+\frac{y}{x+y})^3} = \left[\left(\frac{x^3t_1}{(x+y)^3} + \frac{x^2yt_2}{(x+y)^3}\right) + \left(\frac{x^2yt_3}{(x+y)^3} + \frac{xy^2t_4}{(x+y)^3}\right)\right] +$$
$$+ \left[\left(\frac{x^2yt_5}{(x+y)^3} + \frac{xy^2t_6}{(x+y)^3}\right) + \left(\frac{xy^2t_7}{(x+y)^3} + \frac{y^3t_8}{(x+y)^3}\right)\right]$$

$$f_{(\frac{x}{x+y+z}+\frac{y}{x+y+z}+\frac{z}{x+y+z})^3} = \left[\left(\frac{x^3t_1}{(x+y+z)^3} + \frac{x^2yt_2}{(x+y+z)^3} + \frac{x^2zt_3}{(x+y+z)^3}\right) +\right.$$
$$+ \left(\frac{x^2yt_4}{(x+y+z)^3} + \frac{xy^2t_5}{(x+y+z)^3} + \frac{xyzt_6}{(x+y+z)^3}\right) + \left(\frac{x^2zt_7}{(x+y+z)^3} +\right.$$
$$+ \left.\frac{xyzt_8}{(x+y+z)^3} + \frac{xz^2t_9}{(x+y+z)^3}\right)\right] + \left[\left(\frac{x^2yt_{10}}{(x+y+z)^3} + \frac{xy^2t_{11}}{(x+y+z)^3} +\right.\right.$$
$$+ \left.\frac{xyzt_{12}}{(x+y+z)^3}\right) + \left(\frac{xy^2t_{13}}{(x+y+z)^3} + \frac{y^3t_{14}}{(x+y+z)^3} + \frac{y^2zt_{15}}{(x+y+z)^3}\right) +$$
$$+ \left(\frac{xyzt_{16}}{(x+y+z)^3} + \frac{y^2zt_{17}}{(x+y+z)^3} + \frac{yz^2t_{18}}{(x+y+z)^3}\right)\right] + \left[\left(\frac{x^2zt_{19}}{(x+y+z)^3} +\right.\right.$$
$$+ \frac{xyzt_{20}}{(x+y+z)^3} + \frac{xz^2t_{21}}{(x+y+z)^3}\right) + \left(\frac{xyzt_{22}}{(x+y+z)^3} + \frac{y^2zt_{23}}{(x+y+z)^3} +$$
$$+ \left.\frac{yz^2t_{24}}{(x+y+z)^3}\right) + \left(\frac{xz^2t_{25}}{(x+y+z)^3} + \frac{yz^2t_{26}}{(x+y+z)^3} + \frac{z^3t_{27}}{(x+y+z)^3}\right)\right]$$

$$f_{(\frac{a}{k}+\frac{b}{k}+\frac{c}{k}+\ldots+\frac{m}{k})^3} = \left[\left(\frac{a^3t_1}{k^3} + \frac{a^2bt_2}{k^3} + \frac{a^2ct_3}{k^3} + \ldots + \frac{a^2mt_n}{k^3}\right) + \left(\frac{a^2bt_{n+1}}{k^3} +\right.\right.$$
$$+ \frac{a^2b^2t_{n+2}}{k^3} + \frac{abct_{n+3}}{k^3} + \ldots + \frac{abmt_{2n}}{k^3}\right) + \left(\frac{a^2ct_{2n+1}}{k^3} + \frac{abct_{2n+2}}{k^3} +$$
$$+ \frac{ac^2t_{2n+3}}{k^3} + \ldots + \frac{acmt_{3n}}{k^3}\right) + \ldots + \left(\frac{a^2mt_{3n+1}}{k^3} + \frac{abmt_{3n+2}}{k^3} + \frac{acmt_{3n+3}}{k^3} + \ldots\right.$$

$$\cdots + \frac{a^2m^2t_{4n}}{k^3}\Big)\Big] + \Big[\Big(\frac{a^2bt_{4n+1}}{k^3} + \frac{ab^2t_{4n+2}}{k^3} + \frac{abct_{4n+3}}{k^3} + \cdots + \frac{abmt_{5n}}{k^3}\Big) +$$

$$+ \Big(\frac{ab^2t_{5n+1}}{k^3} + \frac{b^3t_{5n+2}}{k^3} + \frac{b^2ct_{5n+3}}{k^3} + \cdots + \frac{b^2mt_{6n}}{k^3}\Big) + \Big(\frac{abct_{6n+1}}{k^3} +$$

$$+ \frac{b^2ct_{6n+2}}{k^3} + \frac{b^2c^2t_{6n+3}}{k^3} + \cdots + \frac{bcmt_{7n}}{k^3}\Big) + \cdots + \Big(\frac{abmt_{7n+1}}{k^3} +$$

$$+ \frac{b^2mt_{7n+2}}{k^3} + \frac{bcmt_{7n+3}}{k^3} + \cdots + \frac{bm^2t_{8n}}{k^3}\Big)\Big] + \Big[\Big(\frac{a^2ct_{8n+1}}{k^3} + \frac{abct_{8n+2}}{k^3} +$$

$$+ \frac{ac^2t_{8n+3}}{k^3} + \cdots + \frac{acmt_{9n}}{k^3}\Big) + \Big(\frac{abct_{9n+1}}{k^3} + \frac{b^2ct_{9n+2}}{k^3} + \frac{bc^2t_{9n+3}}{k^3} + \cdots$$

$$\cdots + \frac{bcmt_{10n}}{k^3}\Big) + \Big(\frac{ac^2t_{10n+1}}{k^3} + \frac{bc^2t_{10n+2}}{k^3} + \frac{c^3t_{10n+3}}{k^3} + \cdots + \frac{c^2mt_{11n}}{k^3}\Big) + \cdots$$

$$\cdots + \Big(\frac{acmt_{11n+1}}{k^3} + \frac{bcmt_{11n+2}}{k^3} + \frac{c^2mt_{11n+3}}{k^3} + \cdots + \frac{cm^2t_{12n}}{k^3}\Big)\Big] + \cdots$$

$$\cdots + \Big[\Big(\frac{a^2mt_{n^2-4n}}{k^3} + \frac{abmt_{n^2-4n+1}}{k^3} + \frac{acmt_{n^2-4n+2}}{k^3} + \cdots + \frac{am^2t_{n^2-3n-1}}{k^3}\Big) +$$

$$+ \Big(\frac{abmt_{n^2-3n}}{k^3} + \frac{b^2mt_{n^2-3n+1}}{k^3} + \frac{bcmt_{n^2-3n+2}}{k^3} + \cdots + \frac{bm^2t_{n^2-2n-1}}{k^3}\Big) +$$

$$+ \Big(\frac{acmt_{n^2-2n}}{k^3} + \frac{bcmt_{n^2-2n+1}}{k^3} + \frac{c^2mt_{n^2-3n+2}}{k^3} + \cdots + \frac{bm^2t_{n^2-2n-1}}{k^3}\Big) + \cdots$$

$$\cdots + \Big(\frac{am^2t_{n^2-n}}{k^3} + \frac{bm^2t_{n^2-n+1}}{k^3} + \frac{cm^2t_{n^2-n+2}}{k^3} + \cdots + \frac{m^2t_{n^2}}{k^3}\Big)\Big].$$

(4) *Generalization. The nth power.*

$$F_{x^n} = x^n t_1 + x^n t_2 + x^n t_3 + \cdots + x^n t_x + \cdots + x^n t_{x^n}.$$

B. General Treatment of Powers

Besides the technique of raising to powers through consecutive multiplication of all the terms of a polynomial by each term, it is possible to raise any polynomial to any power directly. This problem consists of three items:

1. To find the general number of terms in a given power.
2. To find the progression in which they are distributed.
3. To find the values of the terms.

1. Binomials.

The number of terms of the *nth* power of a binomial equals

$$N_t = 2^n$$

The distribution of terms of the *nth* power in any given order equals

$$D_t = 1 + 2 + 1$$

The coefficients determining the number of groups in their orders respectively are

$$V_c = 1, 2, 2^2, 2^3, \ldots 2^{n-2}$$

These coefficients increase as the orders of groups decrease. The group of the first order has the coefficient 2^{n-2}; the group of the *nth* order (the whole) has the coefficient 1.

Now we can write the complete expression for the distribution of the terms of the *nth* power of a binomial.

$$2^n = \begin{cases} = \dfrac{2^n}{2^2} + \dfrac{2^n}{2} + \dfrac{2^n}{2^2} \\[4pt] = 2\left(\dfrac{2^n}{2^3} + \dfrac{2^n}{2^2} + \dfrac{2^n}{2^3}\right) \\[4pt] = 2^2\left(\dfrac{2^n}{2^4} + \dfrac{2^n}{2^3} + \dfrac{2^n}{2^4}\right) \\[4pt] = 2^3\left(\dfrac{2^n}{2^5} + \dfrac{2^n}{2^4} + \dfrac{2^n}{2^5}\right) \\[4pt] = 2^{n-2}\left(\dfrac{2^n}{2^n} + \dfrac{2^n}{2^{n-1}} + \dfrac{2^n}{2^n}\right) = 2^{n-2}(1 + 2 + 1) \end{cases}$$

Let x and y be the terms of a binomial and a, b, c . . . the terms of the *nth* power. The major term, in the formula for distribution, doubles a single term or a group of the power polynomial at any given order.

Thus if n = 2

$$(x + y)^2 = a + \overline{b + b} + a$$

Here a and b are distributed in the progression $1 + 2 + 1$, where b taken twice represents the major term; $2^{n-2} = 1$ and $\dfrac{2^2}{2^2} + \dfrac{2^2}{2} + \dfrac{2^2}{2^4} = 1 + 2 + 1$

DISTRIBUTIVE INVOLUTION

Likewise if n = 3

$$(x + y)^3 = a + \overline{b + b} + c + \underline{\overline{b + c} + c} + d$$

$2^3 = \dfrac{2^3}{2^2} + \dfrac{2^3}{2} + \dfrac{2^3}{2^2} = 2 + 4 + 2$, where the major term is a binomial b + c taken twice.

$2^3 = 2\left(\dfrac{2^3}{2^3} + \dfrac{2^3}{2^2} + \dfrac{2^3}{2^3}\right) = (1 + 2 + 1) + (1 + 2 + 1)$, representing the groups of the first order.

If n = 4, then

$$(x + y)^4 = a + \overline{b + b} + c + \overline{b + c + c + d} + \overline{b + c + c + d} + \overline{c + d + d} + e$$

$2^4 = \dfrac{2^4}{2^2} + \dfrac{2^4}{2} + \dfrac{2^4}{2^2} = 4 + 8 + 4$

$2^4 = 2\left(\dfrac{2^4}{2^3} + \dfrac{2^4}{2^2} + \dfrac{2^4}{2^3}\right) = (2 + 4 + 2) + (2 + 4 + 2)$

$2^4 = 2^2\left(\dfrac{2^4}{2^4} + \dfrac{2^4}{2^3} + \dfrac{2^4}{2^4}\right) = (1 + 2 + 1) + (1 + 2 + 1) + (1 + 2 + 1)$

If n = 5, then

$$(x + y)^5 = a + \overline{b + b} + c + \overline{b + c + c + d} + \overline{b + c + c + d + c + d + d} +$$

$$+ e + \overline{b + c + c + d + c + d + d + e} + \overline{c + d + d + e} + \overline{d + e} + e + f$$

$= \dfrac{2^5}{2^2} + \dfrac{2^5}{2} + \dfrac{2^5}{2^2} = 8 + 16 + 8$

$= 2\left(\dfrac{2^5}{2^3} + \dfrac{2^5}{2^2} + \dfrac{2^5}{2^3}\right) = (4 + 8 + 4) + (4 + 8 + 4)$

$2^5 = 32$

$= 2^2\left(\dfrac{2^5}{2^4} + \dfrac{2^5}{2^3} + \dfrac{2^5}{2^4}\right) = (2+4+2) + (2+4+2) + (2+4+2) + (2+4+2)$

$= 2^3\left(\dfrac{2^5}{2^5} + \dfrac{2^5}{2^4} + \dfrac{2^5}{2^5}\right) = (1+2+1) + (1+2+1) + (1+2+1) +$

$+ (1+2+1) + (1+2+1) + (1+2+1) + (1+2+1) + (1+2+1)$

THEORY OF REGULARITY AND COORDINATION

When the distribution of the terms at the *nth* power has been found, their value can be obtained directly from the following formula.

$$(x + y)^n = x^n + x^{n-1}y + x^{n-2}y^2 + x^{n-3}y^3 + \ldots$$
$$\ldots + x^3y^{n-3} + x^2y^{n-2} + xy^{n-1} + y^n.$$

Here x^n, $x^{n-1}y$, etc. correspond to the terms of distribution a, b, c, . . .

$$a = x^n;\ b = x^{n-1}y;\ c = x^{n-2}y^2;\ \ldots\ldots$$

Examples:

$(x + y)^2 = a + b + b + c$

$\quad a = x^2;\ b = xy;\ c = y^2$

$(x + y)^2 = x^2 + xy + xy + y^2$

$(x + y)^3 = a + b + b + c + b + c + d$

$\quad a = x^3;\ b = x^2y;\ c = xy^2;\ d = y^3$

$(x + y)^3 = x^3 + x^2y + x^2y + xy^2 + x^2y + x^2y + xy^2 + xy^2 + y^3$

$(x + y)^4 = a + b + b + c + b + c + c + d + b + c + c + d + c +$
$\qquad + d + d + e$

$\quad a = x^4;\ b = x^3y;\ c = x^2y^2;\ d = xy^3;\ e = y^4$

$(x + y)^4 = x^4 + x^3y + x^3y + x^2y^2 + x^3y + x^2y^2 + x^2y^2 + xy^3 +$
$\qquad + x^3y + x^2y^2 + x^2y^2 + xy^3 + x^2y^2 + xy^3 + xy^3 + y^4.$

$x = 3;\ y = 2;\ n = 5$

$(3 + 2)^5 = a + b + b + c + b + c + c + d + b + c + c + d + c + d +$
$\qquad + d + e + b + c + c + d + c + d + d + e + c + d + d +$
$\qquad + e + d + e + e + f$

$\quad a = 243;\ b = 162;\ c = 108;\ d = 72;\ e = 48;\ f = 32$

$(3 + 2)^5 = 243 + 162 + 162 + 108 + 162 + 108 + 108 + 72 + 162 + 108 +$
$\qquad + 108 + 72 + 108 + 72 + 72 + 48 + 162 + 108 + 108 + 72 +$
$\qquad + 108 + 72 + 72 + 48 + 108 + 72 + 72 + 48 + 72 + 48 + 48 + 32.$

DISTRIBUTIVE INVOLUTION

The fractional form of a power polynomial can be obtained from the same formula by using values for the numerators and the sum of the binomial at its *nth* power as the denominator.

$$\left(\frac{x}{x+y} + \frac{y}{x+y}\right)^n = \frac{x^n}{(x+y)^n} + \frac{x^{n-1}y}{(x+y)^n} + \frac{x^{n-2}y^2}{(x+y)^n} + \frac{x^{n-3}y^3}{(x+y)^n} + \cdots$$

$$\cdots + \frac{x^3 y^{n-3}}{(x+y)^n} + \frac{x^2 y^{n-2}}{(x+y)^n} + \frac{xy^{n-1}}{(x+y)^n} + \frac{y^n}{(x+y)^n}$$

Example:

$$\left(\frac{3}{3+2} + \frac{2}{3+2}\right)^5 = \left(\frac{3}{5} + \frac{2}{5}\right)^5 = \frac{243}{3125} + \frac{162}{3125} + \frac{162}{3125} + \frac{108}{3125} + \frac{162}{3125} +$$

$$+ \frac{108}{3125} + \frac{108}{3125} + \frac{72}{3125} + \frac{162}{3125} + \frac{108}{3125} + \frac{108}{3125} + \frac{72}{3125} + \frac{108}{3125} + \frac{72}{3125} +$$

$$+ \frac{72}{3125} + \frac{48}{3125} + \frac{162}{3125} + \frac{108}{3125} + \frac{108}{3125} + \frac{72}{3125} + \frac{108}{3125} + \frac{72}{3125} + \frac{72}{3125} +$$

$$+ \frac{48}{3125} + \frac{108}{3125} + \frac{72}{3125} + \frac{72}{3125} + \frac{48}{3125} + \frac{72}{3125} + \frac{48}{3125} + \frac{48}{3125} + \frac{32}{3125}$$

> *The discovery of new laws of unstable equilibrium is the end that will be attained by the innovator who reveals new paths of art."*
>
> From The Biological Bases of the Evolution of Music, by Ivan Kryzhanovsky.

CHAPTER 6

BALANCE, UNSTABLE EQUILIBRIUM AND CRYSTALLIZATION OF EVENT

A STRUCTURE is in a state of stable (stationary, static) equilibrium when the ratio of its two major components equals one:

$$R = 1$$

The density of such a structure becomes its potential, bearing the tendency of unstabilization. The density set of a ratio can be expressed as a fractional form of unity:

$$R = \frac{n}{n}$$

The greater the value of n, the denser the set. And the denser the set, the greater the stability of its potential.

To illustrate, let us take $R = \frac{2}{2}$. In this case stable equilibrium of the two major components can be expressed as $\frac{1}{2} + \frac{1}{2} = SE$. The potential of unstabilization in this case equals $\frac{1}{2}$, *i.e.*, the fractional determinant of the $\frac{2}{2}$ series. As each of the components in this particular case has the value of its potential (*i.e.*, the value of the component equals that of the potential), or $\frac{1}{2} = \frac{1}{2}$, the potential may be regarded as one that is insufficient to change the state of stable equilibrium.

To increase the power of the potential, we must increase the density of the set. This can be accomplished by means of involution. By squaring the general determinant of a series, we may produce a denser set and a more powerful potential (the fractional determinant).

In the case of $\frac{2}{2}$ series, where $\frac{2}{2}$ is the general determinant and $\frac{1}{2}$ its fractional determinant, (the latter being also its potential of unstabilization), the squaring of its general determinant yields:

$$\left(\frac{2}{2}\right)^2 = \frac{4}{4} = \frac{2}{4} + \frac{2}{4}.$$

The potential of unstabilization becomes $\frac{1}{4}$, and unstable equilibrium acquires the following form:

$$UE'_{1^\circ} = \frac{2+1}{4} + \frac{2-1}{4} = \frac{3}{4} + \frac{1}{4} \text{ and}$$

$$UE''_{1^\circ} = \frac{2-1}{4} + \frac{2+1}{4} = \frac{1}{4} + \frac{3}{4}.$$

E'_{1° is achieved by adding the potential to the first term and subtracting it from the second (*i.e.*, by shifting the potential in the negative direction of its own value), and E''_{1°—by reversal of the first operation. The resulting unstable equilibrium fluctuates between E' and E'':

$$R_{1^\circ} (E'_{1^\circ} \longleftrightarrow E''_{1^\circ}) = \left(\frac{3}{4} + \frac{1}{4}\right) t_1 + \left(\frac{1}{4} + \frac{3}{4}\right) t_2.$$

Still further refinement of the form of unstable equilibrium can be accomplished by greater condensation of the original set $\frac{2}{2}$. Third power involution, *i.e.*, *cubing of the general determinant*, provides such a means. In our case, unstable equilibrium of the second order assumes the following form:

$$SE = \left(\frac{2}{2}\right)^3 = \frac{8}{8} = \frac{4}{8} + \frac{4}{8}.$$ As the potential of stabilization is the fractional determinant, it equals $\frac{1}{8}$. Then:

$$UE'_{2^\circ} = \frac{4+1}{8} + \frac{4-1}{8} = \frac{5}{8} + \frac{3}{8} \text{ and}$$

$$UE''_{2^\circ} = \frac{4-1}{8} + \frac{4+1}{8} = \frac{3}{8} + \frac{5}{8} \text{ and}$$

$$R_{2^\circ} = (E'_{2^\circ} \longleftrightarrow E''_{2^\circ}) = \left(\frac{5}{8} + \frac{3}{8}\right) + \left(\frac{3}{8} + \frac{5}{8}\right)$$

This procedure can be extended ad infinitum. A set can be made as dense as desired, and the potential of stabilization—as powerful as desired. The set (general determinant of the series) becomes saturated quickly and reaches a limit beyond which instability becomes imperceptible. The general form of the resulting unstable equilibrium appears as follows:

$$R_{n^\circ} = (E'_{n^\circ} \longleftrightarrow E''_{n^\circ}) = \left(\frac{n+1}{2} + \frac{n-1}{2}\right) + \left(\frac{n-1}{2} + \frac{n+1}{2}\right).$$

At the average velocity of the last 100,000 years, the chronological life-span of the power-index must be counted in the tens of thousands of years. For example, the $\frac{8}{8}$ series, as it manifests itself in the form of unstable equilibrium of the duration-groups in music, is at present at least 30,000 years old. Unstable equilibrium of the $\frac{3}{3}$ series, as it manifests itself in the field of graphic symbols, as well as duration-groups in music, can be traced several millennia back: to the "mogen dovid" of the Hebrews, and the triadic measure, "Divine Trinity" symbol of early Catholic liturgical music.

We can look upon the involution, and in some cases factoring, technique as a hidden mechanism behind the morphological evolution of the forms of unstable equilibrium. As in the field of organic phenomena, this evolution is slow in pace. Once an adaptable form has been developed, it acquires a better chance for long survival, like the "fittest" of bio-zoological forms.

An event may be considered crystallized when it reaches its optimum during the period of observation. Crystallization means that the organizational tendency has acquired its maximum of realization. Crystallized events are rational; they are a part of the continuum of eventuality and produce a series of isolated terms between which are all the events undergoing the process of crystallization. The latter are irrational and may be considered as "eventual" states of the process of crystallization. The irrational polygon forms that gravitate between the form of a triangle and the form of a square may be looked upon as a "would be" triangle (eventual triangle), or a "would be" square (eventual square), depending on the form of tendency of the eventuality (in this case, variation of the angle value).

An event spends itself during the period of crystallization. An identical event can be duplicated only if all specifications are scientifically integrated. The record of an event, however, can duplicate the event to the degree of precision that physical conditions permit. Up to the present, scientific specifications cannot be worked out for the "fatal" event. These can be reproduced only from the record, which is either physical (optical, acoustical) or psychological (mnemonic).

A. FORMULAE

$$1 \quad \text{unity}$$

$$1 = \frac{t}{t} \quad \text{fractional equivalent of unity}$$

$$SE = \frac{\frac{1}{2}t}{t} + \frac{\frac{1}{2}t}{t} \quad \text{stable equilibrium}$$

$$\frac{u}{t} \quad \text{unstabilizer}$$

$$UE = \frac{\frac{1}{2}t + u}{t} + \frac{\frac{1}{2}t - u}{t} \quad \text{unstable equilibrium}$$

$$R = \frac{\frac{1}{2}t + u}{\frac{1}{2}t - u} \neq 1 \quad \text{ratio of UE}$$

BALANCE, UNSTABLE EQUILIBRIUM ...

Summation:

$$\frac{t-2u}{2t}, \frac{t+2u}{2t}, \frac{t}{t}, \frac{3t+2u}{2t}, \frac{5t+2u}{2t}, \frac{8t+4u}{2t}, \frac{13t+6u}{2t},$$

$$\frac{21t+10u}{2t}, \frac{34t+16u}{2t}, \frac{55t+26u}{2t}, \frac{89t+42u}{2t}, \frac{144t+68u}{2t}, \ldots$$

SE—stable equilibrium

UE—unstable equilibrium

$$\frac{2}{2} \text{ SE} = \frac{1}{2} + \frac{1}{2}$$

No UE

$$\left(\frac{2}{2}\right)^2 = \frac{4}{4}; \frac{2}{4} + \frac{2}{4} = \text{SE}$$

$$\text{UE} = \frac{3}{4} + \frac{1}{4}$$

$$\left(\frac{2}{2}\right)^3 = \frac{8}{8}; \quad \text{SE} = \frac{4}{8} + \frac{4}{8}$$

$$\text{UE} = \frac{5}{8} + \frac{3}{8}$$

$$\left(\frac{2}{2}\right)^4 = \frac{16}{16}; \quad \text{SE} = \frac{8}{16} + \frac{8}{16}$$

$$\text{UE} = \frac{9}{16} + \frac{7}{16}$$

$$\left(\frac{2}{2}\right)^5 = \frac{32}{32}; \quad \text{SE} = \frac{16}{32} + \frac{16}{32}$$

$$\text{UE} = \frac{17}{32} + \frac{15}{32}$$

$$\left(\frac{2}{2}\right)^6 = \frac{64}{64}; \quad \text{SE} = \frac{32}{64} + \frac{32}{64}$$

$$\text{UE} = \frac{33}{64} + \frac{31}{64}$$

$\dfrac{3}{3}$ No SE

$$UE = \dfrac{2}{3} + \dfrac{1}{3}$$

$\left(\dfrac{3}{3}\right)^2 = \dfrac{9}{9};$ No SE

$$UE = \dfrac{5}{9} + \dfrac{4}{9}$$

$\left(\dfrac{3}{3}\right)^3 = \dfrac{27}{27}$ No SE

$$UE = \dfrac{14}{27} + \dfrac{13}{27}$$

$\left(\dfrac{3}{3}\right)^4 = \dfrac{81}{81}$ No SE

$$UE = \dfrac{41}{81} + \dfrac{40}{81}$$

$\dfrac{5}{5};$ No SE

$$UE = \dfrac{3}{5} + \dfrac{2}{5}$$

$\left(\dfrac{5}{5}\right)^2 = \dfrac{25}{25};$ No SE

$$UE = \dfrac{13}{25} + \dfrac{12}{25}$$

$\left(\dfrac{5}{5}\right)^3 = \dfrac{125}{125}$ No SE

$$UE = \dfrac{63}{125} + \dfrac{62}{125}$$

$\dfrac{7}{7};$ No SE

$$UE = \dfrac{4}{7} + \dfrac{3}{7}$$

$\left(\dfrac{7}{7}\right)^2 = \dfrac{49}{49};$ No SE

$$UE = \dfrac{25}{49} + \dfrac{24}{49}$$

$\frac{10}{10}$; $\quad$ SE $= \frac{5}{10} + \frac{5}{10}$

$\quad\quad\quad\quad$ UE $= \frac{7}{10} + \frac{3}{10}$

$\frac{11}{11}$; $\quad$ No SE

$\quad\quad\quad\quad$ UE $= \frac{6}{11} + \frac{5}{11}$

$\frac{12}{12}$; $\quad$ SE $= \frac{6}{12} + \frac{6}{12}$

$\quad\quad\quad\quad$ UE $= \frac{7}{12} + \frac{5}{12}$

B. Unstable Equilibrium in Factorial Composition of Duration-Groups

First Form: in composing the resultants of interference of two synchronized biners, select two identical major generators (a) and two non-identical minor generators (b and b') which are the successive terms of one summation-series, and the sum of which equals the value of major generator (a); the minor generator (b) of the first biner is greater than the minor generator (b') of the second biner, or vice-versa.

Formula:

$E' = r_{a \div b} + r_{a \div b'}$ and/or

$E'' = r_{a \div b'} + r_{a \div b}$, where $b + b' = a$

Examples:

$\frac{3}{3}$ series: $E = r_{3 \div 2} + r_{3 \div 1} = 3T$

$\frac{4}{4}$ series: $E = r_{4 \div 3} + r_{4 \div 1} = 4T$

$\frac{5}{5}$ series: $E = r_{5 \div 3} + r_{5 \div 2} = 5T$

$\frac{6}{6}$ series: $E = r_{6 \div 5} + r_{6 \div 1} = 6T$

$\frac{7}{7}$ series: $E = r_{7 \div 4} + r_{7 \div 3} = 7T$

$\frac{8}{8}$ series: $E = r_{8 \div 5} + r_{8 \div 3} = 8T$

$\frac{9}{9}$ series: $E = r_{9 \div 5} + r_{9 \div 4} = 9T$

$\frac{10}{10}$ series: $E = r_{10 \div 7} + r_{10 \div 3} = 10T$

$\frac{11}{11}$ series: $E = r_{11 \div 6} + r_{11 \div 5} = 11T$

$\frac{12}{12}$ series: $E = r_{12 \div 7} + r_{12 \div 5} = 12T$

$\frac{13}{13}$ series: $E = r_{13 \div 7} + r_{13 \div 6} = 13T$

$\frac{14}{14}$ series: $E = r_{14 \div 9} + r_{14 \div 5} = 14T$

$\frac{15}{15}$ series: $E = r_{15 \div 8} + r_{15 \div 7} = 15T$

$\frac{16}{16}$ series: $E = r_{16 \div 9} + r_{16 \div 7} = 16T$

$\frac{17}{17}$ series: $E = r_{17 \div 9} + r_{17 \div 8} = 17T$

$\frac{18}{18}$ series: $E = r_{18 \div 11} + r_{18 \div 7} = 18T$

$\frac{19}{19}$ series: $E = r_{19 \div 10} + r_{19 \div 9} = 19T$

$\frac{20}{20}$ series: $E = r_{20 \div 11} + r_{20 \div 9} = 20T$

Second Form: in composing the resultants of interference of two synchronized biners, select two non-identical major generators (a and a′) which are the successive terms of one summation-series, and whose sum equals the value of the determinant of the series (and, hence, the value of T); then select the minor generators (b), which are identical and whose value equals the difference between a and a′; the value of a is greater than that of a′, or vice-versa.

BALANCE, UNSTABLE EQUILIBRIUM ...

Formula:

$$E' = r_{a \div b} + r_{a' \div b} \text{ and/or}$$

$$E'' = r_{a' \div b} + r_{a \div b}, \text{ where } a' - a = b$$

The value of coefficients of $r_{a' \div b}$ and $r_{a \div b}$ equals the value of T divided by 2; each half becomes the coefficient of recurrence of either r; when T is an odd number, the value of the first coefficient is $\frac{T+1}{2}$, and the value of the second $\frac{T-1}{2}$, or vice versa.

Complete formula for $T = 2n$:

$$E' = \frac{T}{2} r_{a \div b} + \frac{T}{2} r_{a' \div b} \text{ and}$$

$$E'' = \frac{T}{2} r_{a' \div b} + \frac{T}{2} r_{a \div b}$$

Complete formula for $T = 2n + 1$:

$$E' = \frac{T+1}{2} r_{a \div b} + \frac{T-1}{2} r_{a' \div b} \text{ and}$$

$$E'' = \frac{T+1}{2} r_{a' \div b} + \frac{T+1}{2} r_{a \div b} \text{ and}$$

$$E''' = \frac{T-1}{2} r_{a \div b} + \frac{T+1}{2} r_{a' \div b} \text{ and}$$

$$E'''' = \frac{T-1}{2} r_{a' \div b} + \frac{T+1}{2} r_{a \div b}$$

Examples:

$\frac{8}{8}$ series: $E = 4r_{5 \div 2} + 4r_{3 \div 2} = 5T + 3T$

$\frac{9}{9}$ series: $E = 5r_{5 \div 3} + 4r_{4 \div 3} = 5T + 4T$

$\frac{10}{10}$ series: $E = 5r_{7 \div 2} + 5r_{3 \div 2} = 7T + 3T$

$\frac{12}{12}$ series: $E = 6r_{7 \div 2} + 6r_{5 \div 2} = 7T + 5T$

$\dfrac{13}{13}$ series: $E = 7r_{7 \div 5} + 6r_{6 \div 5} = 7T + 6T$

$\dfrac{14}{14}$ series: $E = 7r_{9 \div 4} + 7r_{5 \div 4} = 9T + 5T$

$\dfrac{15}{15}$ series: $E = 8r_{8 \div 3} + 7r_{7 \div 3} = 8T + 7T$

$\dfrac{15}{15}$ series: $E = 8r_{8 \div 5} + 7r_{7 \div 5} = 8T + 7T$

$\dfrac{16}{16}$ series: $E = 8r_{9 \div 2} + 8r_{7 \div 2} = 9T + 7T$

$\dfrac{16}{16}$ series: $E = 8r_{9 \div 4} + 8r_{7 \div 4} = 9T + 7T$

$\dfrac{16}{16}$ series: $E = 8r_{9 \div 5} + 8r_{7 \div 5} = 9T + 7T$

CHAPTER 7

RATIO AND RATIONALIZATION

A. Ratio. Rational. Relation. Relational

RATIONAL behavior—behavior according to a ratio. *Rational composition*—composition based on a ratio. *Rational thinking*—thinking in terms of ratios. When the ratio is established, involution (power-differentiation) takes its course. Cutting a portion of space by simple (monomial) or complex (polynomial) periodic motion establishes an area. Thus, enclosing an unbounded space in a rational boundary *ipso facto* introduces regulations that are the inherent laws within the boundary. The act of limiting converts potentiality into a tendency (intent).

This process can be defined as *ratio-realization of space*. Viewed in this way, the inscribing of a structure (trajectory) in a boundary, which structure was originally evolved in an unbounded space, corresponds (is equivalent) to rationalization of structure.

These are the two fundamental procedures, mutually compensating each other. In terms of logic, one corresponds to relativity and the other to quanta, as the first works from an enclosed (bounded) all-inclusive whole, while the second operates on a unit (st). The combination of the two bases of departure establishes a unified system of interacting tendencies of the unit and the whole.

B. Rationalization of the Second Order

(Rationalization of a ratio).

Introduction of a new or of an identical ratio rationalizes a given ratio, thus introducing symmetry into a given ratio. If a given ratio is 2, and its extensional equivalent is 2, then the introduction of a 2:1 ratio gives $\frac{2}{3} + \frac{1}{3}$; likewise, $1 \div 2$ ratio yields $\frac{1}{3} + \frac{2}{3}$. The combination or interference of the two yields $\frac{1}{3}$ as a unit of symmetric distribution.

To demonstrate that there is symmetric correspondence between the extension and the ratio, let us take a frequency $2 \div 1$ ratio. In a ratio of two frequencies, the extraction of the root (evolution) corresponds to division as applied to quantitative extension. Then, taking the ratio 2, which is the equivalent of $2 \div 1$, we extract $\sqrt[3]{2}$. This makes one third of the ratio value, as $\sqrt[3]{2^3} = 2$. Therefore $\sqrt[3]{2}$ is the equivalent of one third in an extension. Or to put it mathematically: $\log_{\sqrt[3]{2}} \sqrt[3]{2} = 1$, $\log_{\sqrt[3]{2}} \sqrt[3]{2^2} = 2$, $\log_{\sqrt[3]{2}} \sqrt[3]{2^3} = 3$.

Further evolution is achieved by squaring the exponent of the radical. Thus, to establish a new $2 \div 1$ ratio within each of the $\sqrt[3]{2}$, it is necessary to extract $\sqrt[9]{2}$. Then, $\sqrt[9]{2^2}$ gives the equivalent of $\frac{2}{9}$ of the extensional value, *i.e.*, $\sqrt[9]{2^2} + \sqrt[9]{2} \equiv \frac{2}{9} + \frac{1}{9}$.

As an illustration, we can take the 2÷1 ratio of sound frequencies. This forms an octave. By extracting the $\sqrt[3]{2}$ and $\sqrt[3]{2^2}$, we obtain the following three symmetric intervals between the frequencies:

(1) from 1 to $\sqrt[3]{2}$

(2) from $\sqrt[3]{2}$ to $\sqrt[3]{2^2}$

(3) from $\sqrt[3]{2^2}$ to 2

Assuming 1 to be 256 cycles, we obtain the following values for the symmetric points:

(1) $\sqrt[3]{512}$

(2) $\sqrt[3]{512^2}$

(3) $\sqrt[3]{512^3} = 512$

The musical names of these points are: c, e, g♯, c'.

The limit frequency between 1 and $\sqrt[3]{2}$ in the temperament of $\sqrt[12]{2}$ is $\frac{\sqrt[12]{2^3}}{\sqrt[3]{2}} = \sqrt[4]{2}$. Thus $\sqrt[4]{2}$ becomes the limit of frequency units in the symmetry $\sqrt[3]{2}$. As $\sqrt[4]{2} = 3\sqrt[12]{2}$, we can establish a binomial ratio within this limit. In our original reasoning, we established the symmetry of three intervals to the octave as the $\frac{3}{3}$ series, in which we have developed the 2÷1 ratio. Now we assume the coefficient 3 of the $3\sqrt[12]{2}$ to belong to $\frac{3}{3}$ series as well. Thus we establish a binomial symmetry of 2÷1 within $\sqrt[12]{2}$. It appears in the following form: $2\sqrt[12]{2} + \sqrt[12]{2}$, which in logarithms to the base $\sqrt[12]{2}$ becomes 2 + 1. As 1 is the equivalent of a musical semitone, we obtain the following pitch-scale, which carries out 2÷1 ratio from all three symmetric points:

c — d — e♭ — e♮ — f♯ — g — g♯ — a♯ — b — (c).

Similar ratios can be carried out within assigned ratios of wave-length for the projection of spectral colors. Color-scales may be constructed in any form of symmetry within the assigned range (ratio of wave-length) and from any symmetric points.

C. Rationalization of a Rectangle

The rationalization of a rectangle consists of the process of subjecting the rectangle to the tendency of its own ratio (*i.e.*, the ratio of two sides of the rectangle: a:b) within a new ratio. The new ratio must be one of the simplest. In the following discussion, the rectangle itself may have any ratio, but the new (assumed) ratio is 2, and will be used as a constant. The ratio of 2 represents an octave and is a selective operation of the most general kind. It is applicable to frequencies, wavelengths, etc.; now the application will be extended to a rectangle.

The rationalization of a rectangle has the purpose of extending the original ratio to half of the sum of two sides of the rectangle and to the remaining portion of the longer side. After such an operation has been applied to the two above-mentioned segments, each of the two segments becomes subdivided into two shorter segments which are related to each other as a:b. Thus we acquire four extensions on the longer side of the rectangle.

Assuming that the original rectangle has sides a and b, and assuming further that the sum of the segments of proportionate subdivision x_1, x_2, x_3 and x_4 equal a, we find that the above segments bear the following relation to each other:

$$\frac{a}{b} = \frac{x_1}{x_2} = \frac{x_3}{x_4}$$

By dropping perpendiculars from the points between x_1 and x_2, x_2 and x_3, and x_3 and x_4, we isolate four areas: A_1, A_2, A_3, A_4. These areas bear the same relation to one another as the segments x_1, x_2, x_3 and x_4:

$$\frac{A_1}{A_2} = \frac{A_3}{A_4} = \frac{a}{b}$$

According to our conditions, $x_1 + x_2 = \dfrac{a+b}{2}$.

Then, $x_3 + x_4 = (a+b) - \dfrac{a+b}{2} - b = \dfrac{2a+2b}{2} - \dfrac{a+b}{2} - \dfrac{2b}{2} = \dfrac{2a+2b-a-b-2b}{2} = \dfrac{a-b}{2}$. Thus, side a is subdivided into two extensions: $\dfrac{a+b}{2}$ and $\dfrac{a-b}{2}$.

Any area of four sides, that is, a rectangle, can be expressed as the following trajectory:

A4 = 2(sat 180° + 90° + sbt 180° + 90°). In other words, the boundary of a four-sided rectangular area equals two times the rectilinear extension whose period is at, turning under 90° into another rectilinear extension whose period is bt. Disregarding 180° as a general constant of rectilinear extension and 90° as a rectangular constant, we can express the boundary of any specified rectangle as:

A4 = 2(at + bt), where a and b are the periods producing a 90° angle, and therefore—the ratio of the two sides of A4.

Assuming the sum of two sides of a rectangle to be: A4 = a+b, we shall designate such a sum as the determinant of the $\frac{a+b}{a+b}$ series. For example: A4 = 2(2t + t) expresses the rectangle whose area is 2×1. This particular rectangle belongs to $\frac{3}{3}$ series, as its determinant is: 2 + 1 = 3.

1. Formulae

(a) Extensions of the Major Term.

$$x_1 = \frac{a+b}{2} \cdot \frac{a}{a+b} = \frac{a^2+ab}{2a+2b} = \frac{a(a+b)}{2(a+b)} = \frac{a}{2}$$

$$x_2 = \frac{a+b}{2} \cdot \frac{b}{a+b} = \frac{ab+b^2}{2a+2b} = \frac{b(a+b)}{2(a+b)} = \frac{b}{2}$$

$$x_3 = \frac{a-b}{2} \cdot \frac{a}{a+b} = \frac{a^2-ab}{2a+2b} = \frac{a(a-b)}{2(a+b)}$$

$$x_4 = \frac{a-b}{2} \cdot \frac{b}{a+b} = \frac{ab-b^2}{2a+2b} = \frac{b(a-b)}{2(a+b)}$$

(b) Ratios of the Extensions.

$$\frac{x_1}{x_2} = \frac{a^2+ab}{2a+2b} \div \frac{ab+b^2}{2a+2b} = \frac{a^2+ab}{ab+b^2} = \frac{a}{b}$$

$$\frac{x_3}{x_4} = \frac{a^2-ab}{2a+2b} \div \frac{ab-b^2}{2a+2b} = \frac{a^2-ab}{ab-b^2} = \frac{a(a-b)}{b(a-b)} = \frac{a}{b}$$

$$\frac{x_1}{x_3} = \frac{a^2+ab}{2a+2b} \div \frac{a^2-ab}{2a+2b} = \frac{a^2+ab}{a^2-ab} = \frac{a(a+b)}{a(a-b)} = \frac{a+b}{a-b}$$

$$\frac{x_2}{x_4} = \frac{ab+b^2}{2a+2b} \div \frac{ab-b^2}{2a+2b} = \frac{ab+b^2}{ab-b^2} = \frac{b(a+b)}{b(a-b)} = \frac{a+b}{a-b}$$

$$\frac{x_1}{x_2} = \frac{x_3}{x_4} = \frac{a}{b}; \quad \frac{x_1}{x_3} = \frac{x_2}{x_4} = \frac{a+b}{a-b}$$

RATIO AND RATIONALIZATION

(c) Minor Areas (Subareas) of a Rectangle.

$$A_1 = \frac{a^2 + ab}{2a + 2b} \cdot b = \frac{a^2b + ab^2}{2a + 2b} = \frac{ab(a + b)}{2(a + b)} = \frac{ab}{2}$$

$$A_2 = \frac{ab + b^2}{2a + 2b} \cdot b = \frac{ab^2 + b^3}{2a + 2b} = \frac{b^2(a + b)}{2(a + b)} = \frac{b^2}{2}$$

$$A_3 = \frac{a^2 - ab}{2a + 2b} \cdot b = \frac{a^2b - ab^2}{2a + 2b} = \frac{ab(a - b)}{2(a + b)}$$

$$A_4 = \frac{ab - b^2}{2a + 2b} \cdot b = \frac{ab^2 - b^3}{2a + 2b} = \frac{b^2(a - b)}{2(a + b)}$$

(d) Ratios of the Minor Areas of a Rectangle.

$$\frac{A_1}{A_2} = \frac{a^2b + ab^2}{2a + 2b} \div \frac{ab^2 + b^3}{2a + 2b} = \frac{a^2b + ab^2}{ab^2 + b^3} = \frac{ab(a + b)}{b^2(a + b)} = \frac{ab}{b^2} = \frac{a}{b}$$

$$\frac{A_3}{A_4} = \frac{a^2b - ab^2}{2a + 2b} \div \frac{ab^2 - b^3}{2a + 2b} = \frac{a^2b - ab^2}{ab^2 - b^3} = \frac{ab(a - b)}{b^2(a - b)} = \frac{ab}{b^2} = \frac{a}{b}$$

$$\frac{A_1}{A_3} = \frac{a^2b + ab^2}{2a + 2b} \div \frac{a^2b - ab^2}{2a + 2b} = \frac{a^2b + ab^2}{a^2b - ab^2} = \frac{ab(a + b)}{ab(a - b)} = \frac{a + b}{a - b}$$

$$\frac{A_2}{A_4} = \frac{ab^2 + b^3}{2a + 2b} \div \frac{ab^2 - b^3}{2a + 2b} = \frac{ab^2 + b^3}{ab^2 - b^3} = \frac{b^2(a + b)}{b^2(a - b)} = \frac{a + b}{a - b}$$

$$\frac{A_1}{A_2} = \frac{A_3}{A_4} = \frac{a}{b}; \quad \frac{A_1}{A_3} = \frac{A_2}{A_4} = \frac{a + b}{a - b}$$

(e) Major Areas (Subareas) of a Rectangle.

$$A_1 + A_2 = \frac{a^2b + ab^2}{2a + 2b} + \frac{ab^2 + b^3}{2a + 2b} = \frac{a^2b + 2ab^2 + b^3}{2a + 2b} = \frac{(ab + b^2)(a + b)}{2(a + b)} = \frac{ab + b^2}{2}$$

$$A_3 + A_4 = \frac{a^2b - ab^2}{2a + 2b} + \frac{a^2b - b^3}{2a + 2b} = \frac{a^2b - ab^2 + ab^2 - b^3}{2a + 2b} = \frac{a^2b - b^3}{2a + 2b} = \frac{b(a^2 - b^2)}{2(a + b)} =$$

$$= \frac{b(a + b)(a - b)}{2(a + b)} = \frac{b(a - b)}{2} = \frac{ab - b_2}{2}$$

Ratio of the Major Areas of a Rectangle.

$$\frac{A_1 + A_2}{A_3 + A_4} = \frac{ab + b^2}{2} \div \frac{ab - b^2}{2} = \frac{ab + b^2}{ab - b^2} = \frac{b(a + b)}{b(a - b)} = \frac{a + b}{a - b}$$

(f) Summary.

$$\frac{x_1}{x_2} = \frac{A_1}{A_2} = \frac{x_3}{x_4} = \frac{A_3}{A_4} = \frac{a}{b}$$

$$\frac{x_1}{x_3} = \frac{A_1}{A_3} = \frac{x_2}{x_4} = \frac{A_2}{A_4} = \frac{A_1 + A_2}{A_3 + A_4} = \frac{a+b}{a-b}$$

$$\frac{A_1}{A_3} = \frac{A_2}{A_4} = \frac{a+b}{a-b}$$

2. Application

2×1 rectangle

$$2 + 1 = 3 \quad \frac{3}{2} \quad \text{Ratio of two sides} = 2$$

$$\frac{2}{3} + \frac{1}{3} = \frac{3}{3}$$

$$\frac{3}{3 \cdot 2} = \frac{3}{6}$$

$$A_1 = \frac{3 \cdot 2}{6 \cdot 6} = \frac{6}{36} \;;\; A_3 = \frac{1 \cdot 2}{6 \cdot 6} = \frac{2}{36}$$

$$\frac{A_1}{A_3} = \frac{6}{36} \div \frac{2}{36} = \frac{6}{2} = \frac{3}{1}$$

Ratio of two sub-areas (A_1 and A_3) = 3

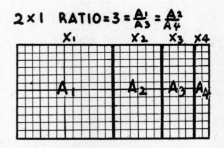

Figure 1. Rationalization of rectangle 2×1.

3×1 rectangle

$$\frac{3}{4} + \frac{1}{4} = \frac{4}{4}$$

$$\frac{4}{4 \cdot 2} = \frac{4}{8}$$

$$A_1 = \frac{4 \cdot 2}{8 \cdot 8} = \frac{8}{64}; \quad A_3 = \frac{2 \cdot 2}{8 \cdot 8} = \frac{4}{64}$$

$$\frac{A_1}{A_3} = \frac{8}{64} \div \frac{4}{64} = \frac{8 \cdot 64}{4 \cdot 64} = 2$$

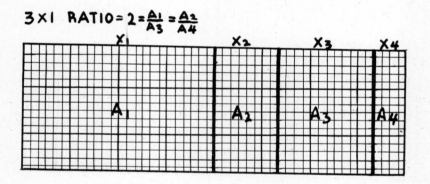

Figure 2. *Rationalization of rectangle* 3×1.

4×1 rectangle

$$\frac{4}{5} + \frac{1}{5} = \frac{5}{5}$$

$$\frac{5}{5 \cdot 2} = \frac{5}{10}$$

$$A_1 = \frac{5 \cdot 2}{10 \cdot 10} = \frac{10}{100}; \quad A_3 = \frac{3 \cdot 2}{10 \cdot 10} = \frac{6}{100}$$

$$\frac{A_1}{A_3} = \frac{10}{100} \div \frac{6}{100} = \frac{10 \cdot 100}{6 \cdot 100} = \frac{5}{3}$$

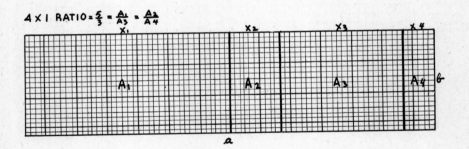

Figure 3. Rationalization of rectangle 4×1.

RATIO AND RATIONALIZATION

5 × 1 rectangle

$$\frac{5}{6} + \frac{1}{6} = \frac{6}{6}$$

$$\frac{6}{6 \cdot 2} = \frac{6}{12}$$

$$A_1 = \frac{6 \cdot 2}{12 \cdot 12} = \frac{12}{144}; \quad A_3 = \frac{4 \cdot 2}{12 \cdot 12} = \frac{8}{144}$$

$$\frac{A_1}{A_3} = \frac{12}{144} \div \frac{8}{144} = \frac{12 \cdot 144}{8 \cdot 144} = \frac{3}{2}$$

6 × 1 rectangle

$$\frac{6}{7} + \frac{1}{7} = \frac{7}{7}$$

$$\frac{7}{7 \cdot 2} = \frac{7}{14}$$

$$A_1 = \frac{7 \cdot 2}{14 \cdot 14} = \frac{14}{196}; \quad A_3 = \frac{2 \cdot 5}{14 \cdot 14} = \frac{10}{196}$$

$$\frac{A_1}{A_3} = \frac{14}{196} \div \frac{10}{196} = \frac{14 \cdot 196}{10 \cdot 196} = \frac{7}{5}$$

7 × 1 rectangle

$$\frac{7}{8} + \frac{1}{8} = \frac{8}{8}$$

$$\frac{8}{8 \cdot 2} = \frac{8}{16}$$

$$A_1 = \frac{8 \cdot 2}{16 \cdot 16} = \frac{16}{256}; \quad A_3 = \frac{2 \cdot 6}{16 \cdot 16} = \frac{12}{256}$$

$$\frac{A_1}{A_3} = \frac{16}{256} \div \frac{12}{256} = \frac{16 \cdot 256}{12 \cdot 256} = \frac{4}{3}$$

3 × 2 rectangle

$$\frac{3}{5} + \frac{2}{5} = \frac{5}{5}$$

$$\frac{5}{5 \cdot 2} = \frac{5}{10}$$

$$A_1 = \frac{5 \cdot 4}{10 \cdot 10} = \frac{20}{100}; \quad A_3 = \frac{1 \cdot 4}{10 \cdot 10} = \frac{4}{100}$$

$$\frac{A_1}{A_3} = \frac{20}{100} \div \frac{4}{100} = \frac{20 \cdot 100}{4 \cdot 100} = 5$$

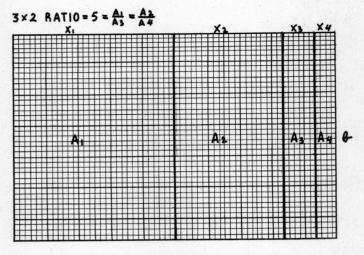

Figure 4. Rationalization of rectangle 3 × 2.

5 × 2 rectangle

$$\frac{5}{7} + \frac{2}{7} = \frac{7}{7}$$

$$\frac{7}{7 \cdot 2} = \frac{7}{14}$$

$$A_1 = \frac{7 \cdot 4}{14 \cdot 14} = \frac{28}{196}; \; A_3 = \frac{3 \cdot 4}{14 \cdot 14} = \frac{12}{196}$$

$$\frac{A_1}{A_3} = \frac{28}{196} \div \frac{12}{196} = \frac{28 \cdot 196}{12 \cdot 196} = \frac{7}{3}$$

7 × 3 rectangle

$$\frac{7}{10} + \frac{3}{10} = \frac{10}{10}$$

$$\frac{10}{10 \cdot 2} = \frac{10}{20}$$

$$A_1 = \frac{10 \cdot 6}{20 \cdot 20} = \frac{60}{400}; \; \frac{6 \cdot 4}{20 \cdot 20} = \frac{24}{400}$$

$$\frac{A_1}{A_3} = \frac{60}{24} = \frac{5}{2}$$

4 × 3 rectangle

$$\frac{4}{7} + \frac{3}{7} = \frac{7}{7}$$

$$\frac{7}{7 \cdot 2} = \frac{7}{14}$$

$$A_1 = \frac{7 \cdot 6}{14 \cdot 14} = \frac{142}{196}; \; \frac{7}{14} - \frac{6}{14} = \frac{1}{14}$$

$$A_3 = \frac{1 \cdot 6}{14 \cdot 14} = \frac{6}{196}; \; \frac{A_1}{A_3} = \frac{42}{6} = 7$$

5 × 3 rectangle

$$\frac{5}{8} + \frac{3}{8} = \frac{8}{8}$$

$$\frac{8}{8 \cdot 2} = \frac{8}{16}$$

$$A_1 = \frac{8 \cdot 6}{16 \cdot 16} = \frac{48}{256}; \quad \frac{8}{16} - \frac{6}{16} = \frac{2}{16}$$

$$A_3 = \frac{2 \cdot 6}{16 \cdot 16} = \frac{12}{256}; \quad \frac{A_1}{A_3} = \frac{48}{12} = 4$$

5 × 4 rectangle

$$\frac{5}{9} + \frac{4}{9} = \frac{9}{9}$$

$$\frac{9}{9 \cdot 2} = \frac{9}{18}$$

$$A_1 = \frac{9 \cdot 8}{18 \cdot 18} = \frac{72}{324}; \quad \frac{9}{18} - \frac{8}{18} = \frac{1}{18}$$

$$A_3 = \frac{1 \cdot 8}{18 \cdot 18} = \frac{8}{324}; \quad \frac{A_1}{A_3} = \frac{72 \cdot 324}{324 \cdot 8} = 9$$

6 × 5 rectangle

$$\frac{6}{11} + \frac{5}{11} = \frac{11}{11}$$

$$\frac{11}{11 \cdot 2} = \frac{11}{22}$$

$$A_1 = \frac{11 \cdot 10}{22 \cdot 22} = \frac{110}{484}; \quad \frac{11}{22} - \frac{10}{22} = \frac{1}{22}$$

$$A_3 = \frac{1 \cdot 10}{22 \cdot 22} = \frac{10}{484}; \quad \frac{A_1}{A_3} = \frac{110 \cdot 484}{484 \cdot 10} = 11$$

RATIO AND RATIONALIZATION

7 × 2 rectangle

$$\frac{7}{9} + \frac{2}{9} = \frac{9}{9}$$

$$\frac{9}{9\cdot 2} = \frac{9}{18}$$

$$A_1 = \frac{9\cdot 4}{18\cdot 18} = \frac{36}{324}; \quad \frac{9}{18} - \frac{4}{18} = \frac{5}{18}$$

$$A_3 = \frac{5\cdot 4}{18\cdot 18} = \frac{20}{324}; \quad \frac{A_1}{A_3} = \frac{36}{20} = \frac{9}{5}$$

7 × 3 rectangle

$$\frac{7}{10} + \frac{3}{10} = \frac{10}{10}$$

$$\frac{10}{10\cdot 2} = \frac{10}{20}$$

$$A_1 = \frac{10\cdot 6}{20\cdot 20} = \frac{60}{400}; \quad \frac{10}{20} - \frac{6}{20} = \frac{4}{20}$$

$$A_3 = \frac{4\cdot 6}{20\cdot 20} = \frac{24}{400}; \quad \frac{A_1}{A_3} = \frac{60}{24} = \frac{10}{4} = \frac{5}{2}$$

7 × 5 rectangle

$$\frac{7}{12} + \frac{5}{12} = \frac{12}{12}$$

$$\frac{12}{12\cdot 2} = \frac{12}{24}$$

$$A_1 = \frac{12\cdot 10}{24\cdot 24} = \frac{120}{576}; \quad \frac{12}{24} - \frac{10}{24} = \frac{2}{24}$$

$$A_3 = \frac{2\cdot 10}{24\cdot 24} = \frac{20}{576}; \quad \frac{A_1}{A_3} = \frac{120}{20} = 6$$

7 × 6 rectangle

$$\frac{7}{13} + \frac{6}{13} = \frac{13}{13}$$

$$\frac{13}{13\cdot 2} = \frac{13}{26}$$

$$A_1 = \frac{13\cdot 12}{26\cdot 26} = \frac{156}{676}; \frac{13}{26} - \frac{12}{26} = \frac{1}{26}$$

$$A_3 = \frac{1\cdot 12}{26\cdot 26} = \frac{12}{26}; \frac{A_1}{A_3} = \frac{156}{12} = 13$$

8 × 3 rectangle

$$\frac{8}{11} + \frac{3}{11} = \frac{11}{11}$$

$$\frac{11}{11\cdot 2} = \frac{11}{22}$$

$$A_1 = \frac{11\cdot 6}{22\cdot 22} = \frac{66}{484}; \frac{11}{22} - \frac{6}{22} = \frac{5}{22}$$

$$A_3 = \frac{5\cdot 6}{22\cdot 22} = \frac{30}{484}; \frac{A_1}{A_3} = \frac{66}{30} = \frac{11}{5}$$

8 × 5 rectangle

$$\frac{8}{13} + \frac{5}{13} = \frac{13}{13}$$

$$\frac{13}{13\cdot 2} = \frac{13}{26}$$

$$A_1 = \frac{13\cdot 10}{26\cdot 26} = \frac{130}{676}; \frac{13}{26} - \frac{10}{26} = \frac{3}{26}$$

$$A_3 = \frac{3\cdot 10}{26\cdot 26} = \frac{30}{676}; \frac{A_1}{A_3} = \frac{130}{30} = \frac{13}{3}$$

8×7 rectangle

$$\frac{8}{15} + \frac{7}{15} = \frac{15}{15}$$

$$\frac{15}{15 \cdot 2} = \frac{15}{30}$$

$$A_1 = \frac{15 \cdot 14}{30 \cdot 30} = \frac{210}{900}; \quad \frac{15}{30} - \frac{14}{30} = \frac{1}{30}$$

$$A_3 = \frac{1 \cdot 14}{30 \cdot 30} = \frac{14}{900}; \quad \frac{A_1}{A_3} = \frac{210}{14} = \frac{30}{2} = 15$$

9×2 rectangle: $\dfrac{A_1}{A_3} = \dfrac{9+2}{9-2} = \dfrac{11}{7}$

$\dfrac{11}{11}$ series

9×4 rectangle: $\dfrac{A_1}{A_3} = \dfrac{9+4}{9-4} = \dfrac{13}{5}$

$\dfrac{13}{13}$ series

9×5 rectangle: $\dfrac{A_1}{A_3} = \dfrac{9+5}{9-5} = \dfrac{14}{4} = \dfrac{7}{2}$

$\dfrac{14}{14}$ series

9×7 rectangle: $\dfrac{A_1}{A_3} = \dfrac{9+7}{9-7} = \dfrac{16}{8} = 2$

$\dfrac{16}{16}$ series

9×8 rectangle: $\dfrac{A_1}{A_3} = \dfrac{9+8}{9-8} = 17$

$\dfrac{17}{17}$ series

10×3 rectangle: $\dfrac{A_1}{A_3} = \dfrac{10+3}{10-3} = \dfrac{13}{7}$

$\dfrac{13}{13}$ series

10×7 rectangle: $\dfrac{A_1}{A_3} = \dfrac{10+7}{10-7} = \dfrac{17}{3}$

$\dfrac{17}{17}$ series

10×9 rectangle: $\dfrac{A_1}{A_3} = \dfrac{10+9}{10-9} = 19$

$\dfrac{19}{19}$ series

11×2 rectangle: $\dfrac{A_1}{A_3} = \dfrac{11+2}{11-2} = \dfrac{13}{9}$

$\dfrac{13}{13}$ series

11×3 rectangle: $\dfrac{A_1}{A_3} = \dfrac{11+3}{11-3} = \dfrac{14}{8} = \dfrac{7}{4}$

$\dfrac{14}{14}$ series

11×4 rectangle: $\dfrac{A_1}{A_3} = \dfrac{11+4}{11-4} = \dfrac{15}{7}$

$\dfrac{15}{15}$ series

11×5 rectangle: $\dfrac{A_1}{A_3} = \dfrac{11+5}{11-5} = \dfrac{16}{6} = \dfrac{8}{3}$

$\dfrac{16}{16}$ series

11×6 rectangle: $\dfrac{A_1}{A_3} = \dfrac{11+6}{11-6} = \dfrac{17}{5}$

$\dfrac{17}{17}$ series

11×7 rectangle: $\dfrac{A_1}{A_3} = \dfrac{11+7}{11-7} = \dfrac{18}{4} = \dfrac{9}{2}$

$\dfrac{18}{18}$ series

11×8 rectangle: $\dfrac{A_1}{A_3} = \dfrac{11+8}{11-8} = \dfrac{19}{3}$

$\dfrac{19}{19}$ series

11×9 rectangle: $\dfrac{A_1}{A_3} = \dfrac{11+9}{11-9} = \dfrac{20}{2} = 10$

$\dfrac{20}{20}$ series

11×10 rectangle: $\dfrac{A_1}{A_3} = \dfrac{11+10}{11-10} = \dfrac{21}{1} = 21$

$\dfrac{21}{21}$ series

12×5 rectangle: $\dfrac{A_1}{A_3} = \dfrac{12+5}{12-5} = \dfrac{17}{7}$

$\dfrac{17}{17}$ series

12×7 rectangle: $\dfrac{A_1}{A_3} = \dfrac{12+7}{12-7} = \dfrac{19}{5}$

$\dfrac{19}{19}$ series

3. Charts.

Series: $a+b$	Rectangle: A_4	Ratio of $A_1 : A_3$
$\dfrac{3}{3}$	2×1	3
$\dfrac{4}{4}$	3×1	2
$\dfrac{5}{5}$	3×2	5
	4×1	$\dfrac{5}{3}$
$\dfrac{6}{6}$	5×1	$\dfrac{3}{2}$
$\dfrac{7}{7}$	4×3	7
	5×2	$\dfrac{7}{3}$
	6×1	$\dfrac{7}{5}$
$\dfrac{8}{8}$	5×3	4
	7×1	$\dfrac{4}{3}$
$\dfrac{9}{9}$	5×4	9
	7×2	$\dfrac{9}{5}$
	8×1	$\dfrac{9}{7}$
$\dfrac{10}{10}$	7×3	$\dfrac{5}{2}$
	9×1	$\dfrac{5}{4}$

RATIO AND RATIONALIZATION

Series: a + b	Rectangle: A4	Ratio of $A_1 : A_3$
$\frac{11}{11}$	6 × 5	11
	7 × 4	$\frac{11}{3}$
	8 × 3	$\frac{11}{5}$
	9 × 2	$\frac{11}{7}$
	10 × 1	$\frac{11}{9}$
$\frac{12}{12}$	7 × 5	$\frac{12}{5}$
	11 × 1	$\frac{6}{5}$
$\frac{13}{13}$	7 × 6	13
	8 × 5	$\frac{13}{3}$
	9 × 4	$\frac{13}{5}$
	10 × 3	$\frac{13}{7}$
	11 × 2	$\frac{13}{9}$
	12 × 1	13
$\frac{14}{14}$	9 × 5	$\frac{7}{2}$
	11 × 3	$\frac{7}{4}$
	13 × 1	$\frac{7}{6}$

Series: $a+b$	Rectangle: $A4$	Ratio of $A_1 : A_3$
$\dfrac{15}{15}$	8×7	15
	11×4	$\dfrac{15}{7}$
	13×2	$\dfrac{15}{11}$
	14×1	15
$\dfrac{16}{16}$	9×7	2
	11×5	$\dfrac{8}{3}$
	13×3	$\dfrac{8}{5}$
	15×1	$\dfrac{8}{7}$
$\dfrac{17}{17}$	9×8	17
	10×7	$\dfrac{17}{3}$
	11×6	$\dfrac{17}{5}$
	12×5	$\dfrac{17}{7}$
	13×4	$\dfrac{17}{9}$
	14×3	$\dfrac{17}{11}$
	15×2	$\dfrac{17}{13}$
	16×1	$\dfrac{17}{15}$

RATIO AND RATIONALIZATION

Series: a + b	Rectangle: A4	Ratio of $A_1 : A_3$
$\frac{18}{18}$	11 × 7	$\frac{9}{2}$
	13 × 5	$\frac{9}{4}$
	17 × 1	$\frac{9}{8}$
$\frac{19}{19}$	10 × 9	19
	11 × 8	$\frac{19}{3}$
	12 × 7	$\frac{19}{5}$
	13 × 6	$\frac{19}{7}$
	14 × 5	$\frac{19}{9}$
	15 × 4	$\frac{19}{11}$
	16 × 3	$\frac{19}{13}$
	17 × 2	$\frac{19}{15}$
	18 × 1	$\frac{19}{17}$
$\frac{20}{20}$	11 × 9	10
	13 × 7	$\frac{10}{3}$
	17 × 3	$\frac{10}{7}$
	19 × 1	$\frac{10}{9}$

Series: a + b	Rectangle: A4	Ratio of $A_1 : A_3$
$\dfrac{21}{21}$	11 × 10	21
	13 × 8	$\dfrac{21}{5}$
	16 × 5	$\dfrac{21}{11}$
	17 × 4	$\dfrac{21}{13}$
	19 × 2	$\dfrac{21}{17}$
	20 × 1	$\dfrac{21}{19}$

D. Ratios of the Rational Continuum [1]

$$\begin{array}{cccccccc}
 & 5 \div 3 & 7 \div 5 & 9 \div 7 & 11 \div 9 & 13 \div 11 & 15 \div 13 & 17 \div 15 \\
2 \div 1 & 3 \div 2 & 4 \div 3 & 5 \div 4 & 6 \div 5 & 7 \div 6 & 8 \div 7 & \\
\end{array}$$

$$\begin{array}{cccccc}
 & 19 \div 17 & 21 \div 19 & 23 \div 21 & 25 \div 23 & 27 \div 25 \\
9 \div 8 & 10 \div 9 & 11 \div 10 & 12 \div 11 & 13 \div 12 & 14 \div 13 \\
\end{array}$$

$$\begin{array}{cccccc}
29 \div 27 & 31 \div 29 & 33 \div 31 & 35 \div 33 & 37 \div 35 \\
 & 15 \div 14 & 16 \div 15 & 17 \div 16 & 18 \div 17 & 19 \div 18 \\
\end{array}$$

$$\begin{array}{cccccc}
39 \div 37 & 41 \div 39 & 43 \div 41 & 45 \div 43 & 47 \div 45 \\
 & 20 \div 19 & 21 \div 20 & 22 \div 21 & 23 \div 22 & 24 \div 23 \\
\end{array}$$

$$\begin{array}{cccccc}
49 \div 47 & 51 \div 49 & 53 \div 51 & 55 \div 53 & 57 \div 55 \\
 & 25 \div 24 & 26 \div 25 & 27 \div 26 & 28 \div 27 & 29 \div 28 \\
\end{array}$$

$$\begin{array}{cccccc}
59 \div 57 & 61 \div 59 & 63 \div 61 & 65 \div 63 & 67 \div 65 \\
 & 30 \div 29 & 31 \div 30 & 32 \div 31 & 33 \div 32 & 34 \div 33 \\
\end{array}$$

$$\begin{array}{cccccc}
69 \div 67 & 71 \div 69 & 73 \div 71 & 75 \div 73 & 77 \div 75 \\
 & 35 \div 34 & 36 \div 35 & 37 \div 36 & 38 \div 37 & 39 \div 38 \\
\end{array}$$

$$\begin{array}{cccccc}
79 \div 77 & 81 \div 79 & 83 \div 81 & 85 \div 83 & 87 \div 85 \\
 & 40 \div 39 & 41 \div 40 & 42 \div 41 & 43 \div 42 & 44 \div 43 \\
\end{array}$$

$$\begin{array}{ccccc}
89 \div 87 & 91 \div 89 & 93 \div 91 & 95 \div 93 \\
 & 45 \div 44 & 46 \div 45 & 47 \div 46 & 48 \div 47. \\
\end{array}$$

[1] The reader is referred to Appendix B, which contains the relative dimensions resulting from these ratios. (Ed.)

CHAPTER 8

POSITIONAL ROTATION

POSITIONAL rotation (p. r.) is equivalent to circular permutation with the implication of the concept of sequence. The use of this term signifies a sequent group evolved by means of circular permutation: for example, positional rotation of coordinate phases of music: melody evolved by means of p. r.; harmony evolved by means of p.r.; and correlated melodies (counterpoint) evolved by means of p. r.[1]

In design: composition of superimposed images obtained by means of p.r.; kinetic sequence of images (cinema) in p.r.; and color sequence obtained through p.r. In the coordination of the groups of components, such as scores of homogeneous or heterogeneous arts, p.r. can control simultaneity (ordinate phases) and continuity (abscissa phases). This device permits variation of a pre-set coordinated group automatically, both in simultaneity and in continuity, either in the positive (↻ ↻) or in the negative (↺ ↺) direction. Two-dimensional (two coordinate) positional rotation may be expressed in terms of the phasic and the directional relations.

A. Dimensionality of Positional Rotation

Positional rotation can be defined as *two-dimensional* (x, y) and *two directional* (positive: ↻ x, ↻ y, and negative: ↺ x, ↺ y). Each phase of positional rotation is defined by the dimensional and directional relations of x and y. Expressing a simultaneous structural group as S, its units— as s, general time period as T and its duration units as t, we may elaborate the four fundamental forms of positional rotation.

(1)   Sms ↻ Tnt ↻;   (2) Sms ↺ Tnt ↻;
(3)   Sms ↻ Tnt ↺;   (4) Sms ↺ Tnt ↺.

Here m and n represent the phasic values. As ↻ and ↻ represent the positive values of t and s respectively, and ↺ and ↺ represent the negative values respectively, any desired st phase of positional rotation may be computed directly. The definition of structure (S) in its original position is as follows:

$S = ms_o nt_o$. The limit position of m is Sm, and the limit position of t is t_n.

The change of position S_o in a positive phase results in S_a position. The change of position t_o in a positive phase results in t_a position. The change of position S_o in b negative phases results in S_{-b} position. The change of position t_o in b negative phases results in t_{-b} position.

---

[1] These processes are described in detail in *The Schillinger System of Musical Composition*. (Ed.)

ST	$\mathop{ST}\limits_{oo}^{mn} \equiv mynx$
S ↻ T ↻	$S = ms$
	$T = nt$
S ↻ T ↻	$ST = msnt$
S ↻ T ↻	The original position:
	$S_o T_o = ms_o nt_o$
S ↻ T ↻	$S_1 T_1 = s_1 t_1$
	$S_k T_{-b} = s_k t_{-b}$

$$\mathop{ST}\limits_{oo}^{mn} = S_o t_o, S_1 t_o, S_o t_1, S_1 t_1, S_2 t_o,$$

$$S_2 t_1, S_o t_2, S_1 t_2, S_2 t_2, \ldots$$

$$S_m t_o, S_m t_1, S_m t_2, \ldots$$

$$S_o t_n, S_1 t_n, S_2 t_n,$$

$$S_n t_n.$$

46
$$\mathop{ST}\limits_{oo}^{} = S_o t_o, S_1 t_o, S_1 t_1, \ldots S_4 t_4, \ldots S_2 t_6$$

If a given phasic position is $S_3 t_2$, the addition of $2S - 2t$ gives: $S_3 t_2 + (2S - 2t) = S_5 t_o$

Let the structure be:

48
$$\mathop{ST}\limits_{oo}^{}$$

(1) $S_o t_o + (5s - 3t) = S_1 t_5$

(2) $S_o t_o + (-2S + 11t) = S_3 t_3$

(3) $S_o t_o + (3S + 5t) = S_3 t_5$

(4) $S_o t_o + (3t + t + 2t + 2t + t + 3t) = S_o t_4$

(5) $S_o t_o + (3t - t + 2t - 2t + t - 3t) = S_o t_o$

(6) $S_o t_o + (3S - t + 2S - 2t + S - 3t) = S_2 t_2$

POSITIONAL ROTATION

Let the structure be: 66
                    ST
                    oo

then: $S_o = S_6$, $t_o = t_6$

(1) $S_o t_o + (2S - t + s - 2t) = S_3 t_3$

(2) $S_o t_o + (3S - t + 2S - 2T + S - 3t) = S_o t_o$

The addition of a resultant or of any group with a compensating form of symmetry, at the end of the group, restores the latter to its original position when the signs of consecutive terms alternate, and the number of terms is even.

It is possible to compute directly any phase either from zero phase or from any other phase for both s and t.

Example: mn
                    ST
                    oo

(1) $S_o t_o + (5S + 2t - S - 3t + 2S + t) = S_o t_o + (6S + Ot) = S_6 t_o$

(2) $S_2 t_2 + (5S + 2t - S + 3t + 2S + t) = S_2 t_2 + (6S + Ot) = S_8 t_2$

In general:

    if $t_o = tn$

    $t_o + nt = nt - nt = t_o$

    $t_1 + nt = nt - 1 = t_1$

    $t_2 + nt = nt - 2 = t_2$

6
T
o

    $t_o = t_6$

    $t_o + t = t_1$

    $t_o + 2t = t_2$

   . . . .

    $t_o + 6t = t_o$

    $t_o + 7t = t_o + (7t - 6t) = t_1$

    $t_o + 8t = t_o + (8t - 6t) = t_2$

If $m > n$, then:

$$t_o + mt = t_o + (mt - nt) = tm - n$$

If $m > 2n$, then:

$$t_o + mt = t_o + (mt - 2nt) = tm - 2n$$

If $m > pn$, then:

$$t_o + mt = t_o + (mt - pnt) = tm - pn$$

Example: $\overset{3}{\underset{o}{T}}$

$t_o + t = t_1$

$t_o + 2t = t_2$

$t_o + 3t = t_3 = t_o$

$t_o + 4t = (4t - 3t) = t_1$

$t_o + 5t = (5t - 3t) = t_2$

$t_o + 6t = (6t - 2 \cdot 3t) = t_o$

$t_o + 7t = (7t - 6t) = t_1$

$t_o + 8t = (8t - 6t) = t_2$

$t_o + 9t = (9t - 3 \cdot 3t) = t_o$

$t_o + 10t = (10t - 9t) = t_1$

. . . .

As temporal units (t) of the structural temporal group (T) move back and forth, controlled by definite positive and negative coefficients, physical time (during which perception takes place) evolves in one direction. Thus, each temporal sub-group of the entire T corresponds to t with some coefficient. We shall indicate it as mt, nt, pt, ... The positional zero of any subgroup therefore corresponds to P, which is the term of physical time (the positive non-reversible time). If P_1 is the original term of physical time, P_2 is the second term of physical time, etc.; then: $P_1 = mt$, $P_2 = nt$, $P_3 = pt$, ... All these considerations hold true when t jumps back and forth with any coefficient, but never actually moves backwards. Positional rotation presupposes phasic variations of T or of its subgroups mt, nt, pt, ... where T, mt, nt, pt, ... are treated as *discontinuity*. In order to produce continuity of phasic rotation, it is necessary to introduce the dense set from t_o to t_1, from t_1 to t_2, etc., and to use such dense sets in the positive as well as in the negative direction.

CHAPTER 9

SYMMETRY

A. SYMMETRIC PARALLELISMS

(Configurational Identities)

(1) $\frac{3}{3}$ series

(a) Six-pointed star (Hebrew mogen dovid) represents binomial periodicity of the $\frac{3}{3}$ series: $2 + 1 = \angle 120° + \angle 60°$; $6(120° + 60°)$ in alternating direction, usually represented as two superimposed equilateral triangles; in combination with the cross ($\frac{12}{12}$) series: $4(5+5+2)$ in constant direction under $\angle 90°$ and heart (bifold symmetry) ($\frac{2}{2}$ series) used by the Franciscan order. It is of importance that $120° + 60°$ produce $180°$, *i.e.*, an infinite rectilinear extension.

(b) All members of $\frac{3}{3}$ series participate in the temporal configurations of Catholic liturgical music, especially in the 14th to the 16th centuries. This was an intentional selection of the symbol of Divine Trinity. This form influenced all the subsequent forms of European music written in $\frac{3}{4}$ time, which forms evolved later in the 18th and the 19th centuries into hexagonal forms of symmetry ($\frac{6}{6}$ series).

(c) A secondary selective system evolved by the Arabs in the 7th century A.D.: a pitch-scale known as "zer ef kend." (literally: string of pearls). This scale was conceived as an alternation of a large step (bead) and a small step (bead):

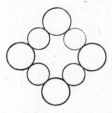

Figure 1. String of pearls. "Zer ef kend."

Zer ef kend approximates with sufficient precision the alternation of a whole tone and a half tone of the $\sqrt[12]{2}$ temperament:

$$c - d - e\flat - f - f\sharp - g\sharp - a - b - c^1$$
$$2 + 1 + 2 + 1 + 2 + 1 + 2 + 1$$

It means that the binomial of the $\frac{3}{3}$ series, after performing four cycles (as in the case of beads), closes in an octave, *i.e.*, in $\frac{2}{1}$ ratio: $4(2+1) = 12$, which links it with the hybrid $\frac{12}{12}$ series, *i.e.*,

$$\frac{3}{3} \cdot \frac{4}{4} = \frac{12}{12}.$$

The latter, in turn, links it with the "blues", which is partly hybrid, partly pure $\frac{12}{12}$ series.

(d) Temporal configurations of the music of Oklahoma Indians:

$$\frac{1}{9} \ldots \frac{1}{3} \ldots \frac{3}{3} \ldots 3 \ldots 9$$

(2) $\frac{5}{5}$ series

(a) The natural pentagonal and pentaclic developments: sea urchin, starfish. The starfish is $r_{(5 \div 2)}$, where the $\frac{2}{2}$ series is the primary genetic factor (as the starfish at first represents bifoldedness without signs of pentagonality) and the $\frac{5}{5}$ series is the secondary genetic factor (possessing the pentagonal tendency).

$r_{(5 \div 2)} = 2+2+1+1+2+2$, and refers to the sequence of angles:

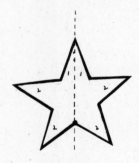

Figure 2. *Sequence of angles in pentagonal form.*

---

[1]The numbers given here and elsewhere to denote pitch scales refer to semitones. The number 2 denotes two semitones or a whole tone (c—d). Three (3) denotes three semitones (c—e♭), etc. (Ed.)

(b) The primary selective system of Javanese tuning:

$$\sqrt[5]{2} \text{ ("slendro")}.$$

(c) The pentameter of verse coupled with the $\frac{5}{5}$ temporal series of music of Asia and European Russia.

(d) The Caucasian rug-patterns, where the width of partial rectangles is arranged as the sequence of terms of $r_{(5 \div 3)}$, *i.e.*, $3+2+1+3+1+2+3$.

(e) Distribution of the visible spectral circle into 5 uniform sectors: $\sqrt[5]{\pi r^2}$

(3) $\frac{6}{6}$ series

(a) Hexagonal symmetry of honeycombs.

(b) Primary selective system of tuning: $\sqrt[6]{2}$, *i.e.*, the "whole-tone scale".

(c) Adaptations of hexagons in tiles, floors (wood inlay) in Persia, Rome, etc.

Figure 3. Practical form of $\frac{3}{3}$ series (star), $\frac{2}{2}$ series (heart), and $\frac{12}{12}$ series (cross).

Courtesy National Geographic Magazine and Buffalo Museum of Science

Figure 4. Snowflakes are all based on $\frac{6}{6}$ series.

SYMMETRY

B. Esthetic Evaluation on the Basis of Symmetry

Beauty as a form of Regularity or Symmetry.

A perfect case of symmetry is the Arabian pitch-scale "Zer ef kend" (String of Pearls): $4(2+1)$. Its derivative (d_1) is: $4(1+2)$.

Among European scales (ecclesiastic modes), we find the following structures:[2]

$(2 + 2 + 1) + (2 + 2 + 2)$	Nat. major	d_0
$(2 + 1 + 2) + (2 + 2 + 1)$	Dorian	d_1
$(1 + 2 + 2) + (2 + 1 + 2)$	Phrygian	d_2
$(2 + 2 + 2) + (1 + 2 + 2)$	Lydian	d_3
$(2 + 2 + 1) + (2 + 2 + 1)$	Mixolydian	d_4
$(2 + 1 + 2) + (2 + 1 + 2)$	Aeolian	d_5
$(1 + 2 + 2) + (1 + 2 + 2)$	Locrian	d_6

Esthetically, grade A are: d_4, d_5, d_6
  "   " B " d_1, d_2
  "   " C " d_0, d_3

Grade A are symmetric
  "   B " modified symmetric
  "   C " assymetric

It is interesting to note that in Russian folk music, only d_4 and d_5 (both symmetric) are commonly used, d, being an exception. In ancient Greek music d_4 (which was known as Mixolydian) was dedicated to the Sun, and was therefore regarded as the royal scale. All other scales were dedicated to different planets. d_6 is symmetric.

The fundamental structure of the so-called "Chinese pentatonic" scale is: $2+2+3+2$ or $2+3+2+2$. It is interesting to note that $2+2+2=6$ and the remaining interval is 3, thus forming a $\frac{6}{3} = \frac{2}{1}$ ratio between the sums of the two kinds of intervals employed. Perfect bifold symmetry appears in d_5 of the original scale: a — c — d — e — g, i.e., $3+2+2+3$, a frequently used scale.

One of the prominent Javanese scales is constructed downward: a—g—e —d—b, i.e., $2+3+2+3$. This scale is also prominent in Madagascar.[3] It is known

---

[2] The symbol d is used to denote displacement or circular permutation. d_0 is the zero displacement scale; d_1 is the first displacement scale that may be derived from it by circular permutation; d_2 is the second displacement, etc. Schillinger shows how the various modal scales may be derived by displacement from natural major. (Ed.)

[3] Compare Ravel's *Chansons Madecasses*.

as "Slendro". The other fundamental Javanese scale is "Pelog" ($\sqrt{2}$). Reading downward: f — e — c — b, *i.e.*, 1+4+1, which is the central trinomial of the $\frac{6}{6}$ series. The Balinese scale "Selenders" (derives from the Javanese $\sqrt[5]{2}$ tuning Slendro): (3+2+3+2) + (2+3+2+3) +2 (reading downward). The Balinese dance "Djanger" is based on the scale: (2+1+4) + (2+1+4), reading downward: d—c—b—g—f—e—c. Persian popular songs are based on (reading downward): g—f—e—d—c—b—a, *i.e.*, (2+1+2) + (2+1+2), a scale identical with the Aeolian mode of Glareanus.

Our pitch discrimination is conditioned by $\sqrt{2}$, our time discrimination (in music and movement) by 2.

C. RECTANGULAR SYMMETRY OF EXTENSIONS IN SERIAL DEVELOPMENT

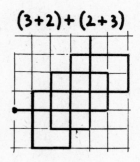

Figure 5. (3 + 2) + (2 + 3).

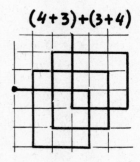

Figure 6. (4 + 3) + (3 + 4) + ...

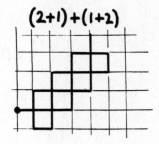

Figure 7. $(2 + 1) + (1 + 2)$.

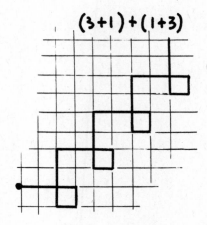

Figure 8. $(3 + 1) + (1 + 3)$.

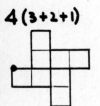 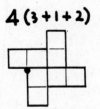 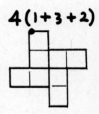

Figure 9. $4(3 + 2 + 1)\ 4(3 + 1 + 2)\ 4(1 + 3 + 2)$

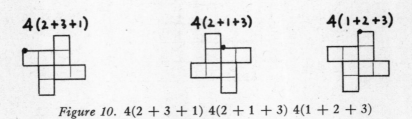

Figure 10. 4(2 + 3 + 1) 4(2 + 1 + 3) 4(1 + 2 + 3)

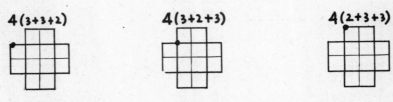

Figure 11. 4(2 + 1 + 1) 4(1 + 2 + 1) 4(1 + 1 + 2)

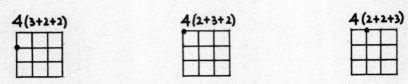

Figure 12. 4(3 + 3 + 2) 4(3 + 2 + 3) 4(2 + 3 + 3)

Figure 13. 4(3 + 2 + 2) 4(2 + 3 + 2) 4(2 + 2 + 3)

SYMMETRY

Figure 14. $4(4 + 3 + 3)\ 4(3 + 4 + 3)\ 4(3 + 3 + 4)$

 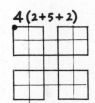 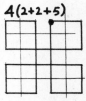

Figure 15. $4(5 + 2 + 2)\ 4(2 + 5 + 2)\ 4(2 + 2 + 5)$

$$\frac{12}{12}$$

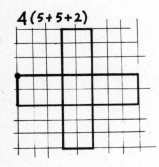 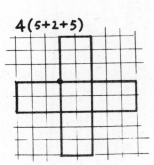 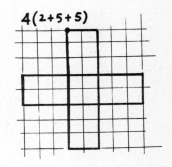

Figure 16. $4(5 + 5 + 2)\ 4(5 + 2 + 5)\ 4(2 + 5 + 5)$

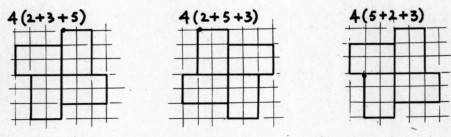

Figure 17. 4(2 + 3 + 5) 4(2 + 5 + 3) 4(5 + 2 + 3)

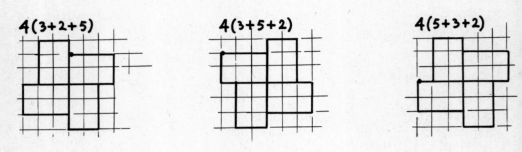

Figure 18. 4(3 + 2 + 5) 4(3 + 5 + 2) 4(5 + 3 + 2)

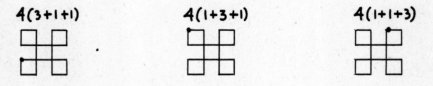

Figure 19. 5(3 + 1 + 1) 4(1 + 3 + 1) 4(1 + 1 + 3)

SYMMETRY

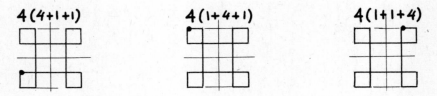

Figure 20. 4(4 + 1 + 1) 4(1 + 4 + 1) 4(1 + 1 + 4)

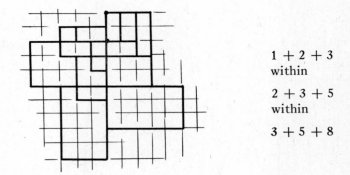

1 + 2 + 3
within

2 + 3 + 5
within

3 + 5 + 8

Figure 21. Common origin.

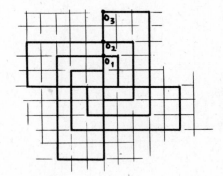

Figure 22. Origins uncommon; superimposition of the three trajectories is centered.

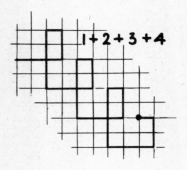

Figure 23. $1 + 2 + 3 + 4$.

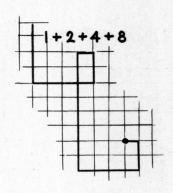

Figure 24. $1 + 2 + 4 + 8$.

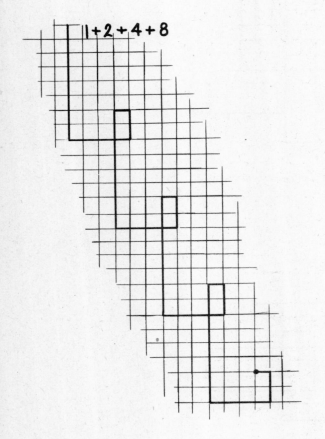

Figure 25. $1 + 2 + 4 + 8$.

SYMMETRY

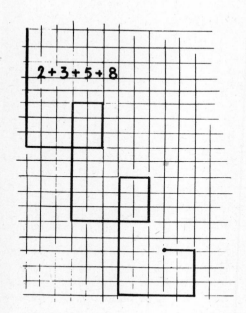

Figure 26. 2 + 3 + 5 + 8

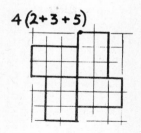

Figure 27. 2 + 3 + 5

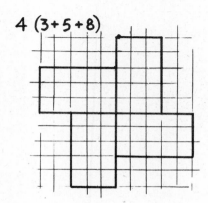

Figure 28. 3 + 5 + 8

CHAPTER 10

QUADRANT ROTATION[1]

MUSIC in any equal temperament, when it is recorded graphically in rectangular projection, expresses the equivalent of musical *notation* in equal temperament. Such a geometrical projection of music is expressed on a plane, and as such is subject to quadrant rotation of the plane through three dimensional space. Rotation may be either *clockwise* or *counterclockwise*.[2]

The conception of time, which is based on the common denominator and not on the logarithmic series, implies two possible positions: (1) the original, under zero degrees to the field of vision (parallel to the eyes); (2) the 180° position derived from the first one through rotation around the ordinate axis. Such an ordinate axis is either the starting or the ending limit of the vertical cross-section of the graph (duration limits). If the original (zero degree position) is conceived as a forward motion of music in time continuity, then the respective variation of it (180° position) is the backward motion of the original, when the ordinate is the ending limit in time.

The *logarithmic* contraction of time corresponds to the logarithmic contraction of space on the graph—and if our music were not bound to a common denominator system of measurement, it would be possible to apply such projection practically. This same form of variation has been known in *visual* art since about 1533 A.D., in skillful paintings made by German and Italian artists. They are based on the principle of angle-perspective and have to be looked at (that is, held at an angle) from right to left, instead of under the zero angle to the field of vision.

---

[1] From *The Schillinger System of Musical Composition*, Copyright, 1941, by Carl Fischer, Inc. Reprinted by permission. This is an abbreviated version of Book III, Chapter 1.

[2] It may be helpful to add at this point the following: geometric inversion of music consists of "a" of the original form of the music, to start with; then, as the "b" inversion, the same thing *backwards*; as the "c" inversion, the original but backwards and *upside down*; and the "d" version, forwards and upside down. (Ed.)

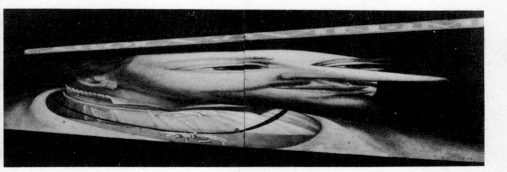

Figure 1. German School, 16th Century: Charles V, 1533.
Collection, Jacques Lipchitz, Paris

Figure 2. Unknown Master, 16th Century: St. Anthony of Padua.
Courtesy Museum of Modern Art. Collection, Jacques Lipchitz, Paris

234 THEORY OF REGULARITY AND COORDINATION

By revolving the second position of a musical graph through the abscissa (which becomes the axis of rotation) 180° in a clockwise direction, we obtain the third position of the original. The axis of rotation must represent a *pt* (pitch-time) maximum and the direction of the third position is backwards upside-down of the original, and forward upside-down of the second position. Further 180° clockwise rotation of the third position about its ordinate produces the fourth position, which is the backwards of the third position, the backwards upside-down of the second position and the forward upside-down of the original. The respective four positions will be expressed in the following exposition through ⓐ, ⓑ, ⓒ and ⓓ.

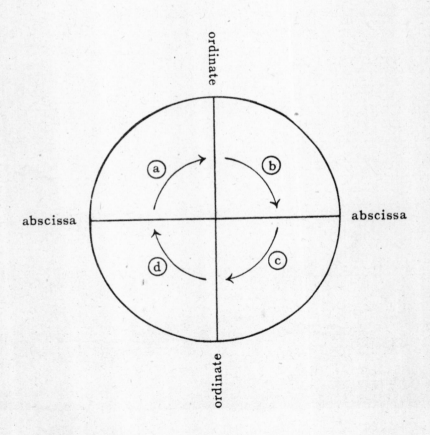

Figure 3. The four positions of geometric inversion.

QUADRANT ROTATION

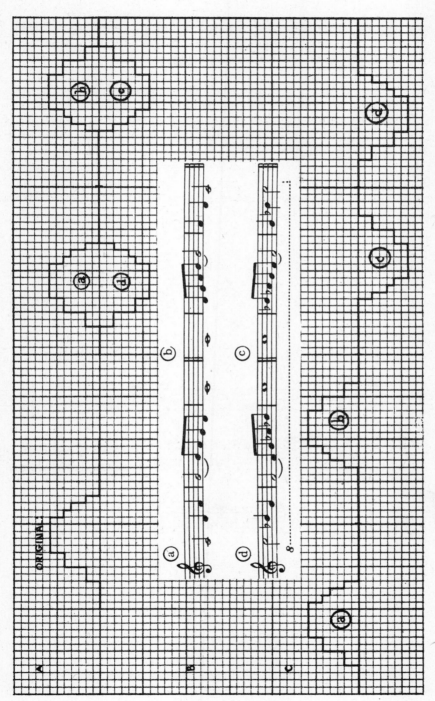

Figure 4. Evolving the four geometrical inversions of a given melody.

These four geometrical inversions may be used individually as variations of a given melody. They may also be developed into a continuity in which the different positions are given different coefficients. Under such conditions the recurrence of the different positions is subject to rhythm.

This method of geometrical inversion, when applied to the composition of melodic continuity, offers much greater versatility—yet preserves the unity more—than any composer in the past was able to achieve. For example, by comparing the music of J. S. Bach with the following illustrations, the full range of what he could have done by using the method of geometrical inversions becomes clear.

In Invention No. 8, from his *Two-Part Inventions*, during the first 8 bars of the leading voice (upper part after the theme ends), the first 2 bars fall into the triple repetition of an insignificant melodic pattern lasting one and one-half times longer than the entire theme.

Figure 5. J. S. Bach, *Two-Part Inventions*, No. 8.

Using the method of geometrical inversion (even with a compromise of the recurrence of the original position), we obtain the following version of thematic continuity.

QUADRANT ROTATION

Inversion of J. S. Bach, Two-Part Inventions, No. 8

Figure 6. Expansions of J. S. Bach, Two-Part Inventions, No. 8

In some cases geometrical inversions of music give new and often more interesting character to the original. When a composer feels dissatisfied with his theme, he may try out some of the inversions—and he may possibly find them more suitable for his purpose, discarding the original. Such was the case when George Gershwin[3] wrote a theme for his opera *Porgy and Bess*, where position ⓒ was used instead of the original which was not as expressive and lacked the character of the later version.

An analysis of well-known works of the composers of the past often throws new light upon them, revealing hidden characteristics that become more apparent in the geometrical inversions. For example, the harmonic minor scale combined with certain rhythmic forms produces an effect of Hungarian dance music. In L. van Beethoven's Piano Sonata No. 8, the first theme of the finale in its position ⓑ reveals a decidedly Hungarian character which is not as noticeable in its original form. This analysis also discloses that position ⓓ of the same theme has a more archaic character than the original, linking Beethoven's music with that of Joseph Haydn.

Figure 7. Geometric inversions of L. van Beethoven, Piano Sonata, No. 8, Finale.
(*continued*)

---

[3] In the *Musical Courier* of Nov. 1, 1940 Leonard Liebling, editor, wrote: "After George Gershwin had written over 700 songs, he felt at the end of his inventive resources and went to Schillinger for advice and study. He must have valued both, for he remained a pupil of the theorist for four and a half years." *Porgy and Bess*, which took Gershwin more than two years of work under his teacher's supervision, was composed according to the Schillinger System. (Ed.)

Figure 7. Geometric inversions of L. van Beethoven, Piano Sonata, No. 8, Finale.
(concluded)

It is possible to plan in advance the composition of melodic continuity through combining geometrical inversions of the original material with a rhythmic group pre-selected for the coefficients of recurrence of the different positions.

Rhythm of Coefficients: $r_{4\div3}$

Geometrical Positions: ⓐ, ⓓ, ⓑ

Continuity: 3 ⓐ + ⓓ + 2 ⓑ + 2 ⓐ + ⓓ + 3 ⓑ

The actual technique of transcribing music from one position to another may be worked out in three different ways. The student may take his choice.

1. Direct transcription of the inverted positions from the *graph* into musical notation.

2. Direct transcription from a complete manifold of chromatic tables representing ⓐ and ⓓ positions for all the 12 axes.

Figure 8. Manifold of chromatic tables for ⓐ and ⓓ.

3. Step by step (melodic) transcription from the original.

The unconscious urge toward geometrical inversions was actually realized in music of the past through those backward and contrary motions of the original pattern which may be found in abundance in the works of the contrapuntalists of the 16th, 17th and 18th centuries. As they did not do it geometrically but tonally, they often *misinterpreted* the tonal structure of a theme appearing in an upside-down position. They tried to preserve the tonal unity instead of preserving the original pattern. Besides these thematic inversions of melodies, evidence of the tendency toward unconscious geometrical inversions may be observed in the juxtaposition of major and minor as the *psychological* poles. In reality, the commonly used harmonic minor is simply an erroneous geometrical inversion of the natural major scale. The correct position ⓓ of the natural major scale is the Phrygian scale and not the harmonic minor. The difference appears in the 2d and 7th degrees of that scale.

In the following examples, dⓑ indicates the upward reading of the ⓓ scale.

Figure 9. Inversion of natural major.

QUADRANT ROTATION

The effect of psychological contrasts, to which I have referred with regard to scales, takes place with chord structures and their progressions as well. The most obvious illustration is a major triad (c — e — g; 4 + 3) with its reciprocal structure minor triad (c — e♭ — g; 3 + 4). When such a chord is to be inverted from c as an axis, all pitch-units take corresponding places in the opposite direction, *i.e.*, c remains constant (the invariant of inversion), e becomes a♭, and g becomes f. Here is a comparative chart of positions ⓐ and ⓓ of the chords commonly known as triads [S(5)] and 7th chords [S(7)].

Figure 10. ⓐ and ⓓ *positions of the triads.*

This method of inverting chords as well as scales in order to find the psychological reciprocal is particularly useful in cases *where there is doubt* as to what the reciprocal chord structure or progression may be. It also provides an exact way of finding the reciprocal structures and progressions in those cases in which the latter are entirely unknown—and the trial and error method does not bring any satisfactory result.

The technique of transcribing any harmonic continuity into different geometrical positions can be greatly simplified by using the method of *enumeration of each voice* of the harmony. Each voice becomes a melody and it is only necessary to know the entire chord (*i.e.*, the starting-points of such melodies) for the starting-point, after which all voices may be transcribed horizontally (as melodies).

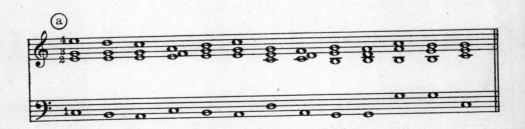

Figure 11. ⓐ *and* ⓓ *of melody with chords.*

The above-mentioned operations make it clear that any of the variations in the original distribution of voices of a chord may serve as a starting-point for any harmonic continuity. Thus, a 4-part harmony offers 24 versions in each of the four geometrical positions. This device is superior to the ingenuity of any composer using an intuitive method in order to achieve variety of instrumental forms of the same harmonic continuity.

The following chart represents 24 original forms of distribution of the starting chord (according to the 24 permutations of 4 elements), for the harmonic continuity offered in the preceding figure 11. When the starting chord has the same structure but different distribution, the resulting sonority of each version also becomes different.

QUADRANT ROTATION

d c c c	d b b b	d b b b	d c c c	d a a a	d a a a
c d b b	b d c c	b d a a	c d a a	a d c c	a d b b
b b d a	c c d a	a a d c	a a d b	c c d b	b b d c
a a a d	a a a d	c c c d	b b b d	b b b d	c c c d
4 3 3 3	4 2 2 2	4 2 2 2	4 3 3 3	4 1 1 1	4 1 1 1
3 4 2 2	2 4 3 3	2 4 1 1	3 4 1 1	1 4 3 3	1 4 2 2
2 2 4 1	3 3 4 1	1 1 4 3	1 1 4 2	3 3 4 2	2 2 4 3
1 1 1 4	1 1 1 4	3 3 3 4	2 2 2 4	2 2 2 4	3 3 3 4

Figure 12. Twenty-four original forms of distribution of starting chord.

CHAPTER 11

COORDINATE EXPANSION[1]

HAVING DISCUSSED the technique of geometrical *inversions*, we may now consider an additional set of techniques, those leading to geometrical *expansions*.

On an ordinary graph, the unit of measurement is equivalent to $\frac{1}{12}$ of an inch, and it represents, in this system of notation, the standard pitch-unit, *i.e.*, $\sqrt[12]{2}$ (a semitone). Such units are expressible in arithmetical integers as logarithms to the base of $\sqrt[12]{2}$. Thus, a semitone consists of one unit, a whole tone of two units, etc., along the ordinate.

A melodic graph may be translated into different absolute pitch values by substituting different coefficients for the original p.

To translate a musical graph into $\sqrt[6]{2}$ we would simply use double units on the ordinate for the original single units, while preserving all the other relations within a given melodic continuity. In this case, p = 2p. By using greater coefficients such as 3, 4, 5, 6 or 7 ($\sqrt[4]{2}$, $\sqrt[3]{2}$, $\sqrt[12]{2^5}$, $\sqrt{2}$, $\sqrt[12]{2^7}$), we obtain the respective units for the pitch intervals.

This form of projection is known as an *optical projection through extension of the ordinate*. It is one of the natural tendencies in visual arts. When artists attempt to produce a distortion (variation) of the original proportions, they are unconsciously attempting to achieve one or another form of geometrical projection.

These variations, when executed geometrically and in accordance with optics, give a greater amount of esthetic satisfaction because they are more natural.

On the next page you will find an example of the translation of one system of proportions into another, as applied to linear design.

[1] Reprinted with permission, from *The Schillinger System of Musical Composition*, Copyright, 1941, by Carl Fischer, Inc. This is an abbreviated form of Chapter 2 of Book III. (Ed.)

COORDINATE EXPANSION

Figure 1. Translating one system of proportions into another.

In the illustration above, the same configuration is presented under different coordinate ratios. The technique of such translation consists of producing a network on the original drawing (with as many units as is desirable with regard to precision) and then transcribing this network into a differently proportioned area, preserving the same number of lines on both coordinates of the network. Then all points of the drawing acquire their respective positions in the corresponding places of the network.

Compare these geometrical projections with the distortions in these and other paintings by El Greco and Modigliani.

Figure 2. El Greco.

*Figure 3. Modigliani.***

* *Metropolitan Museum of Art, New York.*
** *Collection, The Museum of Modern Art, New York.*

As each coefficient of expansion is applied to music, the original is translated into a different style, a style often separated by centuries. It is sufficient to translate music written in the 18th century by the coefficient 2 in order to obtain music of greater consistency than an original of the early 20th century style. For example, a higher quality Debussy-like music may be derived by translation of Bach or Handel into the coefficient 2.

The coefficient 3 is characteristic of any music based on $\sqrt[4]{2}$ (*i.e.*, the "diminished 7th" chord). Any high-quality piece of music of the past exhibits, *under such projection*, a greater versatility than any of the known samples that would stylistically correspond to it in the past. For the sake of comparing the intuitive patterns with the corresponding forms of geometrical projection, it is advisable to analyze such works as J. S. Bach's *Chromatic Fantasy and Fugue*, Liszt's *B Minor Sonata*, L. van Beethoven's *Moonlight Sonata*, first movement.

The coefficient 4, being a multiple of 2, gives too many recurrences of the same pitch-units since it is actually confined to but 3 units in an octave. Naturally, such music is thereby deprived of flexibility.

But the 5p expansion is characteristic of the modern school which utilizes the interval of the 4th—such as Hindemith, Berg, Krenek, etc. Music corresponding to further expansions, such as 7p, has some resemblance to the music written by Anton von Webern. Drawing comparisons between the music of Chopin and Hindemith, under the same coefficient of expansion, *i.e.*, either by expanding Chopin into the coefficient 5, or by contracting Hindemith into the coefficient 1, we find that the versatility of Chopin is much greater than that of Hindemith. Such a comparison may be made between any waltz of Chopin and the waltz written by Hindemith from his piano suite, *1922*.

Comparative study of music under various coefficients of expansion reveals that often we are more impressed by the raw material of intonation than by the actual quality of the composition.

The opposite of this procedure of expansion of pitch is *contraction* of pitch. Any pitch interval-unit may be contracted twice, three times, etc., which is expressible in $\sqrt[24]{2}$, $\sqrt[36]{2}$, etc., providing that instruments with corresponding tuning are devised. Those esthetes who usually love to talk about the "economy of material" and "maximum of expression" will perhaps be delighted to learn that an entire 4-part fugue of Bach occupying a range of $3\frac{1}{2}$ octaves would require only one whole tone if the pitch interval-unit were $\frac{1}{18}$ of a tone ($\sqrt[216]{2}$). Applying the same principle to the contraction of the absolute time duration-unit, we could hear this fugue in a few seconds instead of several minutes!

The natural pitch-scale, *i.e.*, the series of harmonics, does not produce uniform ratios but gives a natural logarithmic contraction. The intervals between the pitch-units decrease, while the absolute frequencies increase. This phenomenon is analogous to the *perspective* contraction in space *as we see it*. If music were devised on natural harmonic series, the relative group-coefficients of expansion and contraction could be used. But it seems that the natural harmonic series does not, in fact, provide any flexible material for musical intonation but merely for building up various tone qualities—for the fact is that a group of harmonics sounded at the same intensity produces one saturated unison rather than harmony. This phenomenon is somewhat similar to that of *white light*, in which all sprectral hues merge—becoming noticeable only when the beam is broken up. Logarithmic contraction of pitch combined with the logarithmic contraction of time *may* come into existence in the remote future in connection with the development of automatic instruments for composition and execution of music.

The technique of pitch-expansion may be executed directly from a graph or from a corresponding chromatic scale of expansion. In such a case, 2p will produce a whole tone scale progressing through 2 octaves instead of a full chromatic scale progressing through one octave (when $p = 1$.) While expansion of time extends the graph along the abscissa, the increase of the absolute time unit is not noticeable unless compared with the original. When we hear a musical continuity, we do not know (unless it is extremely exaggerated) whether it is the original velocity or a derivative thereof. The difference becomes apparent only when different coefficients of velocity of the same musical continuity are brought close together. Thus, time extension produces a different pattern on a graph without producing a difference detectable in the absence of comparison.

Pitch expansion works under the same conditions. It is only through comparison that we can learn that a certain musical continuity has been expanded or contracted from its original. This is apparent in the process of *tonal* expansion (which preserves all the pitch-units while the range increases).

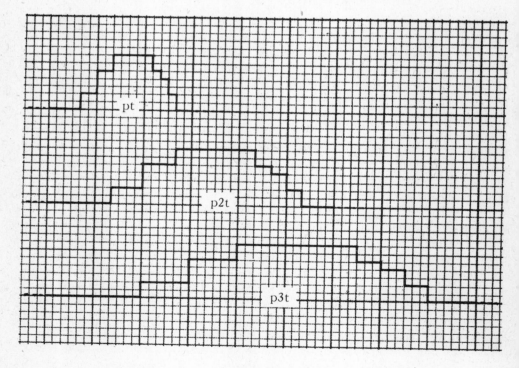

Figure 4. Time and pitch expansions (continued).

COORDINATE EXPANSION

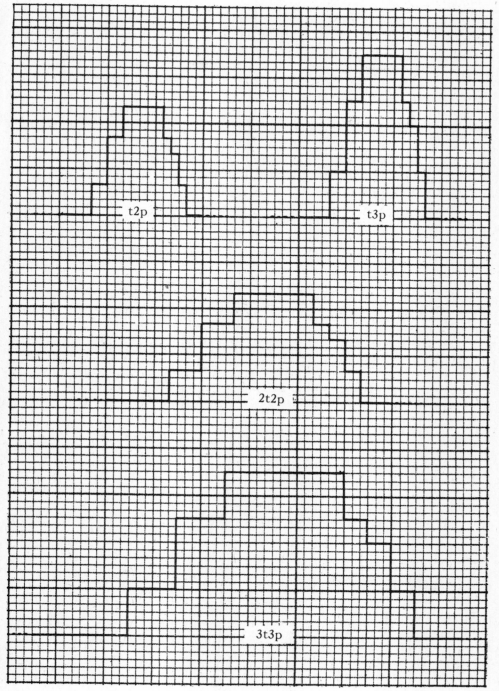

Figure 4. Time and pitch expansions (concluded).

250 THEORY OF REGULARITY AND COORDINATION

If pt represents the original, 2t and 3t produce the corresponding time expansions. Likewise, 2p and 3p produce the corresponding pitch expansions. The expansion through two coordinates preserves the absolute form of the configuration, merely magnifying it (2p2t and 3t3p).

It might seem at first that the ordinary enlargement or reduction of an original image—such as that effected by any natural optical projection (lantern slide projector, motion picture projector, magnifying glass, etc.)—does not change the appearance of the image. Yet when carried to an extreme, it does in fact transform the image to a great extent. For example, an ordinary close-up of a human head seen on the screen does not change our impression of the image. But when a human head is subjected to a several hundred power magnification, the original image is changed beyond recognition. A photograph of the skin surface of the human arm occupying only 1/100th of a square inch produces an image which is not easily associated with the human arm.

Thus, the difference in the actual sound of music (like the magnification of Haydn into von Webern) is only quantitatively different from the enlarging of visual images. Even with coefficients as low as 5, a melody is transformed beyond recognition. But the magnification of visual images requires at least one-hundred power magnification in order to achieve a similar effect.

It is interesting to note that bizarre effects of optical magnification are often due to the fact that such images are merely *hypothetic* and have no actual correspondences in the physical world of our planet. An image of a chicken can be magnified to the size of the Empire State building (for example, by being projected on an outdoor smoke screen), yet no real live chicken could exist on this planet even the size of an ostrich, because—as the volume grows in cubes—the legs of such a chicken could not support the weight of its body.

The following chart represents pitch expansions of the melody: graphed in Figure 4.

Figure 5. Pitch expansions of the melody of Figure 4 (continued).

Figure 5. Pitch expansions of the melody of Figure 4 (concluded).

Geometrical expansions of melody may also serve the purpose of modifying motifs through the method of geometrical projection. The original melodic pattern becomes entirely modified—yet the system of pitch-units is the outcome of a consistent translation from one system of pitch relations to another. The technique of such modification is equivalent to the contraction of the general pitch range emphasized by the geometrically expanded form. Some melodies, especially those with big coefficients of expansion, permit several different versions (degrees) of contraction.

The following example presents the exact geometrical expansions with the respective contractions of their ranges:

Figure 6. *Geometrical expansions with readjusted (contracted) range.*

The process of *range-contraction* often introduces new characteristics into geometrically expanded forms. For example, in the case of 5p in the preceding example: in its readjusted form, it seems to be more "conservative" than in its respective geometrical expansion. In the case of 7p, the contracted form is reminiscent of the music of Prokofief rather than that of von Webern.

Geometrical expansion of the harmony which accompanies melody expanded through the same coefficient (whether with readjusted range or not), must be performed from the pitch axis of the entire system (usually the root-tone).

COORDINATE EXPANSION

Figure 7. Geometrical expansion of a harmonized melody.

This translation of harmony may be accomplished either through transcription of a graph or through step by step translation from the original. One may also prepare in advance chromatic scales from the respective pitch axes where all the pitch-units may be found directly in the corresponding expansions.

Figure 8. Scale of pitch-units and their corresponding expansions.

All geometrical *expansions* are subject to geometrical *inversions* as well. A consistent musical continuity may be evolved through the variation of inversions under the same coefficient of expansion. Thus the two methods of mathematical variation of music, based on geometrical projection, bring an effective solution to two very important technical problems:

1. Composition of infinite melodic or harmonic continuity containing organically related contrasts.
2. Translation of music of one epoch into another, "modernization" and "antiquation."

CHAPTER 12

COMPOSITION OF DENSITY[1]

The behavior of sounding texture in any musical composition is such that it fluctuates between stability and instability, and so remains perpetually in a state of unstable equilibrium. The latter is characteristic of albumen which is chemically basic to all organic forms of nature. For this reason, unstable equilibrium is a manifestation of life itself, and, being applied to the field of musical composition as a formal principle, contributes the quality of life to music.

NOMENCLATURE:

d — density unit $\equiv$ p, S
D — simultaneous density-group $\equiv$ S, 2S, . . . Σ.
$\vec{D}$ — sequent density-group (consecutive D)
Δ (delta) — compound density-group representing density limit in a given score (simultaneous $\Delta \equiv \Sigma$)
$\vec{\Delta}$ (delta) — sequent compound density group: general symbol for the entire consecutive composition of density: $\vec{\Delta} = \Sigma$
$\vec{\Delta}(\vec{\Delta})$ — the delta of a delta: sequent compound delta.
ϕ (phi) — individual rotation-phase:
    $\phi \backsim$ and $\phi \backsim$ in reference to t or T
    ϕ ( ) and ϕ ( ) in reference to p or P, or d or D
θ (theta) — compound rotation-phase, general symbol of the continuity of rotary groups in a given score; it includes both forms of ϕ.

---

[1] From *The Schillinger System of Musical Composition*, Copyright, 1941, by Carl Fischer, Inc., by whose permission it is reprinted. This is an abbreviated version of Chapter 15 of Book IX. (Ed.)

COMPOSITION OF DENSITY

A. Technical Premise

Depending on the degree of refinement with which the composition of density is to be reflected in a score, d may equal p or S. In scores predominantly using individual parts, either as melodic or harmonic parts, it is possible and advisable to make d = p. In scores of predominantly contrapuntal type, where each melody is obtained from a complete S, d = S is a more practical form of assignment.

One of the fundamental forms of variation of the density-groups is rotation of phases.

The *abscissa* (horizontal) *rotation* follows the sequence of harmony (↻ or ↺); in it, all pitch-units (neutral or directional) follow the progression originally pre-set by harmony.

The *ordinate* (vertical) *rotation* does not refer to vertical displacement of p or S, but to thematic textures (melody, counterpoint, harmonic accompaniment) only; therefore there is *no vertical rearrangement of harmonic parts at any time*. Such displacement of simultaneously correlated S would completely change the harmonic meaning and the sounding characteristics of the original. Technically such schemes are possible only under the following conditions:

(1) identical interval of symmetry between all strata;
(2) identical structures with identical number of parts in all strata.

The above requirements impose limitations which are unnecessary in orchestral writing, as it means that each orchestral group would have to be represented by the same number of instruments, which is seldom practical.

The idea of *bi-coordinate rotation* (*i.e.*, through the abscissa and through the ordinate) implies that the whole scheme of density in a composition first appears as a graph on a plane, then is folded into a cylindrical (tubular) shape in such a fashion that the starting and the ending duration-units meet, *i.e.*, $\Delta^{\rightarrow} = \lim t_1 \leftrightarrow t_m$. Under such conditions the cylinder is the result of bending the graph through ordinate, and the cylinder itself appears in a vertical position. Variations are obtained by rotating this cylinder through abscissa, which corresponds to $\phi \circlearrowright$ and $\phi \circlearrowleft$.

Therefore: $\Delta^{\rightarrow} = \phi \circlearrowright (t_1 \rightarrow t_m), \phi \circlearrowleft (t_m \rightarrow t_1)$.

Folding the scheme of density (as it appears on the graph) in such a fashion that the lowest and the highest parts of the score meet, we obtain the limits for p, *i.e.*, $\Delta = \lim p_1 \leftrightarrow p_m$. Under such conditions the cylinder is the result of bending the graph through abscissa, and the cylinder itself appears in horizontal position. Variations are obtained by rotating this cylinder through ordinate, which corresponds to $\phi \circlearrowright$ and $\phi \circlearrowleft$. Therefore: $\Delta^{\rightarrow} = \phi \circlearrowright \begin{pmatrix} p_m \\ \uparrow \\ p_1 \end{pmatrix}, \phi \circlearrowleft \begin{pmatrix} p_m \\ \downarrow \\ p_1 \end{pmatrix}$.

Here delta is consecutive as physical time exists during the period of rotation.

B. Composition of Density-Groups

As we have mentioned before, the choice of p and t, or of S and T as density units, depends on the degree of refinement which is to be attributed to a certain particular score. For the sake of convenience and economy of space, we shall express dt as one square unit of cross-section paper. In each particular case, d may equal p or S, and T may equal t or mt. Yet we shall retain the dt unit of the graph in its general form.

Under such conditions a scale of density-time relations can be expressed as follows:

$$D = d, \quad D = 2d, \quad \ldots \quad D = md$$
$$\overrightarrow{D} = dt, \quad \overrightarrow{D} = d2t, \quad \ldots \quad \overrightarrow{D} = dmt$$
$$\overrightarrow{D} = dt, \quad \overrightarrow{D} = 2dt, \quad \ldots \quad \overrightarrow{D} = mdt$$
$$\overrightarrow{D} = dt, \quad \overrightarrow{D} = 2d2t, \quad \ldots \quad \overrightarrow{D} = mdnt$$

The above are monomial density-groups. On the graph they appear as follows:

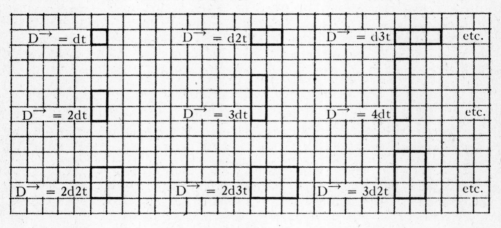

Figure 1. Monomial density-groups.

Binomial density-groups can be evolved in a similar way:

$$\overrightarrow{\Delta} = \overrightarrow{D_1} + \overrightarrow{D_2}; \quad \overrightarrow{D_1} = dt; \quad \overrightarrow{D_2} = 2d2t;$$
$$\overrightarrow{\Delta} = dt + 2d2t$$

Figure 2. Binomial density-groups (continued.)

$$\Delta^{\rightarrow} = D_1^{\rightarrow} + D_2^{\rightarrow}; \quad D_1^{\rightarrow} = d2t; \quad D_2^{\rightarrow} = 2dt;$$
$$\Delta^{\rightarrow} = d2t + 2dt$$

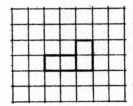

$$\Delta^{\rightarrow} = D_1^{\rightarrow} + D_2^{\rightarrow}; \quad D_1^{\rightarrow} = 2d3t; \quad D_2^{\rightarrow} = 5d3t;$$
$$\Delta^{\rightarrow} = 2d3t + 5d3t$$

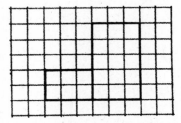

Figure 2. Binomial density-groups (concluded).

Polynomial density-groups may be evolved, depending on the purpose, from rhythmic resultants, permutation-groups, involution-groups, series of variable velocities, etc.

$$\Delta^{\rightarrow} = D_1^{\rightarrow} + D_2^{\rightarrow} + D_3^{\rightarrow}; \quad D_1^{\rightarrow} = 3d3t; \quad D_2^{\rightarrow} = dt; \quad D_3^{\rightarrow} = 2d2t;$$
$$\Delta^{\rightarrow} = 3d3t + dt + 2d2t$$

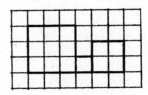

$$\Delta^{\rightarrow} = 4D^{\rightarrow}; \quad D_1^{\rightarrow} = 2d4t + 2d2t + 2d2t + 4d2t$$

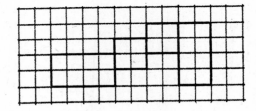

Figure 3. Polynomial density-groups.

As it follows from the above arrangement of density-groups, the latter may be distributed in any desirable fashion, preferably in a symmetric one within the range of D.

$\vec{\Delta} = 5\vec{D}$; $\vec{D_1} = d8t$; $\vec{D_2} = 2d5t$; $\vec{D_3} = 3d3t$; $\vec{D_4} = 5d2t$; $\vec{D_5} = 8dt$;
$\vec{\Delta} = d8t + 2d5t + 3d3t + 5d2t + 8dt$

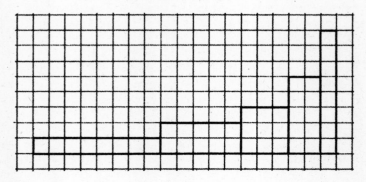

Another variant of the same scheme:

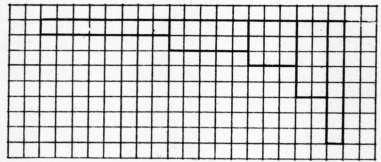

Another variant of the same scheme:

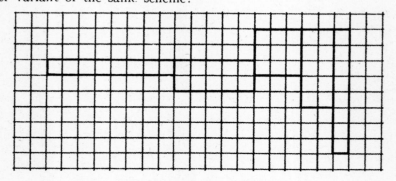

Figure 4. Variants of $\vec{\Delta} = 5\vec{D}$

In all the above cases $\Delta \not> D$, *i.e.*, the compound density-group is not greater than any of the component density groups.

COMPOSITION OF DENSITY

Density groups may be considerably smaller than Δ, in which case there are many more possibilities for the distribution of D's.

$$\Delta = 6D;\ \vec{\Delta} = 4D;\ \vec{D_1} = 2d2t;\ \vec{D_2} = dt;\ \vec{D_3} = dt;\ \vec{D_4} = 2d2t;$$
$$\vec{\Delta} = 2d2t + dt + dt + 2d2t$$

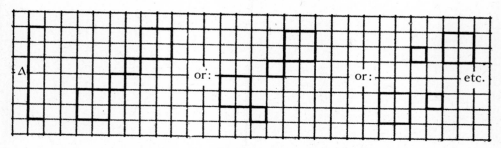

Figure 5. Density groups smaller than Δ.

The different distributions as in the above Figure can be specified by means of their phasic positions.

If we assume that the lowest d of Δ designates ϕ_0, *i.e.*, the zero phase, then $\phi_1, \phi_2, \ldots$ designate all the consecutive phases. Thus the first variant of Figure 5 can be expressed as follows:

$$\vec{\Delta} = (2d2t)\phi_0 + (dt)\phi_2 + (dt)\phi_3 + (2d2t)\phi_4.$$

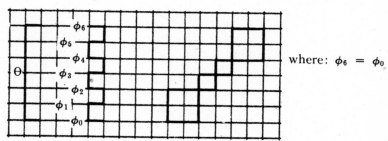

where: $\phi_6 = \phi_0$

Figure 6. First variant of Figure 5.

It follows from the above that the first (ϕ_0) and the last (ϕ_6) phases are identical.

C. Permutation of Sequent Density-Groups within the Compound Sequent Density-Group

(Permutations of $\vec{D}$ within $\vec{\Delta}$)

Continuity where permutations of $\vec{D}$'s take place can be designated as a compound sequent group consisting of several other compound sequent density groups, the latter being permutations of the original compound group. Then such a compound density-group yielding n permutations of the original compound sequent density group can be expressed as follows: $\vec{\Delta}(\vec{\Delta}) = \vec{\Delta_0} + \vec{\Delta_1} + \vec{\Delta_2} + \ldots \vec{\Delta_n}$.

$$\vec{\Delta_0} = (3d3t)\vec{D_1}\phi_0 + (dt)\vec{D_2}\phi_0 + (2d2t)\vec{D_3}\phi_0 + (2d2t)\vec{D_4}\phi_1 + \\ + (dt)\vec{D_5}\phi_2 + (3d3t)\vec{D_6}\phi_0, \text{ where } \Delta = 3d.$$

$$\vec{\Delta}(\vec{\Delta}) \subset = \vec{\Delta_0} + \vec{\Delta_1} + \vec{\Delta_2} + \vec{\Delta_3} + \vec{\Delta_4} + \vec{\Delta_5} = \\ = (\vec{D_1} + \vec{D_2} + \vec{D_3} + \vec{D_4} + \vec{D_5} + \vec{D_6}) + \\ + (\vec{D_2} + \vec{D_3} + \vec{D_4} + \vec{D_5} + \vec{D_6} + \vec{D_1}) + \\ + (\vec{D_3} + \vec{D_4} + \vec{D_5} + \vec{D_6} + \vec{D_1} + \vec{D_2}) + \\ + (\vec{D_4} + \vec{D_5} + \vec{D_6} + \vec{D_1} + \vec{D_2} + \vec{D_3}) + \\ + (\vec{D_5} + \vec{D_6} + \vec{D_1} + \vec{D_2} + \vec{D_3} + \vec{D_4}) + \\ + (\vec{D_6} + \vec{D_1} + \vec{D_2} + \vec{D_3} + \vec{D_4} + \vec{D_5})$$

See Figure 7 on opposite page.

COMPOSITION OF DENSITY

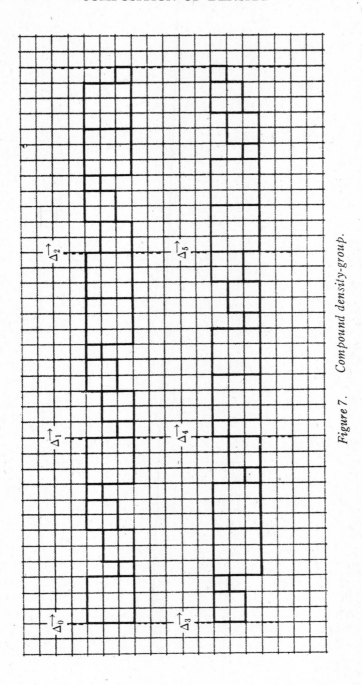

Figure 7. Compound density-group.

The same technique is applicable to all cases where $\Delta > D$, *i.e.*, where delta is greater than any of the simultaneous density-groups.

262 THEORY OF REGULARITY AND COORDINATION

D. Phasic Rotation of Δ and $\vec{\Delta}$ through t and d

Assuming $\vec{\Delta} = \vec{D} = dt$, we can subject it to rotation:

(1) $\vec{\Delta}\theta = t\phi_0 + t\phi_1 + \ldots$ and

(2) $\vec{\Delta}\theta = d\phi_0 + d\phi_1 + \ldots$

The following represents scales of rotation for $\vec{\Delta}\vec{T} = \vec{D}\vec{T} = dt$; $\Delta = 4d$; $\vec{T} = 4t$; $\phi_4 = \phi_0$. $\vec{T}$ symbolizes the range of duration of D.

The original position: $d\phi_0 \; t\phi_0$
The sequence of rotary phases of d:

$$\vec{\Delta}\theta = d\phi_0 \, t\phi_0 + d\phi_1 \, t\phi_0 + d\phi_2 \, t\phi_0 + d\phi_3 t\phi_0:$$

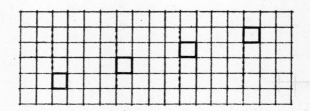

The sequence of rotary phases of t:

$$\vec{\Delta}\theta = d\phi_0 \, t\phi_0 + d\phi_0 \, t\phi_1 + d\phi_0 \, t\phi_2 + d\phi_0 \, t\phi_3:$$

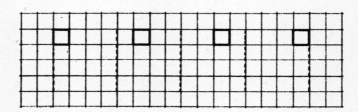

The sequence of rotary phases of dt:

$$\vec{\Delta}\theta = d\phi_0 \, t\phi_0 + d\phi_1 \, t\phi_1 + d\phi_2 \, t\phi_2 + d\phi_3 \, t\phi_3:$$

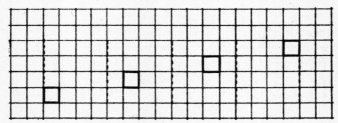

Figure 8. Phasic rotation of Δ and $\vec{\Delta}$

COMPOSITION OF DENSITY

The same technique is applicable to a $\vec{\Delta}$ of any desirable structure.

For example: $\vec{\Delta} = 3\vec{D}$; $\vec{D_1} = 3d3t$; $\vec{D_2} = dt$; $\vec{D_3} = 2d2t$;

$\Delta = 3d$ $\vec{\Delta_0} = (3d3t + dt + 2d2t)\phi_0$;
$\vec{T} = 6t$ $\vec{\Delta_1} = \vec{\Delta_0}\phi_1$; $\vec{\Delta_2} = \vec{\Delta_0}\phi_2$; $\vec{\Delta_3} = \vec{\Delta_0}\phi_3$; ...
$\Theta \subset$ and $()$ $\Delta_0 = (3d + d + 2d)\phi_0$;
     $\Delta_1 = \Delta_0\phi_1$; $\Delta_2 = \Delta_0\phi_2$; $\Delta_3 = \Delta_0\phi_3$; ...

Let $\vec{\Delta}(\vec{\Delta})\Theta = \vec{\Delta_0}(3d\phi_0 3t\phi_0 + d\phi_0 t\phi_0 + 2d\phi_0 2t\phi_0) +$
     $+ \vec{\Delta_1}(3d\phi_1 3t\phi_0 + d\phi_1 t\phi_0 + 2d\phi_1 2t\phi_0) +$
     $+ \vec{\Delta_2}(3d\phi_2 3t\phi_0 + d\phi_2 t\phi_0 + 2d\phi_2 t\phi_0)$.

Then, $\vec{\Delta}(\vec{\Delta})\Theta = \vec{\Delta_0} + \vec{\Delta_1} + \vec{\Delta_2}$ appears as follows:

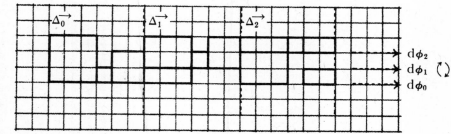

Let further $\vec{\Delta}(\vec{\Delta})\Theta = \vec{\Delta_0}(3d\phi_0 3t\phi_0 + d\phi_0 t\phi_0 + 2d\phi_0 2t\phi_0) +$
     $+ \vec{\Delta_1}(3d\phi_0 3t\phi_1 + d\phi_0 t\phi_1 + 2d\phi_0 2t\phi_1) +$
     $+ \vec{\Delta_2}(3d\phi_0 3t\phi_2 + d\phi_0 t\phi_2 + 2d\phi_0 2t\phi_2) +$
     $+ \vec{\Delta_3}(3d\phi_0 3t\phi_3 + d\phi_0 t\phi_3 + 2d\phi_0 2t\phi_3) +$
     $+ \vec{\Delta_4}(3d\phi_0 3t\phi_4 + d\phi_0 t\phi_4 + 2d\phi_0 2t\phi_4) +$
     $+ \vec{\Delta_5}(3d\phi_0 3t\phi_5 + d\phi_0 t\phi_5 + 2d\phi_0 2t\phi_5)$.

Then, $\vec{\Delta}(\vec{\Delta})\Theta = \vec{\Delta_0} + \vec{\Delta_1} + \vec{\Delta_2} + \vec{\Delta_3} + \vec{\Delta_4} + \vec{\Delta_5}$ appears as follows:

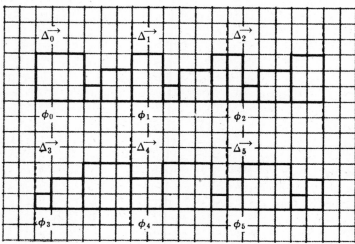

Figure 9.   Phasic rotation of $\vec{\Delta} = 3\vec{D}$ (continued).

Now we shall combine the $\theta()$ and the $\theta\backsim$.

Let $\vec{\Delta}(\vec{\Delta})\cdot\theta = \vec{\Delta_0}D\theta_0\vec{T}\theta_0 + \vec{\Delta_1}D\theta_1\vec{T}\theta_1 + \vec{\Delta_2}D\theta_2\vec{T}\theta_2 +$
$+ \vec{\Delta_3}D\theta_0\vec{T}\theta_3 + \vec{\Delta_4}D\theta_1\vec{T}\theta_4 + \vec{\Delta_5}D\theta_2\vec{T}\theta_5$

Then $\vec{\Delta}(\vec{\Delta})\theta = \vec{\Delta_0} + \vec{\Delta_1} + \vec{\Delta_2} + \vec{\Delta_3} + \vec{\Delta_4} + \vec{\Delta_5}$ appears as follows:

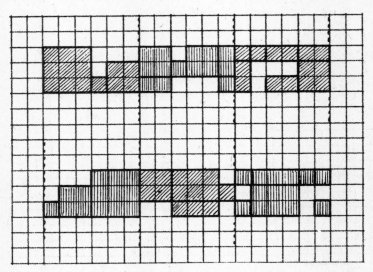

Figure 9. *Phasic rotation of $\vec{\Delta} = 3\vec{D}$ (concluded).*

The diagonal and vertical lines are inserted for clarity.

The addition of positive or negative phases of rotation to any given position of $\vec{\Delta}$ follows the rules of algebraic addition. Thus if the given position is ϕ_0, the addition of one $\phi\backsim$ or $()$ brings the density-group into position ϕ_1, or: $\phi_0 + \phi = \phi_1$. Likewise $\phi_0 + 2\phi = \phi_2$, $\phi_0 + m\phi = \phi_m$.

As the last phase equals the first phase, or $\phi_n = \phi_0$, negative quantities of phases, or the counterclockwise phases, *i.e.*, $\phi\backsim$ or $\phi()$, must be added with the sign minus to the last phase. Thus if the given position is ϕ_0 and the number of phases is n, the addition of one negative phase brings the density-group into position ϕ_{n-1}; or, $\phi_n - \phi = \phi_{n-1}$. Likewise, $\phi_n - 2\phi = \phi_{n-2}$, $\phi_n - m\phi = \phi_{n-m}$.

Problem: find the phase ϕ after the following forms of rotation have been performed from the original ϕ_0, where $\theta = 8\phi$: $2\phi - 3\phi + 5\phi + \phi - 4\phi + 3\phi - \phi$.

Solution: $\phi_x = \phi_0 + 2\phi - 3\phi + 5\phi + \phi - 4\phi + 3\phi - \phi = \phi_0 + 11\phi - 8\phi = \phi_0 + 3\phi = \phi_3$, *i.e.*, the density group appears in its third phase.

This is applicable to both ordinate and abscissa. It follows from the above reasoning that in order to obtain the original position θ_0, after performing a group of phasic rotations, the sum of the coefficients of ϕ must equal zero. As

COMPOSITION OF DENSITY

we know from the theory of rhythm, all resultants with an *even number of terms* have identical terms in both halves of the resultants. If such terms, used as coefficients of ϕ, are supplied with alternating "plus" and "minus", the sum of the whole resultant would be zero. This gives a perfect solution for the cases of variation of density groups, because resultants, being symmetric, produce a perfect form of continuity.

Examples:

$r_{4 \div 3} = 3+1+2+2+1+3$; changing the signs, we obtain:
$3-1+2-2+1-3 = 6-6 = 0$.

$r_{5 \div 4} = 4+1+3+2+2+3+1+4$; changing the signs, we obtain:
$4-1+3-2+2-3+1-4 = 10-10 = 0$.

$\theta(r_{7 \div 2}) = \phi_0 + 2\phi - 2\phi + 2\phi - \phi + \phi - 2\phi + 2\phi - 2\phi =$
$= \phi_0 + 7\phi - 7\phi = \phi_0 + 0 = \phi_0$.

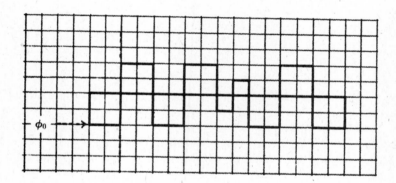

Figure 10. Applying resultants from the theory of rhythm.

Computation of the phasic position θ_x, which is the outcome of a group of phasic rotation, can be applied to any position θ_m to which such rotations have been applied. The computation is performed through the use of same technique as before, *i.e.*, through algebraic addition.

The technique of phasic rotation of the density-groups can be pursued to any desirable degree of refinement. The phases of d and t can be synchronized when they are subjected to independent rotary groups, in which case we follow the usual formula:

$$\frac{\theta d}{\theta t} = \frac{\theta^1 d}{\theta^1 t} ; \qquad \frac{\theta^1 t \, (\theta d)}{\theta^1 d \, (\theta t)}$$

In composing the original density-group ($\overrightarrow{\Delta_0}\overrightarrow{T_0}$), it is important to take into consideration the character of $\frac{D}{T}$ relations with regard to the effects such relations produce. In this respect we can rely on the three fundamental forms of correlation, *i.e.*, the parallel, the oblique and the contrary.

When they are applied to density-groups, these three forms must be interpreted in the following way:

(1) parallel: identical ratios of the coefficients of ϕd and ϕt;
(2) oblique: non-identical ratios of the coefficients of ϕd and ϕt, where—
   (a) partial coincidence of the coefficients takes place, and/or
   (b) the coefficient of one of the components (either d or t) remains constant;
(3) contrary: identical ratios arranged in inverted symmetry. When the number of coefficients in both coefficient-groups is odd, such case should be classified as *oblique*, due to partial coincidence of coefficients.

E. PRACTICAL APPLICATION OF $\overrightarrow{\Delta}$ TO $\overrightarrow{\Sigma}$.

(Composition of Variable Density from Strata)

In its complete form, this subject belongs to the field of Textural Composition and will be treated in this chapter only to the extent necessary in order to make the whole subject more tangible.

The first consideration is that $\overrightarrow{\Delta}$ can be composed to a given $\overrightarrow{\Sigma}$, or $\overrightarrow{\Sigma}$ can be composed to a given $\overrightarrow{\Delta}$. This means that either a progression of chords in strata or a density-group may be the origin of a whole composition. One harmonic progression may be combined with more than one density-group; the opposite is also true, *i.e.*, more than one harmonic progression can be written to the same group of density. For this reason the composer's work on such a scheme may start either with $\overrightarrow{\Sigma}$ or with $\overrightarrow{\Delta}$.

It is practical to consider d = S as the most general form of the density-unit, leaving d = p for cases of particular refinement with regard to density. If d = S it means that one density-unit may consist of p, 2p, 3p or 4p. In actuality, however, harmonic strata acquire instrumental forms, in which case even S4p may sound like rapidly moving melodies. On the other hand, S may be transformed into melody, in which case we also hear one part. The implication is that, in the average case, the density of a melodic line and the density of harmony subjected to instrumental figuration are about the same. Physically and physiologically, and therefore psychologically, *density is in direct proportion to mobility*. This means, for example, that a rapidly moving instrumental form of successive single attacks, which derives from S4p, is nearly as dense as a sustained chord of S4p; the extreme frequency of attacks makes an arpeggio sound like a chord, *i.e.*, in our perception, lines aggregate into an assemblage.

COMPOSITION OF DENSITY

The technique of superimposition of $\vec{\Delta}$ upon $\vec{\Sigma}$ consists of establishing correspondences between ϕd and p, and between ϕt and H, *i.e.*, between the density-phase, or density-unit, and the number of harmonic parts; and between the duration-phase, or duration-unit, of the sequent density-group and the number of successive chords.

All subsequent techniques pertain to composition of continuity, *i.e.*, to coordination of attacks and durations, instrumental forms, etc.

We shall now evolve an illustration of $\vec{\Delta}$ correlated with $\vec{\Sigma}$. To demonstrate this technique beyond doubt, we shall use the most refined form of it, where $d = p$ and $t = H$.

If $Nt = NH$, then the cycle of $\vec{\Delta}$ and $\vec{\Sigma}$ are synchronized a priori; otherwise, (*i.e.*, if $\frac{Nt}{NH} \neq 1$) they have to be synchronized. This shows that with just a few chords and a relatively brief scheme of density, one can evolve a composition of considerable length, since $\vec{\Delta}$ itself, in addition to interference with $\vec{H}$ of $\vec{\Sigma}$, can be subjected to rotational variations.

Let the original $\vec{\Delta_0} = \Delta = 8d$.

Let $\vec{\Delta_0} = \Delta\ \phi_0 2t + d\phi_0 t + 5d\phi_0 2t + 3d\phi_3 t + 2d\phi_0 2t$.

As $\Delta = D = 8d$, Σ must equal 8p.

$\vec{T} = 8t$ and would require $\vec{H} = 8H$, unless we wish to introduce a case in which $\frac{\vec{T}}{\vec{H}} \neq 1$.

We shall introduce such a case.
Let $\vec{H} = 5H$. Then $\frac{\vec{T}}{\vec{H}} = \frac{8}{5}$; $\frac{5}{8}\frac{(8)}{(5)}$.
Hence, $\vec{T}^1 = 8t \cdot 5 = 40t$.

As we intend to use 5 variations of $\vec{\Delta}$, the entire cycle will be synchronized (completed) in the form: $\vec{\Delta}(\vec{\Delta}) = 40t\ 40H$, where $\vec{H}\ (= 5H)$ appears 8 times.

For the sake of greater pliability of thematic textures, it is desirable to pre-set a directional sigma.

We shall choose the following sigma: $\Sigma = S_I 2p + S_{II} 3p + S_{III} 3p$ and $\vec{\Sigma} = 5H$.

Let $\vec{I} = 3i + 2i + 3i + 5i$ and $I(\Sigma) = \dfrac{\sqrt[12]{2}''}{\sqrt[6]{2}}$

THEORY OF REGULARITY AND COORDINATION

We shall now subject the $\overrightarrow{\Delta_0}, \Theta_0, \overrightarrow{\Sigma_0}$ to variations of density

$\overrightarrow{\Delta} \Theta = 8d\,2t + dt + 5d\,2t + 3dt + 2d\,2t$

Figure 11. $\overrightarrow{\Delta}$ correlated with $\overrightarrow{\Sigma}$

COMPOSITION OF DENSITY

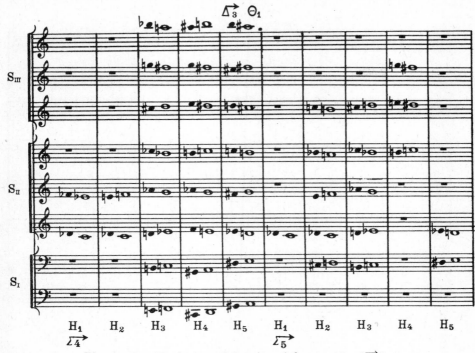

Figure 12. Variation of density of figure 11. $\vec{\Delta_3}\Theta_1$.

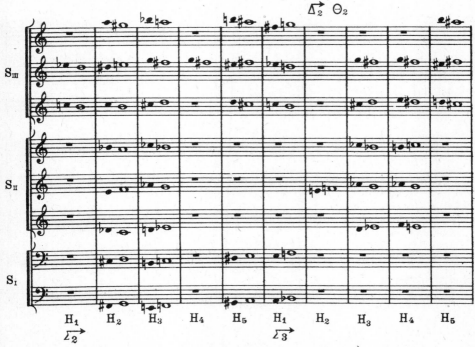

Figure 13. Variation of density of figure 11. $\vec{\Delta_2}\Theta_2$.

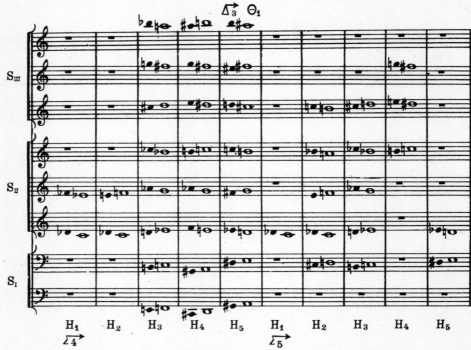

Figure 14. Variation of density of figure 11. $\vec{\Delta_3}\,\Theta_1$.

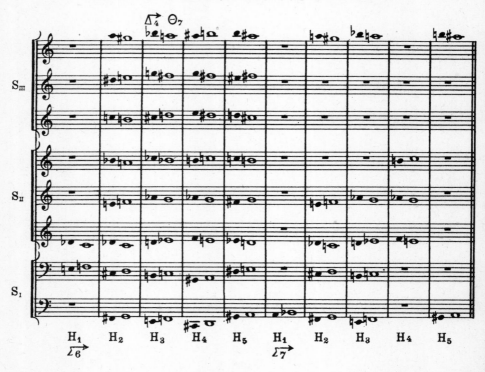

Figure 15. Variation of density of figure 11. $\vec{\Delta_4}\,\Theta_1$.

PART THREE
TECHNOLOGY OF ART PRODUCTION

Consecutive Selective Processes

1. Construction of a *rational set* from a *real set* by selecting the rational numbers and omitting the real.
2. Selection of a limit between two numbers of a rational set (cycles, wavelengths [Angstrom units]).
3. Logarithmic selection (exponent scale) of points within the given limits.
4. Establishing symmetric points within the logarithmic scale.
5. Establishing the forms of group-symmetry (binomial, trinomial, polynomial) within the logarithmic scale.
6. Developing the secondary selection: operand groups ≡ artistic scales (assymetric and symmetric).
7. Variation of the operand group
    (a) phase displacement (circular permutation of elements);
    (b) general permutation of elements;
    (c) quadrant rotation;
    (d) θ ↻ and θ ↺, θ ↶ and θ ↷ and their combinations (variation through coordinate rotation) applied to a compound operand group.
8. Composition of components of the compound operand groups through the process of two-coordinate development.

CHAPTER 1

SELECTIVE SYSTEMS

A. PRIMARY AND SECONDARY SELECTIVE SYSTEMS[1]

THE logarithmic dependence of ratios within a given limit-ratio is one of the more complex forms of dependence in the field of uniformity. Within the limit of $\frac{a}{b}$ ratio, for example, we can establish a scale of uniform n ratios. These may be represented on a straight line, the extension of which would correspond in frequencies to the ratio of $\frac{a}{b}$, by equidistant symmetric points. The first point of such linear extension would correspond to b, and the last point, to a. All the intermediate points of uniform symmetry would thereupon be represented as follows:

$$\sqrt[n]{\frac{a}{b}},\ \sqrt[n]{\left(\frac{a}{b}\right)^2},\ \sqrt[n]{\left(\frac{a}{b}\right)^3},$$

$$\sqrt[n]{\left(\frac{a}{b}\right)^n}\ \ldots\ldots\ldots\ldots,\ \frac{a}{b}$$

Such sets of uniform ratios may be regarded as *primary selective systems*. The number of points in a straight line, which is finite in itself, is infinite. The possible points in a line would thus be represented by irrational as well as rational number values. Another way of expressing this concept is to say that all of the possible number values in a line constitute a "dense set." Primary selective systems, as equidistant symmetric points, are not "dense sets."

Non-uniform forms of regularity, as we have previously shown, are the resultants of the interference of two or more uniform periodic waves of different frequencies brought into synchronization. These resultants, the parent shapes of all rhythms and the source of configurations, may be obtained either by direct computation or through graphs. When these resultants are applied in direct sequence to any of the primary selective systems, they, in turn, produce *secondary selective systems*.

As in the case of primary systems, secondary selective systems vary in density. When a secondary system reaches a point of saturation, it becomes identical with the parent primary system. When the primary system becomes saturated, it merges with the continuum. Thus, we may state that secondary selective series are the result of rarefying primary selective series, which constitute the dense set of the secondary.

---

[1] See Appendix C, which presents a primary selective system (tuning system) in music worked out by Schillinger for the execution of intonations not possible in our present tuning system. Schillinger called it Double Equal Temperament to distinguish it from our equal temperament system. (Ed.)

In linear design, secondary selective systems are the source of sequences of linear, plane, or solid motion, which result either in static configurations like spirals or polygons, or in trajectories of various types that include time as a component. In music, secondary selective systems produce sequences known as the rhythm of durations, of pitch (*i.e.*, pitch scales), of chord progressions, of intensities, of qualities, and of attacks.

The refinement of primary selective systems depends on the discriminatory capacities of perception. As sight is a more developed form of sensation and orientation, it permits the construction of primary selective systems that are denser sets than in the case of auditory orientation.

The primary selective system, which dominates the music of the western world, is known as equal temperament and consists of a series of 12 semitones. It is possible to construct a tuning system which would permit execution of other systems of intonation, such as mean temperament, just intonation, and the inflections of special types of intonation. The author has devised a system of tuning, "double equal temperament," which successfully unifies these systems of intonation. (See Appendix C.)

The material presented in the succeeding pages comprises secondary systems that are of fundamental importance in the various arts. The *temporal scales*, as the parent shapes of all rhythms, are useful in all the arts. The *pitch scales* are primarily of importance in music. The *scales of linear configuration* and the *color scales* apply to the graphic arts.

B. Temporal Scales

We measure time through the use of a clock system, which is based on a sequence of uniform moments corresponding to the set of natural integers. In the field of auditory association, as aroused by the ticking of a clock or metronome, such uniform time intervals, or periodicities, may be regarded as a primary selective system. Secondary selective systems may be constructed by abstracting durations from the uniform set, or by causing interference between two different uniform series. The rhythmic resultants constitute temporal scales, which are the basic forms of regularity and coordination, *i.e.*, of rhythm, in all the arts.

The reader is referred to Appendix A, which contains a detailed presentation of the forms of regularity and coordination. Among the rhythmic resultants presented in Appendix A are the following: (1) binary and ternary synchronization; (2) distributive involution groups; and (3) groups of variable velocity.

In Appendix C the reader will find a description of the Rhythmicon, the first modern instrument for composing music, or temporal scales, automatically. The rhythmic resultants produced by this instrument are based on the interference of generators from one to sixteen.

1. Rhythmic Resultants.

$r_{3 \div 2} = 2 + 1 + 1 + 2$

$r_{4 \div 3} = 3 + 1 + 2 + 2 + 1 + 3$

$r_{5 \div 2} = 2 + 2 + 1 + 1 + 2 + 2$

$r_{5 \div 3} = 3 + 2 + 1 + 3 + 1 + 2 + 3$

$r_{5 \div 4} = 4 + 1 + 3 + 2 + 2 + 3 + 1 + 4$

$r_{6 \div 5} = 5 + 1 + 4 + 2 + 3 + 3 + 2 + 4 + 1 + 5$

$r_{7 \div 2} = 2 + 2 + 2 + 1 + 1 + 2 + 2 + 2$

$r_{7 \div 3} = 3 + 3 + 1 + 2 + 3 + 2 + 1 + 3 + 3$

$r_{7 \div 4} = 4 + 3 + 1 + 4 + 2 + 2 + 4 + 1 + 3 + 4$

$r_{7 \div 5} = 5 + 2 + 3 + 4 + 1 + 5 + 1 + 4 + 3 + 2 + 5$

$r_{7 \div 6} = 6 + 1 + 5 + 2 + 4 + 3 + 3 + 4 + 2 + 5 + 1 + 6$

$r_{8 \div 3} = 3 + 3 + 2 + 1 + 3 + 3 + 1 + 2 + 3 + 3$

$r_{8 \div 5} = 5 + 3 + 2 + 5 + 1 + 4 + 4 + 1 + 5 + 2 + 3 + 5$

$r_{8 \div 7} = 7 + 1 + 6 + 2 + 5 + 3 + 4 + 4 + 3 + 5 + 2 + 6 + 1 + 7$

$r_{9 \div 2} = 2 + 2 + 2 + 2 + 1 + 1 + 2 + 2 + 2 + 2$

$r_{9 \div 4} = 4 + 4 + 1 + 3 + 4 + 2 + 2 + 4 + 3 + 1 + 4 + 4$

$r_{9 \div 5} = 5 + 4 + 1 + 5 + 3 + 2 + 5 + 2 + 3 + 5 + 1 + 4 + 5$

$r_{9 \div 7} = 7 + 2 + 5 + 4 + 3 + 6 + 1 + 7 + 1 + 6 + 3 + 4 + 5 + 2 + 7$

$r_{9 \div 8} = 8 + 1 + 7 + 2 + 6 + 3 + 5 + 4 + 4 + 5 + 3 + 6 + 2 + 7 + 1 + 8$

2. Rhythmic Resultants with Fractioning.

$r_{\underline{3 \div 2}} = 2 + 1 + 1 + 1 + 1 + 1 + 2$

$r_{\underline{4 \div 3}} = 3 + 1 + 2 + 1 + 1 + 1 + 1 + 2 + 1 + 3$

$r_{\underline{5 \div 2}} = 2 + 2 + 8(1) + 1 + 8(1) + 2 + 2$

$r_{5+3} = 3 + 2 + 1 + 2 + 4(1) + 1 + 4(1) + 2 + 1 + 2 + 2$

$r_{5+4} = 4 + 1 + 3 + 1 + 1 + 2 + 1 + 2 + 1 + 1 + 3 + 1 + 4$

$r_{6+5} = 5 + 1 + 4 + 1 + 1 + 3 + 1 + 2 + 2 + 1 + 3 + 1 + 1 + 4 + 1 + 5$

$r_{7+2} = 2 + 2 + 2 + 18(1) + 1 + 18(1) + 2 + 2 + 2$

$r_{7+3} = 3 + 3 + 1 + 2 + 1 + 2 + 12(1) + 1 + 12(1) + 2 + 1 + 2 + 1 + \\ + 3 + 3$

$r_{7+4} = 4 + 3 + 1 + 3 + 1 + 2 + 1 + 1 + 2 + 6(1) + 1 + 6(1) + 2 + \\ + 1 + 1 + 2 + 1 + 3 + 1 + 3 + 4$

$r_{7+5} = 5 + 2 + 3 + 2 + 2 + 1 + 2 + 2 + 1 + 1 + 1 + 2 + 1 + 2 + 1 + \\ + 1 + 1 + 2 + 2 + 1 + 2 + 2 + 3 + 2 + 5$

$r_{7+6} = 6 + 1 + 5 + 1 + 1 + 4 + 1 + 2 + 3 + 1 + 3 + 2 + 1 + 4 + 1 + \\ + 1 + 5 + 1 + 6$

$r_{8+3} = 3 + 3 + 2 + 1 + 2 + 1 + 2 + 18(1) + 18(1) + 2 + 1 + 2 + 1 + \\ + 2 + 3 + 3$

$r_{8+5} = 5 + 3 + 2 + 3 + 2 + 1 + 2 + 2 + 1 + 2 + 1 + 1 + 1 + 1 + 2 + \\ + 1 + 1 + 1 + 1 + 1 + 1 + 1 + 1 + 2 + 1 + 1 + 1 + 2 + 1 + \\ + 2 + 2 + 1 + 2 + 3 + 2 + 3 + 5$

$r_{8+7} = 7 + 1 + 6 + 1 + 1 + 5 + 1 + 2 + 4 + 1 + 3 + 3 + 1 + 4 + 2 + \\ + 1 + 5 + 1 + 1 + 6 + 1 + 7$

$r_{9+2} = 2 + 2 + 2 + 2 + 32(1) + 1 + 32(1) + 2 + 2 + 2 + 2$

$r_{9+4} = 4 + 4 + 1 + 3 + 1 + 3 + 1 + 1 + 2 + 1 + 1 + 2 + 16(1) + 1 + \\ + 16(1) + 2 + 1 + 1 + 2 + 1 + 1 + 3 + 1 + 3 + 1 + 4 + 4$

$r_{9+5} = 5 + 4 + 1 + 4 + 1 + 3 + 1 + 1 + 3 + 1 + 1 + 2 + 1 + 1 + 1 + \\ + 2 + 8(1) + 1 + 8(1) + 2 + 1 + 1 + 1 + 2 + 1 + 1 + 3 + 1 + \\ + 1 + 3 + 1 + 4 + 1 + 4 + 5$

$r_{9+7} = 7 + 2 + 5 + 2 + 2 + 3 + 2 + 2 + 2 + 1 + 2 + 2 + 3 + 1 + 1 + \\ + 2 + 3 + 2 + 1 + 1 + 3 + 2 + 2 + 1 + 2 + 2 + 2 + 3 + 2 + \\ + 2 + 5 + 2 + 7$

$r_{9+8} = 8 + 1 + 7 + 1 + 1 + 6 + 1 + 2 + 5 + 1 + 3 + 4 + 1 + 4 + 3 + \\ + 1 + 4 + 3 + 1 + 5 + 2 + 1 + 6 + 1 + 1 + 7 + 1 + 8$

C. Pitch Scales[2]

In the field of music, our equal temperament system of tuning constitutes the primary selective system. Equal temperament tuning involves 12 semitone units—c, c#, d, d#, e, f, f#, g, g#, a, a#, b and c. These units became established over a period of centuries, as reference points among all the possible frequencies that constitute the audible continuum.

Tuning is a problem of pitch, and pitch is a matter of frequency. In equal temperament tuning, the twelve tonal units are so related that the second c in the series above is an octave higher than the first c—*i.e.*, its frequency is twice the frequency of the first c. The intervening units comprise a series of uniform ratios, which are related as logarithms to the base $\sqrt[12]{2}$.

The entire series takes the following form:

$$C = 2^{\frac{0}{12}} = 1$$

$$C = 2^{\frac{1}{12}} = \sqrt[12]{2}$$

$$D = 2^{\frac{2}{12}} = \sqrt[6]{2}$$

$$D = 2^{\frac{3}{12}} = \sqrt[4]{2}$$

$$E = 2^{\frac{4}{12}} = \sqrt[3]{2}$$

$$F = 2^{\frac{5}{12}} = \sqrt[12]{2^5}$$

$$F = 2^{\frac{6}{12}} = \sqrt{2}$$

$$G = 2^{\frac{7}{12}} = \sqrt[12]{2^7}$$

$$G = 2^{\frac{8}{12}} = \sqrt{2^3}$$

$$A = 2^{\frac{9}{12}} = \sqrt[4]{2^3}$$

$$A = 2^{\frac{10}{12}} = \sqrt[6]{2^5}$$

$$B = 2^{\frac{11}{12}} = \sqrt[12]{2^{11}}$$

$$C = 2^{\frac{12}{12}} = 2$$

From this tuning system, certain sequences of tones, or scales, may be abstracted. These constitute secondary selective systems and provide the raw material for musical composition. A summary of the pitch scales possible in our tuning system follows.

[2] A comprehensive analysis of pitch scales will be found in Book II, *Theory of Pitch Scales* of *The Schillinger System of Musical Composition*. (Ed.)

1. Number of Pitch Scales in all Groups of One Axis[3]

Group I	2048
Group II	9217
Group III	48
Group IV	2000
Grand Total	13,313

2. Number of Pitch Scales in All Four Groups of Equal Temperament ($\sqrt[12]{2}$)

The First Group: $2'' = 2048$ scales
The Second Group: 9217
The Third Group: 48 scales
The Fourth Group: $2^7 + 2^8 + 2^9 + 2^4 + 2^6 + 2^{10} =$
$+ 128 + 256 + 512 + 16 + 64 + 1024 = 2000$

3. The Number of Pitch Scales in the First Group.

1 unit	(p)	1
2 units	(2p)	11
3 units	(3p)	55
4 units	(4p)	165
5 units	(5p)	330
6 units	(6p)	462
7 units	(7p)	462
8 units	(8p)	330
9 units	(9p)	165
10 units	(10p)	55
11 units	(11p)	11
12 units	(12p)	1
	$2'' =$	2048

---

[3]Schillinger does not follow the traditional system of classifying scales simply as major, minor and chromatic, for these classifications are patently not broad enough to encompass all possible scales, or to embrace even those modern scale forms that have, in recent years, become commonplaces in our musical vocabulary. The four large groups into which Schillinger divides pitch scales may be described as follows. Group I: Assymetric scales with on root-tone and a range of one octave. This group includes the seven-unit diatonic scales (known as "major" and "minor"), which constitute our traditional musical language and serve as the basis of traditional harmony. Group II: Expanded scales with one tonic and a range of more than one octave. These are obtained by tonally expanding the scales of the first group, i.e., by rearranging the mutual positions of the specific pitch-units. Group III: Symmetric scales with more than one root-tone and a range of one octave. Derived from roots of the number 2, these scales are based on pitch-units which constitute the symmetric points between the root-tones. Group IV: Symmetrical scales with more than one root-tone and a range exceeding an octave. Like the symmetric scales in Group III, these scales contain an equal number of semitones between pitch-units. (Ed.)

SELECTIVE SYSTEMS

GROUP I. ONE ROOT-TONE. RANGE: 11

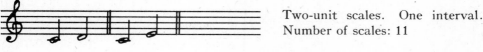
Two-unit scales. One interval.
Number of scales: 11

Three-unit scales. Two intervals.
Number of scales: 55

Four-unit scales. Three intervals.
Number of scales: 165

Seven-unit scales. Six intervals.
Number of scales: 462

Figure 1. Group I. Pitch-Scales.

4. The Number of Pitch-Scales in the Second Group

3p	1_E	× 55	=	55 scales
4p	2_E	× 165	=	330 scales
5p	3_E	× 330	=	990 scales
6p	4_E	× 462	=	1848 scales
7p	5_E	× 462	=	2310 scales
8p	6_E	× 330	=	1980 scales
9p	7_E	× 165	=	1155 scales
10p	8_E	× 55	=	440 scales
11p	9_E	× 11	=	99 scales
12p	10_E	× 1	=	10 scales

GROUP II. ONE ROOT-TONE. RANGE: OVER 12

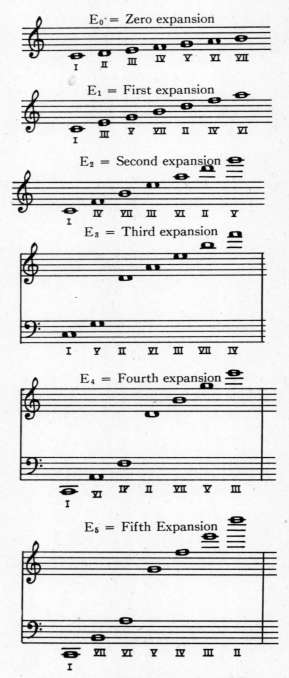

Figure 2. *Pitch-Scales of Group II*.

SELECTIVE SYSTEMS

5. The Third Group of Pitch-Scales (Symmetric System)

(1) 1 0 (4) $\sqrt[4]{2}$ $\frac{12}{4} = 3$

(2) $\sqrt{2}$ $\frac{12}{2} = 6$ (5) $\sqrt[6]{2}$ $\frac{12}{6} = 2$

(3) $\sqrt[3]{2}$ $\frac{12}{3} = 4$ (6) $\sqrt[12]{2}$ $\frac{12}{12} = 1$

GROUP III. MORE THAN ONE ROOT-TONE. RANGE: 12

Figure 3. Pitch-Scales of Group III.

TECHNOLOGY OF ART PRODUCTION

6. *The Fourth Group of Pitch-Scales (Symmetric System)*

(1) $\sqrt[3]{4}$ $\frac{24}{3} = 8$ (4) $\sqrt[12]{32}$ $\frac{60}{12} = 5$

(2) $\sqrt[4]{8}$ $\frac{36}{4} = 9$ (5) $\sqrt[12]{128}$ $\frac{84}{12} = 7$

(3) $\sqrt[6]{32}$ $\frac{60}{6} = 10$ (6) $\sqrt[12]{2048}$ $\frac{132}{12} = 11$

GROUP IV. MORE THAN ONE ROOT-TONE
RANGES: 24, 36, 60, 132

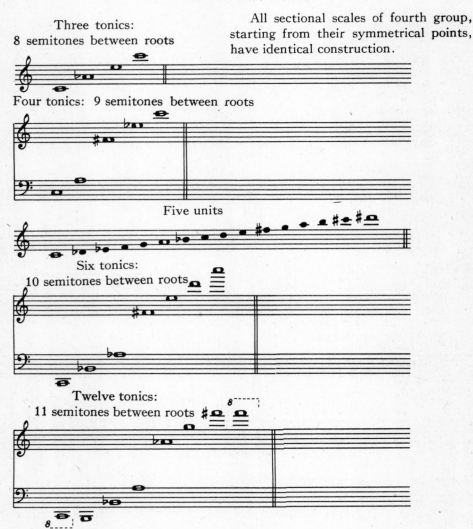

The final C is 11 octaves above beginning C.

Figure 4. Pitch-Scales of Group IV.

7. Symmetric Scales of Equal Temperament

Group	Range of Symmetry	Interval of Symmetry	No. of Semitones	Number of Tonics	Number of Scales	Value
III.	1	1	0	1	1	1
III.	2	$\sqrt[12]{2}$	$\frac{12}{12} = 1$	12	1	1
III.	2	$\sqrt[6]{2}$	$\frac{12}{6} = 2$	6	2	2
III.	2	$\sqrt[4]{2}$	$\frac{12}{4} = 3$	4	4	2^2
III.	2	$\sqrt[3]{2}$	$\frac{12}{3} = 4$	3	8	2^3
IV.	32	$\sqrt[12]{32}$	$\frac{60}{12} = 5$	12	16	2^4
III.	2	$\sqrt{2}$	$\frac{12}{2} = 6$	2	32	2^5
IV.	128	$\sqrt[12]{128}$	$\frac{84}{12} = 7$	12	64	2^6
IV.	4	$\sqrt[3]{4}$	$\frac{24}{3} = 8$	3	128	2^7
IV.	8	$\sqrt[4]{8}$	$\frac{36}{4} = 9$	4	256	2^8
IV.	32	$\sqrt[6]{32}$	$\frac{60}{6} = 10$	6	512	2^9
IV.	2048	$\sqrt[12]{2048}$	$\frac{132}{12} = 11$	12	1024	2^{10}

Figure 5. Summary Analysis of Symmetric Scales.

Six forms of symmetry in group III., and six forms of symmetry in group IV.

TECHNOLOGY OF ART PRODUCTION

D. SCALES OF LINEAR CONFIGURATION AND AREA.[4]

1. *Periodicity of Dimensions.*

A. Monomial Periodicity of Rectilinear Segments Moving Under a Constant Angle.

a. Rectilinear Segments.
   1. 0°, 30°, 60°, 90°, 120°, 150°, 180°, 210°, 240°, 270°, 300°, 330°, 360°. (See Figure 1.)

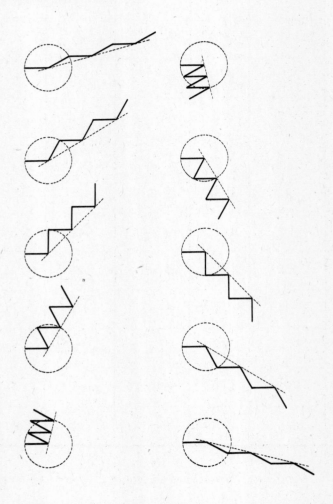

Figure 1. Rectilinear segments moving under a constant angle.

---

[4]Students will find that some of these scales are illustrated. It was Schillinger's intention that the student work out the others as exercises. (Ed.)

SELECTIVE SYSTEMS

b. 180° Arcs Moving in a Constant Clockwise Direction.
   1. 0°, 30°, 60°, 90°, 120°, 150°, 180°, 210°, 240°, 270°, 300°, 330°, 360°. (See Figure 2.)

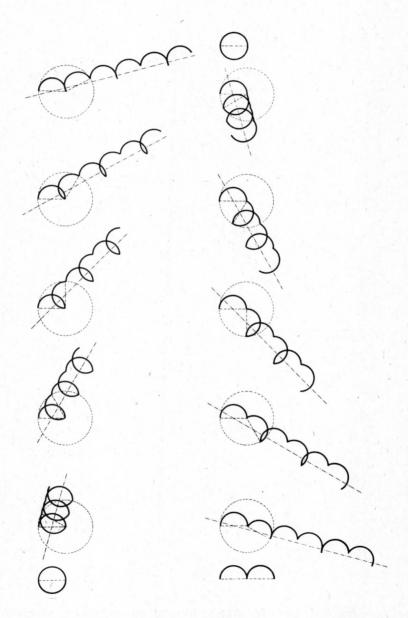

Figure 2. 180° arcs moving in a constant clockwise direction.

c. 180° Arcs Moving in a Constant Alternating Direction.
   1. 0°, 30°, 60°, 90°, 120°, 150°, 180°, 210°, 240°, 270°, 300°, 330°, 360°. (See Figure 3.)

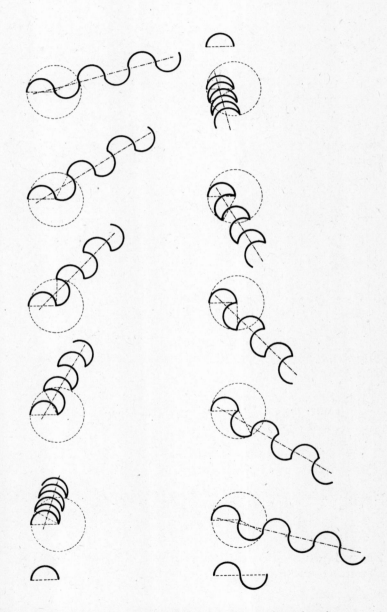

Figure 3. 180° arcs moving in a constant alternating direction.

SELECTIVE SYSTEMS 287

B. **Binomial Periodicity of Rectilinear Segments Moving Under a Constant Angle. (2 + 1) + . . .**

  a. Rectilinear Segments.
    1. 0°, 30°, 60°, 90°, 120°, 150°, 180°, 210°, 240°, 270°, 300°, 330°, 360°.

  b. 180° Arcs Moving in a Constant Clockwise Direction.
    1. 0°, 30°, 60°, 90°, 120°, 150°, 180°, 210°, 240°, 270°, 300°, 330°, 360°.

  c. 180° Arcs Moving in a Constant Alternating Direction.
    1. 0°, 30°, 60°, 90°, 120°, 150°, 180°, 210°, 240°, 270°, 300°, 330°, 360°.

C. **Binomial Periodicity of Rectilinear Segments Moving Under a Constant Angle. (3 + 1) + . . .**

  a. Rectilinear Segments.
    1. 0°, 30°, 60°, 90°, 120°, 150°, 180°, 210°, 240°, 270°, 300°, 330°, 360°.

  b. 180° Arcs Moving in a Constant Clockwise Direction.
    1. 0°, 30°, 60°, 90°, 120°, 150°, 180°, 210°, 240°, 270°, 300°, 330°, 360°.

  c. 180° Arcs Moving in a Constant Alternating Direction.
    1. 0°, 30°, 60°, 90°, 120°, 150°, 180°, 210°, 240°, 270°, 300°, 330°, 360°.

D. **Binomial Periodicity of Rectilinear Segments Moving Under a Constant Angle. (3 + 2) + . . .**

  a. Rectilinear Segments.
    1. 0°, 30°, 60°, 90°, 120°, 150°, 180°, 210°, 240°, 270°, 300°, 330°, 360°.

  b. 180° Arcs Moving in a Constant Clockwise Direction.
    1. 0°, 30°, 60°, 90°, 120°, 150°, 180°, 210°, 240°, 270°, 300°, 330°, 360°.

  c. 180° Arcs Moving in a Constant Alternating Direction.
    1. 0°, 30°, 60°, 90°, 120°, 150°, 180°, 210°, 240°, 270°, 300°, 330°, 360°.

E. **Binomial Periodicity of Rectilinear Segments Moving Under a Constant Angle. (4 + 3) + . . .**

  a. Rectilinear Segments.
    1. 0°, 30°, 60°, 90°, 120°, 150°, 180°, 210°, 240°, 270°, **300°**, 330°, 360°.

b. 180° Arcs Moving in a Constant Clockwise Direction.
    1. 0°, 30°, 60°, 90°, 120°, 150°, 180°, 210°, 240°, 270°, 300°, 330°, 360°.
  c. 180° Arcs Moving in a Constant Alternating Direction.
    1. 0°, 30°, 60°, 90°, 120°, 150°, 180°, 210°, 240°, 270°, 300°, 330°, 360°.

F. **Trinomial Periodicity of Rectilinear Segments Moving Under a Constant Angle. (3 + 2 + 1) + . . .**
  a. Rectilinear Segments.
    1. 0°, 30°, 60°, 90°, 120°, 150°, 180°, 210°, 240°, 270°, 300°, 330°, 360°.
  b. 180° Arcs Moving in a Constant Clockwise Direction.
    1. 0°, 30°, 60°, 90°, 120°, 150°, 180°, 210°, 240°, 270°, 300°, 330°, 360°.
  c. 180° Arcs Moving in a Constant Alternating Direction.
    1. 0°, 30°, 60°, 90°, 120°, 150°, 180°, 210°, 240°, 270°, 300°, 330°, 360°.

INFINITE SERIES

G. **Infinite Series of Rectilinear Segments Moving Under a Constant Angle—Series I: 1 + 2 + 3 + 4 + 5 + 6 + . . .**
  a. Rectilinear Segments.
    1. 0°, 30°, 60°, 90°, 120°, 150°, 180°, 210°, 240°, 270°, 300°, 330°, 360°.
  b. 180° Arcs Moving in a Constant Clockwise Direction.
    1. 0°, 30°, 60°, 90°, 120°, 150°, 180°, 210°, 240°, 270°, 300°, 330°, 360°.
  c. 180° Arcs Moving in a Constant Alternating Direction.
    1. 0°, 30°, 60°, 90°, 120°, 150°, 180°, 210°, 240°, 270°, 300°, 330°, 360°.

H. **Infinite Series of Rectilinear Segments Moving Under a Constant Angle—Series II: 1 + 3 + 5 + 7 + 9 + . . .**
  a. Rectilinear Segments.
    1. 0°, 30°, 60°, 90°, 120°.
    2. 150°, 180°, 210°.
    3. 240°, 270°, 300°, 330°, 360°.

SELECTIVE SYSTEMS

  b. 180° Arcs Moving in a Constant Clockwise Direction.
  1. 0°, 30°, 60°, 90°, 120°.
  2. 150°, 180°, 210°.
  3. 240°, 270°, 300°, 330°, 360°.

  c. 180° Arcs Moving in a Constant Alternating Direction.
  1. 0°, 30°, 60°, 90°, 120°.
  2. 150°, 180°, 210°.
  3. 240°, 270°, 300°, 330°, 360°.

I. **Infinite Series of Rectilinear Segments Moving Under a Constant Angle—Series III:** $1 + 2 + 3 + 5 + 8 + 13 + \ldots$

  a. Rectilinear Segments.
  1. 0°, 30°, 60°, 90°, 120°.
  2. 150°, 180°, 210°.
  3. 240°, 270°, 300°, 330°, 360°.

  b. 180° Arcs Moving in a Constant Clockwise Direction.
  1. 0°, 30°, 60°, 90°, 120°.
  2. 150°, 180°, 210°.
  3. 240°, 270°, 300°, 330°, 360°.

  c. 180° Arcs Moving in a Constant Alternating Direction.
  1. 0°, 30°, 60°, 90°, 120°.
  2. 150°, 180°, 210°.
  3. 240°, 270°, 300°, 330°, 360°.

J. **Infinite Series of Rectilinear Segments Moving Under a Constant Angle—Series IV:** $1 + 2 + 4 + 7 + 11 + 16 + \ldots$

  a. Rectilinear Segments.
  1. 0°, 30°, 60°, 90°, 120°.
  2. 150°, 180°, 210°.
  3. 240°, 270°, 300°, 330°, 360°.

  b. 180° Arcs Moving in a Constant Clockwise Direction.
  1. 0°, 30°, 60°, 90°, 120°.
  2. 150°, 180°, 210°.
  3. 240°, 270°.
  4. 300°, 330°, 360°.

c. 180° Arcs Moving in a Constant Alternating Direction.
   1. 0°, 30°, 60°, 90°, 120°.
   2. 150°, 180°, 210°.
   3. 240°, 270°.
   4. 300°, 330°, 360°.

K. **Infinite Series of Rectilinear Segments Moving Under a Constant 90° Angle.**
   a. Rectilinear Segments.
      1. Series I, Series II (See Figure 4).
      2. Series III, Series IV.

1, 2, 3, 4, 5 17   1, 3, 5, 7, 9, 11, 13, 15, 17, 19, 21, 23, 25

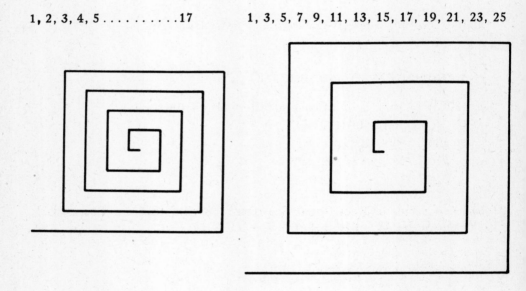

Figure 4. Infinite series of rectilinear segments moving under a constant angle.

SELECTIVE SYSTEMS

b. 180° Arcs Moving in a Constant Clockwise Direction.
   1. Series I, Series II (See Figure 5).
   2. Series III, Series IV.

Figure 5. 180° arcs moving in a constant clockwise direction.

TECHNOLOGY OF ART PRODUCTION

c. 180° Arcs Moving in a Constant Alternating Direction.
  1. Series I, Series II (See Figure 6).
  2. Series III, Series IV.

1, 2, 3, 4, 5, 6, 7 15   1, 3, 5, 7, 9 21

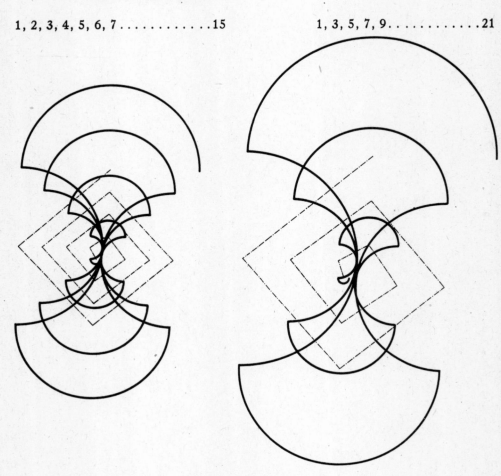

Figure 6. 180° arcs moving in a constant alternating direction.

L. **180° Arcs Moving in a Constant Clockwise Direction, Using 180° As Diameters.**
  1. Series I, Series II.
  2. Series III, Series IV.

2. *Periodicity of Angles.*

A. Monomial Periodicity of Angles, with Constant Dimensions of Rectilinear Segments.

a. Rectilinear Segments.
   1. 10°, 20°, 30°, 40°, 60°, 90°. (See Figure 7.)

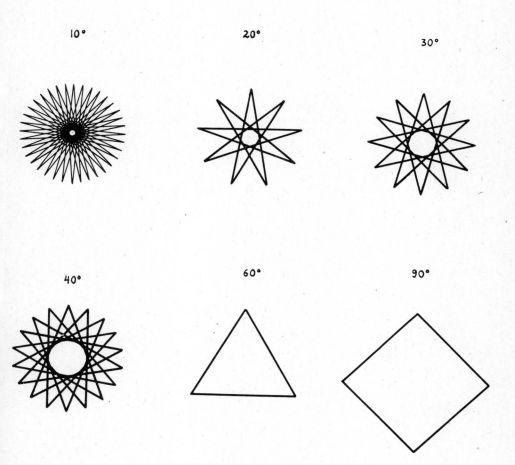

Figure 7. Monomial periodicity of angles, with constant dimensions of rectilinear segments.

b. 180° Arcs Moving in a Constant Clockwise Direction.
   (Rectilinear Segments used as Diameters.)
   1. 20°, 30°, 40°, 60°, 90°. (See Figure 8.)

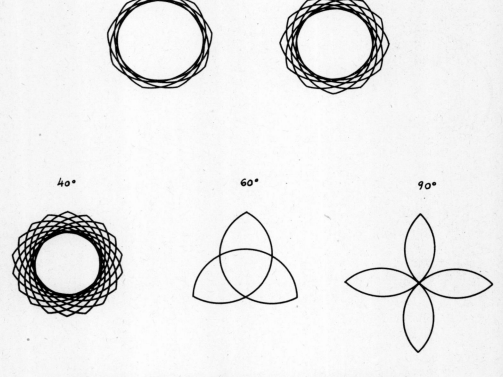

Figure 8. 180° arcs moving in a constant clockwise direction.

SELECTIVE SYSTEMS

c. 180° Arcs Moving in a Constant Alternating Direction.
   (Rectilinear Segments Used as Diameters.)
   1. 20°, 30°, 40°, 60°, 90°. (See Figure 9.)

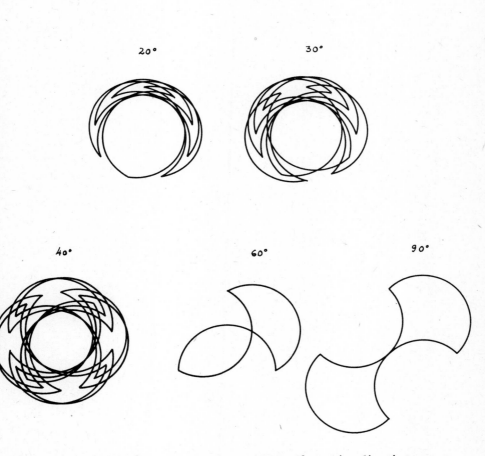

Figure 9. 180° arcs moving in a constant alternating direction.

d. Rectilinear Segments.
   1. 120°, 140°, 150°.
e. 180° Arcs Moving in a Constant Clockwise Direction.
   1. 120°, 140°, 150°.
f. 180° Arcs Moving in a Constant Alternating Direction.
   1. 120°, 140°, 150°.

296   TECHNOLOGY OF ART PRODUCTION

g. Rectilinear Segments.
  1. 150°, 160°, 170°. (See Figure 10.)

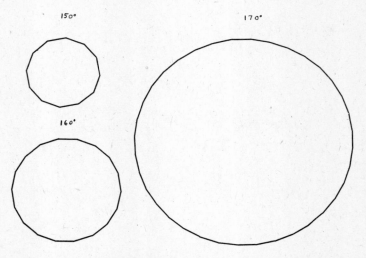

Figure 10. Monomial periodicity of angles with constant dimensions of rectilinear segments.

h. 180° Arcs Moving in a Constant Clockwise Direction.
  1. 150°, 160°, 170°. (See Figure 11.)

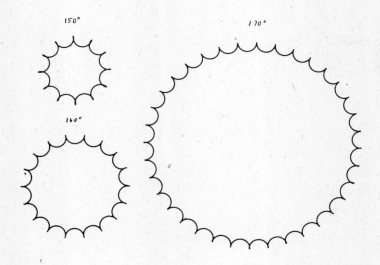

Figure 11. 180° arcs moving in a constant clockwise direction.

SELECTIVE SYSTEMS

i. 180° Arcs Moving in a Constant Alternating Direction.
   1. 150°, 160°, 170°. (See Figure 12.)

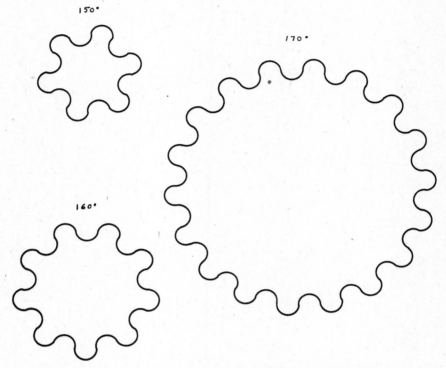

Figure 12. *180° arcs moving in a constant alternating direction.*

B. **Binomial Periodicity of Angles, with Constant Dimensions of Rectilinear Segments.**
   a. Rectilinear Segments.
      1. (10° + 20°) + . . . (10° + 30°) + . . . (20° + 40°) + . . .
         . . . + (30° + 60°) + . . . (30° + 90°) + . . . (See Figure 13.)

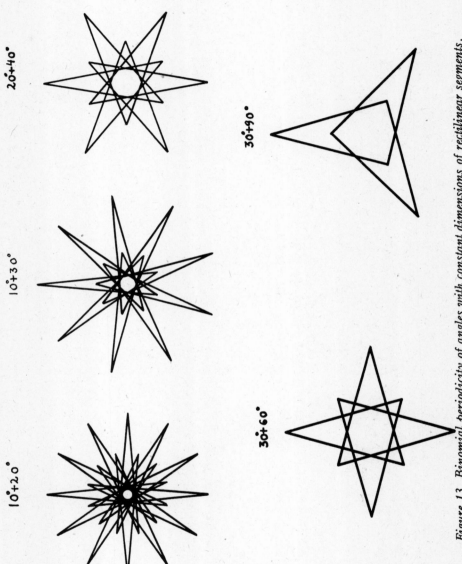

Figure 13. Binomial periodicity of angles with constant dimensions of rectilinear segments.

SELECTIVE SYSTEMS 299

b. 180° Arcs Moving in a Constant Clockwise Direction.
   (Rectilinear Segments Used as Diameters.)
   1. (10° + 20°) + . . . (10° + 30°) + . . . (20° + 40°) + . . .
      . . . + (30° + 60°) + . . . (30° + 90°) + . . . (See Figure 14.)

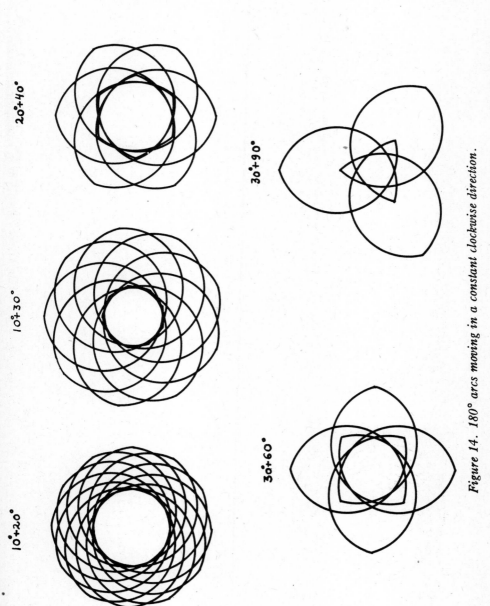

Figure 14. 180° arcs moving in a constant clockwise direction.

300 TECHNOLOGY OF ART PRODUCTION

c. 180° Arcs Moving in a Constant Alternating Direction.
   (Rectilinear Segments Used as Diameters.)
   1. (10° + 20°) + . . . (10° + 30°) + . . . (20° + 40°) + . . .
      . . . + (30° + 60°) + . . . (30° + 90°) + . . . (See Figure 15.)

Figure 15. 180° arcs moving in a constant alternating direction.

SELECTIVE SYSTEMS 301

C. Trinomial Periodicity of Angles.

a. Rectilinear Segments; 180° Arcs Moving in a Constant Counter-Clockwise Direction; 180° Arcs Moving in a Constant Clockwise Direction. 1 + 2 + 6.

1. 10° + 20° + 60°. (See Figure 16.)

180° arcs moving in a constant clockwise direction

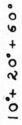
10° + 20° + 60°

180° arcs moving in a constant counter-clockwise direction

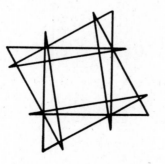
Rectilinear segments

Figure 16. Trinomial periodicity of angles.

b. Rectilinear Segments. 1 + 2 + 3.
   1. (10° + 20° + 30°) + . . . (15° + 30° + 45°) + . . .
      . . . + (18° + 36° + 54°) + . . .
   2. (20° + 40° + 60°) + . . . (30° + 60° + 90°) + . . .
c. 180° Arcs Moving in a Constant Clockwise Direction. 1 + 2 + 3.
   1. (10° + 20° + 30°) + . . . (15° + 30° + 45°) + . . .
      . . . + (18° + 36° + 54°) + . . .
   2. (20° + 40° + 60°) + . . . (30° + 60° + 90°) + . . .
d. 180° Arcs Moving in a Constant Alternating Direction. 1 + 2 + 3.
   1. (10° + 20° + 30°) + . . . (15° + 30° + 45°) + . . .
      . . . + (18° + 36° + 54°) + . . .
   2. (20° + 40° + 60°) + . . . (30° + 60° + 90°) + . . .

3. Rectilinear Segments Forming Angles in Alternating Directions

A. Rectilinear Segments.

a. Binomials by the Sum.
   1. S =   30°, 5 + 1. 4 + 1, 3 + 1, 2 + 1, 3 + 2 (See Figure 17.)
   2. S =   60°, 5 + 1, 4 + 1, 3 + 1, 2 + 1, 3 + 2, 5 + 4 (See Figure 18.)
   3. S =   90°, 5 + 1, 3 + 1, 2 + 1, 3 + 2, 5 + 4
   4. S = 120°, 4 + 1, 3 + 1, 2 + 1, 3 + 2, 7 + 5
   5. S = 150°, 4 + 1, 3 + 2, 8 + 7
   6. S = 180°, 5 + 1, 3 + 1, 2 + 1, 11 + 7, 3 + 2, 5 + 4
   (See Figure 19.)

SELECTIVE SYSTEMS

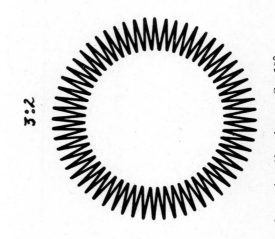

Figure 17. *Rectilinear segments forming angles in alternating directions.* $S = 30°$.

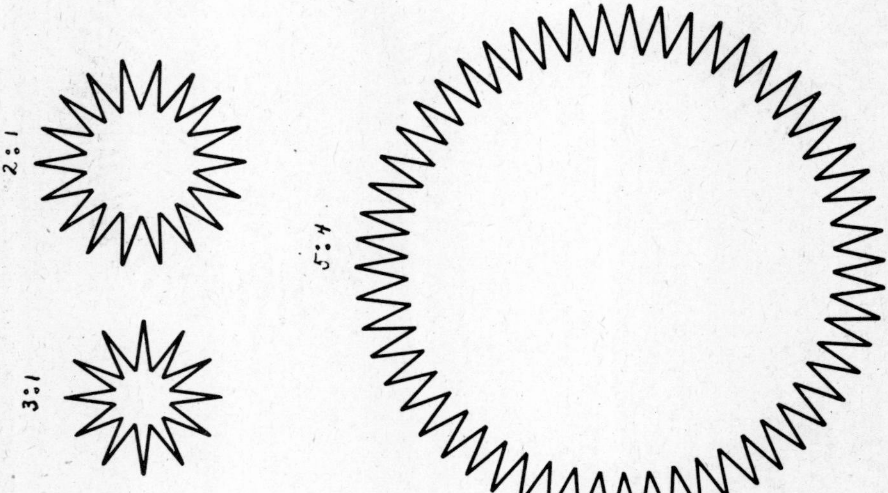

Figure 18. *Rectilinear segments forming angles in an alternating direction.* $S = 60°$

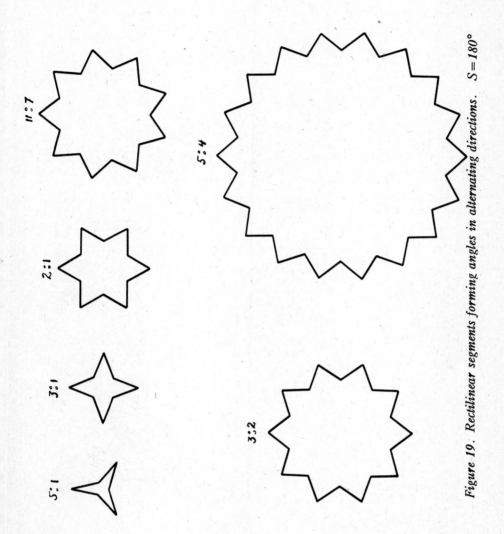

Figure 19. Rectilinear segments forming angles in alternating directions. $S = 180°$

b. Binomials by the Initial Unit.
  1. $10° = 1, 7 + 1, 5 + 1, 4 + 1, 3 + 1, 2 + 1$
  2. $10° = 1, 5 + 2, 3 + 2, 7 + 3, 5 + 3, 4 + 3$
  3. $10° = 1, 7 + 4, 5 + 4$
  4. $10° = 1, 8 + 5, 7 + 5, 6 + 5$
  5. $10° = 1, 7 + 6, 11 + 7, 9 + 7$
  6. $10° = 1, 8 + 7$
  7. $10° = 1, 9 + 8$

B. 180° Arcs Moving in a Constant Direction.

(Rectilinear Segments Serve as Diameters.)

a. Binomials by the Sum.
   1. S =   30°, 5 + 1, 4 + 1, 3 + 1, 2 + 1, 3 + 2   (See Figure 20.)
   2. S =   60°, 5 + 1, 4 + 1, 3 + 1, 2 + 1, 3 + 2, 5 + 4
      (See Figure 21.)
   3. S =   90°, 5 + 1, 3 + 1, 2 + 1, 3 + 2, 5 + 4
   4. S = 120°, 4 + 1, 3 + 1, 2 + 1, 3 + 2, 7 + 5
   5. S = 150°, 4 + 1, 3 + 2, 8 + 7
   6. S = 180°, 5 + 1, 3 + 1, 2 + 1, 11 + 7, 3 + 2, 5 + 4
      (See Figure 22.)

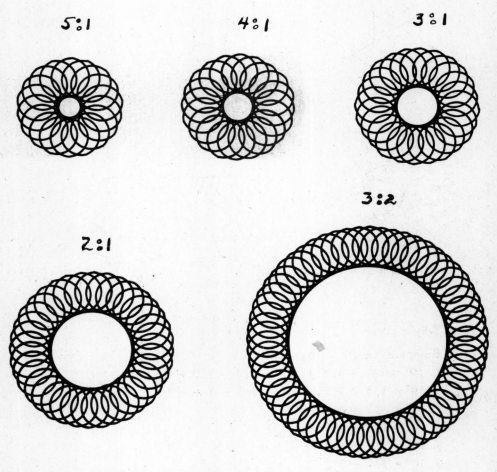

Figure 20. *180° arcs moving in a constant direction. S = 30°*

SELECTIVE SYSTEMS

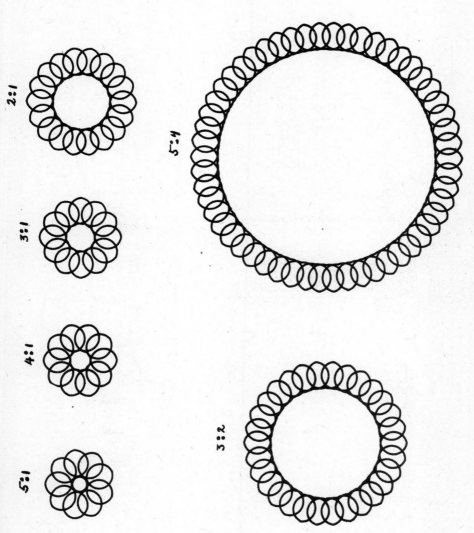

Figure 21. 180° arcs moving in a constant direction. $S = 60°$

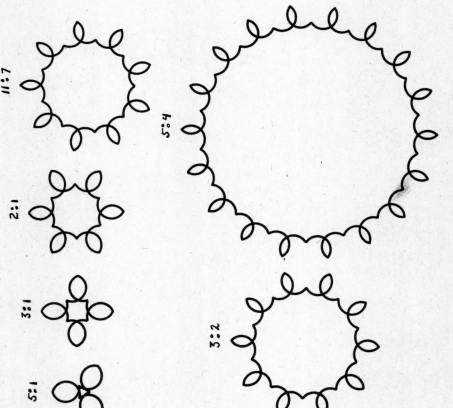

Figure 22. 180° arcs moving in a constant direction. S = 180°

b. Binomials by the Initial Unit.
  1. 10° = 1, 7 + 1, 5 + 1, 4 + 1, 3 + 1, 2 + 1
  2. 10° = 1, 5 + 2, 3 + 2, 7 + 3, 5 + 3, 4 + 3
  3. 10° = 1, 7 + 4, 5 + 4
  4. 10° = 1, 8 + 5, 7 + 5, 6 + 5
  5. 10° = 1, 7 + 6, 11 + 7, 9 + 7
  6. 10° = 1, 8 + 7
  7. 10° = 1, 9 + 8

C. 180° Arcs Moving in a Constant Alternating Direction.

a. Binomials by the Sum.
1. S =   30°, 5 + 1, 4 + 1, 3 + 1, 2 + 1, 3 + 2 (See Figure 23.)
2. S =   60°, 5 + 1, 4 + 1, 3 + 1, 2 + 1, 3 + 2, 5 + 4 (See Figure 24.)
3. S =   90°, 5 + 1, 3 + 1, 2 + 1, 3 + 2, 5 + 4
4. S = 120°, 4 + 1, 3 + 1, 2 + 1, 3 + 2, 7 + 5
5. S = 150°, 4 + 1, 3 + 2, 8 + 7
6. S = 180°, 5 + 1, 3 + 1, 2 + 1, 11 + 7, 3 + 2, 5 + 4
(See Figure 25.)

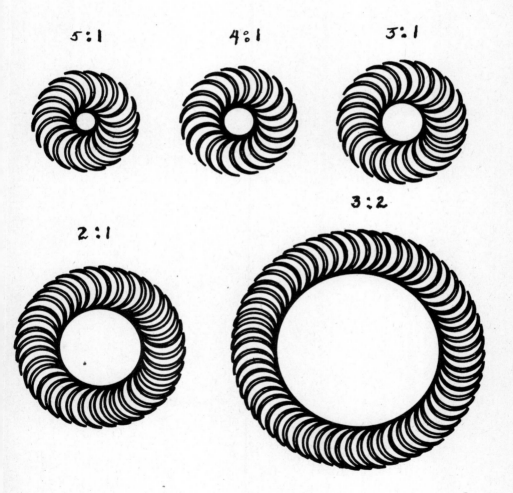

Figure 23. *180° arcs moving in constant alternating directions.*   S = 30°

Figure 24. 180° arcs moving in a constant alternating direction. $S = 60°$

SELECTIVE SYSTEMS

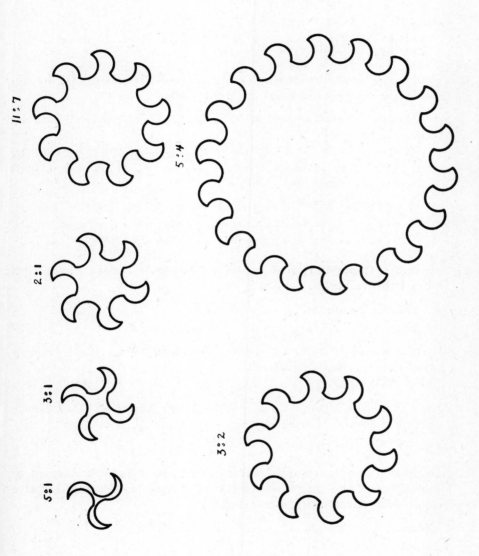

Figure 25. 180° arcs moving in constant alternating directions. S = 180°

312   TECHNOLOGY OF ART PRODUCTION

    b. Binomials by the Initial Unit.
       1. $10° = 1, 7 + 1, 5 + 1, 4 + 1, 3 + 1, 2 + 1$
       2. $10° = 1, 5 + 2, 3 + 2, 7 + 3, 5 + 3, 4 + 3$
       3. $10° = 1, 7 + 4, 5 + 4$
       4. $10° = 1, 8 + 5, 7 + 5, 6 + 5$
       5. $10° = 1, 7 + 6, 11 + 7, 9 + 7$
       6. $10° = 1, 8 + 7$
       7. $10° = 1, 9 + 8$

D. Trinomial Periodicity of Angles Moving in Alternating Direction. (Rectilinear Segments Serve as Diameters.)

    a. Rectilinear Segments—Arcs Move in One Direction and in Alternating Direction.
       1. $1 \div 2 \div 3$, $15° + 30° + 45°$, Rectilinear Segments, 180° Arcs moving in a constant direction, and 180° arcs moving in a constant alternating direction. $30° + 60° + 90°$.
       2. $3 \div 4 \div 5$, $15° + 20° + 25°$, $30° + 40° + 50°$, $60° + 80° + 100°$. Rectilinear Segments, 180° arcs moving in a constant direction, and 180° arcs moving in a constant alternating direction.

E. Symmetric Construction of Angles Moving in Alternating Direction Under Infinite Series.

    a. Rectilinear Segments.
       1. Series I, II, III, IV.
    b. 180° Arcs Moving in a Constant Clockwise Direction (Rectilinear Segments Serve as Diameters).
       1. Series I, II, III, IV.
    c. 180° Arcs Moving in a Constant Alternating Direction. (Rectilinear Segments Serve as Diameters).
       1. Series I, II, III, IV.

F. Symmetric Construction of Rectilinear Segments with Respect to their Dimensions and Periodicity of Angles Following Infinite Series, with Alternating Direction.

    a. Rectilinear Segments.
       1. Series I and II.
       2. Series III and IV.
    b. 180° Arcs Moving in a Constant Clockwise Direction. (Rectilinear Segments Serve as Diameters).
       1. Series I and II.
       2. Series III and IV.

SELECTIVE SYSTEMS

c. 180° Arcs Moving in a Constant Alternating Direction (Rectilinear Segments Serve as Diameters).
  1. Series I and II.
  2. Series III and IV.

4. Monomial, Binomial and Trinomial Periodicity of Sector Radii.

A. Discontinuous Counter-Clockwise 180° Arcs.
180° Arcs, 10° Sectors; 30° Sectors; 30° + 60° Sectors; 30° + 60° + 90° Sectors.

B. Discontinuous 180° Arcs in Alternating Direction.
180° Arcs, 30° Sectors; 30° + 60° Sectors; 30° + 60° + 90° Sectors.

C. Infinite Series of the Sector Radii, Angle Variation of Values, Counter-clockwise 180° discontinuous Arcs.
Series I, II, III, IV.

D. Discontinuous 180° Arcs in Alternating Direction.
Series I, II, III, IV. (See Figure 26.)

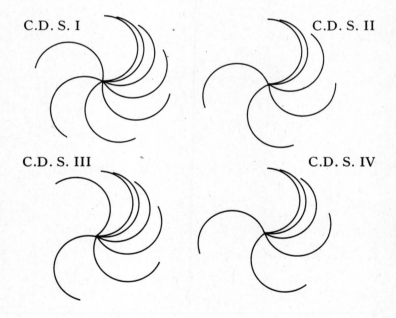

Figure 26. Discontinuous 180° arcs moving in alternating direction.

5. Periodicity of Radii and Angle Values.

A. Arcs Moving in a Constant Direction.

1. 90° Arc, Constant Clockwise Direction.
   $S_r = 1, 2, 3, 4, 5, 6 \ldots$

2. 90° Arc, Constant Clockwise Direction.
   $S_r = 1, 2, 3, 5, 8, 13 \ldots$

3. 90° Arc, Constant Clockwise Direction.
   $S_r = 1, 2, 4, 7, 11, 16, 22, 29 \ldots$

4. 150° Arc, Constant Clockwise Direction.
   $S_r = 1, 2, 3, 4, 5, 6 \ldots$

5. 150° Arc, Constant Clockwise Direction.
   $S_r = 1, 2, 3, 5, 8, 13, 21 \ldots$

6. 150° Arc, Constant Clockwise Direction.
   $S_r = 1, 2, 4, 7, 11, 16, 22 \ldots$

7. 240° Arc, Constant Clockwise Direction.
   $S_r = 1, 2, 3 \ldots$

8. 240° Arc, Constant Clockwise Direction.
   $S_r = 1, 2, 3, 5, 8, 13, 21 \ldots$

9. 240° Arc, Constant Clockwise Direction.
   $S_r = 1, 2, 4, 7, 11, 16, 22 \ldots$

10. 270° Arc, Constant Clockwise Direction.
    $S_r = 1, 2, 3, 4, 5, 6 \ldots$ (See Figure 27.)

11. 270° Arc, Constant Clockwise Direction.
    $S_r = 1, 2, 3, 5, 8, 13, 21 \ldots$

12. 270° Arc, Constant Clockwise Direction.
    $S_r = 1, 2, 4, 7, 11, 16, 22 \ldots$

13. 330° Arc, Constant Clockwise Direction.
    $S_r = 1, 2, 3, 4, 5, 6 \ldots$ (See Figure 28.)

14. 330° Arc, Constant Clockwise Direction.
    $S_r = 1, 2, 3, 5, 8, 13, 21 \ldots$

15. 330° Arc, Constant Clockwise Direction.
    $S_r = 1, 2, 4, 7, 11, 16, 22 \ldots$

SELECTIVE SYSTEMS

16. 360° Arc, Constant Clockwise Direction.
    $S_r = 1, 2, 3, 4, 5, 6 \ldots$

17. 360° Arc, Constant Clockwise Direction.
    $S_r = 1, 2, 3, 5, 8, 13, 21 \ldots$

18. 360° Arc, Constant Clockwise Direction.
    $S_r = 1, 2, 4, 7, 11, 16, 22 \ldots$

Ratios of $\pi \quad \dfrac{3\pi}{2}$

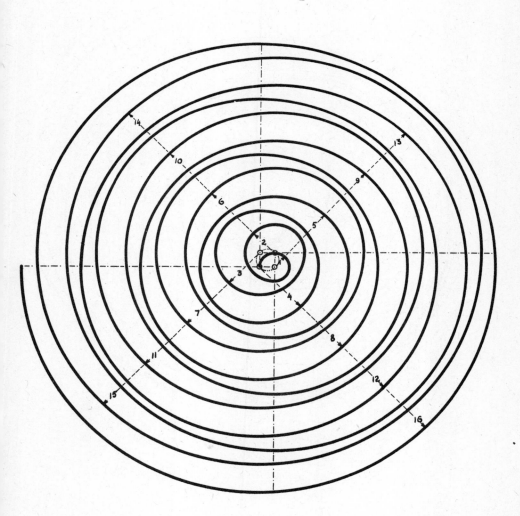

Figure 27. 270° arc moving in a constant clockwise direction.

Ratios of π $\quad \dfrac{11\pi}{12} \angle 330°$

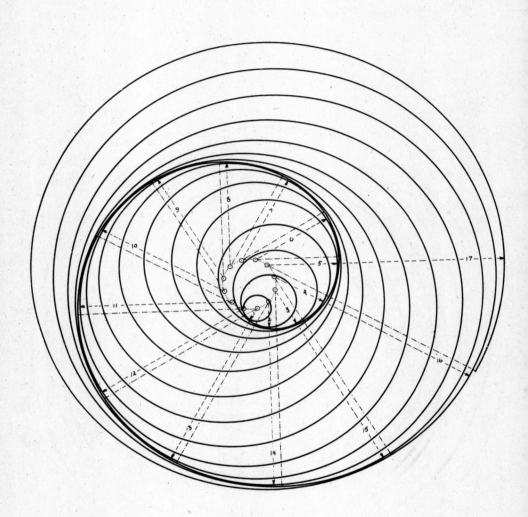

Figure 28. 330° arc moving in a constant clockwise direction.

SELECTIVE SYSTEMS

B. Arcs Moving in a Constant Alternating Direction.
(Sin and cos.)

1. 90° Arc, Constant Alternating Direction.
   $S_r = 1, 2, 3, 4, 5, 6 \ldots$

2. 90° Arc, Constant Alternating Direction.
   $S_r = 1, 2, 3, 5, 8, 13 \ldots$

3. 90° Arc, Constant Alternating Direction.
   $S_r = 1, 2, 4, 7, 11, 16, 22, 29 \ldots$

4. 150° Arc, Constant Alternating Direction.
   $S_r = 1, 2, 3, 4, 5, 6 \ldots$

5. 150° Arc, Constant Alternating Direction.
   $S_r = 1, 2, 3, 5, 8, 13, 21 \ldots$

6. 150° Arc, Constant Alternating Direction.
   $S_r = 1, 2, 4, 7, 11, 16, 22 \ldots$

7. 240° Arc, Constant Alternating Direction.
   $S_r = 1, 2, 3 \ldots$

8. 240° Arc, Constant Alternating Direction.
   $S_r = 1, 2, 3, 5, 8, 13, 21 \ldots$

9. 240° Arc, Constant Alternating Direction.
   $S_r = 1, 2, 4, 7, 11, 16, 22 \ldots$

10. 270° Arc, Constant Alternating Direction.
    $S_r = 1, 2, 3, 4, 5, 6 \ldots$

11. 270° Arc, Constant Alternating Direction.
    $S_r = 1, 2, 3, 5, 8, 13, 21 \ldots$

12. 270° Arc, Constant Alternating Direction.
    $S_r = 1, 2, 4, 7, 11, 16, 22 \ldots$

13. 330° Arc, Constant Alternating Direction.
    $S_r = 1, 2, 3, 4, 5, 6 \ldots$

14. 330° Arc, Constant Alternating Direction.
    $S_r = 1, 2, 3, 5, 8, 13, 21 \ldots$

15. 330° Arc, Constant Alternating Direction.
    $S_r = 1, 2, 4, 7, 11, 16, 22 \ldots$

16. 360° Arc, Constant Alternating Direction.
    $S_r = 1, 2, 3, 4, 5, 6 \ldots$

17. 360° Arc, Constant Alternating Direction.
    $S_r = 1, 2, 3, 5, 8, 13, 21 \ldots$.

18. 360° Arc, Constant Alternating Direction.
    $S_r = 1, 2, 4, 7, 11, 16, 22 \ldots$.

19. Derivative Design. (See Figure 29.)

20. Derivative Design. (See Figure 30.)

$$\frac{\pi}{2} + \frac{5\pi}{6} \qquad \angle 90° + \angle 150°$$

$$r = (1, 2, 3, 4, 5, 6 \ldots) + (1, 2, 3, 5, 8 \ldots) + (1, 2, 4, 7, 11 \ldots)$$

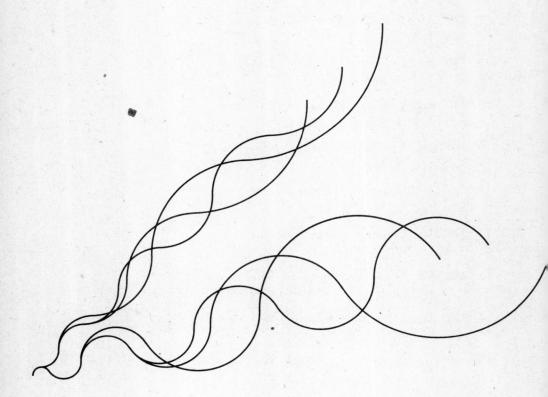

Figure 29. Derivative design.

$270° - 240° - 180° - 150° - 90° - 45° - 0° - 45° - 90° - 150° - 180° - 240° - 270°$

$r = 1, 2, 3, 4, 5, 6\ldots\ldots$

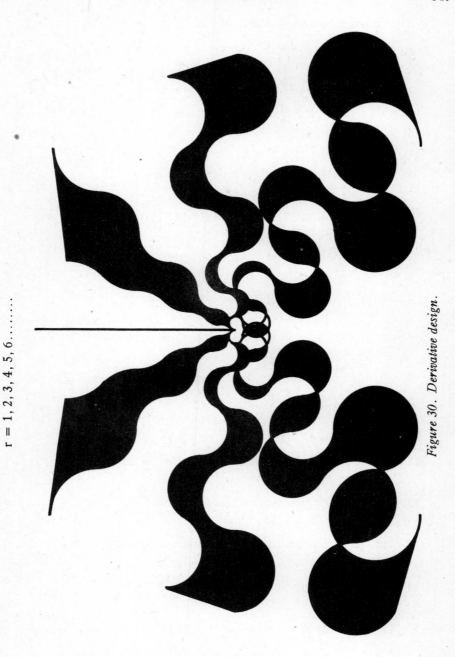

Figure 30. *Derivative design.*

C. Alternating Sin Movement of Arcs.

1. 90° Arc, Alternating Sin Movement of Arcs.
   $S_r = 1, 2, 3, 4, 5, 6 \ldots$ (See Figure 31.)

2. 90° Arc, Alternating Sin Movement of Arcs.
   $S_r = 1, 2, 3, 5, 8, 13 \ldots$

3. 90° Arc, Alternating Sin Movement of Arcs.
   $S_r = 1, 2, 4, 7, 11, 16, 22, 29 \ldots$

4. 150° Arc, Alternating Sin Movement of Arcs.
   $S_r = 1, 2, 3, 4, 5, 6 \ldots$

5. 150° Arc, Alternating Sin Movement of Arcs.
   $S_r = 1, 2, 3, 5, 8, 13, 21 \ldots$

6. 150° Arc, Alternating Sin Movement of Arcs.
   $S_r = 1, 2, 4, 7, 11, 16, 22 \ldots$ (See Figure 32.)

7. 240° Arc, Alternating Sin Movement of Arcs.
   $S_r = 1, 2, 3 \ldots$

8. 240° Arc, Alternating Sin Movement of Arcs.
   $S_r = 1, 2, 3, 5, 8, 13, 21 \ldots$

9. 240° Arc, Alternating Sin Movement of Arcs.
   $S_r = 1, 2, 4, 7, 11, 16, 22 \ldots$

10. 270° Arc, Alternating Sin Movement of Arcs.
    $S_r = 1, 2, 3, 4, 5, 6 \ldots$

11. 270° Arc, Alternating Sin Movement of Arcs.
    $S_r = 1, 2, 3, 5, 8, 13, 21 \ldots$

12. 270° Arc, Alternating Sin Movement of Arcs.
    $S_r = 1, 2, 4, 7, 11, 16, 22 \ldots$

13. 330° Arc, Alternating Sin Movement of Arcs.
    $S_r = 1, 2, 3, 4, 5, 6 \ldots$

14. 330° Arc, Alternating Sin Movement of Arcs.
    $S_r = 1, 2, 3, 5, 8, 13, 21 \ldots$

15. 330° Arc, Alternating Sin Movement of Arcs.
    $S_r = 1, 2, 4, 7, 11, 16, 22 \ldots$

16. Derivative Design. (See Figure 33.)

17. Derivative Design. (See Figure 34.)

18. Derivative Design. (See Figure 35.)

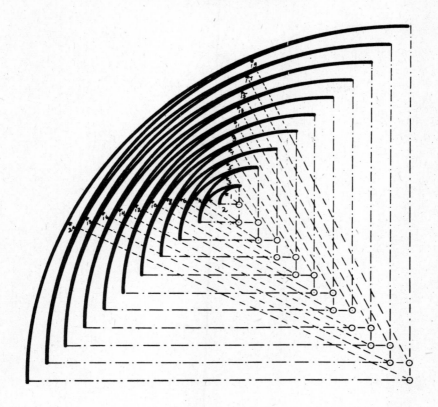

Figure 31. 90° arcs moving in alternating sin movement of arcs.

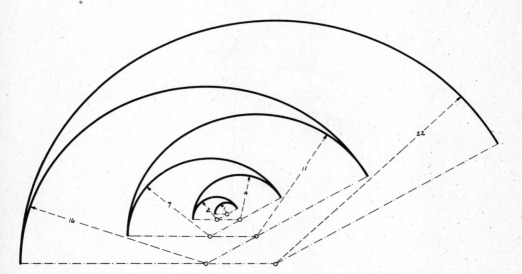

Figure 32. 150° arcs moving in alternating sin movement of arcs.

$$\frac{\pi}{2} \angle 90°$$

$$r = 1, 2, 4, 7, 11, 16\ldots\ldots$$

Figure 33. Derivative design.

$$\frac{\pi}{4} + \frac{\pi}{2} + \frac{3\pi}{2} + \pi + \frac{5\pi}{4} + \frac{3\pi}{2} + \frac{7\pi}{2} + 2\pi$$

$$\angle 45° + 90° + 135° + 180° + 225° + 270° + 315° + 360°$$

$$r = 1, 2, 4.$$

Figure 34. Derivative design.

Figure 35. Derivative design.

TECHNOLOGY OF ART PRODUCTION

6. *Periodicity of Arcs and Radii.*

A. Variable Lengths of Arcs in Constant Clockwise Direction, and Variable Radii.

1. $3 \div 2$, $120° + 60° + 60° + 120°$

   r $2 + 1 + 1 + 2$

   $4 \div 3$, $90° + 30° + 60° + 60° + 30° + 90°$

   r $3 + 1 + 2 + 2 + 1 + 3$

   $5 \div 3$, $72° + 48° + 24° + 72° + 24° + 48° + 72°$

   r $3 + 2 + 1 + 3 + 1 + 2 + 3$ (See Figure 36.)

2. Ellipses

   $2(30° + 60° + 90°)$

   r $2(1+2+3)$

   $2(10° + 20° + 30° + 60° + 30° + 20° + 10°)$

   r $2(1 + 2 + 3 + 6 + 3 + 2 + 1)$

3. $(120° + 60° + 60° + 120°)$ $(120° + 60° + 60° + \ldots)$

   r $(2 + 1 + 1 + 2)$ $(2 + 1 + 1 + \ldots)$

4. Variation of $4 \div 3$ in Table A1 above.

   $90° + 30° + 60° + 60° + 30° + 90° + 30° + 60° + 60° + 30° +$
   $+ 90° + 90° + 60° + 60° + 30° + 90° + 90° + 30° + 60° + 30° +$
   $+ 90° + 90° + 30° + 60° + 30° + 90° + 90° + 30° + 60° + 60° +$
   $+ 90° + 90° + 30° + 60° + 60° + 30° + 90° + 30° + 60° + 60° +$
   $+ 30° + 90°$.

   r $3 + 1 + 2 + 2 + 1 + 3 + 1 + 2 + 2 + 1 + 3 + 3 + 2 + 2 +$
   $+ 1 + 3 + 3 + 1 + 2 + 1 + 3 + 3 + 1 + 2 + 1 + 3 + 3 + 1 +$
   $+ 2 + 2 + 3 + 3 + 1 + 2 + 2 + 1 + 3 + 1 + 2 + 2 + 1 + 3$

   (See Figure 37.)

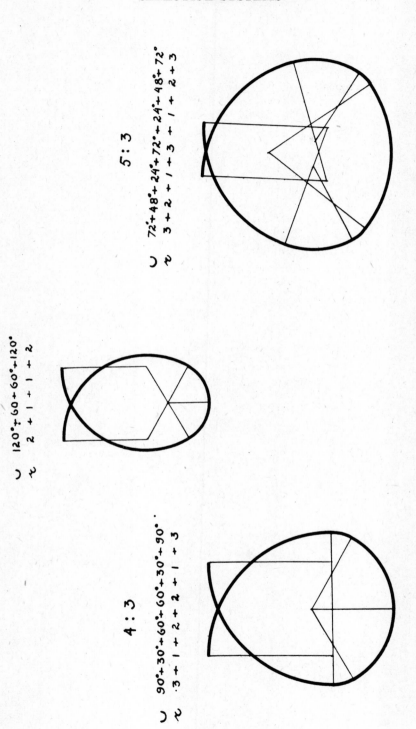

Figure 36. *Variable lengths of arcs in constant clockwise direction, and variable radii.*

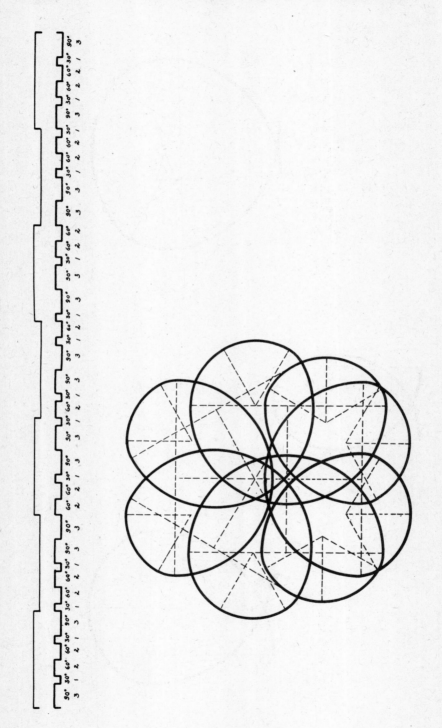

Figure 37. Variation of rhythmic design $4 \div 3$

SELECTIVE SYSTEMS

B. **Variable Lengths of Arcs in a Constant Alternating Direction and Variable Radii.**

1. 10° + 20° + 30° + 60°
   r 1 + 2 + 3 + 6
2. 80° + 50° + 30° + 20° + 10°
   r 1 + 2 + 3 + 5 + 8
3. 160° + 100° + 60° + 40° + 20°
   r 1 + 2 + 3 + 5 + 8 (See Figure 38.)
4. 10° + 20° + 30° + 50° + 80°
   r 1 + 2 + 3 + 5 + 8 (See Figure 39.)

C. **Ratio of the Radii in Relation to the Scale of the Curvature of Arcs.**

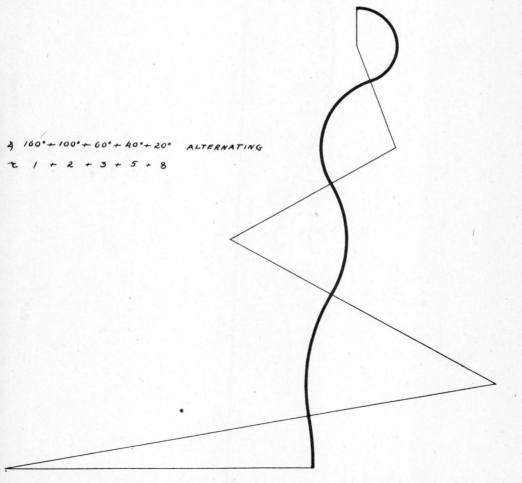

Figure 38. *Variable lengths of arcs in constant alternating direction and variable radii.*

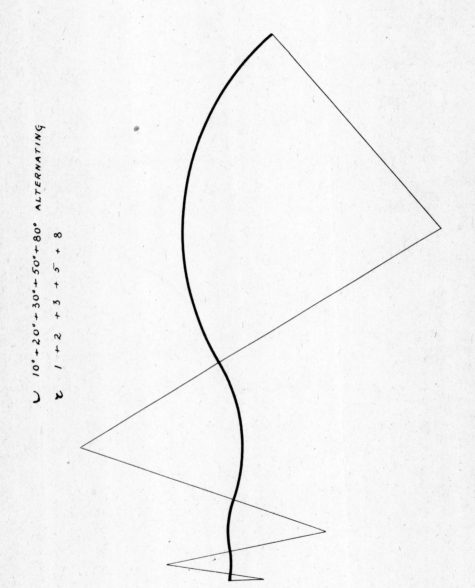

Figure 39. Variable lengths of arcs in constant alternating direction and variable radii.

SELECTIVE SYSTEMS

7. r_{4+3} *Rhythmic Groups in Linear Design* (*Dimensions, Directions, Angles, Constant and Variable*).

A. Angle Unit = 30°.

   a. Rectilinear Segments Moving in Alternating Directions.

   1. r_{4+3} applied to dimensions and angles.

   2. Segments of Table a1 used as diameters.
      r_{4+3} applied to the length of arcs moving in constant clockwise direction.

   3. Segments of Table a1 used as Diameters.
      r_{4+3} applied to the length of arcs moving in constant alternating direction.

   4. Alternating continuity under 90° angle from the groups represented on Tables a1, 2 and 3.

   5. Closed continuity under 90° angle from the groups represented on Tables a1, 2 and 3.

   b. Rectilinear Segments Moving in a Constant Clockwise Direction.

   1. r_{4+3} applied to dimensions and angles.

   2. Segments of Table b1 used as diameters.

   3. Segments of Table b1 used as diameters, r_{4+3} applied to the length of arcs moving in a constant alternating direction.

   4. Alternating continuity under 90° angle from the groups represented on Tables b1, 2 and 3.

   5. Closed continuity under 90° angle from the groups represented on Tables b1, 2 and 3.

   c. Constant Dimensions, Variable Angles.

   1. Rectilinear Segments moving in a constant alternating direction.

   2. 180° Arc moving in a constant clockwise direction through the segments of Table c1, used as diameters.

   3. 180° arc moving in a constant alternating direction through the segments of Table c1 used as diameters.

   4. Continuous alternating patterns under 90° angle from Tables c1, c2 and c3.

   d. Constant Dimensions, Variable Angles moving in a constant clockwise direction.

   1. Rectilinear Segments moving in a constant clockwise direction. (See Figure 40.)

2. 180° Arcs moving in a constant clockwise direction through the segments of Table d1 used as diameters. (See Figure 41.)

3. 180° Arcs moving in a constant alternating direction through the segments of Table d1 used as diameters. (See Figure 42.)

4. Continuous patterns under 180° angle from the Tables d1, 2 and 3. (See Figure 43.)

5. Closed patterns under 90° angle from the Tables d1, 2 and 3. (See Figure 44.)

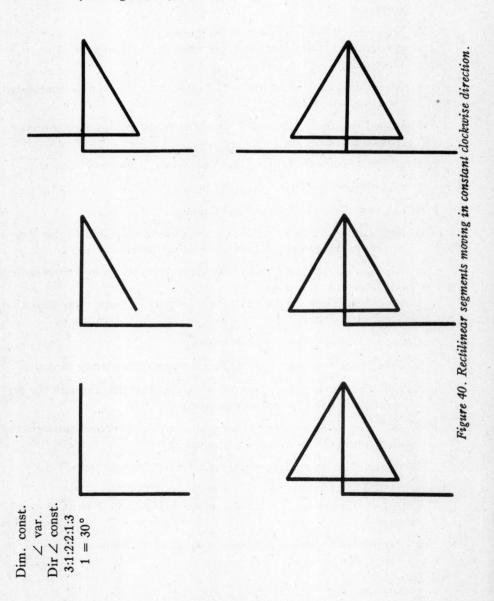

Figure 40. Rectilinear segments moving in constant clockwise direction.

Dim. const.
∠ var.
Dir ∠ const.
3:1:2:2:1:3
1 = 30°

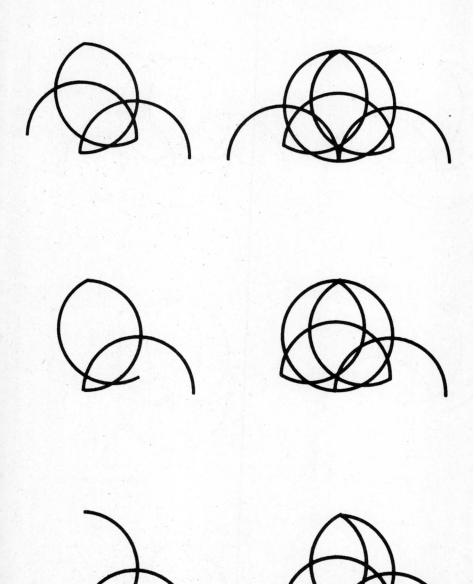

Figure 41. 180° arcs moving in constant clockwise direction.

Dim. const.
∠ var.
Dir. ∠ const.
Dir. ⌒ const.
3:1:2:2:1:3
1 = 30°

Figure 42. 180° arcs moving in constant alternating direction.

SELECTIVE SYSTEMS

3:1:2:2:1:3

Dim. const.
∠ var.
Dir. ∠ const.
1 = 30°

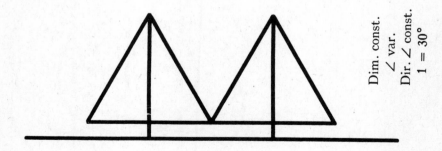

Figure 43. Continuous patterns under 180° angle.

334   TECHNOLOGY OF ART PRODUCTION

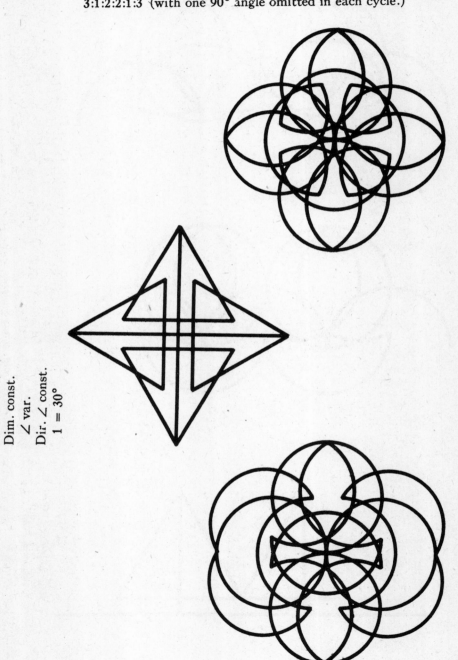

3:1:2:2:1:3 (with one 90° angle omitted in each cycle.)

Dim. const.
∠ var.
Dir. ∠ const.
1 = 30°

Figure 44. Closed patterns under 90° angle.

SELECTIVE SYSTEMS

B. r_{4+3} Rhythmic Groups in Linear Design (Dimensions, Directions, Angles, Constant and Variable). Angle Unit = 10°.

a. Rectilinear Segments moving in alternating direction.

1. r_{4+3} applied to dimensions and angles.

2. Segments of Table a1 used as diameters.
   r_{4+3} applied to the length of arcs moving in a constant clockwise direction.

3. Segments of Table a1 used as diameters.
   r_{4+3} applied to the length of arcs moving in a constant alternating direction.

4. Alternating continuity under 30° angle from group represented in Tables a1, 2 and 3.

5. Closed continuity under 150° angle from group represented in Tables a1, 2 and 3.

b. Rectilinear Segments moving in a constant clockwise direction.

1. r_{4+3} applied to dimensions and angles.

2. Segments of Table b1 used as diameters.
   r_{4+3} applied to the length of arcs moving in a constant clockwise direction.

3. Segments of Table b1 used as diameters.
   r_{4+3} applied to the length of arcs moving in a constant alternating direction.

4. Closed continuity of the patterns of the Tables b1, 2 and 3, under 30° angle.

5. Closed continuity of the patterns of the Tables b1, 2 and 3, under 90° angle.

c. Rectilinear segments moving in a constant alternating direction. Constant dimensions, variable angles.

1. Rectilinear segments moving in a constant alternating direction.

2. 180° Arcs moving in a constant clockwise direction, through the segments of Table c1 used as diameters.

3. 180° Arcs moving in a constant alternating direction, through the segments of Table c1 used as diameters.

4. Continuous patterns from the Tables c1, 2 and 3, under 30° angle.

5. Closed patterns from the Tables c1, 2 and 3, without repeating the 30° angle in the first term of each consecutive group.

d. Constant dimensions, variable angles moving in a constant clockwise direction.

1. Rectilinear segments moving in a constant alternating direction.

2. 180° Arcs moving in a constant clockwise direction through segments of Table c1 used as diameters.

3. 180° Arcs moving in a constant alternating direction through segments of Table c1 used as diameters.

4. Closed continuity of the patterns of Tables c1, 2 and 3 under 30° angle.

C. **Variable Direction of Arcs.**

1. Direction of arcs varying with each term.
   Angle Unit = 10°.

2. Direction of arcs variable with each term.
   Angle Unit = 30°.

3. Direction and dimensions of arcs varying with each term.
   Angle Unit = 30°.

8. Planes.

A. **Rhythmic Centers.**

a. 1×1 area
   1. $(\frac{1}{2} + \frac{1}{2})^2$
   2. $(\frac{1}{2} + \frac{1}{2})^3$
   3. $(\frac{1}{2} + \frac{1}{2})^4$

b. 2×1 area
   1. $(\frac{2}{3} + \frac{1}{3})^2$
   2. $(\frac{2}{3} + \frac{1}{3})^3$
   3. $(\frac{2}{3} + \frac{1}{3})^4$

c. 3×2 area
   1. $(\frac{3}{5} + \frac{2}{5})^2$ (See Figure 45.)
   2. $(\frac{3}{5} + \frac{2}{5})^3$ (See Figure 46.)
   3. $(\frac{3}{5} + \frac{2}{5})^4$ (See Figure 47.)

d. 4×3 area
   1. $(\frac{4}{7} + \frac{3}{7})^2$
   2. $(\frac{4}{7} + \frac{3}{7})^3$
   3. $(\frac{4}{7} + \frac{3}{7})^4$

e. 7×6 area
   1. $(\frac{7}{13} + \frac{6}{13})^5$
   2. $(\frac{6}{13} + \frac{7}{13})^5$

Exterior Sides:

$$\left(\frac{3}{5} + \frac{2}{5}\right)^2 = \frac{9}{25} + \frac{6}{25} + \frac{6}{25} + \frac{4}{25}$$

Areas:

$$\frac{9 \times 6}{25} + \frac{6 \times 6}{25} + \frac{6 \times 4}{25} + \frac{9 \times 4}{25} = \frac{54}{25} + \frac{36}{25} + \frac{24}{25} + \frac{36}{25}$$

3 : 2

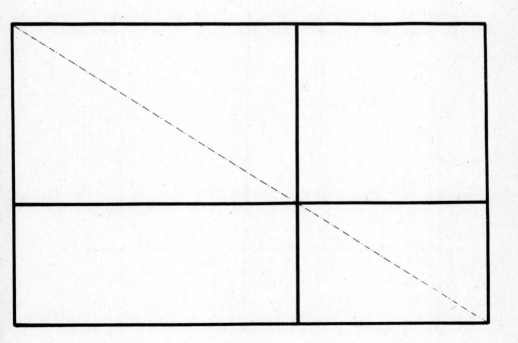

Figure 45. *Rhythmic center of 3×2 area.*

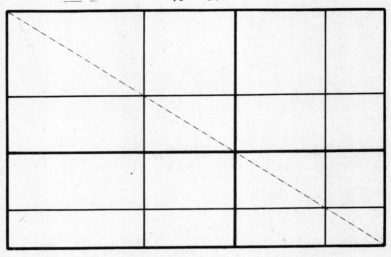

Figure 46. *Rhythmic centers of 3×2 area.*

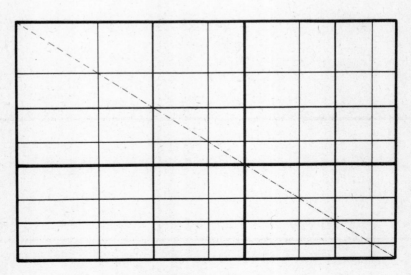

Figure 47. *Rhythmic centers of 3×2 area.*

SELECTIVE SYSTEMS

B. Monomial Periodicity of the Sector Radii Produced from a Rhythmic Center (Sector Angle = 30°).

1. Continuous circular arcs moving in one direction, 2×1 area.

2. Continuous circular arcs moving in one direction, 3×2 area.

3. Continuous circular arcs moving in one direction, 4×3 area.

4. Discontinuous circular arcs moving in alternating direction, 2×1 area.

5. Discontinuous circular arcs moving in alternating direction, 3×2 area.

6. Discontinuous circular arcs moving in alternating direction, 4×3 area.

7. Discontinuous circular arcs moving in one direction, 2×1 area.

8. Discontinuous circular arcs moving in one direction, 3×2 area.

9. Discontinuous circular arcs moving in one direction, 4×3 area.

C. Rhythmic Groups Applied to Rectangular Area. Vertical and Horizontal Circular Permutations.

a. Horizontal extension.

1. r_{3+2} combined coordinates.

2. r_{3+2} phase arrangement and vertical coincidence.

3. r_{4+3} phase arrangement and vertical coincidence.

4. r_{5+3} phase arrangement and vertical coincidence.

5. r_{5+4} phase arrangement and vertical coincidence. (See Figure 48.)

b. Vertical extension.

1. r_{3+2} horizontal coincidence and phase arrangement.

2. r_{4+3} horizontal coincidence and phase arrangement.

3. r_{5+3} horizontal coincidence and phase arrangement.

4. r_{5+4} horizontal coincidence and phase arrangement.

340 TECHNOLOGY OF ART PRODUCTION

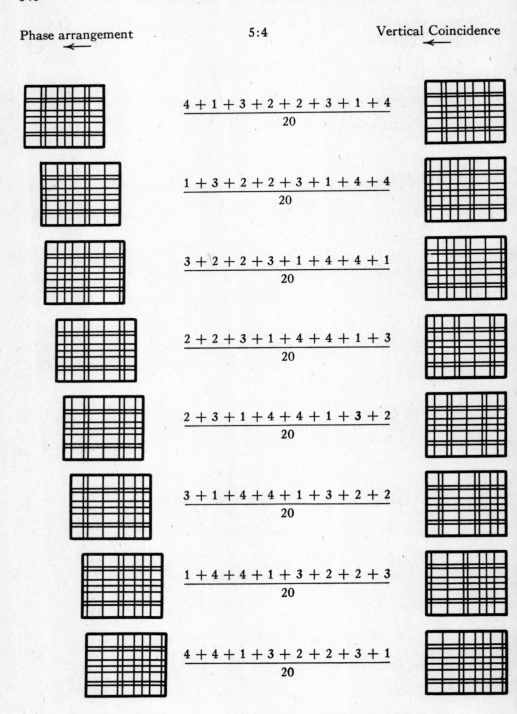

Figure 48. r_{5+4} phase arrangement and vertical coincidence.

SELECTIVE SYSTEMS

D. Automatic Continuity of Arcs in Rectangular Areas.

   a. 4 × 3 area.

1. Arrangement of the radii from rhythmic center. (a, b, c, d, and x, y, z radii) (See Figure 49.)
2. Radius a ↘ 13 arcs.
3. Radius b ↗ 51 arcs.
4. Radius b ↙ 53 arcs. (See Figure 50.)
5. Radius c ↖ 16 arcs. (See Figure 51.)
6. Radius c ↘ 61 arcs.
7. Radius d ↗ 17 arcs.
8. Radius d ↙ 27 arcs.
9. Radius x ⊶→ 22 arcs.
10. Radius x ←⊶ 42 arcs.
11. Radius y ↓ 20 arcs.
12. Radius y ↑ 40 arcs.
13. Radius z ↑ 35 arcs.
14. Rectilinear arc derivatives.

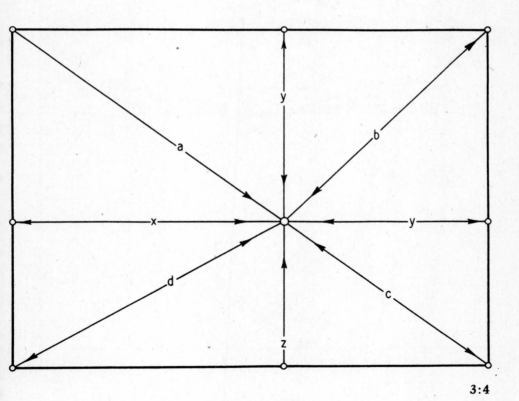

Figure 49. Arrangement of radii from the rhythmic center.

Figure 50. Radius b ∠ 53 arcs.

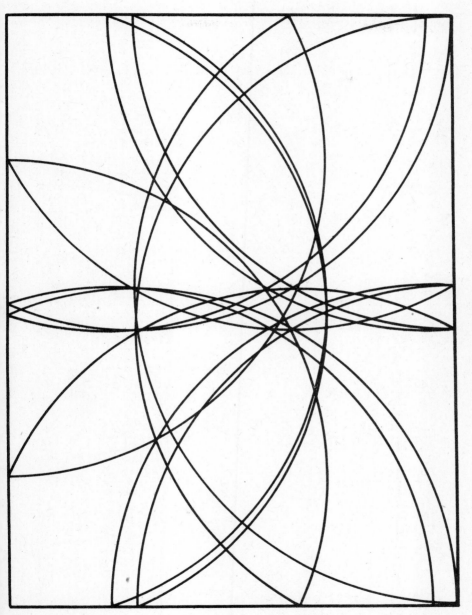

Figure 51. Radius $c \searrow$ 16 arcs.

9. Closed Polygons Conceived as Monomial Periodicity of Angles, Dimensions and Directions.

A. Formula: $t = \dfrac{180° (T-2)}{T}$

Where t is the value of an angle and T the number of angles and sides in a polygon.

$t = \dfrac{180° (3-2)}{3} = 60°$

$t = \dfrac{180° (4-2)}{4} = 90°$

$t = \dfrac{180° (5-2)}{5} = 108°$

$t = \dfrac{180° (6-2)}{6} = 120°$

$t = \dfrac{180° (7-2)}{7} = 128\dfrac{4}{7}°$

$t = \dfrac{180° (8-2)}{8} = 135°$

$t = \dfrac{180° (9-2)}{9} = 140°$

$t = \dfrac{180° (10-2)}{10} = 144°$

$t = \dfrac{180° (11-2)}{11} = 147\dfrac{3}{11}°$

$t = \dfrac{180° (12-2)}{12} = 150°$

$t = \dfrac{180° (13-2)}{13} = 152\dfrac{4}{13}°$

$t = \dfrac{180° (14-2)}{14} = 154\dfrac{2}{7}°$

$t = \dfrac{180° (15-2)}{15} = 156°$

$t = \dfrac{180° (16-2)}{16} = 157\dfrac{1}{2}°$

$t = \dfrac{180° (17-2)}{17} = 158\dfrac{14}{17}°$

$t = \dfrac{180° (18-2)}{18} = 160°$

$t = \dfrac{180° (19-2)}{19} = 161\dfrac{1}{19}°$

$t = \dfrac{180° (20-2)}{20} = 162°$

$t = \dfrac{180° (21-2)}{21} = 162\dfrac{6}{7}°$

$t = \dfrac{180° (22-2)}{22} = 163\dfrac{7}{11}°$

$t = \dfrac{180° (23-2)}{23} = 164\dfrac{8}{23}°$

$t = \dfrac{180° (24-2)}{24} = 165°$

$t = \dfrac{180° (25-2)}{25} = 165\dfrac{3}{5}°$

$t = \dfrac{180° (26-2)}{26} = 166\dfrac{2}{13}°$

$t = \dfrac{180° (27-2)}{27} = 166\dfrac{2}{3}°$

$t = \dfrac{180° (28-2)}{28} = 167\dfrac{1}{7}°$

SELECTIVE SYSTEMS

$$t = \frac{180°\,(29-2)}{29} = 167\frac{17°}{29}$$

$$t = \frac{180°\,(30-2)}{30} = 168°$$

$$t = \frac{180°\,(31-2)}{31} = 168\frac{12°}{31}$$

$$t = \frac{180°\,(32-2)}{32} = 168\frac{3°}{4}$$

$$t = \frac{180°\,(33-2)}{33} = 169\frac{1°}{11}$$

$$t = \frac{180°\,(34-2)}{34} = 169\frac{7°}{17}$$

$$t = \frac{180°\,(35-2)}{35} = 169\frac{5°}{7}$$

$$t = \frac{180°\,(36-2)}{36} = 170°$$

B. Kinetic Geometry.

Formulation of Closed Polygons

$S = 4\,(St\,180° + 90°)$ square

$S = 3\,(St\,180° + 60°)$ triangle

$S = 5\,(St\,180° + 108°)$ pentagon

$S = 6\,(St\,180° + 120°)$ hexagon (any)

$S = \infty\,t\,180° \parallel mt\,180°$ infinite rectilinear extension

$S = s$ point (geometrical)

$S = st$ point (physical)

In one system:

$S = 3\,(smt\,180° + 60°)$ triangle

$S = 4\,(smt\,180° + 90°)$ square

where mt is a given period of time moving under 180°.

C. Geometry

Polygonal Series

$\frac{3}{3}$ Triadic   composition of triangles

$\frac{4}{4}$ Tetradic   composition of squares, rectangles

$\frac{5}{5}$ Pentadic   composition of pentagons

$\frac{6}{6}$ Hexadic   composition of hexagons

$\frac{7}{7}$ Heptadic   composition of heptagons

$\frac{8}{8}$ Octadic   composition of octagons

$\frac{9}{9}$ Enneadic   composition of enneagons

$\frac{10}{10}$ Decadic   composition of decagons

$\frac{m}{m}$ Polyadic   composition of polygons

E. COLOR SCALES

Color origin is determined by the geometrical origin of the circumference.

Color axis is determined by identical colors on the opposite sides of the circumference.

The angle of a color-axis depends on the number of color-sectors in the color range distributed through a circle.

The distribution of color sectors may be either through a circle or an ellipse.

1. Full Spectrum Distributed in Bilateral Symmetry Through 360°

Twenty-four Sectors

Each Sector = 15°

Yellow Origin (*Violet Axis*)		Yellow-Green Origin (*Violet-Red Axis*)		Green Origin (*Red Axis*)	
Yellow	0°	Yellow-Green	0°	Green	0°
Yellow-Green	15°	Green	15°	Green-Blue	15°
Green	30°	Green-Blue	30°	Blue	30°
Green-Blue	45°	Blue	45°	Blue-Violet	45°
Blue	60°	Blue-Violet	60°	Violet	60°
Blue-Violet	75°	Violet	75°	Violet-Red	75°
Violet	90°	Violet-Red	90°	Red	90°
Violet-Red	105°	Red	105°	Red-Orange	105°
Red	120°	Red-Orange	120°	Orange	120°
Red-Orange	135°	Orange	135°	Orange-Yellow	135°
Orange	150°	Orange-Yellow	150°	Yellow	150°
Orange-Yellow	165°	Yellow	165°	Yellow-Green	165°
Yellow	180°	Yellow-Green	180°	Green	180°
Yellow-Green	195°	Green	195°	Green-Blue	195°
Green	210°	Green-Blue	210°	Blue	210°
Green-Blue	225°	Blue	225°	Blue-Violet	225°
Blue	240°	Blue-Violet	240°	Violet	240°
Blue-Violet	255°	Violet	255°	Violet-Red	255°
Violet	270°	Violet-Red	270°	Red	270°
Violet-Red	285°	Red	285°	Red-Orange	285°
Red	300°	Red-Orange	300°	Orange	300°
Red-Orange	315°	Orange	315°	Orange-Yellow	315°
Orange	330°	Orange-Yellow	330°	Yellow	330°
Orange-Yellow	345°	Yellow	345°	Yellow-Green	345°
	(360°)		(360°)		(360°)

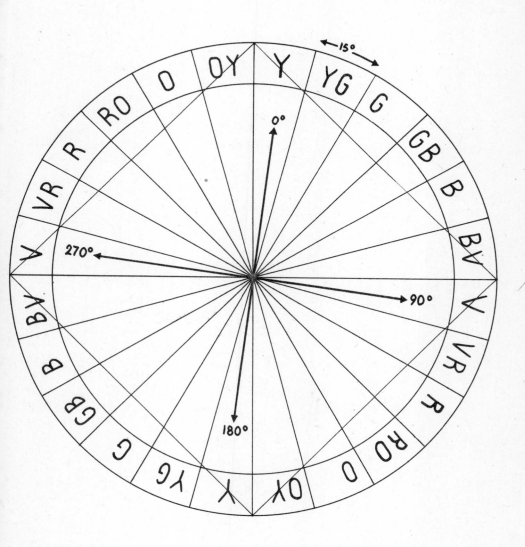

Figure 52. Full spectrum in bilateral symmetry through 360° yellow origin.

Green-Blue Origin
(Red-Orange Axis)

Green-Blue	0°
Blue	15°
Blue-Violet	30°
Violet	45°
Violet-Red	60°
Red	75°
Red-Orange	90°
Orange	105°
Orange-Yellow	120°
Yellow	135°
Yellow-Green	150°
Green	165°
Green-Blue	180°
Blue	195°
Blue-Violet	210°
Violet	225°
Violet-Red	240°
Red	255°
Red-Orange	270°
Orange	285°
Orange-Yellow	300°
Yellow	315°
Yellow-Green	330°
Green	345°
	(360°)

Blue Origin
(Orange Axis)

Blue	0°
Blue-Violet	15°
Violet	30°
Violet-Red	45°
Red	60°
Red-Orange	75°
Orange	90°
Orange-Yellow	105°
Yellow	120°
Yellow-Green	135°
Green	150°
Green-Blue	165°
Blue	180°
Blue-Violet	195°
Violet	210°
Violet-Red	225°
Red	240°
Red-Orange	255°
Orange	270°
Orange-Yellow	285°
Yellow	300°
Yellow-Green	315°
Green	330°
Green-Blue	345°
	(360°)

Blue-Violet Origin
(Orange-Yellow Axis)

Blue-Violet	0°
Violet	15°
Violet-Red	30°
Red	45°
Red-Orange	60°
Orange	75°
Orange-Yellow	90°
Yellow	105°
Yellow-Green	120°
Green	135°
Green-Blue	150°
Blue	165°
Blue-Violet	180°
Violet	195°
Violet-Red	210°
Red	225°
Red-Orange	240°
Orange	255°
Orange-Yellow	270°
Yellow	285°
Yellow-Green	300°
Green	315°
Green-Blue	330°
Blue	345°
	(360°)

Violet Origin
(Yellow Axis)

Violet	0°
Violet-Red	15°
Red	30°
Red-Orange	45°
Orange	60°
Orange-Yellow	75°
Yellow	90°
Yellow-Green	105°
Green	120°
Green-Blue	135°
Blue	150°
Blue-Violet	165°
Violet	180°
Violet-Red	195°
Red	210°
Red-Orange	225°
Orange	240°
Orange-Yellow	255°
Yellow	270°
Yellow Green	285°
Green	300°
Green-Blue	315°
Blue	330°
Blue-Violet	345°
	(360°)

Violet-Red Origin
(Yellow-Green Axis)

Violet-Red	0°
Red	15°
Red-Orange	30°
Orange	45°
Orange-Yellow	60°
Yellow	75°
Yellow-Green	90°
Green	105°
Green-Blue	120°
Blue	135°
Blue-Violet	150°
Violet	165°
Violet-Red	180°
Red	195°
Red-Orange	210°
Orange	225°
Orange-Yellow	240°
Yellow	255°
Yellow-Green	270°
Green	285°
Green-Blue	300°
Blue	315°
Blue-Violet	330°
Violet	345°
	(360°)

Red Origin
(Green Axis)

Red	0°
Red-Orange	15°
Orange	30°
Orange-Yellow	45°
Yellow	60°
Yellow-Green	75°
Green	90°
Green-Blue	105°
Blue	120°
Blue-Violet	135°
Violet	150°
Violet-Red	165°
Red	180°
Red-Orange	195°
Orange	210°
Orange-Yellow	225°
Yellow	240°
Yellow-Green	255°
Green	270°
Green-Blue	285°
Blue	300°
Blue-Violet	315°
Violet	330°
Violet-Red	345°
	(360°)

Red-Orange Origin
(Green-Blue Axis)

Red-Orange	0°
Orange	15°
Orange-Yellow	30°
Yellow	45°
Yellow-Green	60°
Green	75°
Green-Blue	90°
Blue	105°
Blue-Violet	120°
Violet	135°
Violet-Red	150°
Red	165°
Red-Orange	180°
Orange	195°
Orange-Yellow	210°
Yellow	225°
Yellow-Green	240°
Green	255°
Green-Blue	270°
Blue	285°
Blue-Violet	300°
Violet	315°
Violet-Red	330°
Red	345°
	(360°)

Orange Origin
(Blue Axis)

Orange	0°
Orange-Yellow	15°
Yellow	30°
Yellow-Green	45°
Green	60°
Green-Blue	75°
Blue	90°
Blue-Violet	105°
Violet	120°
Violet-Red	135°
Red	150°
Red-Orange	165°
Orange	180°
Orange-Yellow	195°
Yellow	210°
Yellow-Green	225°
Green	240°
Green-Blue	255°
Blue	270°
Blue-Violet	285°
Violet	300°
Violet-Red	315°
Red	330°
Red-Orange	345°
	(360°)

Orange-Yellow Origin
(Blue-Violet Axis)

Orange-Yellow	0°	Yellow	195°
Yellow	15°	Yellow-Green	210°
Yellow-Green	30°	Green	225°
Green	45°	Green-Blue	240°
Green-Blue	60°	Blue	255°
Blue	75°	Blue-Violet	270°
Blue-Violet	90°	Violet	285°
Violet	105°	Violet-Red	300°
Violet-Red	120°	Red	315°
Red	135°	Red-Orange	330°
Red-Orange	150°	Orange	345°
Orange	165°		(360°)
Orange-Yellow	180°		

SELECTIVE SYSTEMS

2. *The Scale of Twelve Hues*

a. FULL SPECTRUM

Each Sector = 30°

Yellow Origin		Yellow-Green Origin		Green Origin	
Yellow	0°	Yellow-Green	0°	Green	0°
Yellow-Green	30°	Green	30°	Green-Blue	30°
Green	60°	Green-Blue	60°	Blue	60°
Green-Blue	90°	Blue	90°	Blue-Violet	90°
Blue	120°	Blue-Violet	120°	Violet	120°
Blue-Violet	150°	Violet	150°	Violet-Red	150°
Violet	180°	Violet-Red	180°	Red	180°
Violet-Red	210°	Red	210°	Red-Orange	210°
Red	240°	Red-Orange	240°	Orange	240°
Red-Orange	270°	Orange	270°	Orange-Yellow	270°
Orange	300°	Orange-Yellow	300°	Yellow	300°
Orange-Yellow	330°	Yellow	330°	Yellow-Green	330°

Green-Blue Origin		Blue Origin		Blue-Violet Origin	
Green-Blue	0°	Blue	0°	Blue-Violet	0°
Blue	30°	Blue-Violet	30°	Violet	30°
Blue-Violet	60°	Violet	60°	Violet-Red	60°
Violet	90°	Violet-Red	90°	Red	90°
Violet-Red	120°	Red	120°	Red-Orange	120°
Red	150°	Red-Orange	150°	Orange	150°
Red-Orange	180°	Orange	180°	Orange-Yellow	180°
Orange	210°	Orange-Yellow	210°	Yellow	210°
Orange-Yellow	240°	Yellow	240°	Yellow-Green	240°
Yellow	270°	Yellow-Green	270°	Green	270°
Yellow-Green	300°	Green	300°	Green-Blue	300°
Green	330°	Green-Blue	330°	Blue	330°

Violet Origin

Violet	0°
Violet-Red	30°
Red	60°
Red-Orange	90°
Orange	120°
Orange-Yellow	150°
Yellow	180°
Yellow-Green	210°
Green	240°
Green-Blue	270°
Blue	300°
Blue-Violet	330°

Violet-Red Origin

Violet-Red	0°
Red	30°
Red-Orange	60°
Orange	90°
Orange-Yellow	120°
Yellow	150°
Yellow-Green	180°
Green	210°
Green-Blue	240°
Blue	270°
Blue-Violet	300°
Violet	330°

Red Origin

Red	0°
Red-Orange	30°
Orange	60°
Orange-Yellow	90°
Yellow	120°
Yellow-Green	150°
Green	180°
Green-Blue	210°
Blue	240°
Blue-Violet	270°
Violet	300°
Violet-Red	330°

Red-Orange Origin

Red-Orange	0°
Orange	30°
Orange-Yellow	60°
Yellow	90°
Yellow-Green	120°
Green	150°
Green-Blue	180°
Blue	210°
Blue-Violet	240°
Violet	270°
Violet-Red	300°
Red	330°

Orange Origin

Orange	0°
Orange-Yellow	30°
Yellow	60°
Yellow-Green	90°
Green	120°
Green-Blue	150°
Blue	180°
Blue-Violet	210°
Violet	240°
Violet-Red	270°
Red	300°
Red-Orange	330°

Orange-Yellow Origin

Orange-Yellow	0°
Yellow	30°
Yellow-Green	60°
Green	90°
Green-Blue	120°
Blue	150°
Blue-Violet	180°
Violet	210°
Violet-Red	240°
Red	270°
Red-Orange	300°
Orange	330°

b. 180° OF THE SPECTRUM DISTRIBUTED THROUGH 360°.

Yellow Origin
Green-Blue Axis

Yellow	0°
Yellow-Green	30°
Green	60°
Green-Blue	90°
Blue	120°
Blue-Violet	150°
Violet	180°
Blue-Violet	210°
Blue	240°
Green-Blue	270°
Green	300°
Yellow-Green	330°

Yellow-Green Origin
Blue Axis

Yellow-Green	0°
Green	30°
Green-Blue	60°
Blue	90°
Blue-Violet	120°
Violet	150°
Violet-Red	180°
Violet	210°
Blue-Violet	240°
Blue	270°
Green-Blue	300°
Green	330°

Green Origin
Blue-Violet Axis

Green	0°
Green-Blue	30°
Blue	60°
Blue-Violet	90°
Violet	120°
Violet-Red	150°
Red	180°
Violet-Red	210°
Violet	240°
Blue-Violet	270°
Blue	300°
Green-Blue	330°

Green-Blue Origin
Violet Axis

Green-Blue	0°
Blue	30°
Blue-Violet	60°
Violet	90°
Violet-Red	120°
Red	150°
Red-Orange	180°
Red	210°
Violet-Red	240°
Violet	270°
Blue-Violet	300°
Blue	330°

Blue Origin
Violet-Red Axis

Blue	0°
Blue-Violet	30°
Violet	60°
Violet-Red	90°
Red	120°
Red-Orange	150°
Orange	180°
Red-Orange	210°
Red	240°
Violet-Red	270°
Violet	300°
Blue-Violet	330°

Blue-Violet Origin
Red Axis

Blue-Violet	0°
Violet	30°
Violet-Red	60°
Red	90°
Red-Orange	120°
Orange	150°
Orange-Yellow	180°
Orange	210°
Red-Orange	240°
Red	270°
Violet-Red	300°
Violet	330°

180° OF THE SPECTRUM DISTRIBUTED THROUGH 360° (*Concluded*)

Violet Origin
Red-Orange Axis

Violet	0°
Violet-Red	30°
Red	60°
Red-Orange	90°
Orange	120°
Orange-Yellow	150°
Yellow	180°
Orange-Yellow	210°
Orange	240°
Red-Orange	270°
Red	300°
Violet-Red	330°

Violet-Red Origin
Orange Axis

Violet-Red	0°
Red	30°
Red-Orange	60°
Orange	90°
Orange-Yellow	120°
Yellow	150°
Yellow-Green	180°
Yellow	210°
Orange-Yellow	240°
Orange	270°
Red-Orange	300°
Red	330°

Red Origin
Orange-Yellow Axis

Red	0°
Red-Orange	30°
Orange	60°
Orange-Yellow	90°
Yellow	120°
Yellow-Green	150°
Green	180°
Yellow-Green	210°
Yellow	240°
Orange-Yellow	270°
Orange	300°
Red-Orange	330°

Red-Orange Origin
Yellow Axis

Red-Orange	0°
Orange	30°
Orange-Yellow	60°
Yellow	90°
Yellow-Green	120°
Green	150°
Green-Blue	180°
Green	210°
Yellow-Green	240°
Yellow	270°
Orange-Yellow	300°
Orange	330°

Orange Origin
Yellow-Green Axis

Orange	0°
Orange-Yellow	30°
Yellow	60°
Yellow-Green	90°
Green	120°
Green-Blue	150°
Blue	180°
Green-Blue	210°
Green	240°
Yellow-Green	270°
Yellow	300°
Orange-Yellow	330°

Orange-Yellow Origin
Green Axis

Orange-Yellow	0°
Yellow	30°
Yellow-Green	60°
Green	90°
Green-Blue	120°
Blue	150°
Blue-Violet	180°
Blue	210°
Green-Blue	240°
Green	270°
Yellow-Green	300°
Yellow	330°

SELECTIVE SYSTEMS

c. 150° OF THE SPECTRUM DISTRIBUTED THROUGH 360°.

Yellow Origin
Green-Green-Blue Axis

Yellow	0°
Yellow-Green	36°
Green	72°
Green-Blue	108°
Blue	144°
Blue-Violet	180°
Blue	216°
Green-Blue	252°
Green	288°
Yellow-Green	324°

Yellow-Green Origin
Grene-Blue-Blue Axis

Yellow-Green	0°
Green	36°
Green-Blue	72°
Blue	108°
Blue-Violet	144°
Violet	180°
Blue-Violet	216°
Blue	252°
Green-Blue	288°
Green	324°

Green Origin
Blue-Blue-Violet Axis

Green	0°
Green-Blue	36°
Blue	72°
Blue-Violet	108°
Violet	144°
Violet-Red	180°
Violet	216°
Blue-Violet	252°
Blue	288°
Green-Blue	324°

Green-Blue Origin
Blue-Violet-Violet Axis

Green-Blue	0°
Blue	36°
Blue-Violet	72°
Violet	108°
Violet-Red	144°
Red	180°
Violet-Red	216°
Violet	252°
Blue-Violet	288°
Blue	324°

Blue Origin
Violet-Violet-Red Axis

Blue	0°
Blue-Violet	36°
Violet	72°
Violet-Red	108°
Red	144°
Red-Orange	180°
Red	216°
Violet-Red	252°
Violet	288°
Blue-Violet	324°

Blue-Violet Origin
Violet-Red-Red Axis

Blue-Violet	0°
Violet	36°
Violet-Red	72°
Red	108°
Red-Orange	144°
Orange	180°
Red-Orange	216°
Red	252°
Violet-Red	288°
Violet	324°

150° OF THE SPECTRUM DISTRIBUTED THROUGH 360° (*Concluded*)

Violet Origin
Red-Red-Orange Axis

Violet	0°
Violet-Red	36°
Red	72°
Red-Orange	108°
Orange	144°
Orange-Yellow	180°
Orange	216°
Red-Orange	252°
Red	288°
Violet-Red	324°

Violet-Red Origin
Red-Orange-Orange Axis

Violet-Red	0°
Red	36°
Red-Orange	72°
Orange	108°
Orange-Yellow	144°
Yellow	180°
Orange-Yellow	216°
Orange	252°
Red-Orange	288°
Red	324°

Red Origin
Orange-Orange-Yellow Axis

Red	0°
Red-Orange	36°
Orange	72°
Orange-Yellow	108°
Yellow	144°
Yellow-Green	180°
Yellow	216°
Orange-Yellow	252°
Orange	288°
Red-Orange	324°

Red-Orange Origin
Orange-Yellow-Yellow Axis

Red-Orange	0°
Orange	36°
Orange-Yellow	72°
Yellow	108°
Yellow-Green	144°
Green	180°
Yellow-Green	216°
Yellow	252°
Orange-Yellow	288°
Orange	324°

Orange Origin
Yellow-Yellow-Green Axis

Orange	0°
Orange-Yellow	36°
Yellow	72°
Yellow-Green	108°
Green	144°
Green-Blue	180°
Green	215°
Yellow-Green	252°
Yellow	288°
Orange-Yellow	324°

Orange-Yellow Origin
Yellow-Green-Green Axis

Orange-Yellow	0°
Yellow	36°
Yellow-Green	72°
Green	108°
Green-Blue	144°
Blue	180°
Green-Blue	216°
Green	252°
Yellow-Green	288°
Yellow	324°

SELECTIVE SYSTEMS

d. 120° OF THE SPECTRUM DISTRIBUTED THROUGH 360°.

Yellow Origin
Green Axis

Yellow	0°
Yellow-Green	45°
Green	90°
Green-Blue	135°
Blue	180°
Green-Blue	225°
Green	270°
Yellow-Green	315°

Yellow-Green Origin,
Green-Blue Axis

Yellow-Green	0°
Green	45°
Green-Blue	90°
Blue	135°
Blue-Violet	180°
Blue	225°
Green-Blue	270°
Green	315°

Green Origin
Blue Axis

Green	0°
Green-Blue	45°
Blue	90°
Blue-Violet	135°
Violet	180°
Blue-Violet	225°
Blue	270°
Green-Blue	315°

Green-Blue Origin
Blue-Violet Axis

Green-Blue	0°
Blue	45°
Blue-Violet	90°
Violet	135°
Violet-Red	180°
Violet	225°
Blue-Violet	270°
Blue	315°

Blue Origin
Violet Axis

Blue	0°
Blue-Violet	45°
Violet	90°
Violet-Red	135°
Red	180°
Violet-Red	225°
Violet	270°
Blue-Violet	315°

Blue-Violet Origin
Violet-Red Axis

Blue-Violet	0°
Violet	45°
Violet-Red	90°
Red	135°
Red-Orange	180°
Red	225°
Violet-Red	270°
Violet	315°

120° OF THE SPECTRUM DISTRIBUTED THROUGH 360° (*Concluded*)

Violet Origin
Red Axis

Violet	0°
Violet-Red	45°
Red	90°
Red-Orange	135°
Orange	180°
Red-Orange	225°
Red	270°
Violet-Red	315°

Violet-Red Origin
Red-Orange Axis

Violet-Red	0°
Red	45°
Red-Orange	90°
Orange	135°
Orange-Yellow	180°
Orange	225°
Red-Orange	270°
Red	315°

Red Origin
Orange Axis

Red	0°
Red-Orange	45°
Orange	90°
Orange-Yellow	135°
Yellow	180°
Orange-Yellow	225°
Orange	270°
Red-Orange	315°

Red-Orange Origin
Orange-Yellow Axis

Red-Orange	0°
Orange	45°
Orange-Yellow	90°
Yellow	135°
Yellow-Green	180°
Yellow	225°
Orange-Yellow	270°
Orange	315°

Orange Origin
Yellow Axis

Orange	0°
Orange-Yellow	45°
Yellow	90°
Yellow-Green	135°
Green	180°
Yellow-Green	225°
Yellow	270°
Orange-Yellow	315°

Orange-Yellow Origin
Yellow-Green Axis

Orange-Yellow	0°
Yellow	45°
Yellow-Green	90°
Green	135°
Green-Blue	180°
Green	225°
Yellow-Green	270°
Yellow	315°

e. 90° OF THE SPECTRUM DISTRIBUTED THROUGH 360°.

Yellow Origin
Yellow-Green-Green Axis

Yellow	0°
Yellow-Green	60°
Green	120°
Green-Blue	180°
Green	240°
Yellow-Green	300°

Yellow-Green Origin
Green-Green-Blue Axis

Yellow-Green	0°
Green	60°
Green-Blue	120°
Blue	180°
Green-Blue	240°
Green	300°

Green Origin
Green-Blue-Blue Axis

Green	0°
Green-Blue	60°
Blue	120°
Blue-Violet	180°
Blue	240°
Green-Blue	300°

Green-Blue Origin
Blue-Blue-Violet Axis

Green-Blue	0°
Blue	60°
Blue-Violet	120°
Violet	180°
Blue-Violet	240°
Blue	300°

Blue Origin
Blue-Violet-Violet Axis

Blue	0°
Blue-Violet	60°
Violet	120°
Violet-Red	180°
Violet	240°
Blue-Violet	300°

Blue-Violet Origin
Violet-Violet-Red Axis

Blue-Violet	0°
Violet	60°
Violet-Red	120°
Red	180°
Violet-Red	240°
Violet	300°

Violet Origin
Violet-Red-Red Axis

Violet	0°
Violet-Red	60°
Red	120°
Red-Orange	180°
Red	240°
Violet-Red	300°

Violet-Red Origin
Red-Red-Orange Axis

Violet-Red	0°
Red	60°
Red-Orange	120°
Orange	180°
Red-Orange	240°
Red	300°

90° OF THE SPECTRUM DISTRIBUTED THROUGH 360° (*Concluded*)

Red Origin
Red-Orange-Orange Axis

Red	0°
Red-Orange	60°
Orange	120°
Orange-Yellow	180°
Orange	240°
Red-Orange	300°

Red-Orange Origin
Orange-Orange-Yellow Axis

Red-Orange	0°
Orange	60°
Orange-Yellow	120°
Yellow	180°
Orange-Yellow	240°
Orange	300°

Orange Origin
Orange-Yellow-Yellow Axis

Orange	0°
Orange-Yellow	60°
Yellow	120°
Yellow-Green	180°
Yellow	240°
Orange-Yellow	300°

Orange-Yellow Origin
Yellow-Yellow-Green Axis

Orange-Yellow	0°
Yellow	60°
Yellow-Green	120°
Green	180°
Yellow-Green	240°
Yellow	300°

f. 60° OF THE SPECTRUM DISTRIBUTED THROUGH 360°.

Yellow Origin
Yellow-Green Axis

Yellow	0°
Yellow-Green	90°
Green	180°
Yellow-Green	270°

Yellow-Green Origin
Green Axis

Yellow-Green	0°
Green	90°
Green-Blue	180°
Green	270°

Green Origin
Green-Blue Axis

Green	0°
Green-Blue	90°
Blue	180°
Green-Blue	270°

Green-Blue Origin
Blue Axis

Green-Blue	0°
Blue	90°
Blue-Violet	180°
Blue	270°

SELECTIVE SYSTEMS

60° OF THE SPECTRUM DISTRIBUTED THROUGH 360° (*Concluded*)

Blue Origin
Blue-Violet Axis

Blue	0°
Blue-Violet	90°
Violet	180°
Blue-Violet	270°

Blue-Violet Origin
Violet Axis

Blue-Violet	0°
Violet	90°
Violet-Red	180°
Violet	270°

Violet Origin
Violet-Red Axis

Violet	0°
Violet-Red	90°
Red	180°
Violet-Red	270°

Violet-Red Origin
Red Axis

Violet-Red	0°
Red	90°
Red-Orange	180°
Red	270°

Red Origin
Red-Orange Axis

Red	0°
Red-Orange	90°
Orange	180°
Red-Orange	270°

Red-Orange Origin
Orange Axis

Red-Orange	0°
Orange	90°
Orange-Yellow	180°
Orange	270°

Orange Origin
Orange-Yellow Axis

Orange	0°
Orange-Yellow	90°
Yellow	180°
Orange-Yellow	270°

Orange-Yellow Origin
Yellow Axis

Orange-Yellow	0°
Yellow	90°
Yellow-Green	180°
Yellow	270°

g. 30° OF THE SPECTRUM DISTRIBUTED THROUGH 360°.

Yellow Origin
Yellow-Yellow-Green Axis

Yellow	0°
Yellow-Green	180°

Yellow-Green Origin
Yellow-Green-Green Axis

Yellow-Green	0°
Green	180°

Green Origin
Green-Green-Blue Axis

Green	0°
Green-Blue	180°

Green-Blue Origin
Green-Blue-Blue Axis

Green-Blue	0°
Blue	180°

Blue Origin
Blue-Blue-Violet Axis

Blue	0°
Blue-Violet	180°

Blue-Violet Origin
Blue-Violet-Violet Axis

Blue-Violet	0°
Violet	180°

Violet Origin
Violet-Violet-Red Axis

Violet	0°
Violet-Red	180°

Violet-Red Origin
Violet-Red-Red Axis

Violet-Red	0°
Red	180°

Red Origin
Red-Red-Orange Axis

Red	0°
Red-Orange	180°

Red-Orange Origin
Red-Orange-Orange Axis

Red-Orange	0°
Orange	180°

Orange Origin
Orange-Orange-Yellow Axis

Orange	0°
Orange-Yellow	180°

Orange-Yellow Origin
Orange-Yellow-Yellow Axis

Orange-Yellow	0°
Yellow	180°

CHAPTER 2

PRODUCTION OF DESIGN

A. Elements of Linear Design

A GEOMETRIC point has no physical extension. An artistic point is the physical expression of a geometric point in a given material medium. This medium renders it visible. We can assume this point to be πr^2, where r approaches zero. Different media require different dimensions for r. A scale moving toward zero may be evolved from a wide flat brush to a pointed one, to a pastel edge, and finally, to the finest pencil point.

A point moving uniformly on a plane produces a visible trajectory. This trajectory is a *linear design*. The time required to evolve such a trajectory and the speed of movement determine its linear dimensions.

A point can move only through an angle. When the angle equals zero degrees, the point does not move. When the angle equals 180°, the point moves indefinitely and the resulting trajectory is an unending straight line. When the time of the movement of a point under 180° is limited, the resulting trajectory is a rectilinear segment of a definite extension. This extension can be measured in terms of linear measurement. Thus, we obtain the first element of a linear design: *rectilinear segment of a definite dimension* (x_1).

Continuous progression of rectilinear segments moving in one direction produces closed forms, or with a tendency to close, according to the arithmetical property of the angles under which they move, each angle being the divisor of a dividend of 180°, or its multiple, with various coefficients.

Continuous progression of rectilinear segments moving in alternating directions produces broken lines when the segments change their direction under one constant angle. Their range of extension is from the dimension of a selected segment unit ($\angle 0°$) to infinity ($\angle 180°$), and back to the dimensions of the unit ($\angle 360°$).

A linear element moving through an angle extends itself through the second coordinate (x_2)[1], thus evolving on a plane. *Angle* and *direction* become the two other elements of linear design. Dimension and angle allow an infinite number of variations. Direction can only be *clockwise* or *counter-clockwise*.

A design evolving through angles moves in the direction opposite to the geometric formation of the angles.

---

[1] x_2 = width.

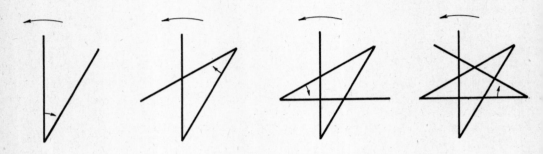

Figure 1. Design evolving through angles.

When angles form in a clockwise direction (⟳), the design moves counter-clockwise (⟲), and vice-versa.

Curvilinear configurations are the derivatives of rectilinear configurations, each *arc* being produced from a rectilinear segment as a radius or a diameter. Dimensions of rectilinear segments determine the curvature of the corresponding arcs. Angular values determine the dimensions of the corresponding arcs. The increase of a radius decreases the curvature of a corresponding arc. The increase of an angle increases the dimension of a corresponding arc.

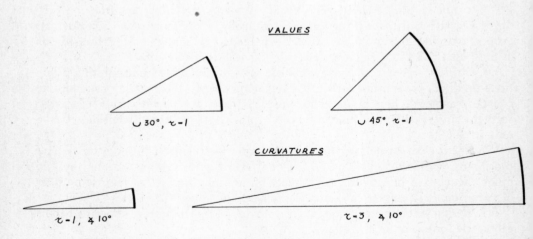

Figure 2. Increase of angle increases dimension of corresponding arc.

B. Rhythmic Design

The human idea of the universe changes through the ages. There have been radical changes from Ptolemy to Eddington. The dominant tendency in modern science is to "re-create" the universe geometrically, *i.e.*, to explain all physical phenomena as derivatives of the properties of space. The idea of the homogeneity of matter and space was introduced first by Democritus in philosophical rather than physical form. Many centuries elapsed until Descartes established this postulate on a physical basis. Nevertheless, Newton challenged it. The mathematical equipment of Descartes' epoch, 39 years before Leibnitz invented differential calculus (1676), was insufficient for the mathematical interpretation of the postulate. Recent progress in science, due to the contributions of Riemann, Minkowsky, Lorentz, Einstein, Michelson, Millikan, Compton and Eddington, makes possible a universe of highly complex space. It is known as a space-time continuum, where time is one of the components of space.

Any art-form is also a derivative of the space-time continuum. Art can be measured and analyzed like any other phenomenon of our universe. Any analytical result can be represented graphically, *i.e.*, geometrically, in short, in terms of space-time relations. The esthetic value, from the analytical point of view, is a function of the space-time relations. Design is geometrically the most obvious art-form, since the idea and the realization in an art medium are both accomplished in empirical space.

A design is rhythmic if analysis reveals the regularity in the sequence of its components and their correlations. The regularity or the irregularity of sequence may be obtained in any finite continuity. The simplest form of regularity can be determined by a constant relation between two consecutive terms. When this relation is unity, we have the simplest case of regularity. In wave theory it is known as "simple harmonic motion"; it is a uniform motion in which the consecutive terms are related as one to one. It may be determined as monomial periodicity and expressed as $a+a+a$ ... More developed forms of rhythm may be observed in a continuity where regularity can be deduced from relations of the groups of terms, for example $(a+b+c) + (a+b+c)$ ... In this case, in order to discover the form of regularity, it is necessary to observe at least six terms. A non-rhythmic continuity may be determined as a sequence where regularity can be observed only in an infinite number of terms. Irregularities producing regular sequences should be considered as rhythmic deviations and variations.

The general method of producing rhythmic sequences is based on the physical phenomenon known as interference. Two harmonic waves of the same period (frequency) acting simultaneously do not produce any new sequence. In order to obtain new sequences from two or more waves, it is necessary that their periods should be different, even when they are harmonic. This requirement will be fulfilled when the two or more periodicities are not related as $1:1:1:\ldots$

A wave resulting from an interference of any periodicities, related as $a:1$, where any value may be attributed to a, does not produce a new sequence of values but only of intensities. In order to obtain new sequences of values as well as of intensities, unity must be excluded from the previous relation. It will assume the form A = B, or A = B = C, etc., where none of the terms equals unity. These rhythmic sequences of values and intensities produce series, which result from interference of the two or more component periodicities. The procedure of obtaining the resulting rhythmic series is based on the synchronization of the component periodicities by means of their common denominator or common product. Taking A and B as the two component periodicities, we obtain the product AB which represents the total of units on which the synchronization can be performed. A is taken B times against B taken A times, *i.e.*, if A = 3 and B = 2, then AB = 6. Thus 3 will follow 2 times against 2 following 3 times.

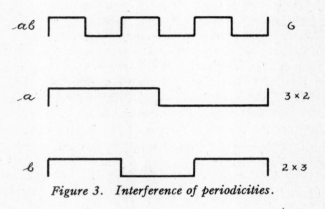

Figure 3. Interference of periodicities.

If we draw the result by dropping perpendiculars to both of the two component periodicities through their consecutive terms, we obtain the following series: 2 + 1 + 1 + 2.

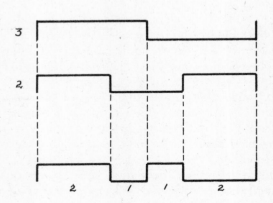

Figure 4. Resulting periodicity.

PRODUCTION OF DESIGN 367

The interference of 4 against 3, obtained by the method described above, will produce 3 + 1 + 2 + 2 + 1 + 3.

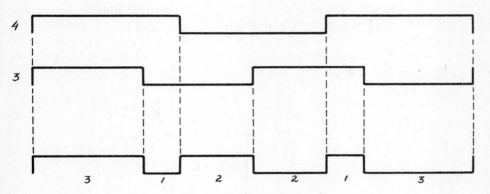

Figure 5. Resulting periodicity of 4 ÷ 3.

The elements of linear design are angles, dimensions, directions and their derivatives (arcs, sectors, etc.), under which a point realized in an art medium moves through an area.

The rhythmic series obtained above, being applied to dimensions of the rectilinear segments moving in a constant direction under a 90° angle, will produce the following results:

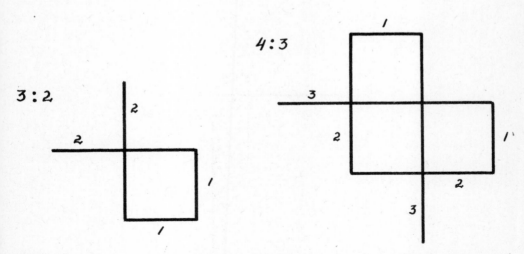

Figure 6. 3 ÷ 2 applied to rectilinear segments moving in a constant direction under 90° angle.

Figure 7. 4 ÷ 3 applied to rectilinear segments moving in a constant direction under 90° angle.

Applying the same series to the progression of angles, where the angle unit equals 30°, we obtain the following results.

The sequence of angles follows the linear periodicity
60° (2) + 30° (1) + 30° (1) + 60° (2)

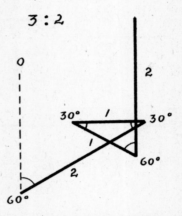

Figure 8. 3 ÷ 2 applied to rectilinear segments moving under angle unit of 30°.

The selection of angles accords with the sequence of periodicities:
90° (3) + 30° (1) + 60° (2) + 60° (2) + 30° (1) + 90° (3)

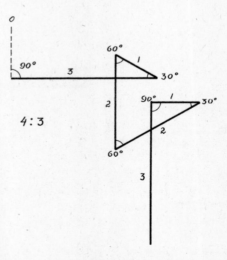

Figure 9. 4 ÷ 3 applied to rectilinear segments moving under angle unit of 30°.

The simple process of evolving circular arcs of 180°, using the previous schemes as diameters and moving in one direction, will result in the following design.

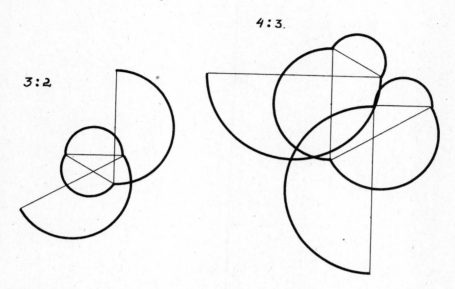

Figure 10. *3 ÷ 2 and 4 ÷ 3 moving in one direction with circular arcs of 180°.*

Different variations of the linear designs, once obtained, may be achieved by means of the displacement of the terms in a rhythmic series. Thus, $2 + 1 + 1 + 2$ (see Figure 4) will produce the following displacements: $1 + 1 + 2 + 2; 1 + 2 + 2 + 1; 2 + 2 + 1 + 1$; while the series $3 + 1 + 2 + 2 + 1 + 3$ (see Figure 5) will produce these: $2 + 2 + 1 + 3 + 3 + 1; 1 + 2 + 2 + 1 + 3 + 3; 2 + 1 + 3 + 3 + 1 + 2; 1 + 3 + 3 + 1 + 2 + 2; 3 + 3 + 1 + 2 + 2 + 1$. These mathematical variations, applied to dimensions or angles, will produce corresponding variations in design.

A more complex form of variation may be obtained by the distributive use of algebraic powers. For instance, $(2 + 1 + 1 + 2)^2 = (4 + 2 + 2 + 4) + (2 + 1 + 1 + 2) + (2 + 1 + 1 + 2) + (4 + 2 + 2 + 4)$. Applying these new periodic series to the components of linear design, we obtain a multiple motive, which harmoniously repeats the relations observed in the primary one.

All these methods may be applied to any linear design evolved on an area without boundaries. In a given area, the relation of its sides, or coordinates, will determine the behavior of linear and angular values, and result in a rhythmic design conditioned by the properties of the area. For example, in a square area, moving clockwise through the 4 : 3 series of angles, we obtain a different result from that of an oblong, where the relation of the two sides is 2 : 1.

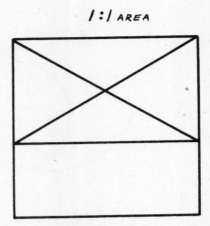

Figure 11. Square and oblong areas contrasted.

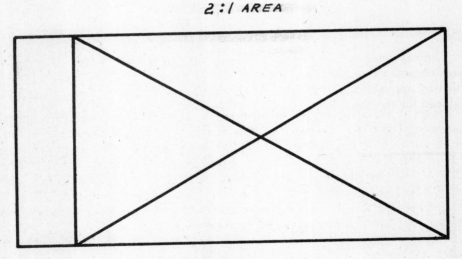

Figure 12. 2 : 1 area.

The effect produced by the powers applied to the sum of the sides of an oblong is to split each of the two sides into partial segments, which are in the same relation to each other as the sides of the oblong. Connecting these partial segments under a right angle, we produce rectangular coordinates. The point of intersection of these is the *rhythmic center* of the area.

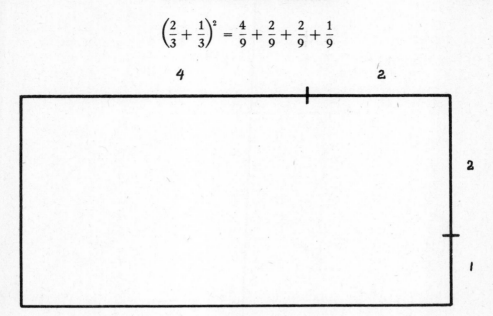

Figure 13. *Powers applied to the sum of the sides of an oblong.*

The rhythmic center determines the origin and the behavior of the linear design obtained through angular relation. This process may be indefinitely continued by application of the higher powers. There will be four rhythmic centers produced for the third power, sixteen for the fourth, etc.

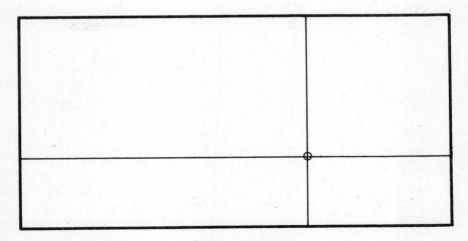

Figure 14. *Rhythmic center of a rectangle.*

There are many other processes in evolving a rhythmic design through its elements, or through the properties of its area. There are some continuous procedures based on the recurrence of one process, such as:

(1) Dropping perpendiculars from the vertices of the two triangles produced by the diagonal of an oblong.

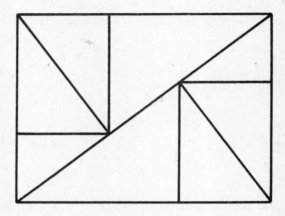

Figure 15. Evolving a rhythmic design on the basis of the diagonal of a rectangle.

(2) Producing continuous sequence of arcs where the radius may be selected from any segment, connecting the rhythmic center with one of the sides of an oblong. The consecutive origins are formed by the tangent points of arcs touching the sides of an oblong.

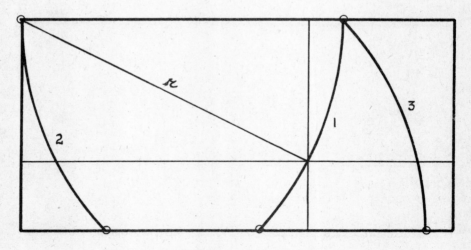

Figure 16. Continuous sequence of arcs.

PRODUCTION OF DESIGN

Geometrical variation of a given design may be obtained by means of:

(1) Consecutive displacement of the partial power areas;

(2) revolving an area around one of the sides;

(3) revolving areas around the rhythmic centers;

(4) displacement in the direction of the coordinates and the rays connecting the sides with the rhythmic center.

Various other forms of areas, such as the circle, the ellipse or any irregular area, may be determined in their proportions by its axes. Any design may be inscribed or described through these axes. Various deviations from symmetry may be obtained by the translation of a motive from an area of one type of structure to that of another.

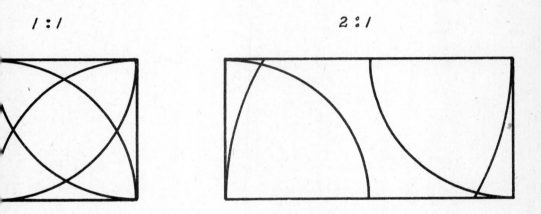

Figure 17. Translating a motive from an area of one type of structure to another.

The same method of rhythmic series is applicable to color in its components (hue, value, intensity). In order to treat one of these color components rhythmically, it is necessary to select a scale within the given limits. This spectral (or value) scale may be inscribed in a circle. The center of the color-wheel will designate the origins of the corresponding angles which are the foundation of a design to which color is to be applied. The rhythmic progression on the twelve step spectral scale may be expressed through the values of the arcs of the circular color chart. For example, assuming a circular scale in the following sequence:

yellow, yellow-green, green, blue-green, blue, blue-violet, violet, red-violet, red, red-orange, orange, yellow-orange,

and assuming yellow to be the initial color (*i. e.*, 0°), the rhythmic series 3 + 1 + 2 + 2 + 1 + 3 will equal 90° + 30° + 60° + 60° + 30° + 90°. This will result in the following series of hues:

Yellow, blue-green, blue-violet, red, red-orange, yellow.

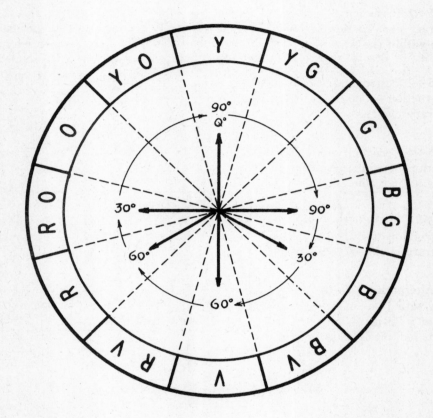

Figure 18. Twelve step spectral scale.

In this way, correlation between the rhythm of design and that of color may be achieved by the identical method of rhythmic sequence of the elements of space.

C. DESIGN ANALYSIS

1. Fish Panel

5 ÷ 3 Rhythmic centers—3rd power.

1.
  5 ÷ 2 Periodicity—clockwise—unit of angle 20°; unit of length of lines 2"—arbitrary construction of circles and a square—the periodicity of 5 ÷ 2 drawn twice within the square form. Arbitrary choice of color.

2.
  5 ÷ 2 Periodicity—clockwise—unit of angle 20°; unit of lines 2"—arbitrary construction of circles and squares, retaining all lines of periodicity. Color arbitrary choice.

3.
  5 ÷ 4 Periodicity—counterclockwise—unit 1/2"; unit of angle 10°—construction of equilateral triangles on each line. Color—three sets of triads—changing in value.

4.
  5 ÷ 3 Periodicity—clockwise—unit of angle 15°; unit of line 1"—arbitrary construction of square, triangle and circles, retaining some lines of the periodicity. Arbitrary choice of color.

5.
  Alternating ∡S—50° and 30° clockwise—length of sides based on 5 ÷ 7 periodicity with 1" as unit. Squares divided into stripes 5 ÷ 3. Color scheme—center unit of 5 ÷ 3 periodicity green and moving from green in both directions—blue, violet, red, orange and yellow.

Figure 19. Fish Panel.

Figure 20. 4÷3 rectangle.

2. 4÷3 Rectangle

Construction: Rhythmic center determined first. Using periodicity 4÷3 (3 + 1 + 2 + 2 + 1 + 3) and 20° as a unit; angles drawn clockwise in each quadrant, using the coordinate (from the rhythmic center to the side of the rectangle) as first side of angle of 60°. In entire rectangle, same angular scheme was developed in each quadrant.

Color: For application of color, each quadrant was developed individually. An angle of 15° used as a unit, and six such angles drawn from rhythmic center to outer boundaries of each quadrant. Each area painted in a different set of two colors. Each set of colors chosen from the color wheel, from points moving clockwise at right angles.

Value: Divisions for values drawn clockwise in each quadrant and parallel to the coordinate of the quadrant. The spacing of value areas determined by the application of the periodicity of 4÷3. Six units to the periodicity, and six values in each quadrant. The lowest value placed nearest the coordinate in each quadrant.

(See preceding page)

PRODUCTION OF DESIGN

Figure 21. 3 ÷ 2 rectangle.

3. 3÷2 Rectangle

Construction: Rhythmic center first determined. Sides of each quadrant divided into 24 units. Then the crossing lines drawn. In the square quadrant, the design is drawn normally. In each of the other 3 quadrants, the design is redrawn in distorted position, as dictated by different spacing.

Color: Four colors in each quadrant. Color wheel with sixteen colors developed.

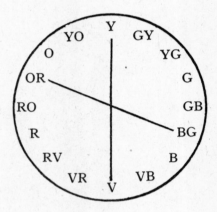

Omitting orange yellow and blue violet, a group of four colors for each quadrant chosen from the four points of 2 diameters that cross at one unit less than a right angle (as shown above). Y, BG, V and OR used in first quadrant. These four colors arranged so that the one of lowest natural value would be placed in the same area in each quadrant. This applied to the placement of all colors.

Value: The value areas are equal; but in each quadrant, there are a different number, according to the proportions of the sides of the 3÷2 rectangle, (9÷6÷6÷4). These areas are arranged perpendicular to the coordinate of each quadrant and move around the rectangle in clockwise direction.

(See preceding page)

PRODUCTION OF DESIGN

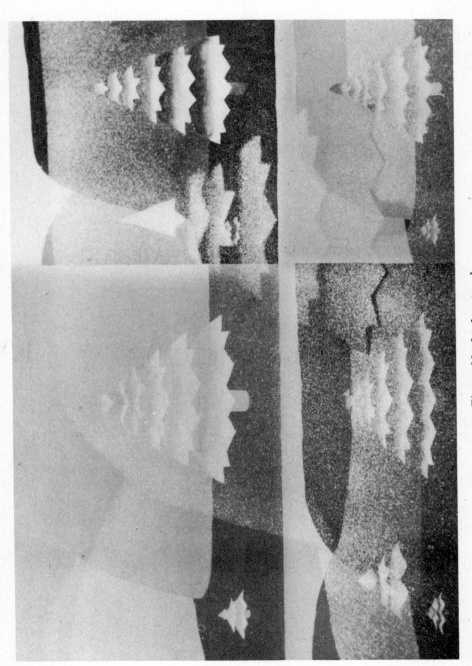

Figure 22. 3 ÷ 2 rectangle.

4. 3÷2 Rectangle

Construction: Division into 576 proportionate areas. Rhythmic center determined. Division of each quadrant into 576 proportionate areas. In square quadrant landscape is drawn, and repeated in other three quadrants and in large rectangle in the first power.

Color: Chose four sets of colors for 4 quadrants.

```
        | yo - Y - yg |
   o    |             |   g
   R  O |             |   GB
   r    |             |   b
        | rv - V - bv |
```

Yellow represents Spring
GB   "   Summer
V   "   Winter
RO   "   Fall

Rendering of landscape in first power in white represents *Snow*. Colors distributed in same succession, a, b, and c, always occurring in same areas. The color to left of dominant color always occurring at—a; the dominant color—b; and color to right of dominant color—c.

Values: There are two values of a and two values of c.

(See preceding page)

Figure 23. Instruments.

5. Instruments

3÷2 rectangle. Rhythmic center determined. Violin drawn normally in square quadrant, divided into 576 spaces. Same subject redrawn in other 3 quadrants in distortion determined by network of spaces. Large instrument drawn on coordinates of rhythmic center with normal section in square area. Bows drawn on diagonal of each quadrant, stemming from the rhythmic center.

12 divisions in each quadrant for values.

15 divisions in each quadrant for intensity.

For background, Red Orange—dominant color, modulated by its complement—Blue Green. For instruments, mixture of Red Orange with small amount of Blue Green of normal value for center of instrument, changing intensity and values as the edges are approached. Ornaments on instrument:—mixture of Red Orange with small amount of Blue Green in a medium value, changing value as the edges are approached.

(See preceding page)

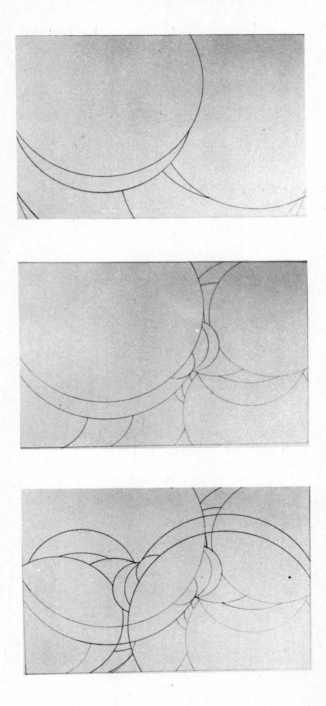

Figure 24. Scale of saturation produced by linear configurations in a confined area.

Figure 25. Scale of saturation produced by linear configurations in a confined area

PRODUCTION OF DESIGN

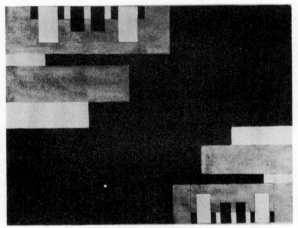

Figure 26. Rug designs.

388 TECHNOLOGY OF ART PRODUCTION

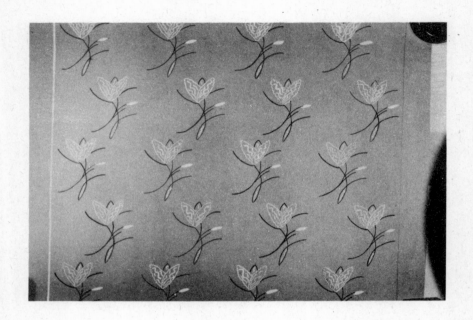

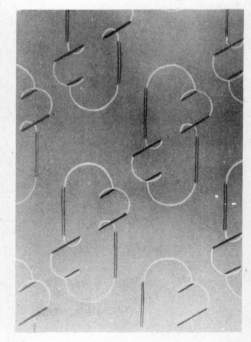

Figure 27. Wall paper designs.

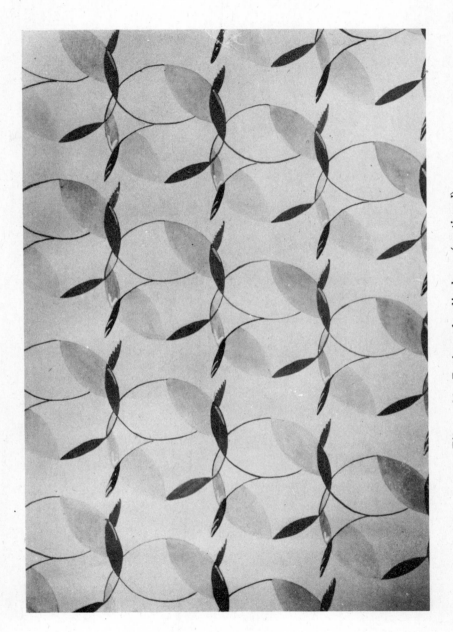

Figure 28. Designs for linoleum (continued).

Figure 28. Designs for linoleum (concluded).

D. Three Compositions in Linear Design

One origin point for two subjects.

Family: $\frac{6}{8}$ series.

Clockwise progression for both subjects. Variation of direction occurs with the recurrence of values. All number-values express radii and the length of arcs. Radii and arcs are in direct relation.

First Subject:

The first power trinomial synchronized with its distributive square. The group: $6 = 4 + 1 + 1$.

Values: $6(4 + 1 + 1) = 24 + 6 + 6$

$6(1 + 4 + 1) = 6 + 24 + 6$

$6(1 + 1 + 4) = 6 + 6 + 24$

Second Subject:

The distributive square of the same group $(4 + 1 + 1)$.

Values: $(4 + 1 + 1)^2 = (16 + 4 + 4) + (4 + 1 + 1) + (4 + 1 + 1)$

$(1 + 4 + 1)^2 = (1 + 4 + 1) + (4 + 16 + 4) + (1 + 4 + 1)$

$(1 + 1 + 4)^2 = (1 + 1 + 4) + (1 + 1 + 4) + (4 + 4 + 16)$

Linear unit for the radii equals $1/4''$.

Arc unit $= 10°$.

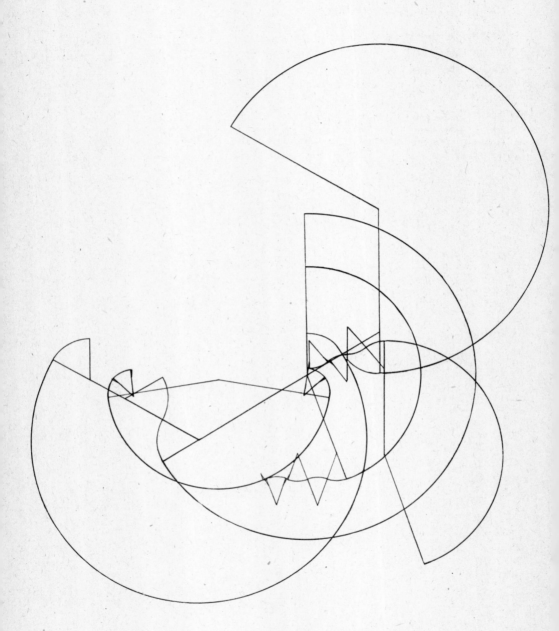

Figure 29. Linear design 6/6 series.

PRODUCTION OF DESIGN

One origin point for all groups: the center of a diameter whose length is 12″.

Family: $\frac{12}{12}$ series.

Clockwise progression for all groups. Variation of direction in each linear group takes place with each term. All number-values express radii and the length of arcs. Radii and arcs are in direct relation.

The following number-values represent binary, ternary and quinternary groups in the order of their appearance. The degrees indicate the origin-axes for each group in relation to the ordinate, which is zero-axis.

Binomials:

$$\left.\begin{array}{ll} 7+5 & 0° \\ 5+7 & 180° \end{array}\right\} \text{two sectors in symmetry}$$

$$\left.\begin{array}{ll} 5+2+5 & 0° \\ 2+5+5 & 120° \\ 5+5+2 & 240° \end{array}\right\} \text{three sectors in symmetry}$$

$$\left.\begin{array}{ll} 2+3+2+3+2 & 0° \\ 2+2+3+2+3 & 72° \\ 3+2+3+2+2 & 144° \\ 2+3+2+2+3 & 216° \\ 3+2+2+3+2 & 288° \end{array}\right\} \text{five sectors in symmetry}$$

Linear unit = 1/2″.

Arc unit = 10°.

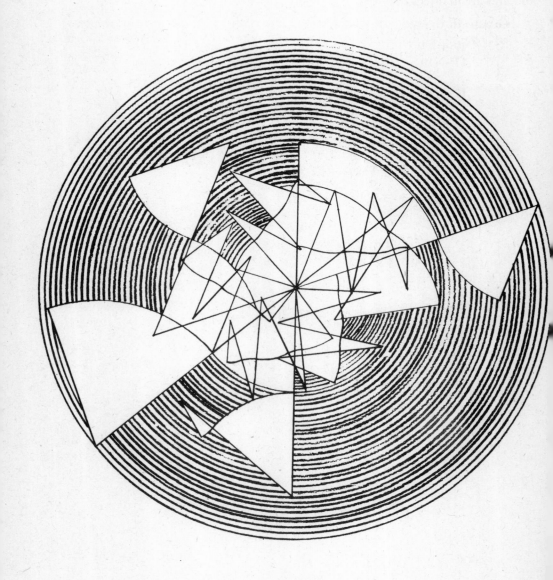

Figure 30. Linear design 12/12 series.

PRODUCTION OF DESIGN

One origin point for all groups.

Family: $\frac{36}{36}$ series.

Clockwise progression for all groups. Variation of direction occurs with the recurrence of values. All number-values express radii and the length of arcs. Radii and arcs are in direct relation.

Groups in the order of their appearance:

$19 + 17$
$17 + 19$

$17 + 2 + 17$
$2 + 17 + 17$
$17 + 17 + 2$

$2 + 15 + 2 + 15 + 2$
$15 + 2 + 15 + 2 + 2$
$2 + 15 + 2 + 2 + 15$
$15 + 2 + 2 + 15 + 2$
$2 + 2 + 15 + 2 + 15$

$2 + 2 + 11 + 2 + 2 + 2 + 11 + 2 + 2$
$2 + 11 + 2 + 2 + 2 + 11 + 2 + 2 + 2$
$11 + 2 + 2 + 2 + 11 + 2 + 2 + 2 + 2$
$2 + 2 + 2 + 11 + 2 + 2 + 2 + 2 + 11$
$2 + 2 + 11 + 2 + 2 + 2 + 2 + 11 + 2$
$2 + 11 + 2 + 2 + 2 + 2 + 11 + 2 + 2$
$11 + 2 + 2 + 2 + 2 + 11 + 2 + 2 + 2$
$2 + 2 + 2 + 2 + 11 + 2 + 2 + 2 + 11$
$2 + 2 + 2 + 11 + 2 + 2 + 2 + 11 + 2$

Linear unit = 1/4″.

Arc unit = 10°.

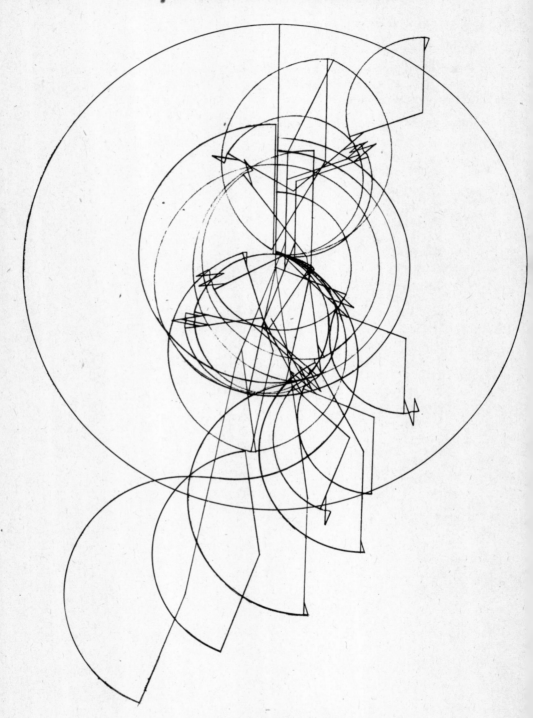

Figure 31. Linear design 36/36 series.

E. Problems in Design

(1) Rhythmic series of angles. Overlapping progressions of striped areas: stripes parallel or perpendicular to one of the sides. Color progressions on same areas.

(2) Rhythmic series of angles used:

        (a) as radii

        (b) as diameters.

(3) The cross section segments used as radii or as diameters, combined with the striping or coloring.

Rhythmic progression of values only (black — white) or combined with hues.

(4) Using halves of rhythmic series of angles in their crossings as determining boundaries for a motive.

(5) Developing the same procedure with the second power of rhythmic series of angles.

(6) Evolving any motive in the second power areas.

(7) Evolving a motive in the first and the second power areas.

(8) Clockwise or counter-clockwise revolving of the four motives in the second power areas.

(9) Various combinations through the four positions in the second power areas

(10) Using a motive twice: once—in the whole area, and once—in one of the second power areas.

Using a motive three times: once in the whole area, and twice in different second power areas.

Using a motive four times: once in the whole area, and three times in the second power areas.

(11) Bithematic Composition

One theme in the area of the first power, one theme (repeated through quadrants) in the second power. Various methods of overlapping the two themes.

(12) Rhythmic series of angles used in the second power through the whole area and in the first power through the second power areas.

(13) Rhythmic distortions of the subject in various second power areas (with revolving, and without).

(14) Consecutive displacement of the subject where rhythm represents the number of the cross section units in the second power areas. Clockwise and counter-clockwise displacement.

(15) Distortion of a subject of the first power through the proportions of the second power areas.

(16) All of the above forms applied to the third power. Bithematic and trithematic composition, using the first, the second and the third power.

CHAPTER 3

PRODUCTION OF MUSIC[1]

A. COORDINATION OF TEMPORAL STRUCTURES[2]

MOTION—that is, changeability in time—is the most important intrinsic property of music. Different cultures of different geographical and historical localities have developed many types and forms of intonation. The latter varies greatly in tuning, in quantity of pitches employed, in quantity of simultaneous parts, and in the ways of treating them.

The types are as diversified as drum-beats, instrumental and vocal monody (one part music), organum, discantus, counterpoint, harmony, combinations of melody and harmony, combinations of counterpoint and harmony, different forms of coupled voices, simultaneous combinations of several harmonies, and many others. Any of these types—as well as any combinations of them—constitute the different musical cultures. In each case, musical culture crystallizes itself into a definite combination of types and forms of intonation. The latter crystallize into habits and traditions.

For example, people belonging to a *harmonic* musical culture want every melody harmonized. But people belonging to a monodic musical culture are disturbed by the very presence of harmony. Music of one culture may be *music* (meaningful sound) to the members of that culture; but the very same music may be *noise* (meaningless sound) to the members of another. The functionality of music is comparable to a great extent to that of a language.

Nevertheless, *all* forms of music have one fundamental property in common: *organized time*. The plasticity of the temporal structure of music, as expressed through its attacks and durations, defines the quality of music. Different types and forms of intonation—as well as different types of musical instruments—come and go like the fashions, while the everlasting strife for *temporal plasticity* remains a *symbol of the "eternal" in music*.

---

[1] This chapter concerns one basic aspect of musical composition, the temporal structure of music as it pertains both to simultaneity and continuity. The problem is one of coordinating the rhythmic elements of individual parts with the structure of the score as a whole. For comprehensive analysis of every phase of the art of composition, the reader is referred to *The Schillinger System of Musical Composition*, published in two volumes by Carl Fischer, Inc. The scope of this work is suggested by the titles of its twelve branches. Book 1: Theory of Rhythm. Book 2: Theory of Pitch Scales. Book 3: Variation of Music by Means of Geometrical Projection. Book 4: Theory of Melody. Book 5: Special Theory of Harmony. Book 6: The Correlation of Melody and Harmony. Book 7: Theory of Counterpoint. Book 8: Instrumental Forms. Book 9: General Theory of Harmony. Book 10: Evolution of Pitch-Families (Style). Book 11: Theory of Composition. Book 12: Theory of Orchestration. (Ed.)

[2] From Book 1, "Theory of Rhythm," of *The Schillinger System of Musical Composition*. Copyright 1941 by Carl Fischer, Inc. Reprinted by permission.

The temporal structure of music, usually known as *rhythm*, pertains to two directions: *simultaneity* and *continuity*. The rhythm of *simultaneity* is a form of coordination among the different components (parts). The rhythm of *continuity* is a form of coordination of the successive moments of one component (part).

People of our civilization have developed the power of reasoning at the price of losing many of the instincts of primitive man. Europeans have never possessed the "instinct of rhythm" with which the Africans are endowed. So-called European "classical music" has never attained the ideal it strived for, that ideal being: the utmost plasticity of the temporal organization. When J. S. Bach, for example, tried to develop a coordinated independence of simultaneous parts, he succeeded in producing only a resultant which is uniformity.[3] We find evidence of the same failures in Mozart and Beethoven. But a score in which the several coordinated parts produce a resultant which has a distinct pattern—has been a "lost art" of the aboriginal African drummers. The age of this art can probably be counted in tens of thousands of years!

Today in the United States, owing to the transplantation of Africans to this continent, there is a *renaissance* of rhythm. Habits form quickly—and the instinct of rhythm in the present American generation surpasses anything known throughout European history. Yet our professional "coordinators of rhythm," specifically in the field of dance music, are slaves to, rather than masters of, rhythm. There is plenty of evidence that the urge for coordination of the whole through individualized parts is growing. The so-called "pyramids" (sustained arpeggio produced by successive entrances of several instruments) is but an incompetent attempt to solve the same problem.

Fortunately, we do not have to feel discouraged or moan over this "lost art." The *power of reasoning* offers us a complete *scientific solution*.

This problem can be formulated as the *distribution of a duration-group through instrumental and attack-groups*.

The entire technique consists of five successive operations with respect to the following:

(1) The number of individual parts in a score;

(2) The quantity of attacks appearing with each individual part in succession;

(3) The rhythmic patterns for each individual part;

(4) The coordination of all parts (which become the resultants of instrumental interference) into a form which, in turn, results in a specified rhythmic pattern (the resultant of interference of all parts); and

(5) The application of such scores to any type of musical measures (bars).

Any part of such a score can be treated as melody, coupled melody, block-harmony, harmony, instrumental figuration—or as a purely percussive (drum)

---

[3]That is to say, when the separate rhythms of the separate parts of a Bach score are "added up," the result tends to be simple uniformity. Schillinger suggests the desirability of scores, and develops a method of scoring, so that the separate parts, while satisfactory rhythmically by themselves, all "add up" to a new rhythm which is *not* uniformity. (Ed.)

part. Aside from the temporal structure of the score, *the practical uses of this technique in intonation* depend on the composer's skill in the respective fields concerned, *i.e.*, melody, harmony, counterpoint and orchestration.

B. DISTRIBUTION OF A DURATION-GROUP (T) THROUGH INSTRUMENTAL (i) AND ATTACK (a) GROUPS

Notation

pli   number of places in the instrumental group.
pla   number of places in the attack-group.
a_a number of attacks in the attack-group.
a_T number of attacks in the duration-group.
PL   the final number of places.
A   the synchronized attack-group (the number of attacks synchronized with the number of places).
A'   the final attack group (number of attacks synchronized with the duration-group).
T   the original duration-group.
T'   the synchronized duration-group.
T"   the final duration-group.
$N_T"$ the number of final duration-groups.

Procedures:

(1) Interference between the number of places in the instrumental group (pli) and the number of places in the attack-group (pla).

$$PL = \frac{pli}{pla}; \quad \begin{array}{l} pla\,(pli) \\ pli\,(pla) \end{array}$$

(2) The product of the number of attacks in the attack group (a_a) by the complementary factor to the number of places in the attack-group (pli after reduction).

$$A = a_a \cdot pli$$

(3) Interference between the synchronized attack-group (A) and the number of attacks in the original duration-group (a_T).

$$A' = \frac{A}{a_T} = \frac{a_a pli}{a_T}$$

(4) The product of the original duration-group (T) by the complementary factor to its number of attacks (A').

$$T' = T \cdot A' = \frac{T \cdot a_a \cdot pli}{a_T}$$

(5) Interference between the synchronized duration-group (T′) and the final duration-group (T″).

$$N_{T''} = \frac{T'}{T''}$$

C. SYNCHRONIZATION OF AN ATTACK-GROUP (a) WITH A DURATION-GROUP (T)

Distribution of attacks of an attack-group (a_a) through the number of attacks of a duration-group (a_T).

First Case: $\dfrac{a_a}{a_T} = 1$

$A = a_T$

$T' = T$

Example:

$a_a = 4a;\ T = r_{3 \div 2} = 6t;\ a_T = 4a$

$A = 4a$

$T' = 6t$

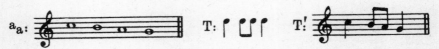

Figure 1. Synchronization of an attack-group with a duration-group.

Second Case: $\dfrac{a_a}{a_T} \neq 1$

$A = a_T \cdot a_a$

$T' = T \cdot a_a$

Example:

$a_a = 5a;\ T = r_{3 \div 2} = 6t;\ a_T = 4a$

$\tfrac{5}{4}$ $\qquad A = 5a \cdot 4 = 20a$

$T' = 6t \cdot 5 = 30t$

Figure 2. *Another illustration of synchronization.*

Third Case: $\dfrac{a_a}{a_T} = \dfrac{a_{a'}}{a_{T'}}$ *i.e.*, a reducible fraction

$A = a_T \cdot a_{a'}$

$T' = T \cdot a_{a'}$

Example:

$a_a = 6a;\ T = r_{3 \div 2} = 6t;\ a_T = 4a$

$\dfrac{6}{4} = \dfrac{3}{2}$

$A = 4a \cdot 3 = 12a$

$T' = 6t \cdot 3 = 18t$

Figure 3. *A third illustration of synchronization.*

D. Distribution of a Synchronized Duration-Group (T′) through the Final Duration-Group (T″)

$$\text{First Case:} \quad \frac{T'}{T''} = 1$$

$$T'' = T'$$

Example:

$T' = 6t; \; T'' = 6t$

$6t = 6t$

Figure 4. Duration-group distribution.

$$\text{Second Case:} \quad \frac{T'}{T''} \neq 1$$

$$N_{T''} = T'$$

Example:

$T' = 6t; \; T'' = 5t$

$N_{5t} = 6$

Figure 5. Another illustration of duration-group distribution.

$$\text{Third Case:} \quad \frac{T'}{T''} = \frac{T'}{T''} \quad \textit{i.e., a reducible fraction}$$

$$N_{T''} = T'$$

Example:

$T' = 6t; \; T'' = 4t$

$\frac{6}{4} = \frac{3}{2}$

$N_{4t} = 3$

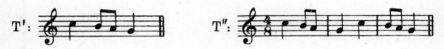

Figure 6. A third illustration of duration-group distribution.

Example:

$a_a = 5a \qquad T = r_{5 \div 2} = 10t \qquad a_T = 6a$

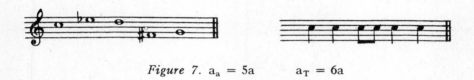

Figure 7. $a_a = 5a \qquad a_T = 6a$

(1) $\frac{6}{5} \quad \begin{matrix} 5 \, (6) \\ 6 \, (5) \end{matrix}$

(2) 6 attacks are equivalent to 10t; $10t \times 5 = 50t$

(3) When $T'' = \frac{8}{8}$, $\frac{50t}{8} = \frac{25 \cdot 4}{4} = 25T$

Figure 8. $T'' = 25T$ *(continued)*

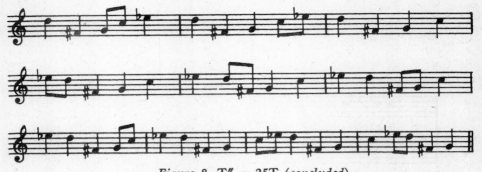

Figure 8. $T'' = 25T$ (*concluded*)

E. Synchronization of an Instrumental Group (pli) with an Attack-Group (pla)

Example:

$$\text{pli} = 4; \quad \text{pla} = 3; \quad a_a = 3 + 2 + 3 = 8; \quad T = r_{5 \div 2} = 10t; \quad 6a$$

(1) $\frac{4}{3}$; $\frac{3}{4}\left(\frac{4}{3}\right)$ \qquad (3) $\frac{32}{6} = \frac{16}{3}$

(2) $8 \times 4 = 32$ \qquad (4) $\frac{16 \cdot 10}{3} = \frac{160}{3}$

(5) $T'' = 8t; \quad \frac{160}{3 \cdot 8} = \frac{20}{3}; \quad \frac{20 \cdot 3}{3} = 20T''$

Figure 9. Synchronization of an instrumental group with an attack-group.

Example:

$$\text{pli} = 3; \quad \text{pla} = 3; \quad a_a = 3 + 2 + 2 + 3 = 10; \quad T = r_{4 \div 3} = 16t; \quad 10a$$

(1) $\frac{3}{3} = 1$ \qquad (3) $\frac{10}{10} = 1$

(2) $10 \cdot 1 = 10$ \qquad (4) $16 \cdot 1 = 16$

(5) $T'' = 8t; \quad \frac{16}{8} = 2T''$

Figure 10. pli = 3; pla = 3; $a_a = 3 + 2 + 2 + 3$.

Example:

pli = 6; pla = 8; $a_a = r_{5 \div 4} = 20$; $T = r_{\underline{4 \div 3}} = 16t$; 10a $T'' = 8t$

(1) PL = $\frac{6}{8} = \frac{3}{4}$; $\frac{4}{3} \binom{6}{8}$ (3) $A' = \frac{60}{10} = 6$

(2) $A = 20 \cdot 3 = 60$ (4) $T' = 16t \cdot 6 = 96t$

(5) $\frac{96}{8} = 12T''$

(*See Fig. 16, p. 411 for an example based on this formula*)

Example of composition of the resultant of instrumental interference.

$$\text{pli} = \text{pla} = 2$$

Form of distribution: $5 + 3$

Figure 11. pli = pla = 2. $5 + 3$

408   TECHNOLOGY OF ART PRODUCTION

(1) $\frac{2}{2} = 1$

(2) 2 is an equivalent of $5 + 3 = 8$

(3) Duration-group: $T = r_{5 \div 2} = 10t$

$$a_T = 6$$

$$\frac{8}{6} = \frac{4}{3} \quad \frac{3}{4} \cdot \binom{8}{6}$$

(4) $10t \times 4 = 40t$

(5) When $T'' = \frac{8}{8}, \frac{40t}{8} = 5T''$

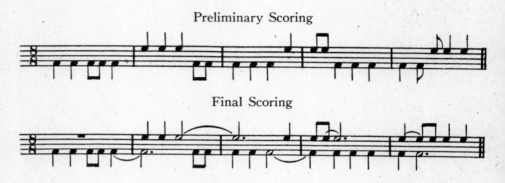

Figure 12. *Preliminary and final score.*

Example of composition of the resultant of instrumental interference.

$$\text{pli} = 3 \quad \text{pla} = 3$$

Form of distribution: $8 + 3 + 5 + 2$.

Figure 13. pli = 3; pla = 3. *Distribution:* $8 + 3 + 5 + 2$.

PRODUCTION OF MUSIC

(1) $\frac{3}{3} = 1$

(2) 4 is an equivalent of $8 + 3 + 5 + 2 = 18$
$$18 \times 1 = 18$$

(3) Duration group: $r_{5 \div 2} = 10t$ $\quad \frac{18}{6} = 3 \quad\quad 3(6)$
$$a_T = 6$$

(4) $10t \times 3 = 30t$

(5) When $T'' = \frac{8}{8}$, $\frac{30t}{8} = \frac{15}{4}$; $\frac{15 \cdot 4}{4} = 15T''$

Preliminary Scoring

Figure 14. Preliminary Score

410 TECHNOLOGY OF ART PRODUCTION

Figure 15. Final score

PRODUCTION OF MUSIC

Example of composition of the resultant of instrumental interference.

pli = 6;   pla = 8;

Form of distribution: $r_{5 \div 4}$

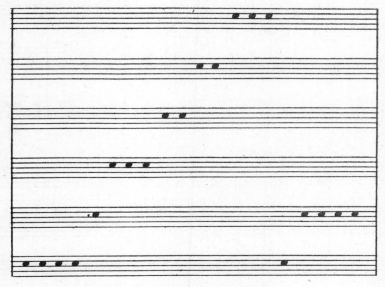

Figure 16. pli = 6;   pla = 8; $r_{5 \div 4}$

(1) $\frac{6}{8} = \frac{3}{4}$ $\begin{matrix}4\\3\end{matrix}\begin{pmatrix}6\\8\end{pmatrix}$

(2) 8 is equivalent to 20 in $r_{5 \div 4}$ $20 \times 3 = 60$

(3) Duration-group = $r_{4 \div 3}$ $a_T = 10$; $\frac{60}{10} = 6$ 6(10)

$r_{4 \div 3} = 16t$

(4) $16t \times 6 = 96t$; a given $T'' = \frac{8}{8}$

(5) $\frac{96t}{8t} = 12T''$

Preliminary Scoring

Figure 17. Preliminary score.

PRODUCTION OF MUSIC

Final Scoring

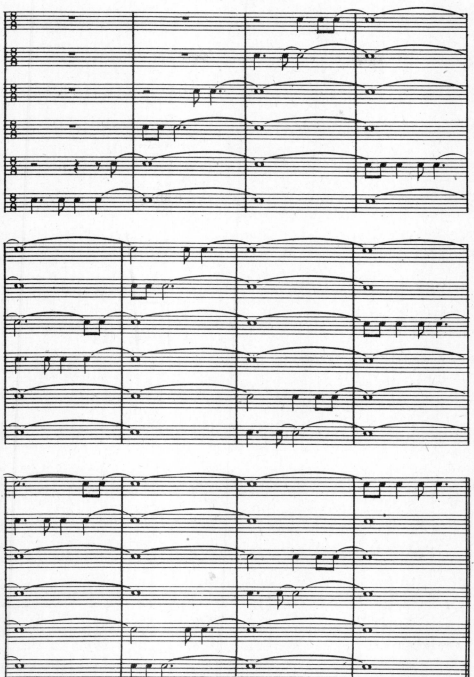

Figure 18. Final score

CHAPTER 4

PRODUCTION OF KINETIC DESIGN

A. Proportionate Distribution within Rectangular Areas

1. Distributive squaring of the sum of two sides of a rectangle.

$$\left(\frac{a}{a+b} + \frac{b}{a+b}\right)^2 = \frac{a^2 + ab + ab + b^2}{(a+b)^2}$$

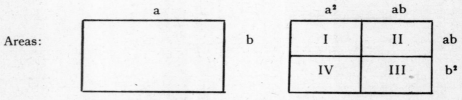

Figure 1. *Distributive squaring of sum of two sides.*

$$ab = \frac{a^2 \cdot ab}{(a+b)^2} + \frac{ab \cdot ab}{(a+b)^2} + \frac{b^2 \cdot ab}{(a+b)^2} + \frac{a^2 \cdot b^2}{(a+b)^2} =$$

$$= \frac{a^3b + a^2b^2 + ab^3 + a^2b^2}{(a+b)^2}$$

$$I = \frac{a^3b}{(a+b)^2};\ II = \frac{a^2b^2}{(a+b)^2};\ III = \frac{ab^3}{(a+b)^2};\ IV = \frac{a^2b^2}{(a+b)^2}$$

Areas identical in square units: II and IV.

Areas non-identical in square units: I and III.

When a = b all areas are identical in square units and in form.

PRODUCTION OF KINETIC DESIGN

2. *Distributive cubing of the sum of two sides of a rectangle.*

$$\left(\frac{a}{a+b} + \frac{b}{a+b}\right)^3 = \frac{a^3 + a^2b + a^2b + ab^2 + a^2b + ab^2 + ab^2 + b^3}{(a+b)^3}$$

	a^3	a^2b	a^2b	ab^2	
	I	II	V	VI	a^2b
Areas:	IV	III	VIII	VII	ab^2
	XIII	XIV	IX	X	ab^2
	XVI	XV	XII	XI	b^3

Figure 2. Distributive cubing of sum of two sides.

$$ab = \frac{a^3 \cdot a^2b + a^2b \cdot a^2b + a^2b \cdot ab^2 + a^3 \cdot ab^2}{(a+b)^3} +$$

$$+ \frac{a^2b \cdot a^2b + ab^2 \cdot a^2b + ab^2 \cdot ab^2 + a^2b \cdot ab^2}{(a+b)^3} +$$

$$+ \frac{a^2b \cdot ab^2 + ab^2 \cdot ab^2 + ab^2 \cdot b^3 + a^2b \cdot b^3}{(a+b)^3} +$$

$$+ \frac{a^3 \cdot ab^2 + a^2b \cdot ab^2 + a^2b \cdot b^3 + a^3 \cdot b^3}{(a+b)^3} =$$

$$= \frac{(a^5b + a^4b^2 + a^3b^3 + a^4b^2) + (a^4b^2 + a^3b^3 + a^2b^4 + a^3b^3)}{(a+b)^3} +$$

$$+ \frac{(a^3b^3 + a^2b^4 + ab^5 + a^2b^4) + (a^4b^2 + a^3b^3 + a^2b^4 + a^3b^3)}{(a+b)^3}$$

Areas identical in square units:

  II, IV, V, XIII.   ——— a^4b^2
  III, VI, VIII, IX, XIV, XVI ——— a^3b^3
  VII, X, XII, XV   ——— a^2b^4

Areas non-identical in square units: I $= a^5b$ and XI $= ab^5$.

3. Distributive involution of the fourth power of the sum of two sides of a rectangle.

$$\left(\frac{a}{a+b} + \frac{b}{a+b}\right)^4 = \frac{a^4 + a^3b + a^3b + a^2b^2 + a^3b + a^2b^2 + a^2b^2 + ab^3}{(a+b)^4} +$$

$$+ \frac{a^3b + a^2b^2 + a^2b^2 + ab^3 + a^2b^2 + ab^3 + ab^3 + b^4}{(a+b)^4}$$

Areas:

a^4	a^3b	a^3b	a^2b^2	a^3b	a^2b^2	a^2b^2	ab^3
I	II	V	VI	XVII	XVIII	XXI	XXII
IV	III	VIII	VII	XX	XIX	XXIV	XXIII
XIII	XIV	IX	X	XXIX	XXX	XXV	XXVI
XVI	XV	XII	XI	XXXII	XXXI	XXVIII	XXVII
XLIX	L	LIII	LIV	XXXIII	XXXIV	XXXVII	XXXVIII
LII	LI	LVI	LV	XXXVI	XXXV	XL	XXXIX
LXI	LXII	LVII	LVIII	XLV	XLVI	XLI	XLII
LXIV	LXIII	LX	LIX	XLVIII	XLVII	XLIV	XLIII

Figure 3. Distributive involution of fourth power of sum of two sides.

$$ab = \left\{\left[\frac{a^4 \cdot a^3b + a^3b \cdot a^3b + a^3b \cdot a^2b^2 + a^4 \cdot a^2b^2}{(a+b)^4}\right] + \right.$$

$$+ \left[\frac{a^3b \cdot a^3b + a^2b \cdot a^3b + a^2b^2 \cdot a^2b^2 + a^3b \cdot a^2b^2}{(a+b)^4}\right] +$$

$$+ \left[\frac{a^3b \cdot a^2b^2 + a^2b^2 \cdot a^2b^2 + a^2b^2 \cdot ab^3 + a^3b \cdot ab^3}{(a+b)^4}\right] +$$

$$+ \left[\frac{a^4 \cdot a^2b^2 + a^3b \cdot a^2b^2 + a^3b \cdot ab^3 + a^4 \cdot ab^3}{(a+b)^4}\right]\right\} +$$

$$+ \left\{\left[\frac{a^3b \cdot a^3b + a^2b^2 \cdot a^3b + a^2b^2 \cdot a^2b^2 + a^3b \cdot a^2b^2}{(a+b)^4}\right] + \right.$$

$$+ \left[\frac{a^2b^2 \cdot a^3b + ab^3 \cdot a^3b + ab^3 \cdot a^2b^2 + a^2b^2 \cdot a^2b^2}{(a+b)^4}\right] +$$

$$+ \left[\frac{a^2b^2 \cdot a^2b^2 + ab^3 \cdot a^2b^2 + ab^3 \cdot ab^3 + a^2b^2 \cdot ab^3}{(a+b)^4}\right] +$$

$$+ \left[\frac{a^3b \cdot a^2b^2 + a^2b^2 \cdot a^2b^2 + a^2b^2 \cdot ab^3 + a^3b \cdot ab^3}{(a+b)^4}\right]\right\} +$$

$$+ \left\{\left[\frac{a^3b \cdot a^2b^2 + a^2b^2 \cdot a^2b^2 + a^2b^2 \cdot ab^3 + a^3b \cdot ab^3}{(a+b)^4}\right] + \right.$$

$$+ \left[\frac{a^2b^2 \cdot a^2b^2 + ab^3 \cdot a^2b^2 + ab^3 \cdot ab^3 + a^2b^2 \cdot ab^3}{(a+b)^4}\right] +$$

$$+ \left[\frac{a^2b^2 \cdot ab^3 + ab^3 \cdot ab^3 + ab^3 \cdot b^4 + a^2b^2 \cdot b^4}{(a+b)^4}\right] +$$

$$+ \left[\frac{a^3b \cdot ab^3 + a^2b^2 \cdot ab^3 + a^2b^2 \cdot b^4 + a^3b \cdot b^4}{(a+b)^4}\right]\right\} +$$

$$+ \left\{\left[\frac{a^4 \cdot a^2b^2 + a^3b \cdot a^2b^2 + a^3b \cdot ab^3 + a^4 \cdot ab^3}{(a+b)^4}\right] + \right.$$

$$+ \left[\frac{a^3b \cdot a^2b^2 + a^2b^2 \cdot a^2b^2 + a^2b^2 \cdot ab^3 + a^3b \cdot ab^3}{(a+b)^4}\right] +$$

$$+ \left[\frac{a^3b \cdot ab^3 + a^2b^2 \cdot ab^3 + a^2b^2 \cdot b^4 + a^3b \cdot b^4}{(a+b)^4}\right] +$$

$$+ \left[\frac{a^4 \cdot ab^3 + a^3b \cdot ab^3 + a^3b \cdot b^4 + a^4 \cdot b^4}{(a+b)^4}\right]\right\} =$$

$$= \left\{\left[\frac{(a^7b + a^6b^2 + a^5b^3 + a^6b^2) + (a^6b^2 + a^5b^3 + a^4b^4 + a^5b^3)}{(a+b)^4}\right. + \right.$$

$$+ \left.\frac{(a^5b^3 + a^4b^4 + a^3b^5 + a^4b^4) + (a^6b^2 + a^5b^3 + a^4b^4 + a^5b^3)}{(a+b)^4}\right] +$$

$$+ \left[\frac{(a^6b^2 + a^5b^3 + a^4b^4 + a^5b^3) + (a^5b^3 + a^4b^4 + a^3b^5 + a^4b^4)}{(a+b)^4} + \right.$$

$$+ \left.\frac{(a^4b^4 + a^3b^5 + a^2b^6 + a^3b^5) + (a^5b^3 + a^4b^4 + a^3b^5 + a^4b^4)}{(a+b)^4}\right] +$$

$$+ \left[\frac{(a^5b^3 + a^4b^4 + a^3b^5 + a^4b^4) + (a^4b^4 + a^3b^5 + a^2b^6 + a^3b^5)}{(a+b)^4} +\right.$$

$$+ \left.\frac{(a^3b^5 + a^2b^6 + ab^7 + a^2b^6) + (a^4b^4 + a^3b^5 + a^2b^6 + a^3b^5)}{(a+b)^4}\right] +$$

$$+ \left[\frac{(a^6b^2 + a^5b^3 + a^4b^4 + a^5b^3) + (a^5b^3 + a^4b^4 + a^3b^5 + a^4b^4)}{(a+b)^4} +\right.$$

$$+ \left.\frac{(a^4b^4 + a^3b^5 + a^2b^6 + a^3b^5) + (a^5b^3 + a^4b^4 + a^3b^5 + a^4b^4)}{(a+b)^4}\right]\Bigg\}$$

Areas identical in square units:

II, IV, V, XIII, XVII, XLIX ——— a^6b^2

III, VI, VIII, IX, XIV, XVI, XVIII, XX, XXI, XXIX —— a^5b^3
XXXIII, L, LII, LIII, LXI

VII, X, XII, XV, XIX, XXII, XXIV, XXV, XXX, XXXII, ——— a^4b^4
XXXIV, XXXVI, XXXVII, XLV, LI, LIV, LVI, LVII, LXII, LXIV

XI, XXIII, XXVI, XXVIII, XXXI, XXXV, XXXVIII, XL, XLI, —— a^3b^5
XLVI, XLVIII, LV, LVIII, LX, LXIII

XXVII, XXXIX, XLII, XLIV, XLVII, XLIX ——— a^2b^6

Areas non-identical in square units:

    I $= a^7b$ and XLIII $= ab^7$

B. Distributive Involution in Linear Design

$$4:3$$

$$\measuredangle 10° \cdots \circlearrowright \cdots 1 \text{ inch.}$$

$(3+1+2+1+1+1+1+2+1+3)^2 = (9+3+6+3+3+3+3+6+3+9) +$
$(3+1+2+1+1+1+1+2+1+3) + (6+2+4+2+2+2+2+4+2+6) +$
$(3+1+2+1+1+1+1+2+1+3) + (3+1+2+1+1+1+1+2+1+3) +$
$(3+1+2+1+1+1+1+2+1+3) + (3+1+2+1+1+1+1+2+1+3) +$
$(6+2+4+2+2+2+2+4+2+6) + (3+1+2+1+1+1+1+2+1+3) +$
$(9+3+6+3+3+3+3+6+3+9)$

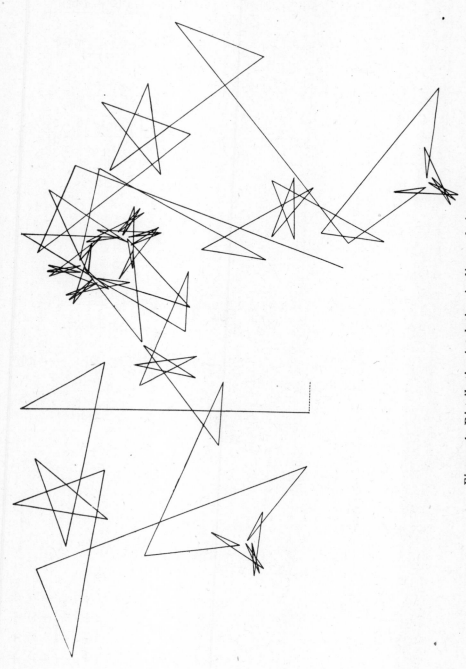

Figure 4. Distributive involution in linear design.

Figure 5. Distributive involution in linear design.

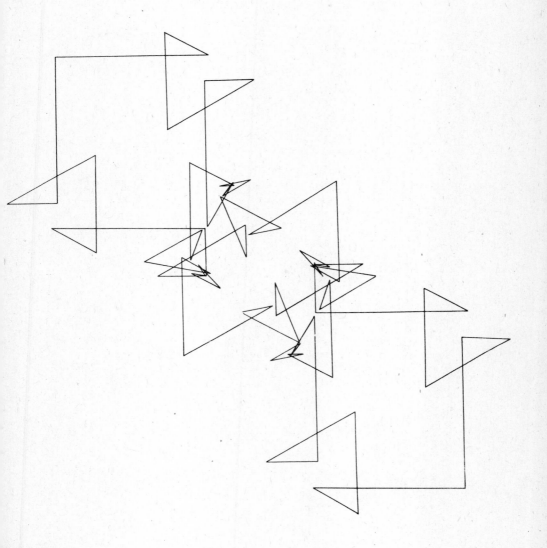

Figure 6. Distributive involution in linear design (continued).

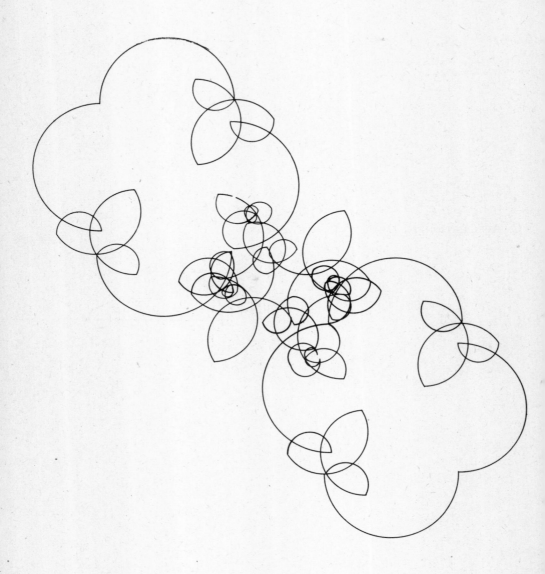

Figure 6. Lines of preceding configuration used as diameters of semi-circles (concluded.)

C. APPLICATION TO DIMENSIONS AND ANGLES

$$r_{(3 \div 2)}$$
$$(2 + 1 + 1 + 2)^3$$

$\angle \text{st} = 5°.$ Direction: ↻

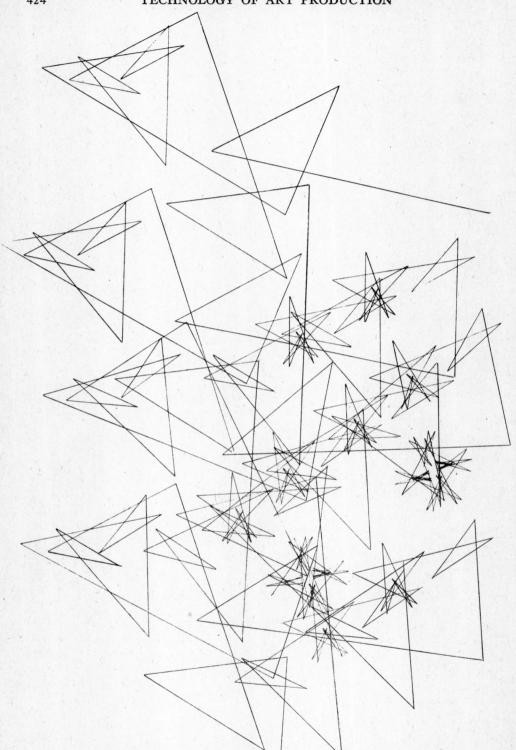

Figure 7. Application to dimensions and angles.

PRODUCTION OF KINETIC DESIGN 425

D. Positional Rotation Applied to Kinetic Design Within Rectangular Area

Φ is equivalent to the proportionate rectangles of the third power.

Figure 8. Positional rotation applied to kinetic design.

Figure 9. Positional rotation applied to kinetic design.

Figure 10. Positional rotation applied to kinetic design.

Figure 11. Positional rotation applied to kinetic design.

CHAPTER 5

PRODUCTION OF COMBINED ARTS

A. THE TIME-SPACE UNIT IN CINEMATIC DESIGN

KINETIC design for cinematic production deals with two elements:

1. the space of the screen
2. the time of the projection

The space of the screen may be used either as a conditionally limitless space, or a definitely proportionate area with boundaries.

In treating the screen space as a conditionally limitless area, it is necessary to assign a certain linear unit which might be a simple partial area of the entire screen. For example, it is possible to divide the length and the width of the screen by 24, thus obtaining the units of horizontal and vertical directions, which through their intersection, form boundary units of the same proportions as the entire screen. The total number of such partial areas is $24^2 = 576$. Such subdivision permits a practically limitless number of designs, possible through various combinations of the partial area-units. If the number of designs to be executed is more limited, one can limit the subdivision of the area of the entire screen into a respectively smaller quantity of area-units.

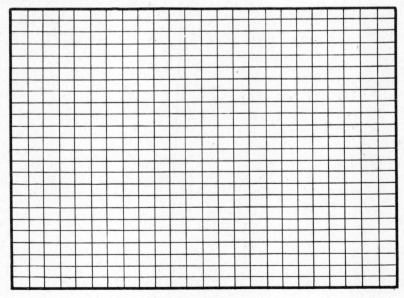

Figure 1. The screen.

Treating the screen space as a definitely proportioned area with boundaries, one has to adhere to its original harmony. The screen space of the silent film is a 4 × 3 rectangle, while the sound-track cuts off one side of it in such a way that the remaining area becomes approximately 8 × 7. The problem of defining partial area-units, harmonically related to the entire area of the screen, is as fundamental as the problem of tuning in music. The solution of this problem can be satisfied by the following formula:

$$\frac{a}{a+b} + \frac{b}{a+b} = \frac{a+b}{a+b} = 1,$$

where a and b are the two sides of the screen area, or any rectangular area in general.

Applying this formula to the silent screen, we obtain:

$$\frac{4}{7} + \frac{3}{7} = \frac{7}{7} = 1$$

and for the sound screen:

$$\frac{8}{15} + \frac{7}{15} = \frac{15}{15} = 1$$

In order to develop the partial proportionate areas of the entire screen area, it is necessary to subject the sum of two sides, as expressed above, to distributive involution (squaring, cubing, etc.). The formula for distributive squaring, through which we obtain 4 partial areas harmonically related to the original area, is:

$$\left(\frac{a}{a+b} + \frac{b}{a+b}\right)^2 = \frac{a^2}{(a+b)^2} + \frac{ab}{(a+b)^2} + \frac{ab}{(a+b)^2} + \frac{b^2}{(a+b)^2}$$

Applying this formula to the silent screen we obtain

$$\left(\frac{4}{7} + \frac{3}{7}\right)^2 = \frac{16}{49} + \frac{12}{49} + \frac{12}{49} + \frac{9}{49}.$$

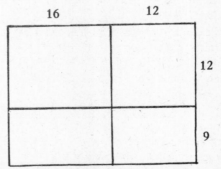

Figure 2. *Distributive squaring of the screen.*

Applying the same formula to the sound screen, we obtain:

$$\left(\frac{8}{15} + \frac{7}{15}\right)^2 = \frac{64}{225} + \frac{56}{225} + \frac{56}{225} + \frac{49}{225}$$

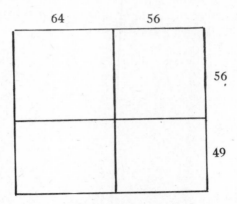

Figure 3. Distributive squaring of sound screen.

Through application of the higher forms of involution through distributive formulae similar to the one above, it is possible to split the entire area into any desired number of partial areas. As squaring of the area produces 4 partial areas, cubing produces 4^2, *i. e.*, 16 partial areas. Raising to the fourth power produces 4^3, *i. e.*, 64 partial areas, and so on. Thus, all the partial areas are in harmony with the entire area of the screen and serve as elements for the spatial cinematic composition.

The time of projection must be approached on a similar basis. As the sound film is projected at 24 frames per second, which at the same time is a desirable speed for silent film, it is most practical to devise the individual time-units from this number 24 as a group-unit of time measurement. Thus, the individual units of time are projected single frames, which correspond to the individual phases of the design made for animation: 24 drawings produce in projection one second of fluent motion.

The distribution of the individual time-units (each equivalent to one phase of drawing) in different quantities and groupings produces temporal rhythm. Temporal rhythm must be coordinated with spatial rhythm in such a fashion that one second of a screen projection, which is equivalent to 24 consecutive frames of the film, must correspond to the formation of one linear or area-unit. For example, in drawing a horizontal line through 24 consecutive phases, which on the screen appears as a line growing to the length of one screen unit during one second of projection, it is necessary for one phase to be $\frac{1}{24^2} = \frac{1}{576}$ of the entire length of the screen. Thus, time and space can be coordinated on the basis of one time-space unit, *i.e.*, $t = \frac{1}{24}$, second $= \frac{1}{576}$ of the screen length.

In making drawings for animation on a sheet 12″ long, t = $\frac{1}{48}$″. In projecting a film made from such drawings on a screen 12′ long, t becomes $\frac{1}{4}$″ in projection, thus giving a 6″ linear growth in the course of one second. This obviously guarantees a maximum graduality of motion.

B. Correlation of Visual and Auditory Forms

1. ELEMENTS OF VISUAL KINETIC COMPOSITION.

1. Linear, plane and solid trajectories
   (distance, dimension, direction, form).
2. Illumination
   (forms and intensity of light).
3. Texture
   (density of matter, quality of surface).
4. General component: time.

2. ELEMENTS OF MUSIC

1. Frequency
   (pitch).
2. Intensity
   (relative dynamics).
3. Quality
   (harmonic composition).
4. Density
   (quantitative aggregation of sound).
5. General component: time.

The correlation of these two art forms may be performed through relative coordination of the different components individually, and in groups, pertaining to both art forms. All values must be determined from an initial axis-point or line of the graph record of the composition.

$$\frac{\text{distance}}{\text{pitch} + \text{relative dynamics}}$$

$$\frac{\text{distance}}{\text{relative dynamics} + \text{harmonic composition}}$$

$$\frac{\text{distance}}{\text{harmonic composition} + \text{quantitative aggregation of sound}}$$

$$\frac{\text{distance}}{\text{quantitative aggregation of sound} + \text{pitch}}$$

PRODUCTION OF COMBINED ARTS

$$\frac{\text{dimension}}{\text{pitch} + \text{relative dynamics}}$$

$$\frac{\text{dimension}}{\text{harmonic composition} + \text{quantitative aggregation of sound}}$$

$$\frac{\text{dimension}}{\text{relative dynamics} + \text{harmonic composition}}$$

$$\frac{\text{dimension}}{\text{quantitative aggregation of sound} + \text{pitch}}$$

$$\frac{\text{direction}}{\text{pitch} + \text{relative dynamics}}$$

$$\frac{\text{direction}}{\text{harmonic composition} + \text{quantitative aggregation of sound}}$$

$$\frac{\text{direction}}{\text{relative dynamics} + \text{harmonic composition}}$$

$$\frac{\text{direction}}{\text{quantitative aggregation of sound} + \text{pitch}}$$

$$\frac{\text{form}}{\text{pitch} + \text{relative dynamics}}$$

$$\frac{\text{form}}{\text{harmonic composition} + \text{quantitative aggregation of sound}}$$

$$\frac{\text{form}}{\text{relative dynamics} + \text{harmonic composition}}$$

$$\frac{\text{form}}{\text{quantitative aggregation of sound} + \text{pitch}}$$

$$\frac{\text{form of light}}{\text{pitch} + \text{relative dynamics}}$$

$$\frac{\text{form of light}}{\text{harmonic composition} + \text{quantitative aggregation of sound}}$$

$$\frac{\text{form of light}}{\text{relative dynamics} + \text{harmonic composition}}$$

$$\frac{\text{form of light}}{\text{quantitative aggregation of sound} + \text{pitch}}$$

$$\frac{\text{intensity of light}}{\text{pitch} + \text{relative dynamics}}$$

$$\frac{\text{intensity of light}}{\text{harmonic composition} + \text{quantitative aggregation of sound}}$$

$$\frac{\text{intensity of light}}{\text{relative dynamics} + \text{harmonic composition}}$$

$$\frac{\text{intensity of light}}{\text{quantitative aggregation of sound} + \text{pitch}}$$

$$\frac{\text{density of matter}}{\text{pitch + relative dynamics}}$$

$$\frac{\text{density of matter}}{\text{relative dynamics + harmonic composition}}$$

$$\frac{\text{density of matter}}{\text{harmonic composition + quantitative aggregation of sound}}$$

$$\frac{\text{density of matter}}{\text{quantitative aggregation of sound + pitch}}$$

$$\frac{\text{quality of matter's surface}}{\text{pitch + relative dynamics}}$$

$$\frac{\text{quality of matter's surface}}{\text{relative dynamics + harmonic composition}}$$

$$\frac{\text{quality of matter's surface}}{\text{harmonic composition + quantitative aggregation of sound}}$$

$$\frac{\text{quality of matter's surface}}{\text{quantitative aggregation of sound + pitch}}$$

$$\frac{\text{distance + dimension}}{\text{pitch + relative dynamics}}$$

$$\frac{\text{distance + dimension}}{\text{relative dynamics + harmonic composition}}$$

$$\frac{\text{distance + dimension}}{\text{harmonic composition + quantitative aggregation of sound}}$$

$$\frac{\text{distance + dimension}}{\text{quantitative aggregation of sound + pitch}}$$

$$\frac{\text{dimension + direction}}{\text{pitch + relative dynamics}}$$

$$\frac{\text{dimension + direction}}{\text{relative dynamics + harmonic composition}}$$

$$\frac{\text{dimension + direction}}{\text{harmonic composition + quantitative aggregation of sound}}$$

$$\frac{\text{dimension + direction}}{\text{quantitative aggregation of sound + pitch}}$$

$$\frac{\text{direction + form}}{\text{pitch + relative dynamics}}$$

$$\frac{\text{direction + form}}{\text{relative dynamics + harmonic composition}}$$

$$\frac{\text{direction + form}}{\text{harmonic composition + quantitative aggregation of sound}}$$

$$\frac{\text{direction + form}}{\text{quantitative aggregation of sound + pitch}}$$

$$\frac{\text{form + distance}}{\text{pitch + relative dynamics}}$$

$$\frac{\text{form + distance}}{\text{relative dynamics + harmonic composition}}$$

$$\frac{\text{form + distance}}{\text{harmonic composition + quantitative aggregation of sound}}$$

$$\frac{\text{form + distance}}{\text{quantitative aggregation of sound + pitch}}$$

$$\frac{\text{forms + intensity of light}}{\text{pitch + relative dynamics}}$$

$$\frac{\text{forms + intensity of light}}{\text{relative dynamics + harmonic composition}}$$

$$\frac{\text{forms + intensity of light}}{\text{harmonic composition + quantitative aggregation of sound}}$$

$$\frac{\text{forms + intensity of light}}{\text{quantitative aggregation of sound + pitch}}$$

$$\frac{\text{density of matter + quality of matter's surface}}{\text{pitch + relative dynamics}}$$

$$\frac{\text{density of matter + quality of matter's surface}}{\text{relative dynamics + harmonic composition}}$$

$$\frac{\text{density of matter + quality of matter's surface}}{\text{harmonic composition + quantitative aggregation of sound}}$$

$$\frac{\text{density of matter + quality of matter's surface}}{\text{quantitative aggregation of sound + pitch}}$$

$$\frac{\text{distance + dimension}}{\text{pitch}}$$

$$\frac{\text{dimension + direction}}{\text{pitch}}$$

$$\frac{\text{direction + form}}{\text{pitch}}$$

$$\frac{\text{form + distance}}{\text{pitch}}$$

$$\frac{\text{distance + dimension}}{\text{relative dynamics}}$$

$$\frac{\text{dimension + direction}}{\text{relative dynamics}}$$

$$\frac{\text{direction + form}}{\text{relative dynamics}}$$

$$\frac{\text{form + distance}}{\text{relative dynamics}}$$

$$\frac{\text{distance + dimension}}{\text{harmonic composition}}$$

$$\frac{\text{dimension + direction}}{\text{harmonic composition}}$$

$$\frac{\text{direction + form}}{\text{harmonic composition}}$$

$$\frac{\text{form + distance}}{\text{harmonic composition}}$$

$$\frac{\text{distance} + \text{dimension}}{\text{quantitative aggregation of sound}} \qquad \frac{\text{dimension} + \text{direction}}{\text{quantitative aggregation of sound}}$$

$$\frac{\text{direction} + \text{form}}{\text{quantitative aggregation of sound}} \qquad \frac{\text{form} + \text{distance}}{\text{quantitative aggregation of sound}}$$

$$\frac{\text{forms} + \text{intensity of light}}{\text{pitch}} \qquad \frac{\text{forms} + \text{intensity of light}}{\text{relative dynamics}}$$

$$\frac{\text{forms} + \text{intensity of light}}{\text{harmonic composition}} \qquad \frac{\text{forms} + \text{intensity of light}}{\text{quantitative aggregation of-sound}}$$

$$\frac{\text{density of matter} + \text{quality of matter's surface}}{\text{pitch}} \qquad \frac{\text{density of matter} + \text{quality of matter's surface}}{\text{relative dynamics}}$$

$$\frac{\text{density of matter} + \text{quality of matter's surface}}{\text{harmonic composition}} \qquad \frac{\text{density of matter} + \text{quality of matter's surface}}{\text{quantitative aggregation of sound}}$$

$$\frac{\text{distance} + \text{dimension} + \text{direction}}{\text{pitch}} \qquad \frac{\text{distance} + \text{dimension} + \text{direction}}{\text{relative dynamics}}$$

$$\frac{\text{distance} + \text{dimension} + \text{direction}}{\text{harmonic composition}} \qquad \frac{\text{distance} + \text{dimension} + \text{direction}}{\text{quantitative aggregation of sound}}$$

$$\frac{\text{dimension} + \text{direction} + \text{form}}{\text{pitch}} \qquad \frac{\text{dimension} + \text{direction} + \text{form}}{\text{relative dynamics}}$$

$$\frac{\text{dimension} + \text{direction} + \text{form}}{\text{harmonic composition}} \qquad \frac{\text{dimension} + \text{direction} + \text{form}}{\text{quantitative aggregation of sound}}$$

$$\frac{\text{direction} + \text{form} + \text{distance}}{\text{pitch}} \qquad \frac{\text{direction} + \text{form} + \text{distance}}{\text{relative dynamics}}$$

$$\frac{\text{direction} + \text{form} + \text{distance}}{\text{harmonic composition}} \qquad \frac{\text{direction} + \text{form} + \text{distance}}{\text{quantitative aggregation of sound}}$$

$$\frac{\text{form} + \text{distance} + \text{dimension}}{\text{pitch}} \qquad \frac{\text{form} + \text{distance} + \text{dimension}}{\text{relative dynamics}}$$

$$\frac{\text{form} + \text{distance} + \text{dimension}}{\text{harmonic composition}} \qquad \frac{\text{form} + \text{distance} + \text{dimension}}{\text{quantitative aggregation of sound}}$$

PRODUCTION OF COMBINED ARTS

$$\frac{\text{distance} + \text{dimension} + \text{direction}}{\text{pitch} + \text{relative dynamics}}$$

$$\frac{\text{distance} + \text{dimension} + \text{direction}}{\text{harmonic composition} + \text{quantitative aggregation of sound}}$$

$$\frac{\text{dimension} + \text{direction} + \text{form}}{\text{pitch} + \text{relative dynamics}}$$

$$\frac{\text{dimension} + \text{direction} + \text{form}}{\text{harmonic composition} + \text{quantitative aggregation of sound}}$$

$$\frac{\text{direction} + \text{form} + \text{distance}}{\text{pitch} + \text{relative dynamics}}$$

$$\frac{\text{direction} + \text{form} + \text{distance}}{\text{harmonic composition} + \text{quantitative aggregation of sound}}$$

$$\frac{\text{form} + \text{distance} + \text{dimension}}{\text{pitch} + \text{relative dynamics}}$$

$$\frac{\text{form} + \text{distance} + \text{dimension}}{\text{harmonic composition} + \text{quantitative aggregation of sound}}$$

$$\frac{\text{distance} + \text{dimension} + \text{direction}}{\text{pitch} + \text{relative dynamics} + \text{harmonic composition}}$$

$$\frac{\text{distance} + \text{dimension} + \text{direction}}{\text{harmonic composition} + \text{quantitative aggregation of sound} + \text{pitch}}$$

$$\frac{\text{distance} + \text{dimension} + \text{direction}}{\text{relative dynamics} + \text{harmonic composition}}$$

$$\frac{\text{distance} + \text{dimension} + \text{direction}}{\text{quantitative aggregation of sound} + \text{pitch}}$$

$$\frac{\text{dimension} + \text{direction} + \text{form}}{\text{relative dynamics} + \text{harmonic composition}}$$

$$\frac{\text{dimension} + \text{direction} + \text{form}}{\text{quantitative aggregation of sound} + \text{pitch}}$$

$$\frac{\text{direction} + \text{form} + \text{distance}}{\text{relative dynamics} + \text{harmonic composition}}$$

$$\frac{\text{direction} + \text{form} + \text{distance}}{\text{quantitative aggregation of sound} + \text{pitch}}$$

$$\frac{\text{form} + \text{distance} + \text{dimension}}{\text{relative dynamics} + \text{harmonic composition}}$$

$$\frac{\text{form} + \text{distance} + \text{dimension}}{\text{quantitative aggregation of sound} + \text{pitch}}$$

$$\frac{\text{distance} + \text{dimension} + \text{direction}}{\text{relative dynamics} + \text{harmonic composition} + \text{quantitative aggregation of sound}}$$

$$\frac{\text{distance} + \text{dimension} + \text{direction}}{\text{quantitative aggregation of sound} + \text{pitch} + \text{relative dynamics}}$$

dimension + direction + form	dimension + direction + form
pitch + relative dynamics + harmonic composition	relative dynamics + harmonic composition + quantitative aggregation of sound
dimension + direction + form	dimension + direction + form
harmonic composition + quantitative aggregation of sound + pitch	quantitative aggregation of sound + pitch + relative dynamics
direction + form + distance	direction + form + distance
pitch + relative dynamics + harmonic composition	relative dynamics + harmonics composition + quantitative aggregation of sound
distance + dimension + direction	distance + dimension + direction
pitch + relative dynamics + harmonic composition	relative dynamics + harmonic composition + quantitative aggregation of sound
distance + dimension + direction	distance + dimension + direction
harmonic composition + quantitative aggregation of sound + pitch	quantitative aggregation of sound + pitch + relative dynamics
dimension + direction + form	dimension + direction + form
pitch + relative dynamics + harmonic composition	relative dynamics + harmonic composition + quantitative aggregation of sound
dimension + direction + form	dimension + direction + form
harmonic composition + quantitative aggregation of sound + pitch	quantitative aggregation of sound + pitch + relative dynamics
direction + form + distance	direction + form + distance
pitch + relative dynamics + harmonic composition	relative dynamics + harmonic composition + quantitative aggregation of sound
direction + form + distance	direction + form + distance
harmonic composition + quantitative aggregation of sound + pitch	quantitative aggregation of sound + pitch + relative dynamics

form + distance + dimension	form + distance + dimension
pitch + relative dynamics + harmonic composition	relative dynamics + harmonic composition + quantitative aggregation of sound

form + distance + dimension	form + distance + dimension
harmonic composition + quantitative aggregation of sound + pitch	quantitative aggregation of sound + pitch + relative dynamics

distance	distance
pitch + relative dynamics + harmonic composition	relative dynamics + harmonic composition + quantitative aggregation of sound

distance	distance
harmonic composition + quantitative aggregation of sound + pitch	quantitative aggregation of sound + pitch + relative dynamics

dimension	dimension
pitch + relative dynamics + harmonic composition	relative dynamics + harmonic composition + quantitative aggregation of sound

dimension	dimension
harmonic composition + quantitative aggregation of sound + pitch	quantitative aggregation of sound + pitch + relative dynamics

direction	direction
pitch + relative dynamics + harmonic composition	relative dynamics + harmonic composition + quantitative aggregation of sound

direction	direction
harmonic composition + quantitative aggregation of sound + pitch	quantitative aggregation of sound + pitch + relative dynamics

form
pitch + relative dynamics + harmonic composition

form
harmonic composition + quantitative aggregation of sound + pitch

forms of light
pitch + relative dynamics + harmonic composition

forms of light
harmonic composition + quantitative aggregation of sound + pitch

intensity of light
pitch + relative dynamics + harmonic composition

intensity of light
harmonic composition + quantitative aggregation of sound + pitch

density of matter
pitch + relative dynamics + harmonic composition

density of matter
harmonic composition + quantitative aggregation of sound + pitch

quality of matter's surface
pitch + relative dynamics + harmonic composition

quality of matter's surface
harmonic composition + quantitative aggregation of sound + pitch

form
relative dynamics + harmonic composition + quantitative aggregation of sound

form
quantitative aggregation of sound + pitch + relative dynamics

forms of light
relative dynamics + harmonic composition + quantitative aggregation of sound

forms of light
quantitative aggregation of sound + pitch + relative dynamics

intensity of light
relative dynamics + harmonic composition + quantitative aggregation of sound

intensity of light
quantitative aggregation of sound + pitch + relative dynamics

density of matter
relative dynamics + harmonic composition + quantitative aggregation of sound

density of matter
quantitative aggregation of sound + pitch + relative dynamics

quality of matter's surface
relative dynamics + harmonic composition + quantitative aggregation of sound

quality of matter's surface
quantitative aggregation of sound + pitch + relative dynamics

PRODUCTION OF COMBINED ARTS 441

distance
pitch + relative dynamics + harmonic composition + quantitative aggregation of sound

dimension
pitch + relative dynamics + harmonic composition + quantitative aggregation of sound

direction
pitch + relative dynamics + harmonic composition + quantitative aggregation of sound

form
pitch + relative dynamics + harmonic composition + quantitative aggregation of sound

forms of light
pitch + relative dynamics + harmonic composition + quantitative aggregation of sound

intensity of light
pitch + relative dynamics + harmonic composition + quantitative aggregation of sound

density of matter
pitch + relative dynamics + harmonic composition + quantitative aggregation of sound

quality of matter's surface
pitch + relative dynamics + harmonic composition + quantitative aggregation of sound

distance + dimension
pitch + relative dynamics + harmonic composition + quantitative aggregation of sound

dimension + direction
pitch + relative dynamics + harmonic composition + quantitative aggregation of sound

direction + form
pitch + relative dynamics + harmonic composition + quantitative aggregation of sound

form + distance
pitch + relative dynamics + harmonic composition + quantitative aggregation of sound

form + intensity of light
pitch + relative dynamics + harmonic composition + quantitative aggregation of sound

density of matter + quality of matter's surface
pitch + relative dynamics + harmonic composition + quantitative aggregation of sound

distance + dimension + direction
pitch + relative dynamics + harmonic composition + quantitative aggregation of sound

dimension + direction + form
pitch + relative dynamics + harmonic composition + quantitative aggregation of sound

direction + form + distance
pitch + relative dynamics + harmonic composition + quantitative aggregation of sound

form + distance + dimension
pitch + relative dynamics + harmonic composition + quantitative aggregation of sound

The correlation of the general component in both art forms may be assigned to different proportionate relations, such as harmonic ratios, distributive powers, series of growth, etc. The entire manifold of synchronized components must be based on a standard space-time unit expressed through a single motion picture frame ($\frac{1}{24}$th of a second) and the common denominator of musical time.

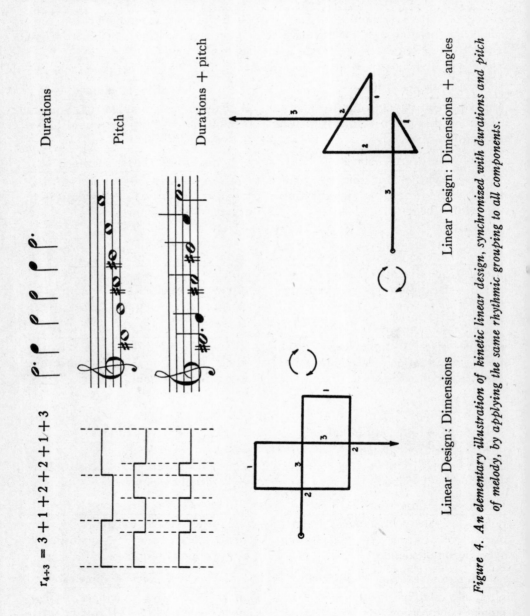

Figure 4. An elementary illustration of kinetic linear design, synchronized with durations and pitch of melody, by applying the same rhythmic grouping to all components.

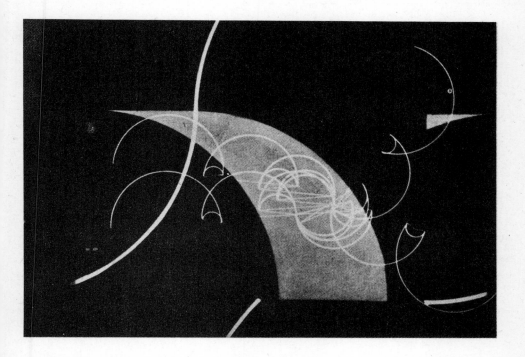

Figure 5. One frame from a composition of kinetic design (cinema) with sound (music), based on r_{8+7} in two reciprocally moving trajectories and on the diagonal rotation of the background.

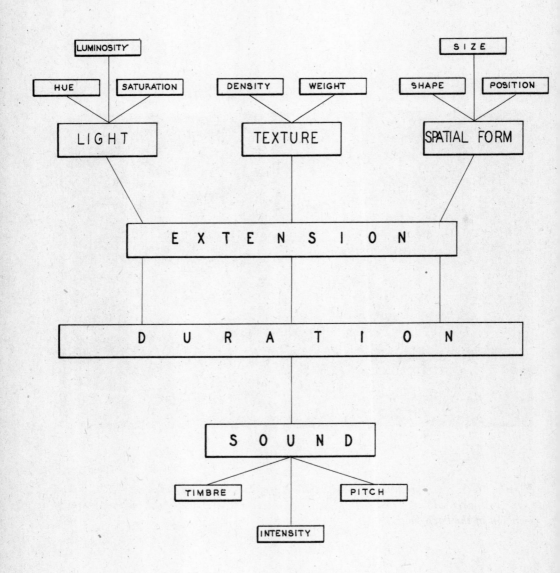

Figure 6. Components of a combined kinetic art form.

APPENDIX A.

BASIC FORMS OF REGULARITY AND COORDINATION[1]

    I. BINARY AND TERNARY SYNCHRONIZATION
    II. DISTRIBUTIVE INVOLUTION—GROUPS
    III. GROUPS OF VARIABLE VELOCITY

---

[1] These basic forms of regularity and coordination constitute a comprehensive elaboration of formulae presented in Part Two, *Theory of Regularity and Coordination*, Chapters 2–5. (Ed.)

I. BINARY AND TERNARY SYNCHRONIZATION OF GENETIC FACTORS (GENERATORS).

Nomenclature:

a—monomial periodic group (equivalent to sine-wave), representing *major genetic factor (major generator)*, whose phases have greater period than that of—

b—monomial periodic group (equivalent to sine-wave), representing *minor genetic factor (minor generator)*, whose phases have smaller period than that of a.

a:b—*a to b ratio*

S—*synchronization* (coordination of durations), *symmetrization* (coordination of extensions), the process in which a and b are coordinated by means of complementary factors.

I—*interference* produced by the interaction of phases a and b.

$r_{a \div b}$—*the resultant of interference* of a to b, a harmonic-symmetric group, possessing an axis of symmetry (therefore reversible), and belonging to the basic forms of regularity.

The components of synchronization or symmetrization, including ab (the product), $\dfrac{1}{ab}$ (the denominator), a, b and r, are applicable to:

(1) identical components of one art-form;

(2) different components of one art-form;

(3) identical components of different art-forms;

(4) different components of different art-forms.

Formulae:

A. BINARY SYNCHRONIZATION

$$a:b; \quad S(a:b) = a \cdot b = ab; \quad S\left(\frac{1}{a:b}\right) = \frac{1}{a \cdot b} = \frac{1}{ab};$$

$$r_{(a \div b)} = I\left[b\left(\frac{a}{ab}\right) \div a\left(\frac{b}{ab}\right)\right]$$

APPENDIX A

1. Rhythmic Resultants

$r_{3 \div 2} = 2 + 1 + 1 + 2$

$r_{4 \div 3} = 3 + 1 + 2 + 2 + 1 + 3$

$r_{5 \div 2} = 2 + 2 + 1 + 1 + 2 + 2$

$r_{5 \div 3} = 3 + 2 + 1 + 3 + 1 + 2 + 3$

$r_{5 \div 4} = 4 + 1 + 3 + 2 + 2 + 3 + 1 + 4$

$r_{6 \div 5} = 5 + 1 + 4 + 2 + 3 + 3 + 2 + 4 + 1 + 5$

$r_{7 \div 2} = 2 + 2 + 2 + 1 + 1 + 2 + 2 + 2$

$r_{7 \div 3} = 3 + 3 + 1 + 2 + 3 + 2 + 1 + 3 + 3$

$r_{7 \div 4} = 4 + 3 + 1 + 4 + 2 + 2 + 4 + 1 + 3 + 4$

$r_{7 \div 5} = 5 + 2 + 3 + 4 + 1 + 5 + 1 + 4 + 3 + 2 + 5$

$r_{7 \div 6} = 6 + 1 + 5 + 2 + 4 + 3 + 3 + 4 + 2 + 5 + 1 + 6$

$r_{8 \div 3} = 3 + 3 + 2 + 1 + 3 + 3 + 1 + 2 + 3 + 3$

$r_{8 \div 5} = 5 + 3 + 2 + 5 + 1 + 4 + 4 + 1 + 5 + 2 + 3 + 5$

$r_{8 \div 7} = 7 + 1 + 6 + 2 + 5 + 3 + 4 + 4 + 3 + 5 + 2 + 6 + 1 + 7$

$r_{9 \div 2} = 2 + 2 + 2 + 2 + 1 + 1 + 2 + 2 + 2 + 2$

$r_{9 \div 4} = 4 + 4 + 1 + 3 + 4 + 2 + 2 + 4 + 3 + 1 + 4 + 4$

$r_{9 \div 5} = 5 + 4 + 1 + 5 + 3 + 2 + 5 + 2 + 3 + 5 + 1 + 4 + 5$

$r_{9 \div 7} = 7 + 2 + 5 + 4 + 3 + 6 + 1 + 7 + 1 + 6 + 3 + 4 + 5 + 2 + 7$

$r_{9 \div 8} = 8 + 1 + 7 + 2 + 6 + 3 + 5 + 4 + 4 + 5 + 3 + 6 + 2 + 7 + 1 + 8$

BINARY AND TERNARY SYNCHRONIZATION

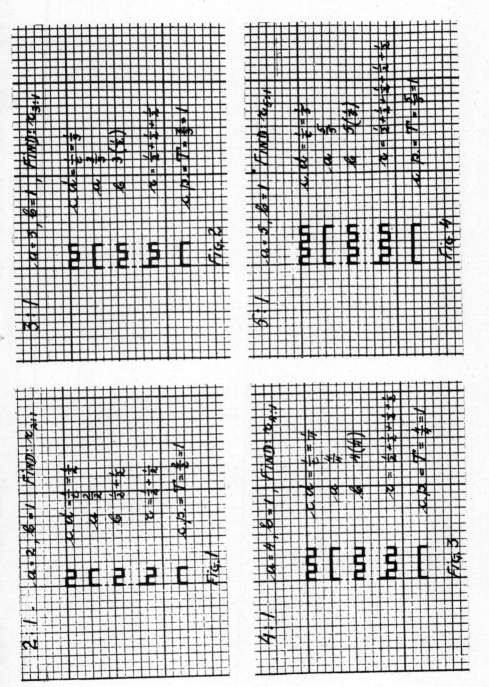

Binary Synchronization

APPENDIX A

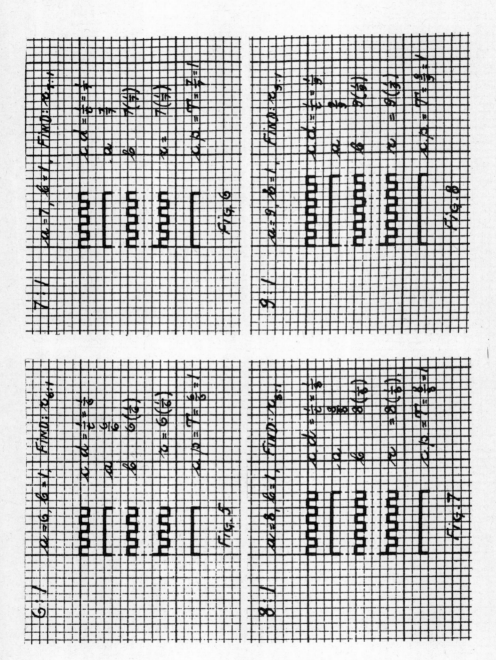

Binary Synchronization

BINARY AND TERNARY SYNCHRONIZATION

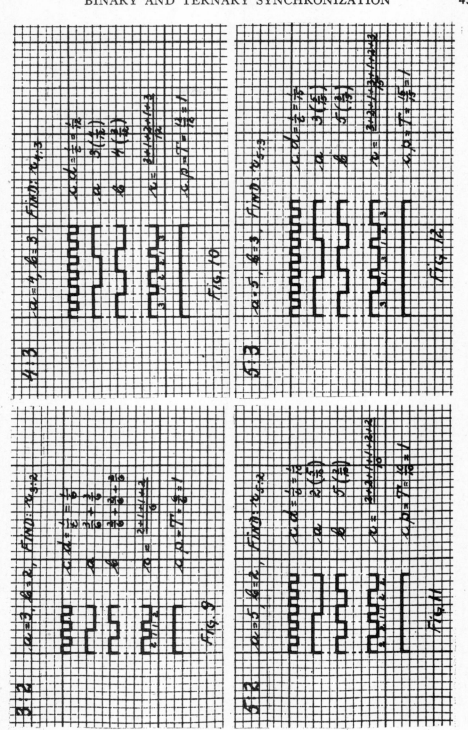

Binary Synchronization

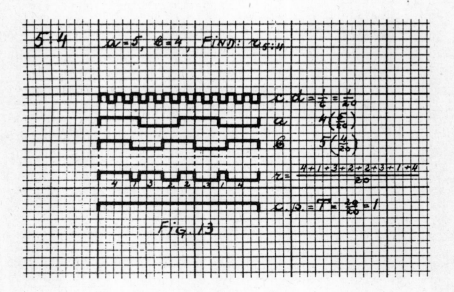

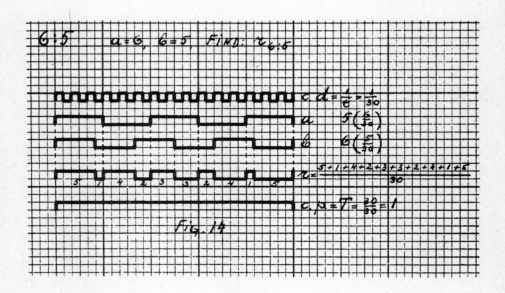

Binary Synchronization

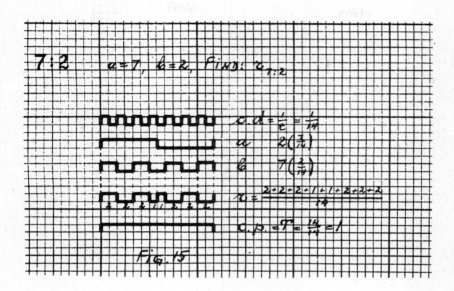

Fig. 15

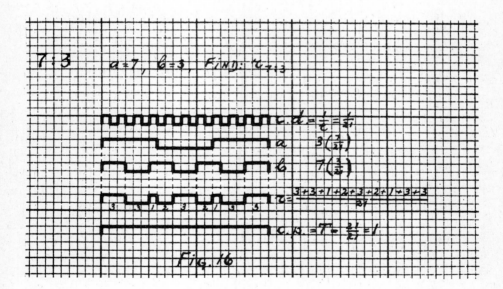

Fig. 16

Binary Synchronization

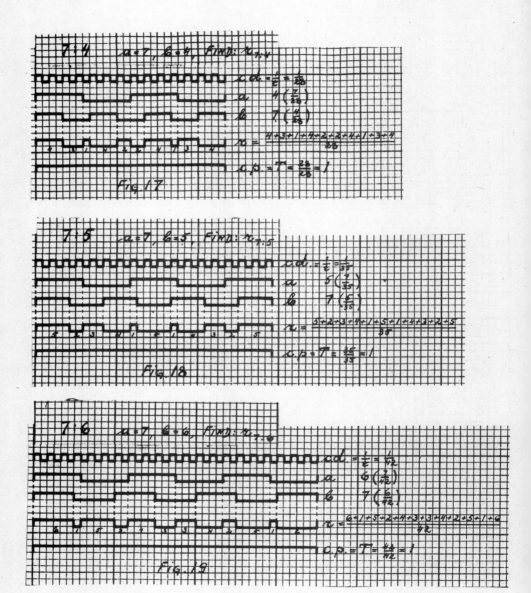

Binary Synchronization

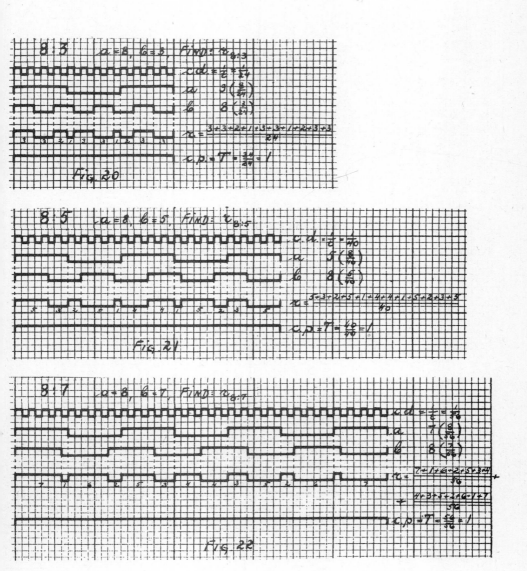

Binary Synchronization

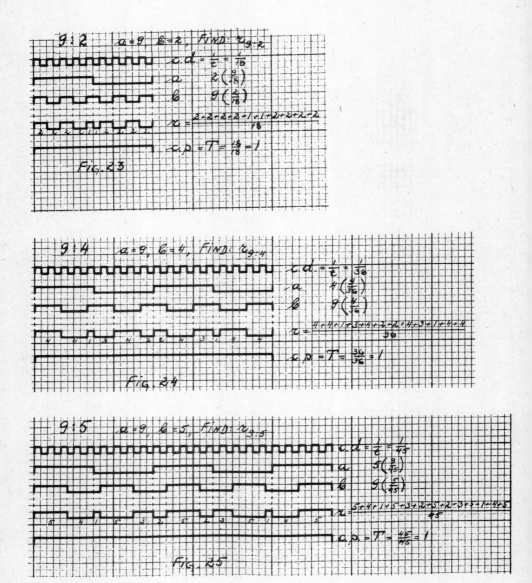

Binary Synchronization

BINARY AND TERNARY SYNCHRONIZATION

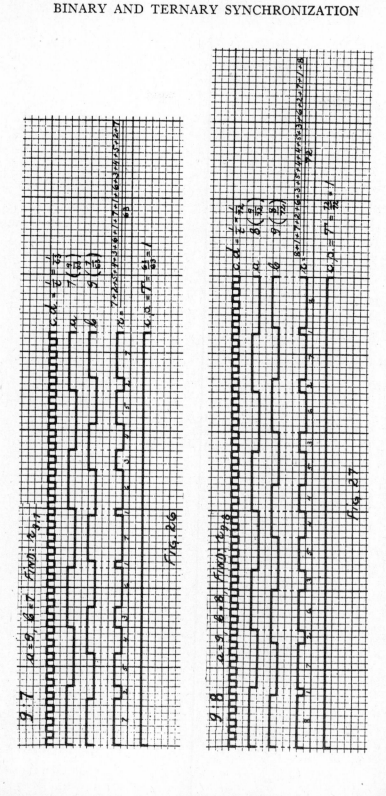

Binary Synchronization

B. Binary Synchronization with Fractioning

(Displacement of b through the phases of a).

$$\underline{a:b};\ S(\underline{a:b}) = a \cdot a = a^2;\ S\left(\frac{1}{\underline{a \div b}}\right) = \frac{1}{a \cdot a} = \frac{1}{a^2};$$

$$r_{\underline{(a \div b)}} = I\left\{a\left(\frac{a}{a^2}\right) \div \left[(a - b + 1) \cdot a\left(\frac{b}{a^2}\right)\right]\right\},\ \text{where}$$

NGb (the number of b groups) $= a - b + 1$;

hence: $b_1 t_0 \equiv a t_0,\ b_2 t_0 \equiv a t_1,\ \ldots\ b_n t_0 \equiv a t_{a-b+1}$, where b_1 is the first Gb, b_2 is the second Gb, etc., and where $T_0, T_1, \ldots$ represent the initial points of respective phases (phasic origins or attacks).

1. Rhythmic Resultants with Fractioning

$r_{\underline{3 \div 2}} = 2 + 1 + 1 + 1 + 1 + 1 + 2$

$r_{\underline{4 \div 3}} = 3 + 1 + 2 + 1 + 1 + 1 + 1 + 2 + 1 + 3$

$r_{\underline{5 \div 2}} = 2 + 2 + 8(1) + 1 + 8(1) + 2 + 2$

$r_{\underline{5 \div 3}} = 3 + 2 + 1 + 2 + 4(1) + 1 + 4(1) + 2 + 1 + 2 + 2$

$r_{\underline{5 \div 4}} = 4 + 1 + 3 + 1 + 1 + 2 + 1 + 2 + 1 + 1 + 3 + 1 + 4$

$r_{\underline{6 \div 5}} = 5 + 1 + 4 + 1 + 1 + 3 + 1 + 2 + 2 + 1 + 3 + 1 + 1 + 4 + 1 + 5$

$r_{\underline{7 \div 2}} = 2 + 2 + 2 + 18(1) + 1 + 18(1) + 2 + 2 + 2$

$r_{\underline{7 \div 3}} = 3 + 3 + 1 + 2 + 1 + 2 + 12(1) + 1 + 12(1) + 2 + 1 + 2 + 1 + 3 + 3$

$r_{\underline{7 \div 4}} = 4 + 3 + 1 + 3 + 1 + 2 + 1 + 1 + 2 + 6(1) + 1 + 6(1) + 2 + 1 + 1 + 2 + 1 + 3 + 1 + 3 + 4$

$r_{\underline{7 \div 5}} = 5 + 2 + 3 + 2 + 2 + 1 + 2 + 2 + 1 + 1 + 1 + 2 + 1 + 2 + 1 + 1 + 1 + 2 + 2 + 1 + 2 + 2 + 3 + 2 + 5$

$r_{\underline{7 \div 6}} = 6 + 1 + 5 + 1 + 1 + 4 + 1 + 2 + 3 + 1 + 3 + 2 + 1 + 4 + 1 + 1 + 5 + 1 + 6$

$r_{8 \div 3} = 3 + 3 + 2 + 1 + 2 + 1 + 2 + 18(1) + 18(1) + 2 + 1 + 2 + 1 +$
$\phantom{r_{8 \div 3} = 3 + 3 + 2 + 1 + 2 + 1 + 2 + 18(1) + 18(1) + 2 + 1} + 2 + 3 + 3$

$r_{8 \div 5} = 5 + 3 + 2 + 3 + 2 + 1 + 2 + 2 + 1 + 2 + 1 + 1 + 1 + 1 + 2 +$
$\phantom{r_{8 \div 5} =} + 1 + 1 + 1 + 1 + 1 + 1 + 1 + 1 + 2 + 1 + 1 + 1 + 2 + 1 +$
$\phantom{r_{8 \div 5} =} + 2 + 2 + 1 + 2 + 3 + 2 + 3 + 5$

$r_{8 \div 7} = 7 + 1 + 6 + 1 + 1 + 5 + 1 + 2 + 4 + 1 + 3 + 3 + 1 + 4 + 2 +$
$\phantom{r_{8 \div 7} =} + 1 + 5 + 1 + 1 + 6 + 1 + 7$

$r_{9 \div 2} = 2 + 2 + 2 + 2 + 32(1) + 1 + 32(1) + 2 + 2 + 2 + 2$

$r_{9 \div 4} = 4 + 4 + 1 + 3 + 1 + 3 + 1 + 1 + 2 + 1 + 1 + 2 + 16(1) + 1 +$
$\phantom{r_{9 \div 4} =} + 16(1) + 2 + 1 + 1 + 2 + 1 + 1 + 3 + 1 + 3 + 1 + 4 + 4$

$r_{9 \div 5} = 5 + 4 + 1 + 4 + 1 + 3 + 1 + 1 + 3 + 1 + 1 + 2 + 1 + 1 + 1 +$
$\phantom{r_{9 \div 5} =} + 2 + 8(1) + 1 + 8(1) + 2 + 1 + 1 + 1 + 2 + 1 + 1 + 3 + 1 +$
$\phantom{r_{9 \div 5} =} + 1 + 3 + 1 + 4 + 1 + 4 + 5$

$r_{9 \div 7} = 7 + 2 + 5 + 2 + 2 + 3 + 2 + 2 + 2 + 1 + 2 + 2 + 3 + 1 + 1 +$
$\phantom{r_{9 \div 7} =} + 2 + 3 + 2 + 1 + 1 + 3 + 2 + 2 + 1 + 2 + 2 + 2 + 3 + 2 + 2 +$
$\phantom{r_{9 \div 7} =} + 5 + 2 + 7$

$r_{9 \div 8} = 8 + 1 + 7 + 1 + 1 + 6 + 1 + 2 + 5 + 1 + 3 + 4 + 1 + 4 + 3 +$
$\phantom{r_{9 \div 8} =} + 1 + 4 + 3 + 1 + 5 + 2 + 1 + 6 + 1 + 1 + 7 + 1 + 8$

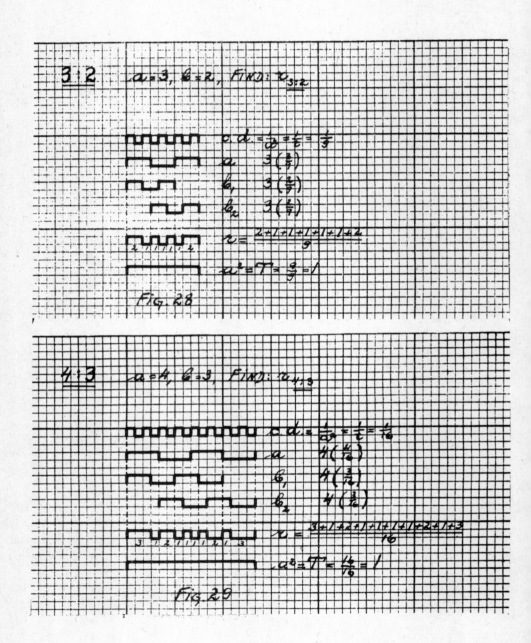

Binary Synchronization with fractioning

BINARY AND TERNARY SYNCHRONIZATION

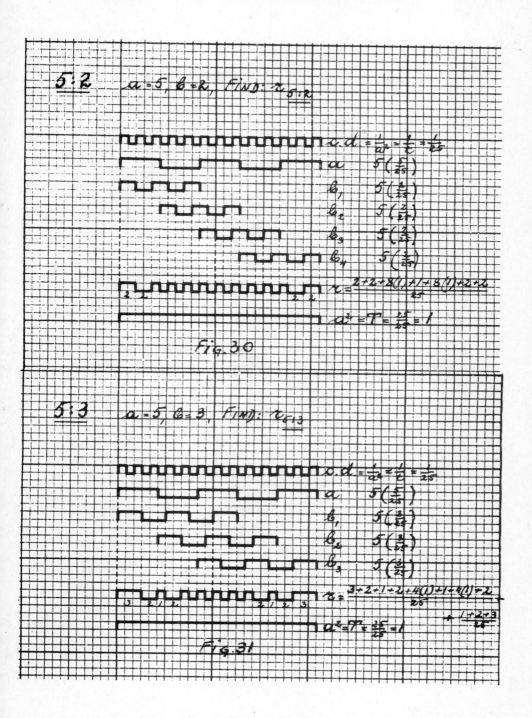

Binary Synchronization with fractioning

APPENDIX A

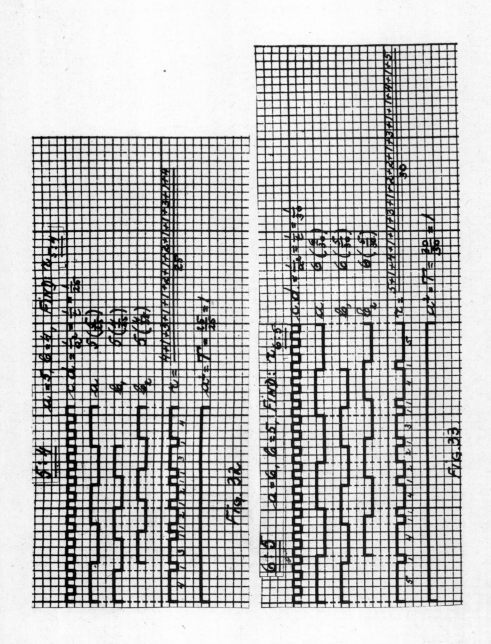

Fig. 32, Fig. 33. Binary Synchronization with fractioning

BINARY AND TERNARY SYNCHRONIZATION

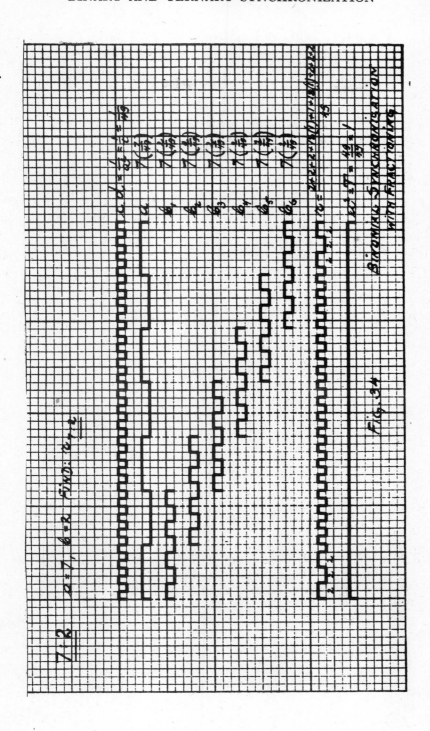

Fig. 34

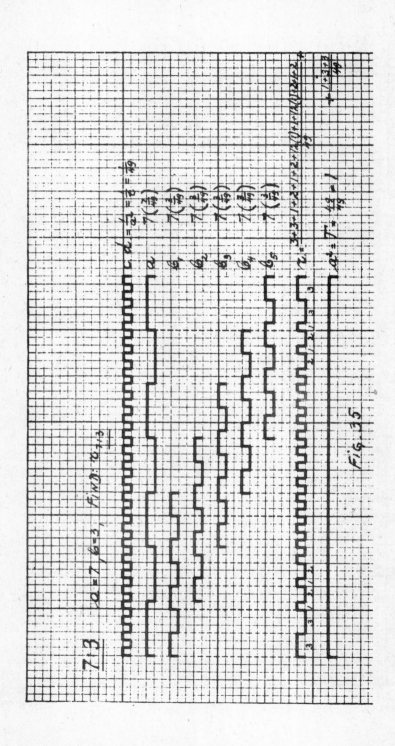

Fig. 35

Binary Synchronization with fractioning

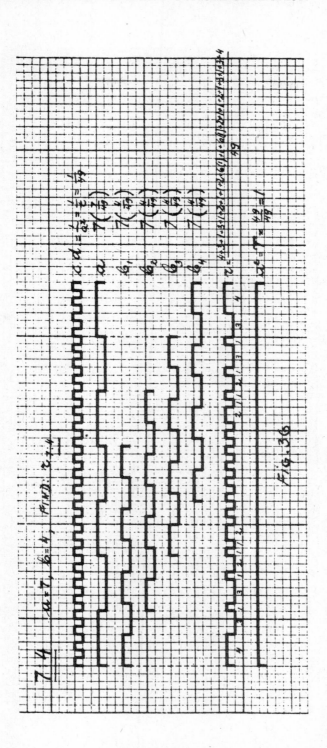

Binary Synchronization with fractioning

APPENDIX A

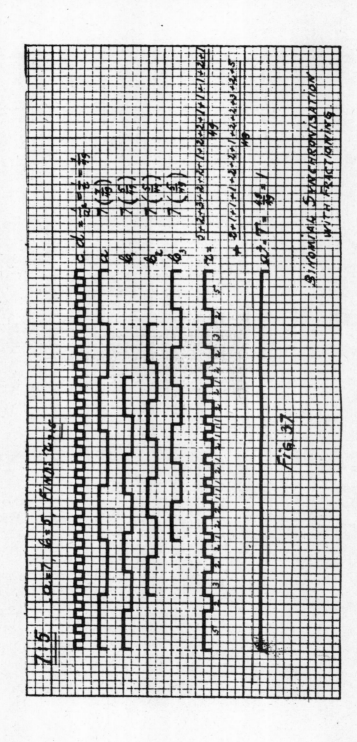

Fig. 37

BINARY AND TERNARY SYNCHRONIZATION

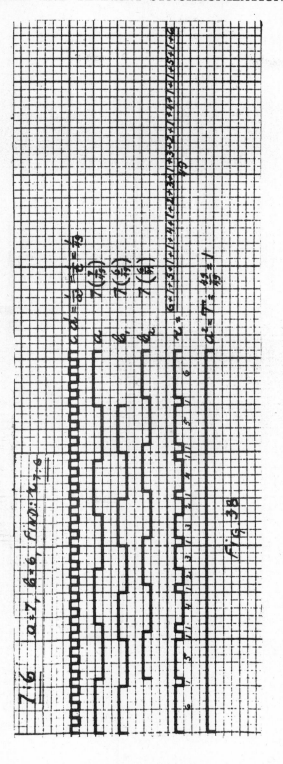

Fig. 38

Binary Synchronization with fractioning

APPENDIX A

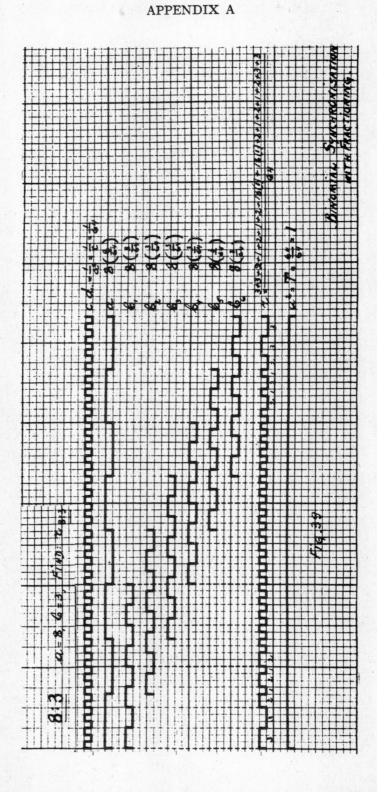

Fig. 39

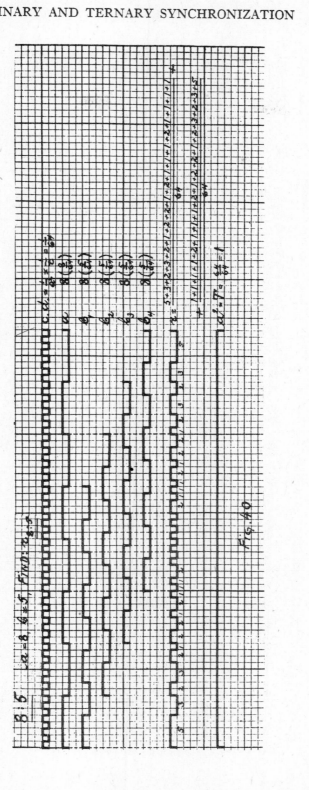

Binary Synchronization with fractioning

APPENDIX A

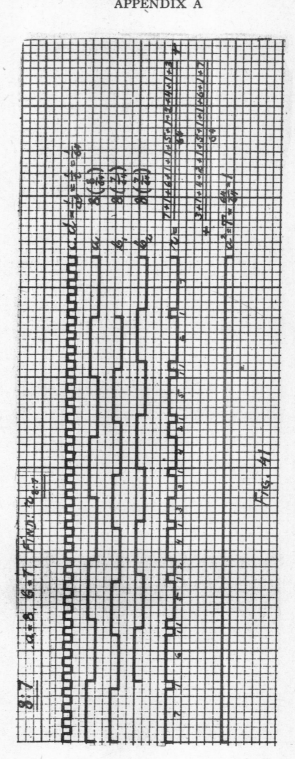

Fig. 41

Binary Synchronization with fractioning

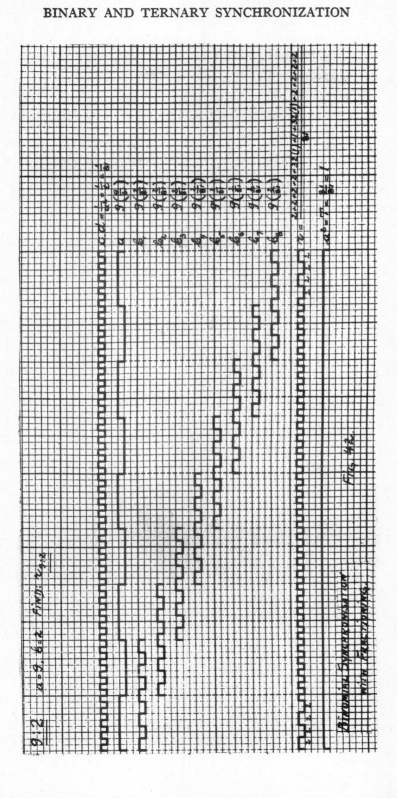

Fig. 42. Binomial Synchronization with Pseudonyms.

APPENDIX A

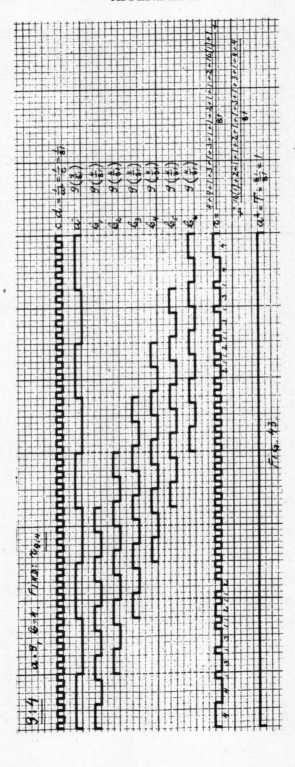

Fig. 43.

Binary Synchronization

BINARY AND TERNARY SYNCHRONIZATION

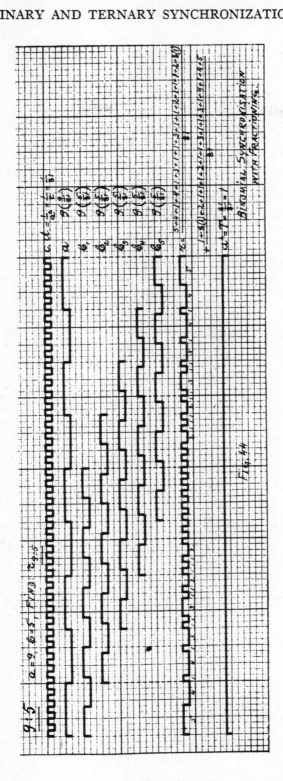

Fig. 44. Binomial Synchronisation with Fractioning.

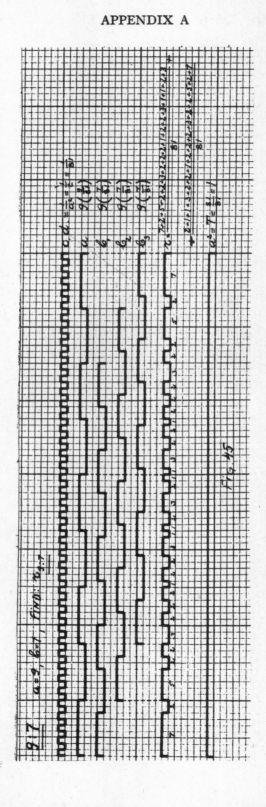

Fig. 45

Binary Synchronization with fractioning

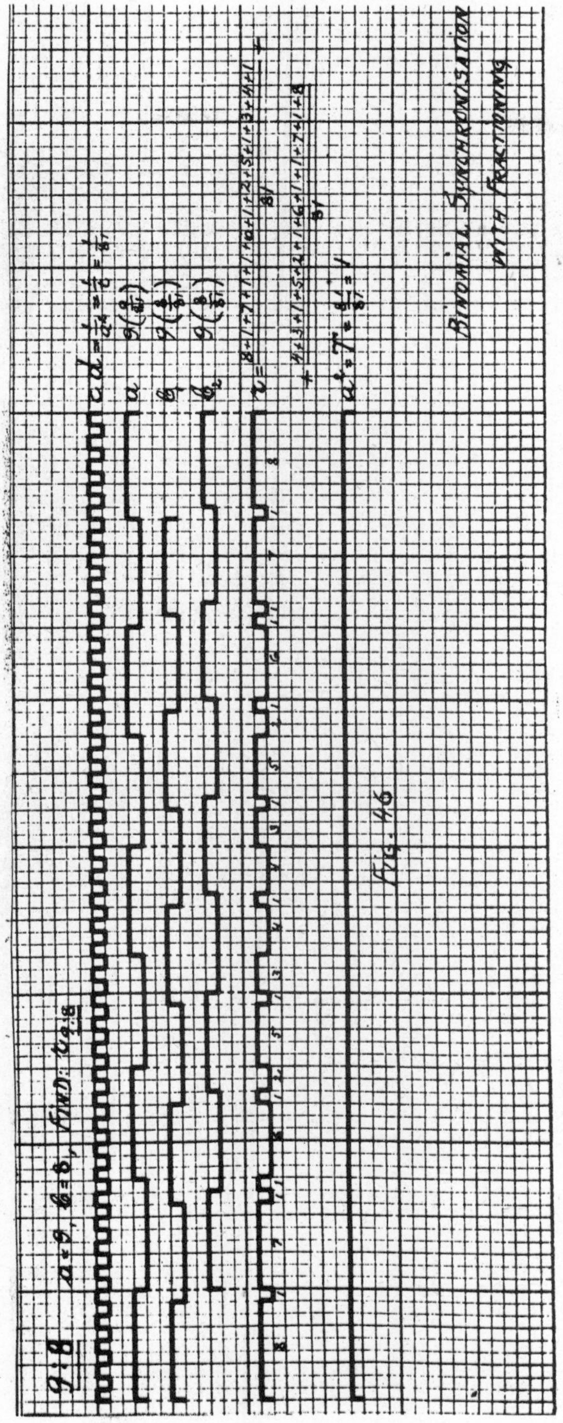

BINARY AND TERNARY SYNCHRONIZATION

$\frac{q}{8}$ $a=9, b=8,$ (fixed) $c=9.8$

$c \cdot d = \frac{c}{a \cdot c} = \frac{1}{b} = \frac{1}{8}$

$a = q\left(\frac{8}{a \cdot c}\right)$

$b_1 = q\left(\frac{8}{8}\right)$

$b_2 = q\left(\frac{8}{8\tau}\right)$

$v = \frac{8 \cdot 1 + 7 + 1 + 7 + 1 + 6 + 1 + 7 + 2 \cdot 5 + 1 + 3 + 7 + 1}{8\tau} + \frac{4 + 3 + 1 + 5 \cdot 2 + 1 + 6 + 1 + 7 + 4 + 8}{8\tau}$

$a \cdot c \cdot \tau = \frac{8 \cdot 1}{8\tau} = 1$

Fig. 46

Binomial Synchronization with Fractioning

C. Ternary Synchronization

Ternary synchronization includes 3 generators: a, b and c, their product (abc), their denominator $\left(\dfrac{1}{abc}\right)$, their complementary factors (bc, ac, ab) and two resultants of interference: r and r'.

r represents the resultant of interference of the generators; it has a number of recurrence-groups;

r' represents the resultant of interference of the complementary factors; it is of the type $r_{(a+b)}$.

$\dfrac{r}{r'}$ produces a perfect coordination of two contrasting forms of regularity, each accompanied by its own generators.

The presence of two resultants is characteristic of all polynomials.

$$a:b:c; \quad S(a:b:c) = a \cdot b \cdot c = abc; \quad S\left(\frac{1}{a:b:c}\right) = \frac{1}{abc};$$

$$r_{(a+b+c)} = I\left[bc\left(\frac{a}{abc}\right) \div ac\left(\frac{b}{abc}\right) \div ab\left(\frac{c}{abc}\right)\right];$$

$$r'_{(a+b+c)} = I\left[a\left(\frac{bc}{abc}\right) \div b\left(\frac{ac}{abc}\right) \div c\left(\frac{ab}{abc}\right)\right].$$

D. Generalization: Synchronization of N Generators.

$$a:b:c: \ldots :m; \quad S(a:b:c: \ldots :m) = abc \ldots m;$$

$$S\left(\frac{1}{a:b:c: \ldots :m}\right) = \frac{1}{abc \ldots m}$$

$$r_{(a+b+c+\ldots+m)} = I\left[bcd \ldots m\left(\frac{a}{abc \ldots m}\right) \div acd \ldots m\left(\frac{b}{abc \ldots m}\right) \div\right.$$
$$\left. \div abd \ldots m\left(\frac{c}{abc \ldots m}\right) \div \ldots \div abc \ldots 1\left(\frac{m}{abc \ldots m}\right)\right];$$

$$r'_{(a+b+c+\ldots+m)} = I\left[a\left(\frac{bcd \ldots m}{abc \ldots m}\right) \div b\left(\frac{acd \ldots m}{abc \ldots m}\right) \div\right.$$
$$\left. \div c\left(\frac{abd \ldots m}{abc \ldots m}\right) \div \ldots \div m\left(\frac{abc \ldots 1}{abc \ldots m}\right)\right].$$

BINARY AND TERNARY SYNCHRONIZATION

1. Rhythmic Resultants from Three Generators

$r_{2 \div 3 \div 5}\ = 2 + 1 + 1 + 1 + 1 + 2 + 1 + 1 + 2 + 2 + 1 + 1 + 2 +$
$\qquad\qquad + 2 + 1 + 1 + 2 + 1 + 1 + 1 + 1 + 2$

$r'_{2 \div 3 \div 5} = 6 + 4 + 2 + 3 + 3 + 2 + 4 + 6$

$r_{3 \div 5 \div 8}\ = 3 + 2 + 1 + 2 + 1 + 1 + 2 + 3 + 1 + 2 + 2 + 1 + 3 +$
$\qquad\qquad + 1 + 2 + 3 + 2 + 1 + 2 + 1 + 3 + 1 + 2 + 3 + 3 +$
$\qquad\qquad + 2 + 1 + 3 + 1 + 1 + 1 + 3 + 3 + 1 + 1 + 1 + 3 +$
$\qquad\qquad + 1 + 2 + 3 + 3 + 2 + 1 + 3 + 1 + 2 + 1 + 2 + 3 +$
$\qquad\qquad + 2 + 1 + 3 + 1 + 2 + 2 + 1 + 3 + 2 + 1 + 1 + 2$
$\qquad\qquad + 1 + 2 + 3$

$r'_{3 \div 5 \div 8} = 15 + 9 + 6 + 10 + 5 + 3 + 12 + 12 + 3 + 5 + 10 + 6 + 9 + 15$

$r_{3 \div 5 \div 13} = 3 + 2 + 1 + 3 + 1 + 2 + 1 + 2 + 3 + 2 + 1 + 3 + 1 +$
$\qquad\qquad + 1 + 1 + 3 + 3 + 2 + 1 + 3 + 1 + 2 + 3 + 3 + 2 + 1 +$
$\qquad\qquad + 1 + 2 + 1 + 2 + 3 + 3 + 2 + 1 + 3 + 1 + 2 + 3 + 3 +$
$\qquad\qquad + 2 + 1 + 3 + 1 + 2 + 3 + 1 + 2 + 2 + 1 + 3 + 1 + 2 +$
$\qquad\qquad + 2 + 1 + 3 + 2 + 1 + 3 + 1 + 2 + 3 + 3 + 2 + 1 + 3 +$
$\qquad\qquad + 1 + 2 + 3 + 3 + 2 + 1 + 2 + 1 + 1 + 2 + 3 + 3 + 2 +$
$\qquad\qquad + 1 + 3 + 1 + 2 + 3 + 3 + 1 + 1 + 1 + 3 + 1 + 2 + 3 +$
$\qquad\qquad + 2 + 1 + 2 + 1 + 3 + 1 + 2 + 3$

$r'_{3 \div 5 \div 13} = 15 + 15 + 9 + 6 + 15 + 5 + 10 + 3 + 12 + 15 + 12 + 3 +$
$\qquad\qquad + 10 + 5 + 15 + 6 + 9 + 15 + 15$

$r_{5 \div 8 \div 13} = 5 + 3 + 2 + 3 + 2 + 1 + 4 + 4 + 1 + 1 + 4 + 2 + 3 +$
$\qquad\qquad + 4 + 1 + 5 + 3 + 2 + 2 + 3 + 1 + 4 + 4 + 1 + 5 +$
$\qquad\qquad + 2 + 3 + 3 + 2 + 5 + 3 + 2 + 1 + 4 + 1 + 4 + 4 +$
$\qquad\qquad + 1 + 5 + 2 + 3 + 2 + 3 + 5 + 3 + 2 + 5 + 1 + 4 +$
$\qquad\qquad + 3 + 1 + 1 + 5 + 2 + 3 + 1 + 4 + 5 + 3 + 1 + 1 +$
$\qquad\qquad + 5 + 1 + 4 + 2 + 2 + 1 + 5 + 2 + 3 + 5 + 5 + 3 +$
$\qquad\qquad + 2 + 5 + 1 + 4 + 1 + 3 + 1 + 5 + 2 + 2 + 1 + 5 +$
$\qquad\qquad + 5 + 2 + 1 + 2 + 5 + 1 + 4 + 4 + 1 + 5 + 2 + 1 +$
$\qquad\qquad + 2 + 5 + 5 + 1 + 2 + 2 + 5 + 1 + 3 + 1 + 4 + 1 +$
$\qquad\qquad + 5 + 2 + 3 + 5 + 5 + 3 + 2 + 5 + 1 + 2 + 2 + 4 +$
$\qquad\qquad + 1 + 5 + 1 + 1 + 3 + 5 + 4 + 1 + 3 + 2 + 5 + 1 +$
$\qquad\qquad + 1 + 3 + 4 + 1 + 5 + 2 + 3 + 5 + 3 + 2 + 3 + 2 +$
$\qquad\qquad + 5 + 1 + 4 + 4 + 1 + 4 + 1 + 2 + 3 + 5 + 2 + 3 +$
$\qquad\qquad + 3 + 2 + 5 + 1 + 4 + 4 + 1 + 3 + 2 + 2 + 3 + 5 +$
$\qquad\qquad + 1 + 4 + 3 + 2 + 4 + 1 + 1 + 4 + 4 + 1 + 2 + 3 +$
$\qquad\qquad + 2 + 3 + 5$

$r'_{5 \div 8 \div 13} = 40 + 25 + 15 + 24 + 16 + 10 + 30 + 35 + 5 + 8 +$
$+ 32 + 20 + 20 + 32 + 8 + 5 + 35 + 30 + 10 + 16 +$
$+ 24 + 15 + 25 + 40$

$r_{3 \div 4 \div 7} = 3 + 1 + 2 + 1 + 1 + 1 + 3 + 2 + 1 + 1 + 2 + 2 +$
$+ 1 + 3 + 3 + 1 + 2 + 2 + 1 + 2 + 1 + 3 + 1 + 2 +$
$+ 2 + 1 + 3 + 1 + 2 + 1 + 2 + 2 + 1 + 3 + 3 + 1 +$
$+ 2 + 2 + 1 + 1 + 2 + 3 + 1 + 1 + 1 + 2 + 1 + 3$

$r'_{3 \div 4 \div 7} = 12 + 9 + 3 + 4 + 8 + 6 + 6 + 8 + 4 + 3 + 9 + 12$

$r_{3 \div 4 \div 11} = 3 + 1 + 2 + 2 + 1 + 2 + 1 + 3 + 1 + 2 + 2 + 1 +$
$+ 1 + 2 + 3 + 1 + 2 + 2 + 1 + 3 + 3 + 1 + 2 +$
$+ 2 + 1 + 3 + 3 + 1 + 2 + 1 + 1 + 1 + 3 + 3 +$
$+ 1 + 2 + 2 + 1 + 3 + 3 + 1 + 1 + 1 + 2 + 1 +$
$+ 3 + 3 + 1 + 2 + 2 + 1 + 3 + 3 + 1 + 2 + 2 +$
$+ 1 + 3 + 2 + 1 + 1 + 2 + 2 + 1 + 3 + 1 + 2 +$
$+ 1 + 2 + 2 + 1 + 3$

$r'_{3 \div 4 \div 11} = 12 + 12 + 9 + 3 + 8 + 4 + 12 + 6 + 6 + 12 + 4 +$
$+ 8 + 3 + 9 + 12 + 12$

$r_{3 \div 7 \div 11} = 3 + 3 + 1 + 2 + 2 + 1 + 2 + 1 + 3 + 3 + 1 + 2 +$
$+ 3 + 1 + 2 + 3 + 2 + 1 + 3 + 3 + 2 + 1 + 3 +$
$+ 1 + 2 + 3 + 1 + 1 + 1 + 3 + 3 + 3 + 3 + 1 +$
$+ 2 + 3 + 2 + 1 + 3 + 3 + 3 + 1 + 2 + 1 + 2 +$
$+ 3 + 2 + 1 + 3 + 3 + 3 + 2 + 1 + 1 + 2 + 3 +$
$+ 2 + 1 + 1 + 2 + 3 + 3 + 3 + 1 + 2 + 3 + 2 +$
$+ 1 + 2 + 1 + 3 + 3 + 3 + 1 + 2 + 3 + 2 + 1 +$
$+ 3 + 3 + 3 + 3 + 1 + 1 + 1 + 3 + 2 + 1 + 3 +$
$+ 1 + 2 + 3 + 3 + 1 + 2 + 3 + 2 + 1 + 3 + 2 +$
$+ 1 + 3 + 3 + 1 + 2 + 1 + 2 + 2 + 1 + 3 + 3$

$r'_{3 \div 7 \div 11} = 21 + 12 + 9 + 21 + 3 + 11 + 7 + 15 + 6 + 21 +$
$+ 6 + 15 + 7 + 11 + 3 + 21 + 9 + 12 + 21$

$r_{4 \div 7 \div 11} = 4 + 3 + 1 + 3 + 1 + 2 + 2 + 4 + 1 + 1 + 2 + 4 +$
$+ 4 + 1 + 2 + 1 + 4 + 2 + 2 + 4 + 1 + 3 + 3 +$
$+ 1 + 4 + 3 + 1 + 2 + 2 + 2 + 2 + 4 + 1 + 3 +$
$+ 4 + 4 + 3 + 1 + 4 + 2 + 1 + 1 + 4 + 1 + 3 +$
$+ 2 + 2 + 4 + 3 + 1 + 1 + 3 + 2 + 2 + 4 + 1 +$
$+ 3 + 4 + 3 + 1 + 3 + 1 + 4 + 2 + 2 + 4 + 1 +$
$+ 3 + 1 + 3 + 4 + 3 + 1 + 4 + 2 + 2 + 3 + 1 +$
$+ 1 + 3 + 4 + 2 + 2 + 3 + 1 + 4 + 1 + 1 + 2 +$
$+ 4 + 1 + 3 + 4 + 4 + 3 + 1 + 4 + 2 + 2 + 2 +$
$+ 2 + 1 + 3 + 4 + 1 + 3 + 3 + 1 + 4 + 2 + 2 +$
$+ 4 + 1 + 2 + 1 + 4 + 4 + 2 + 1 + 1 + 4 + 2 +$
$+ 2 + 1 + 3 + 1 + 3 + 4$

$r'_{4 \div 7 \div 11} = 28 + 16 + 12 + 21 + 7 + 4 + 24 + 20 + 8 + 14 +$
$+ 14 + 8 + 20 + 24 + 4 + 7 + 21 + 12 + 16 + 28$

$r_{4 \div 5 \div 9} = 4 + 1 + 3 + 1 + 1 + 2 + 3 + 1 + 2 + 2 + 4 + 1 +$
$+ 2 + 1 + 2 + 2 + 3 + 1 + 4 + 4 + 1 + 3 + 2 +$
$+ 2 + 2 + 1 + 1 + 4 + 3 + 1 + 1 + 3 + 2 + 2 +$
$+ 3 + 1 + 4 + 1 + 3 + 1 + 3 + 2 + 2 + 3 + 1 +$
$+ 3 + 1 + 4 + 1 + 3 + 2 + 2 + 3 + 1 + 1 + 3 +$
$+ 4 + 1 + 1 + 2 + 2 + 2 + 3 + 1 + 4 + 4 + 1 +$
$+ 3 + 2 + 2 + 1 + 2 + 1 + 4 + 2 + 2 + 1 + 3 +$
$+ 2 + 1 + 1 + 3 + 1 + 4$

$r'_{4 \div 5 \div 9} = 20 + 16 + 4 + 5 + 15 + 12 + 8 + 10 + 10 + 8 +$
$+ 12 + 15 + 5 + 4 + 16 + 20$

$r_{4 \div 5 \div 14} = 4 + 1 + 3 + 2 + 2 + 2 + 1 + 1 + 4 + 4 + 1 + 3 +$
$+ 2 + 2 + 3 + 1 + 4 + 2 + 2 + 1 + 3 + 2 + 2 +$
$+ 3 + 1 + 4 + 4 + 1 + 3 + 2 + 2 + 3 + 1 + 4 +$
$+ 4 + 1 + 3 + 2 + 2 + 3 + 1 + 2 + 2 + 4 + 1 +$
$+ 3 + 2 + 2 + 3 + 1 + 4 + 4 + 1 + 1 + 2 + 2 +$
$+ 2 + 3 + 1 + 4 + 4 + 1 + 3 + 2 + 2 + 2 + 1 +$
$+ 1 + 4 + 4 + 1 + 3 + 2 + 2 + 3 + 1 + 4 + 2 +$
$+ 2 + 1 + 3 + 2 + 2 + 3 + 1 + 4 + 4 + 1 + 3 +$
$+ 2 + 2 + 3 + 1 + 4 + 4 + 1 + 3 + 2 + 2 + 3 +$
$+ 1 + 2 + 2 + 4 + 1 + 3 + 2 + 2 + 3 + 1 + 4 +$
$+ 4 + 1 + 1 + 2 + 2 + 2 + 3 + 1 + 4$

$r'_{4 \div 5 \div 14} = 20 + 20 + 16 + 4 + 10 + 10 + 20 + 12 + 8 + 20 +$
$+ 20 + 8 + 12 + 20 + 10 + 10 + 4 + 16 + 20 + 20$

$r_{4 \div 9 \div 14} = 4 + 4 + 1 + 3 + 2 + 2 + 2 + 2 + 4 + 3 + 1 + 4 +$
$+ 4 + 4 + 2 + 2 + 1 + 3 + 4 + 2 + 2 + 4 + 3 +$
$+ 1 + 4 + 2 + 2 + 4 + 4 + 1 + 3 + 4 + 2 + 2 +$
$+ 4 + 2 + 1 + 1 + 4 + 4 + 4 + 4 + 1 + 3 + 4 +$
$+ 2 + 2 + 4 + 3 + 1 + 4 + 4 + 4 + 4 + 1 + 1 +$
$+ 2 + 4 + 2 + 2 + 4 + 3 + 1 + 4 + 4 + 2 + 2 +$
$+ 4 + 1 + 3 + 4 + 2 + 2 + 4 + 3 + 1 + 2 + 2 +$
$+ 4 + 4 + 4 + 1 + 3 + 4 + 2 + 2 + 2 + 2 + 3 +$
$+ 1 + 4 + 4 + 4 + 4 + 1 + 3 + 2 + 2 + 2 + 2 +$
$+ 4 + 3 + 1 + 4 + 4 + 4 + 2 + 2 + 1 + 3 + 4 +$
$+ 2 + 2 + 4 + 3 + 1 + 4 + 2 + 2 + 4 + 4 + 1 +$
$+ 3 + 4 + 2 + 2 + 4 + 2 + 1 + 1 + 4 + 4 + 4 +$
$+ 4 + 1 + 3 + 4 + 2 + 2 + 4 + 3 + 1 + 4 + 4 +$
$+ 4 + 4 + 1 + 1 + 2 + 4 + 2 + 2 + 4 + 3 + 1 +$
$+ 4 + 4 + 2 + 2 + 4 + 1 + 3 + 4 + 2 + 2 + 4 +$
$+ 3 + 1 + 2 + 2 + 4 + 4 + 4 + 1 + 3 + 4 + 2 +$
$+ 2 + 2 + 2 + 3 + 1 + 4 + 4$

$r'_{4 \div 9 \div 14} = 36 + 20 + 16 + 36 + 4 + 14 + 18 + 24 + 12 +$
$+ 36 + 8 + 28 + 28 + 8 + 36 + 12 + 24 + 18 +$
$+ 14 + 4 + 36 + 16 + 20 + 36$

$$\begin{aligned}
r_{5 \div 9 \div 14} = \; & 5+4+1+4+1+3+2+5+2+1+2+5+ \\
& +1+4+2+3+5+4+1+1+4+3+2+ \\
& +5+2+3+5+1+3+1+5+5+3+1+ \\
& +1+5+3+2+2+3+2+3+5+1+4+ \\
& +5+5+4+1+5+3+1+1+5+2+3+ \\
& +3+2+1+4+5+2+3+4+1+5+1+ \\
& +2+2+5+2+3+5+1+4+4+1+5+ \\
& +4+1+3+2+3+2+5+2+3+5+1+ \\
& +4+1+4+5+4+1+5+3+2+4+1+ \\
& +2+3+5+1+2+2+5+5+2+2+1+ \\
& +5+3+2+1+4+2+3+5+1+4+5+ \\
& +4+1+4+1+5+3+2+5+2+3+2+ \\
& +3+1+4+5+1+4+4+1+5+3+2+ \\
& +5+2+2+1+5+1+4+3+2+5+4+ \\
& +1+2+3+3+2+5+1+1+3+5+1+ \\
& +4+5+5+4+1+5+3+2+3+2+2+ \\
& +3+5+1+1+3+5+5+1+3+1+5+ \\
& +3+2+5+2+3+4+1+1+4+5+3+ \\
& +2+4+1+5+2+1+2+5+2+3+1+ \\
& +4+1+4+5
\end{aligned}$$

$$\begin{aligned}
r'_{5 \div 9 \div 14} = \; & 45+25+20+36+9+5+40+30+15+27+ \\
& +18+10+35+35+10+18+27+15+30+ \\
& 40+5+9+36+20+25+45
\end{aligned}$$

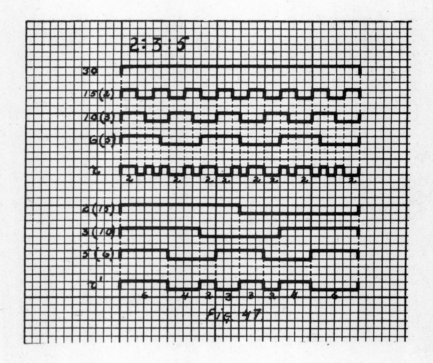

Trinomial Synchronization

BINARY AND TERNARY SYNCHRONIZATION

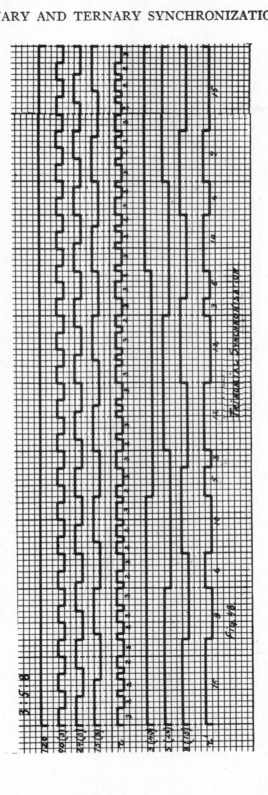

APPENDIX A

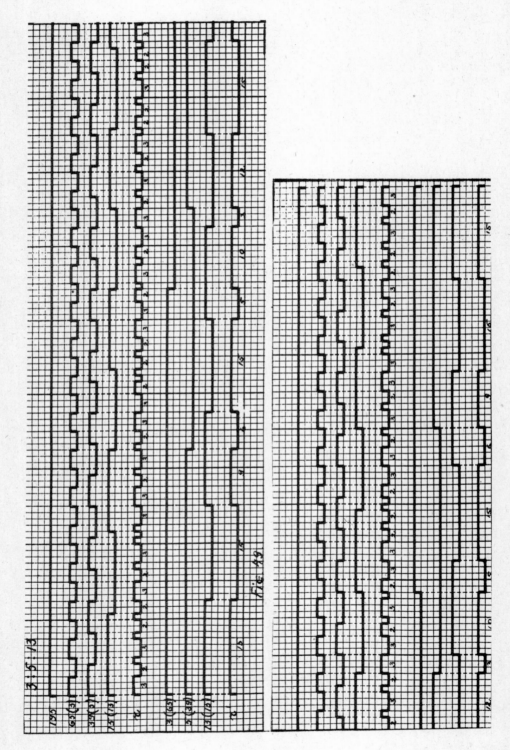

Figure 49. Ternary Synchronization (concluded)

BINARY AND TERNARY SYNCHRONIZATION

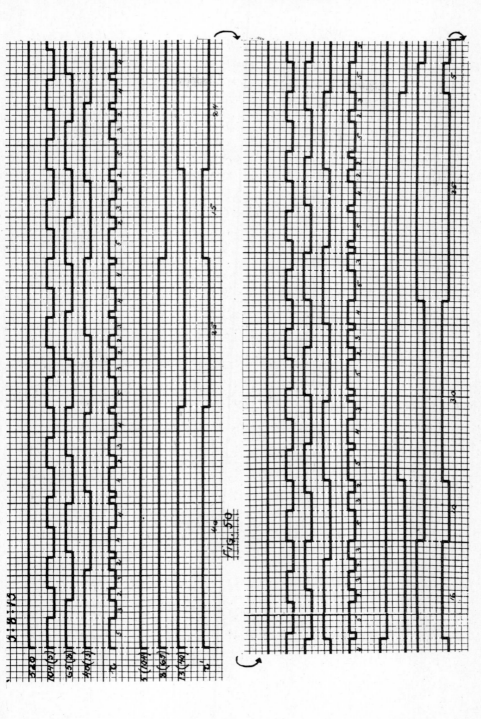

Figure 50. Ternary Synchronization (continued)

APPENDIX A

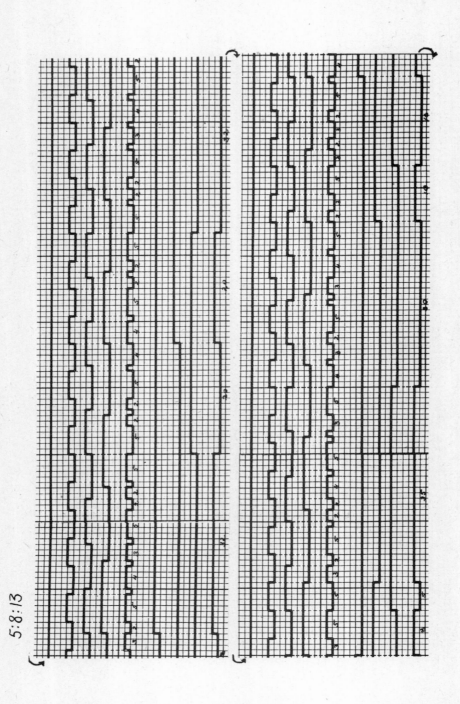

Figure 50. Ternary Synchronization (continued)

BINARY AND TERNARY SYNCHRONIZATION

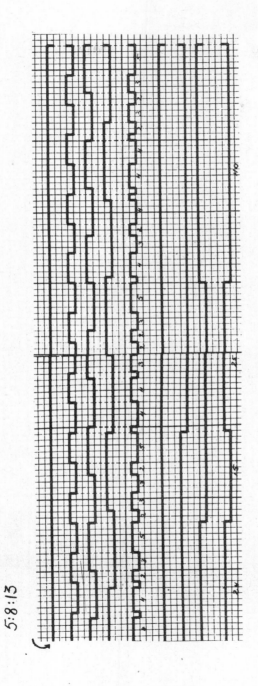

Figure 50. Ternary Synchronization (concluded)

488 APPENDIX A

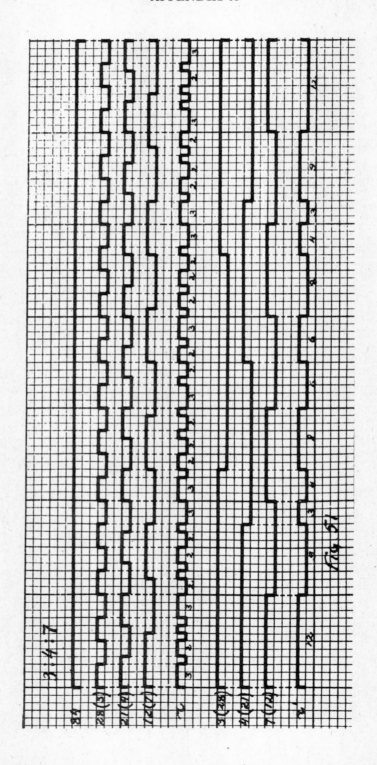

Figure 51. Ternary Synchronization

BINARY AND TERNARY SYNCHRONIZATION

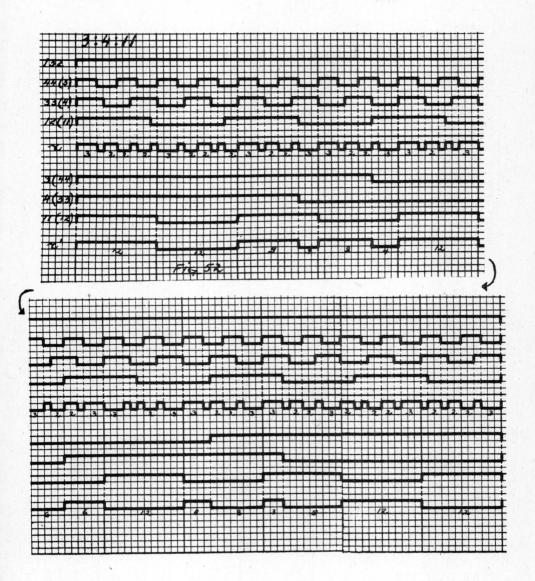

Figure 52. Ternary Synchronization

APPENDIX A

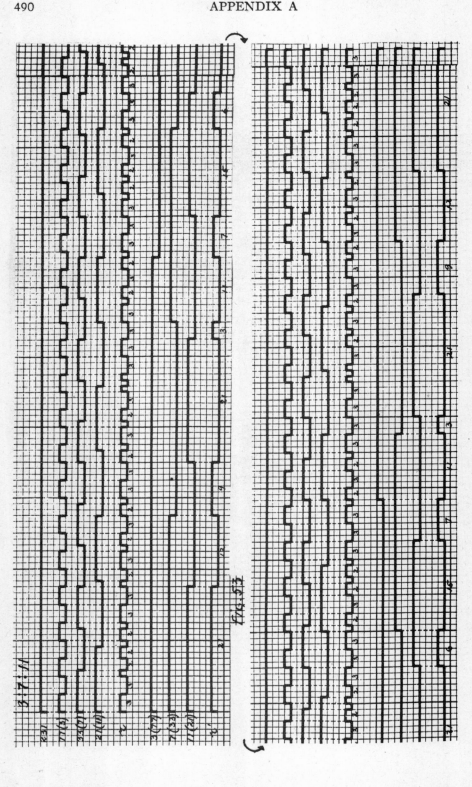

Figure 58. Ternary Synchronization

BINARY AND TERNARY SYNCHRONIZATION

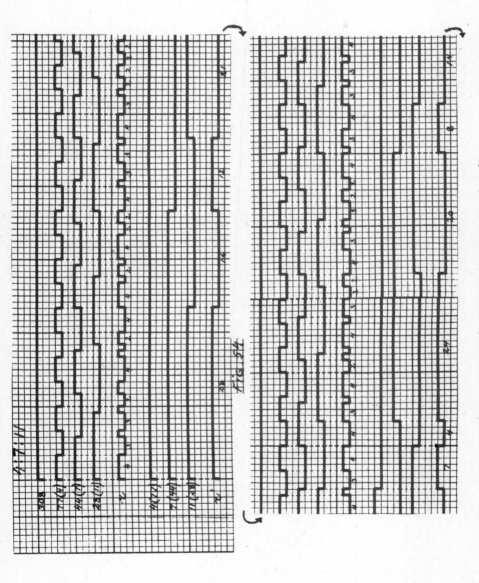

Figure 54. Ternary Synchronization (continued)

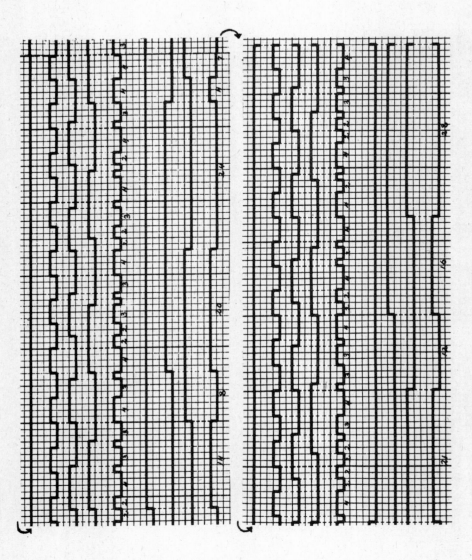

Figure 54. Ternary Synchronization (concluded)

BINARY AND TERNARY SYNCHRONIZATION

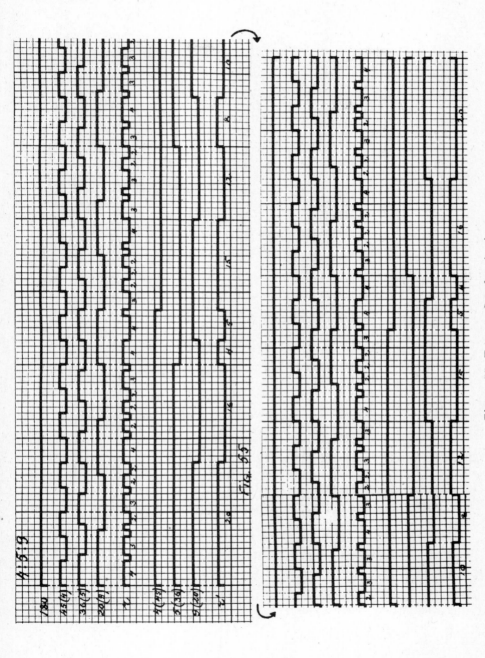

Figure 55. Ternary Synchronization

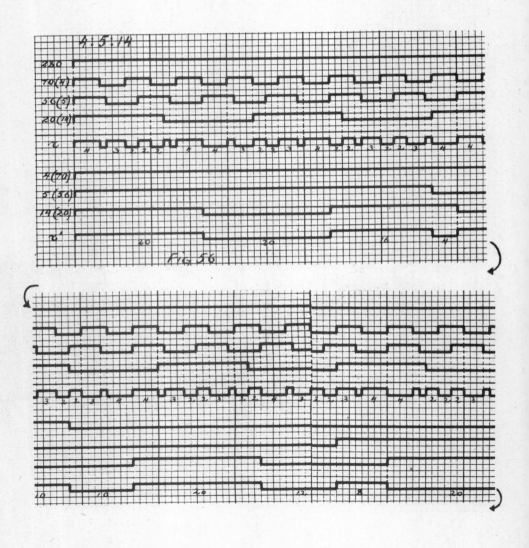

Figure 56. Ternary Synchronization (continued)

4:5:14

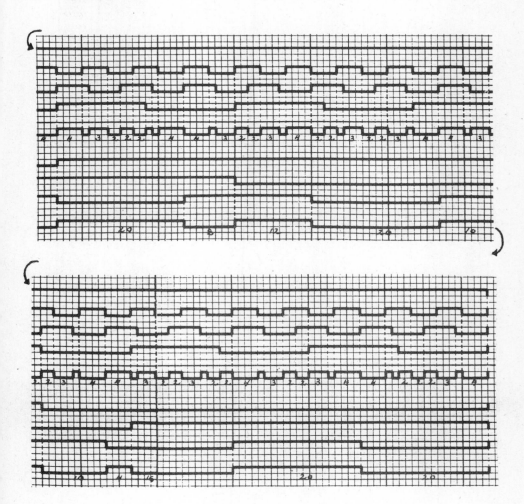

Figure 56. Ternary Synchronization (concluded)

496   APPENDIX A

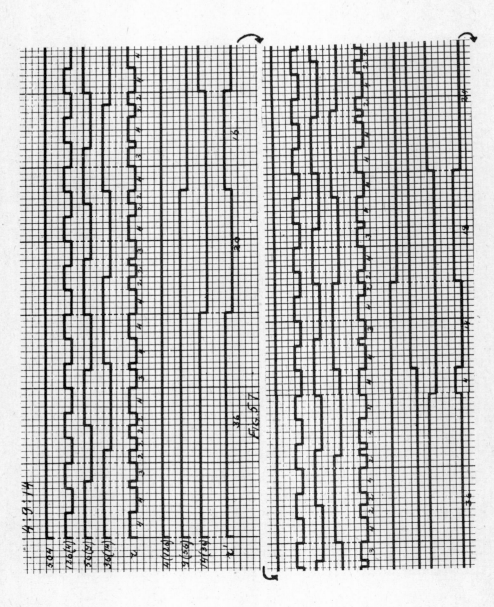

Figure 57. Ternary Synchronization (continued)

BINARY AND TERNARY SYNCHRONIZATION

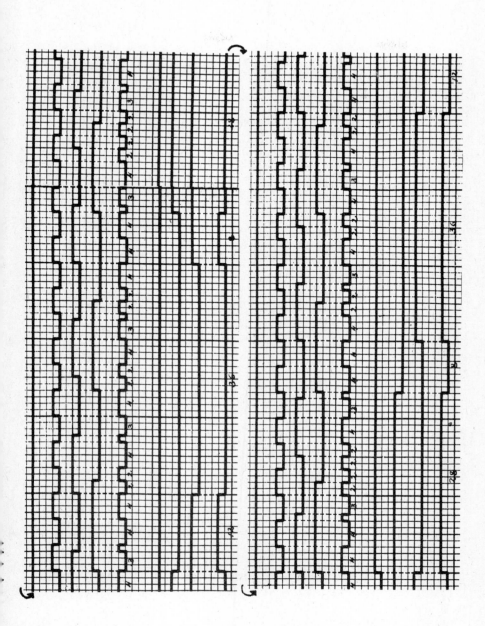

Figure 57. Ternary Synchronization (continued)

498 APPENDIX A

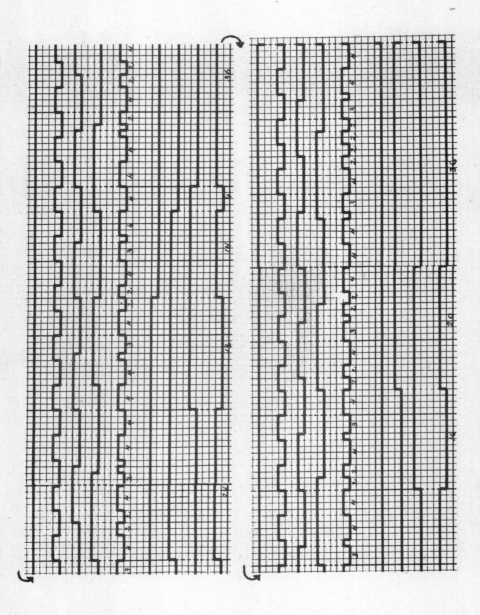

Figure 57. Ternary Synchronization (concluded)

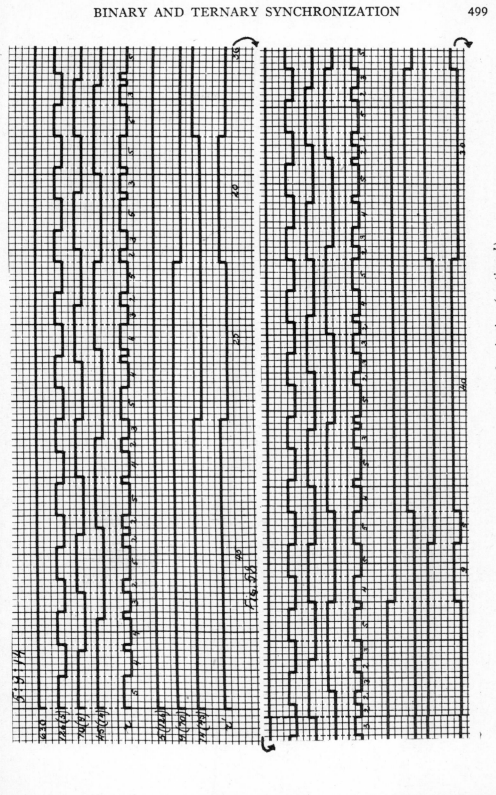

Figure 58. Ternary Synchronization (continued)

500　　APPENDIX A

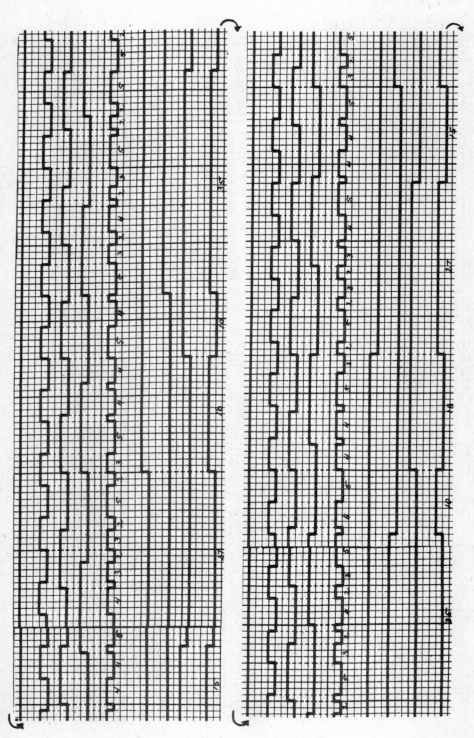

Figure 58. Ternary Synchronization (continued)

BINARY AND TERNARY SYNCHRONIZATION

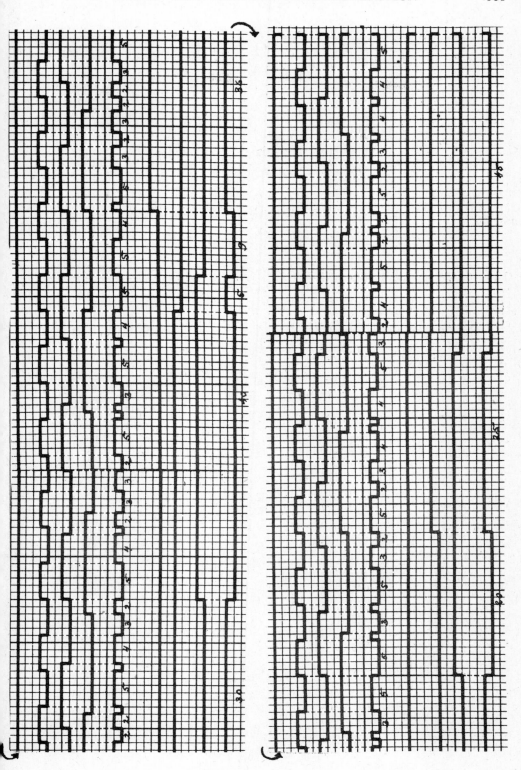

Figure 58. Ternary Synchronization (concluded)

II. DISTRIBUTIVE INVOLUTION—GROUPS.

(Binomials, Trinomials, Polynomials)

Formulae:

A. DISTRIBUTIVE SQUARE OF BINOMIALS.

(1) Factorial: $(a + b)^2 = a^2 + ab + ab + b^2$;

(2) Fractional: $\left(\dfrac{a}{a+b} + \dfrac{b}{a+b}\right)^2 = \dfrac{a^2}{(a+b)^2} + \dfrac{ab}{(a+b)^2} + \dfrac{ab}{(a+b)^2} + \dfrac{b^2}{(a+b)^2}.$

Synchronization of the first power group with the second power group.

(1) Factorial: $S = a(a+b) + b(a+b)$;

(2) Fractional: $S = \dfrac{a}{a+b} \cdot \left(\dfrac{a+b}{a+b}\right) + \dfrac{b}{a+b} \cdot \left(\dfrac{a+b}{a+b}\right).$

$\dfrac{3}{3}$

$\left(\dfrac{2}{3} + \dfrac{1}{3}\right)^2 = \dfrac{4+2+2+1}{9}$

$\left(\dfrac{1}{3} + \dfrac{2}{3}\right)^2 = \dfrac{1+2+2+4}{9}$

$r^2 = \dfrac{1+2+1+1+1+2+1}{9}$

$\dfrac{3}{3}\left(\dfrac{2}{3} + \dfrac{1}{3}\right) = \dfrac{6+3}{9}$

$\dfrac{3}{3}\left(\dfrac{1}{3} + \dfrac{2}{3}\right) = \dfrac{3+6}{9}$

$r = \dfrac{3+3+3}{9}$

$\frac{4}{4}$

$$\left(\frac{3}{4}+\frac{1}{4}\right)^2 = \frac{9+3+3+1}{16}$$

$$\left(\frac{1}{4}+\frac{3}{4}\right)^2 = \frac{1+3+3+9}{16}$$

$$r^2 = \frac{1+3+3+2+3+3+1}{16}$$

$$\frac{4}{4}\left(\frac{3}{4}+\frac{1}{4}\right) = \frac{12+4}{16}$$

$$\frac{4}{4}\left(\frac{1}{4}+\frac{3}{4}\right) = \frac{4+12}{16}$$

$$r = \frac{4+8+4}{16}$$

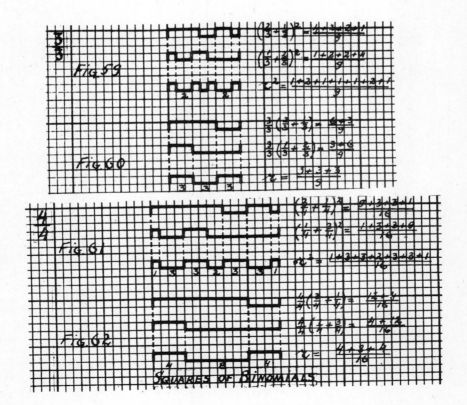

SQUARES OF BINOMIALS

$$\frac{5}{5}\left(\frac{3}{5}+\frac{2}{5}\right)^2 = \frac{9+6+6+4}{25}$$

$$\left(\frac{2}{5}+\frac{3}{5}\right)^2 = \frac{4+6+6+9}{25}$$

$$r_1^2 = \frac{4+5+1+5+1+5+4}{25}$$

$$\left(\frac{4}{5}+\frac{1}{5}\right)^2 = \frac{16+4+4+1}{25}$$

$$\left(\frac{1}{5}+\frac{4}{5}\right)^2 = \frac{1+4+4+16}{25}$$

$$r_2^2 = \frac{1+4+4+7+4+4+1}{25}$$

$$r_1^2 = *$$

$$r_2^2 =$$

$$r_3^2 = \frac{1+3+1+4+1+5+1+4+1+3+1}{25}$$

$$\frac{5}{5}\left(\frac{3}{5}+\frac{2}{5}\right) = \frac{15+10}{25}$$

$$\frac{5}{5}\left(\frac{2}{5}+\frac{3}{5}\right) = \frac{10+15}{25}$$

$$r_1 = \frac{10+5+10}{25}$$

$$\frac{5}{5}\left(\frac{4}{5}+\frac{1}{5}\right) = \frac{20+5}{25}$$

$$\frac{5}{5}\left(\frac{1}{5}+\frac{4}{5}\right) = \frac{5+20}{25}$$

$$r_2 = \frac{5+15+5}{25}$$

$$r_1 =$$

$$r_2 =$$

$$r_3 = \frac{5+5+5+5+5}{25}$$

*This formula and others like it on succeeding pages have been worked out in previous similar examples, and therefore are not repeated. They will also be found on the graphs. (Ed.)

DISTRIBUTIVE INVOLUTION GROUPS

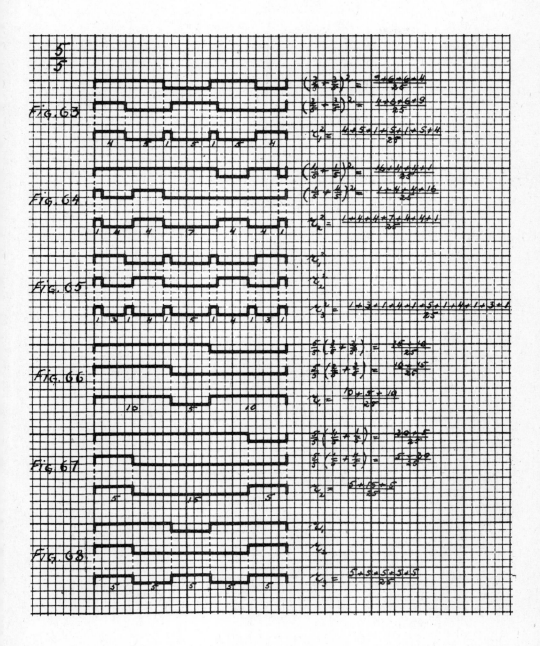

Squares of Binomials

APPENDIX A

$\frac{6}{6}$

$$\left(\frac{5}{6} + \frac{1}{6}\right)^2 = \frac{25 + 5 + 5 + 1}{36}$$

$$\left(\frac{1}{6} + \frac{5}{6}\right)^2 = \frac{1 + 5 + 5 + 25}{36}$$

$$r^2 = \frac{1 + 5 + 5 + 14 + 5 + 5 + 1}{36}$$

$$\frac{6}{6}\left(\frac{5}{6} + \frac{1}{6}\right) = \frac{30 + 6}{36}$$

$$\frac{6}{6}\left(\frac{1}{6} + \frac{5}{6}\right) = \frac{6 + 30}{36}$$

$$r_1 = \frac{6 + 24 + 6}{36}$$

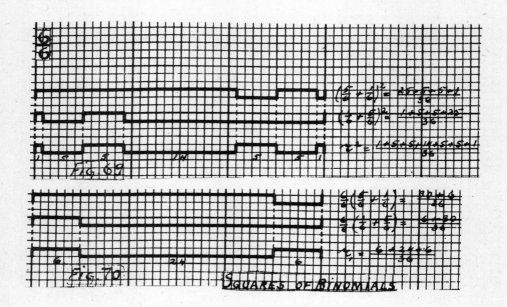

Fig. 69

Fig. 70

SQUARES OF BINOMIALS

$\frac{7}{7}$

$$\left(\frac{4}{7}+\frac{3}{7}\right)^2 = \frac{16+12+12+9}{49}$$

$$\left(\frac{3}{7}+\frac{4}{7}\right)^2 = \frac{9+12+12+16}{49}$$

$$r_1^2 = \frac{9+7+5+7+5+7+9}{49}$$

$$\left(\frac{5}{7}+\frac{2}{7}\right)^2 = \frac{25+10+10+4}{49}$$

$$\left(\frac{2}{7}+\frac{5}{7}\right)^2 = \frac{4+10+10+25}{49}$$

$$r_2^2 = \frac{4+10+10+1+10+10+4}{49}$$

$$\left(\frac{6}{7}+\frac{1}{7}\right)^2 = \frac{36+6+6+1}{49}$$

$$\left(\frac{1}{7}+\frac{6}{7}\right)^2 = \frac{1+6+6+36}{49}$$

$$r_3^2 = \frac{1+6+6+23+6+6+1}{49}$$

$$r_1^2 =$$

$$r_2^2 =$$

$$r_4^2 = \frac{4+5+5+2+5+3+1+3+5+2+5+5+4}{49}$$

$$r_2^2 =$$

$$r_3^2 =$$

$$r_5^2 = \frac{1+3+3+6+1+10+1+10+1+6+3+3+1}{49}$$

$$r_4^2 =$$

$$r_5^2 =$$

$$r_6^2 = \frac{1+3+3+2+4+1+2+5+3+1+3+5+2+1+}{49}$$

$$\frac{+4+2+3+3+1}{49}$$

APPENDIX A

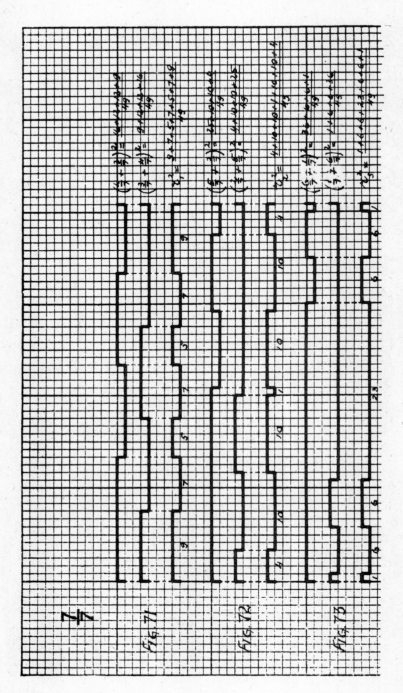

Squares of Binomials

DISTRIBUTIVE INVOLUTION GROUPS

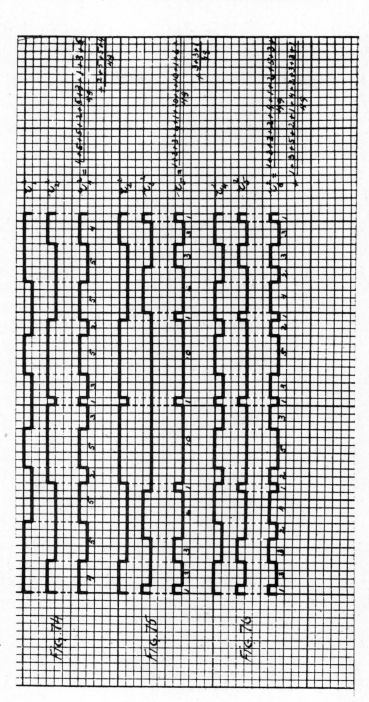

Fig. 74 Fig. 75 Fig. 76

$\frac{7}{7}$

$$\frac{7}{7}\left(\frac{4}{7}+\frac{3}{7}\right) = \frac{28+21}{49}$$

$$\frac{7}{7}\left(\frac{3}{7}+\frac{4}{7}\right) = \frac{21+28}{49}$$

$$r_1 = \frac{21+7+21}{49}$$

$$\frac{7}{7}\left(\frac{5}{7}+\frac{2}{7}\right) = \frac{35+14}{49}$$

$$\frac{7}{7}\left(\frac{2}{7}+\frac{5}{7}\right) = \frac{14+35}{49}$$

$$r_2 = \frac{14+21+14}{49}$$

$$\frac{7}{7}\left(\frac{6}{7}+\frac{1}{7}\right) = \frac{42+7}{49}$$

$$\frac{7}{7}\left(\frac{1}{7}+\frac{6}{7}\right) = \frac{7+42}{49}$$

$$r_3 = \frac{7+35+7}{49}$$

$$r_1 =$$

$$r_2 =$$

$$r_4 = \frac{14+7+7+7+14}{49}$$

$$r_2 =$$

$$r_3 =$$

$$r_5 = \frac{7+7+21+7+7}{49}$$

$$r_4 =$$

$$r_5 =$$

$$r_6 = \frac{7+7+7+7+7+7+7}{49}$$

DISTRIBUTIVE INVOLUTION GROUPS

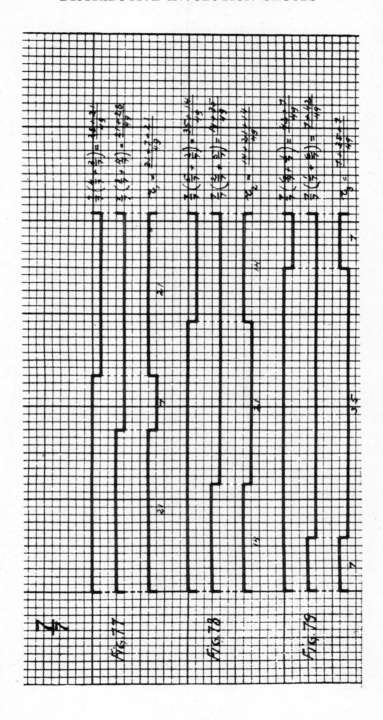

Squares of Binomials

APPENDIX A

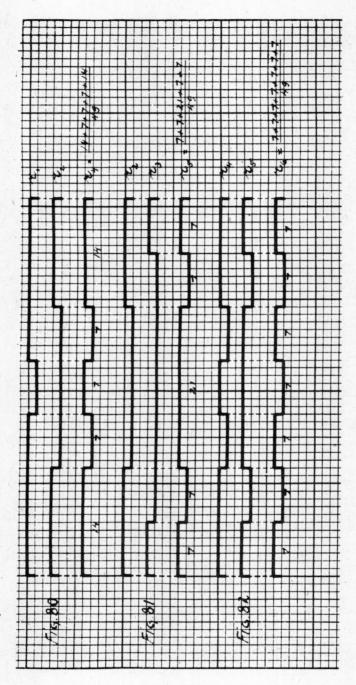

Squares of Binomials

$\frac{8}{8}$

$$\left(\frac{5}{8}+\frac{3}{8}\right)^2 = \frac{25+15+15+9}{64}$$

$$\left(\frac{3}{8}+\frac{5}{8}\right)^2 = \frac{9+15+15+25}{64}$$

$$r_1^2 = \frac{9+15+1+14+1+15+9}{64}$$

$$\left(\frac{7}{8}+\frac{1}{8}\right)^2 = \frac{49+7+7+1}{64}$$

$$\left(\frac{1}{8}+\frac{7}{8}\right)^2 = \frac{1+7+7+49}{64}$$

$$r_2^2 = \frac{1+7+7+34+7+7+1}{64}$$

$$r_1^2 =$$

$$r_2^2 =$$

$$r_3^2 = \frac{1+7+1+6+9+1+14+1+9+6+1+7+1}{64}$$

$$\frac{8}{8}\left(\frac{5}{8}+\frac{3}{8}\right) = \frac{40+24}{64}$$

$$\frac{8}{8}\left(\frac{3}{8}+\frac{5}{8}\right) = \frac{24+40}{64}$$

$$r_1 = \frac{24+16+24}{64}$$

$$\frac{8}{8}\left(\frac{7}{8}+\frac{1}{8}\right) = \frac{56+8}{64}$$

$$\frac{8}{8}\left(\frac{1}{8}+\frac{7}{8}\right) = \frac{8+56}{64}$$

$$r_2 = \frac{8+48+8}{64}$$

$$r_1 =$$

$$r_2 =$$

$$r_3 = \frac{8+16+16+16+8}{64}$$

APPENDIX A

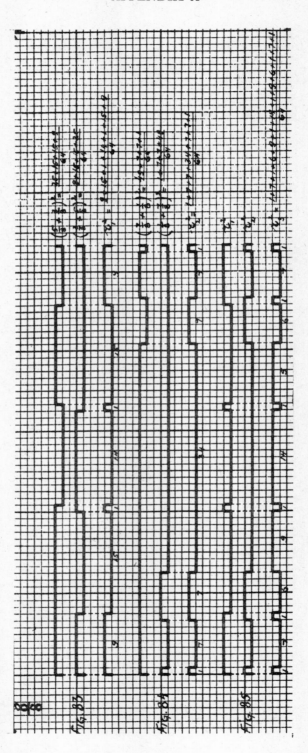

Squares of Binomials

DISTRIBUTIVE INVOLUTION GROUPS 515

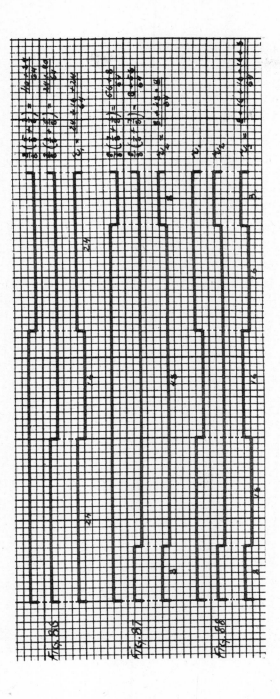

Squares of Binomials

$\frac{9}{9}$

$$\left(\frac{5}{9}+\frac{4}{9}\right)^2 = \frac{25+20+20+16}{81}$$

$$\left(\frac{4}{9}+\frac{5}{9}\right)^2 = \frac{16+20+20+25}{81}$$

$$r_1^2 = \frac{16+9+11+9+11+9+16}{81}$$

$$\left(\frac{7}{9}+\frac{2}{9}\right)^2 = \frac{49+14+14+4}{81}$$

$$\left(\frac{2}{9}+\frac{7}{9}\right)^2 = \frac{4+14+14+49}{81}$$

$$r_2^2 = \frac{4+14+14+17+14+14+4}{81}$$

$$\left(\frac{8}{9}+\frac{1}{9}\right)^2 = \frac{64+8+8+1}{81}$$

$$\left(\frac{1}{9}+\frac{8}{9}\right)^2 = \frac{1+8+8+64}{81}$$

$$r_3^2 = \frac{1+8+8+47+8+8+1}{81}$$

$$r_1^2 =$$

$$r_2^2 =$$

$$r_4^2 = \frac{4+12+2+7+7+4+9+4+7+7+2+12+4}{81}$$

$$r_2^2 =$$

$$r_3^2 =$$

$$r_5^2 = \frac{1+3+5+8+1+14+17+14+1+8+5+3+1}{81}$$

$$r_4^2 =$$

$$r_5^2 =$$

$$r_6^2 = \frac{1+3+5+7+1+1+7+7+4+9+4+7+7+1+}{81}$$

$$\frac{+1+7+5+3+1}{81}$$

DISTRIBUTIVE INVOLUTION GROUPS

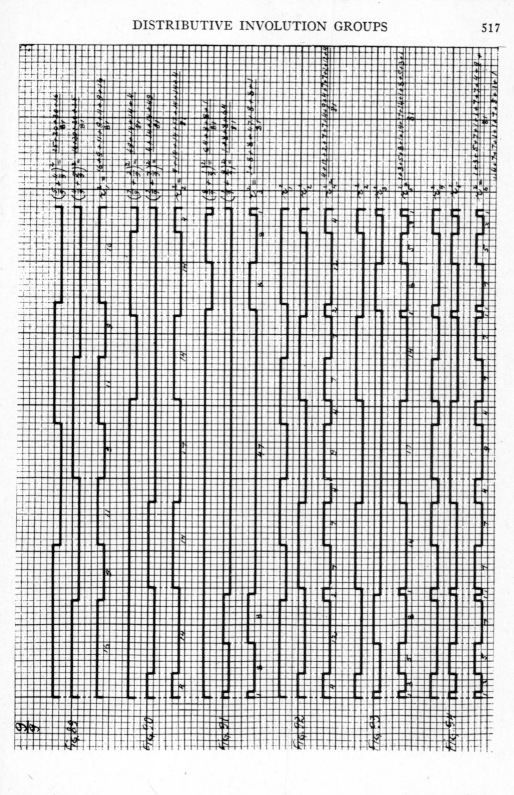

Squares of Binomials

$$\frac{9}{9}\left(\frac{5}{9} + \frac{4}{9}\right) = \frac{45 + 36}{81}$$

$$\frac{9}{9}\left(\frac{4}{9} + \frac{5}{9}\right) = \frac{36 + 45}{81}$$

$$r_1 = \frac{36 + 9 + 36}{81}$$

$$\frac{9}{9}\left(\frac{7}{9} + \frac{2}{9}\right) = \frac{63 + 18}{81}$$

$$\frac{9}{9}\left(\frac{2}{9} + \frac{7}{9}\right) = \frac{18 + 63}{81}$$

$$r_2 = \frac{18 + 45 + 18}{81}$$

$$\frac{9}{9}\left(\frac{8}{9} + \frac{1}{9}\right) = \frac{72 + 9}{81}$$

$$\frac{9}{9}\left(\frac{1}{9} + \frac{8}{9}\right) = \frac{9 + 72}{81}$$

$$r_3 = \frac{9 + 63 + 9}{81}$$

$$r_1 =$$

$$r_2 =$$

$$r_4 = \frac{18 + 18 + 9 + 18 + 18}{81}$$

$$r_2 =$$

$$r_3 =$$

$$r_5 = \frac{9 + 9 + 45 + 9 + 9}{81}$$

$$r_4 =$$

$$r_5 =$$

$$r_6 = \frac{9 + 9 + 18 + 9 + 18 + 9 + 9}{81}$$

DISTRIBUTIVE INVOLUTION GROUPS 519

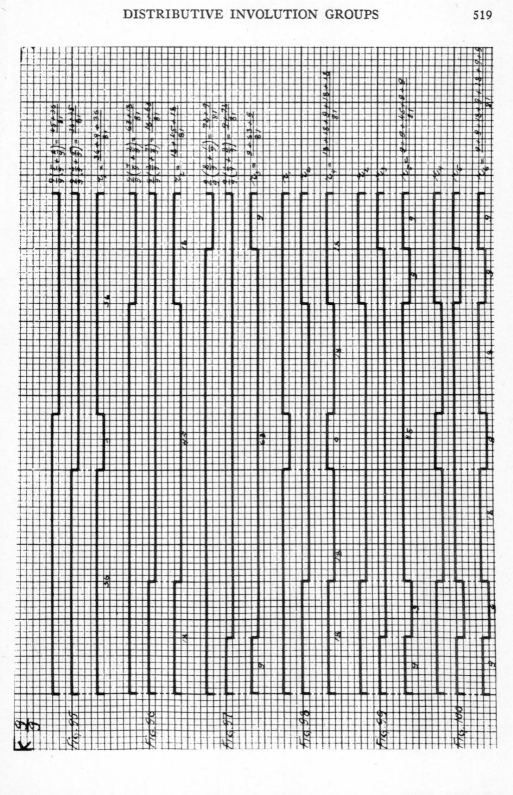

Squares of Binomials

B. Distributive Square of Trinomials.

(1) Factorial: $(a+b+c)^2 = (a^2+ab+ac) + (ab+b^2+bc) + (ac+bc+c^2)$;

(2) Fractional: $\left(\dfrac{a}{a+b+c} + \dfrac{b}{a+b+c} + \dfrac{c}{a+b+c}\right)^2 =$

$= \left[\dfrac{a^2}{(a+b+c)^2} + \dfrac{ab}{(a+b+c)^2} + \dfrac{ac}{(a+b+c)^2}\right] + \left[\dfrac{ab}{(a+b+c)^2} + \dfrac{b^2}{(a+b+c)^2} + \dfrac{bc}{(a+b+c)^2}\right] + \left[\dfrac{ac}{(a+b+c)^2} + \dfrac{bc}{(a+b+c)^2} + \dfrac{c^2}{(a+b+c)^2}\right].$

Synchronization of the first power group with the second power group.

(1) Factorial: $S = a(a+b+c) + b(a+b+c) + c(a+b+c)$;

(2) Fractional: $S = \dfrac{a}{a+b+c} \cdot \left(\dfrac{a+b+c}{a+b+c}\right) + \dfrac{b}{a+b+c} \cdot \left(\dfrac{a+b+c}{a+b+c}\right) +$

$+ \dfrac{c}{a+b+c} \cdot \left(\dfrac{a+b+c}{a+b+c}\right).$

$\dfrac{4}{4}$

$\left(\dfrac{1}{4} + \dfrac{2}{4} + \dfrac{1}{4}\right)^2 = \dfrac{(1+2+1)+(2+4+2)+(1+2+1)}{16}$

$\left(\dfrac{2}{4} + \dfrac{1}{4} + \dfrac{1}{4}\right)^2 = \dfrac{(4+2+2)+(2+1+1)+(2+1+1)}{16}$

$\left(\dfrac{1}{4} + \dfrac{1}{4} + \dfrac{2}{4}\right)^2 = \dfrac{(1+1+2)+(1+1+2)+(2+2+4)}{16}$

$r^2 = \dfrac{1+1+1+1+1+1+2+2+1+1+1+1+1+1}{16}$

$\dfrac{4}{4}\left(\dfrac{1}{4} + \dfrac{2}{4} + \dfrac{1}{4}\right) = \dfrac{4+8+4}{16}$

$\dfrac{4}{4}\left(\dfrac{2}{4} + \dfrac{1}{4} + \dfrac{1}{4}\right) = \dfrac{8+4+4}{16}$

$\dfrac{4}{4}\left(\dfrac{1}{4} + \dfrac{1}{4} + \dfrac{2}{4}\right) = \dfrac{4+4+8}{16}$

$r = \dfrac{4+4+4+4}{16}$

DISTRIBUTIVE INVOLUTION GROUPS

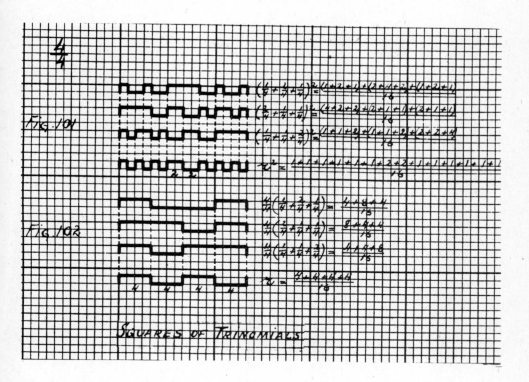

$\dfrac{5}{5}$

$$\left(\dfrac{2}{5}+\dfrac{1}{5}+\dfrac{2}{5}\right)^2 = \dfrac{(4+2+4)+(2+1+2)+(4+2+4)}{25}$$

$$\left(\dfrac{1}{5}+\dfrac{2}{5}+\dfrac{2}{5}\right)^2 = \dfrac{(1+2+2)+(2+4+4)+(2+4+4)}{25}$$

$$\left(\dfrac{2}{5}+\dfrac{2}{5}+\dfrac{1}{5}\right)^2 = \dfrac{(4+4+2)+(4+4+2)+(2+2+1)}{25}$$

$$r_1^2 = \dfrac{1+2+5(1)+2+5(1)+2+5(1)+2+1}{25}$$

$$\left(\dfrac{1}{5}+\dfrac{3}{5}+\dfrac{1}{5}\right)^2 = \dfrac{(1+3+1)+(3+9+3)+(1+3+1)}{25}$$

$$\left(\dfrac{3}{5}+\dfrac{1}{5}+\dfrac{1}{5}\right)^2 = \dfrac{(9+3+3)+(3+1+1)+(3+1+1)}{25}$$

$$\left(\frac{1}{5}+\frac{1}{5}+\frac{3}{5}\right)^2 = \frac{(1+1+3)+(1+1+3)+(3+3+9)}{25}$$

$$r_2^2 = \frac{1+1+2+6(1)+2+1+2+6(1)+2+1+1}{25}$$

$$r_1^2 =$$

$$r_2^2 =$$

$$r_3^2 = \frac{25(1)}{25}$$

$$\frac{5}{5}\left(\frac{2}{5}+\frac{1}{5}+\frac{2}{5}\right) = \frac{10+5+10}{25}$$

$$\frac{5}{5}\left(\frac{1}{5}+\frac{2}{5}+\frac{2}{5}\right) = \frac{5+10+10}{25}$$

$$\frac{5}{5}\left(\frac{2}{5}+\frac{2}{5}+\frac{1}{5}\right) = \frac{10+10+5}{25}$$

$$r_1 = \frac{5+5+5+5+5}{25}$$

$$\frac{5}{5}\left(\frac{1}{5}+\frac{3}{5}+\frac{1}{5}\right) = \frac{5+15+5}{25}$$

$$\frac{5}{5}\left(\frac{3}{5}+\frac{1}{5}+\frac{1}{5}\right) = \frac{15+5+5}{25}$$

$$\frac{5}{5}\left(\frac{1}{5}+\frac{1}{5}+\frac{3}{5}\right) = \frac{5+5+15}{25}$$

$$r_2 = \frac{5+5+5+5+5}{25}$$

$$r_3 = r_2 = r_1$$

DISTRIBUTIVE INVOLUTION GROUPS

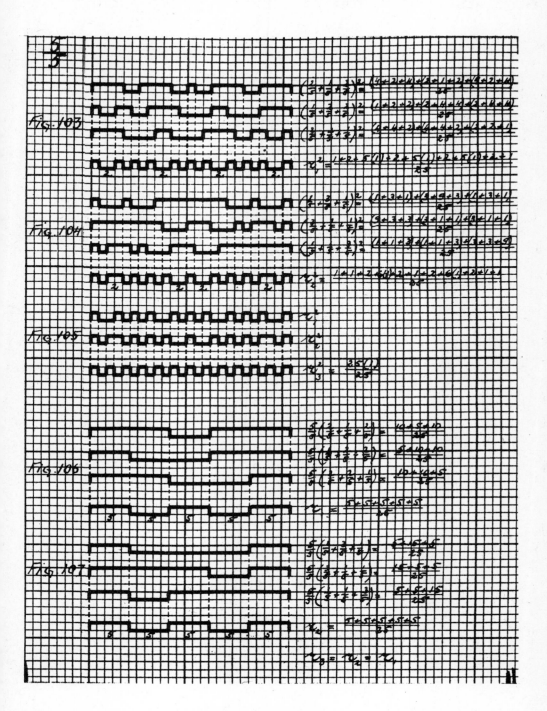

Squares of Trinomials

$$\frac{6}{6}$$

$$\left(\frac{1}{6}+\frac{4}{6}+\frac{1}{6}\right)^2 = \frac{(1+4+1)+(4+16+4)+(1+4+1)}{36}$$

$$\left(\frac{4}{6}+\frac{1}{6}+\frac{1}{6}\right)^2 = \frac{(16+4+4)+(4+1+1)+(4+1+1)}{36}$$

$$\left(\frac{1}{6}+\frac{1}{6}+\frac{4}{6}\right)^2 = \frac{(1+1+4)+(1+1+4)+(4+4+16)}{36}$$

$$r^2 = \frac{1+1+3+3(1)+2+2+4+4+4+2+2+}{36}$$

$$\frac{+3(1)+3+1+1}{36}$$

$$\frac{6}{6}\left(\frac{1}{6}+\frac{4}{6}+\frac{1}{6}\right) = \frac{6+24+6}{36}$$

$$\frac{6}{6}\left(\frac{4}{6}+\frac{1}{6}+\frac{1}{6}\right) = \frac{24+6+6}{36}$$

$$\frac{6}{6}\left(\frac{1}{6}+\frac{1}{6}+\frac{4}{6}\right) = \frac{6+6+24}{36}.$$

$$r = \frac{6+6+12+6+2}{36}$$

DISTRIBUTIVE INVOLUTION GROUPS

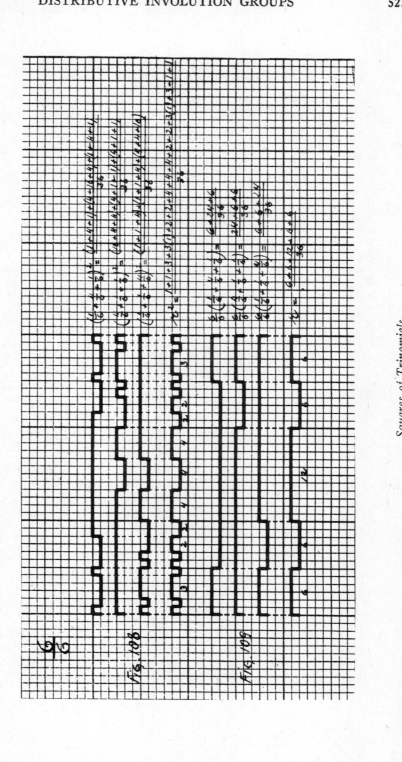

Squares of Trinomials

$\frac{7}{7}$

$$\left(\frac{3}{7}+\frac{1}{7}+\frac{3}{7}\right)^2 = \frac{(9+3+9)+(3+1+3)+(9+3+9)}{49}$$

$$\left(\frac{1}{7}+\frac{3}{7}+\frac{3}{7}\right)^2 = \frac{(1+3+3)+(3+9+9)+(3+9+9)}{49}$$

$$\left(\frac{3}{7}+\frac{3}{7}+\frac{1}{7}\right)^2 = \frac{(9+9+3)+(9+9+3)+(3+3+1)}{49}$$

$$r_1^2 = \frac{1+3+3+2+1+2+6+1+2+3+1+}{49}$$

$$\frac{+3+2+1+6+2+1+2+3+3+1}{49}$$

$$\left(\frac{2}{7}+\frac{3}{7}+\frac{2}{7}\right)^2 = \frac{(4+6+4)+(6+9+6)+(4+6+4)}{49}$$

$$\left(\frac{3}{7}+\frac{2}{7}+\frac{2}{7}\right)^2 = \frac{(9+6+6)+(6+4+4)+(6+4+4)}{49}$$

$$\left(\frac{2}{7}+\frac{2}{7}+\frac{3}{7}\right)^2 = \frac{(4+4+6)+(4+4+6)+(6+6+9)}{49}$$

$$r_2^2 = \frac{4+4+1+1+4+1+3+2+1+1+5+}{49}$$

$$\frac{+1+1+2+3+1+4+1+1+4+4}{49}$$

$$\left(\frac{1}{7}+\frac{5}{7}+\frac{1}{7}\right)^2 = \frac{(1+5+1)+(5+25+5)+(1+5+1)}{49}$$

$$\left(\frac{5}{7}+\frac{1}{7}+\frac{1}{7}\right)^2 = \frac{(25+5+5)+(5+1+1)+(5+1+1)}{49}$$

$$\left(\frac{1}{7}+\frac{1}{7}+\frac{5}{7}\right)^2 = \frac{(1+1+5)+(1+1+5)+(5+5+25)}{49}$$

$$r_3^2 = \frac{1+1+4+1+1+1+3+2+5+5+1+}{49}$$

$$\frac{+5+5+2+3+1+1+1+4+1+1}{49}$$

DISTRIBUTIVE INVOLUTION GROUPS 527

$r_1^2 =$

$r_2^2 =$

$r_4^2 = \dfrac{1 + 3 + 3 + 3(1) + 2 + 2 + 1 + 3 + 4(1) + 2 + 1 +}{49}$
$\dfrac{+ 2 + 4(1) + 3 + 1 + 2 + 2 + 3(1) + 3 + 3 + 1}{49}$

$r_2^2 =$

$r_3^2 =$

$r_5^2 = \dfrac{1 + 1 + 2 + 2 + 4(1) + 2 + 2 + 1 + 3 + 4(1) + 2 + 1 +}{49}$
$\dfrac{+ 2 + 4(1) + 3 + 1 + 2 + 2 + 4(1) + 2 + 2 + 1 + 1}{49}$

$r_4^2 =$

$r_5^2 =$

$r_6^2 = \dfrac{1 + 1 + 2 + 2 + 4(1) + 2 + 2 + 1 + 3 + 4(1) + 2 + 1 +}{49}$
$\dfrac{+ 2 + 4(1) + 3 + 1 + 2 + 2 + 4(1) + 2 + 2 + 1 + 1}{49}$

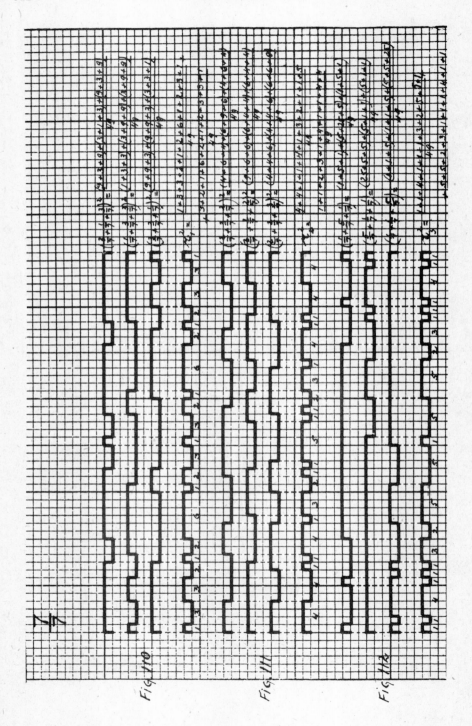

Squares of Trinomials

Fig. 110

Fig. 111

Fig. 112

DISTRIBUTIVE INVOLUTION GROUPS

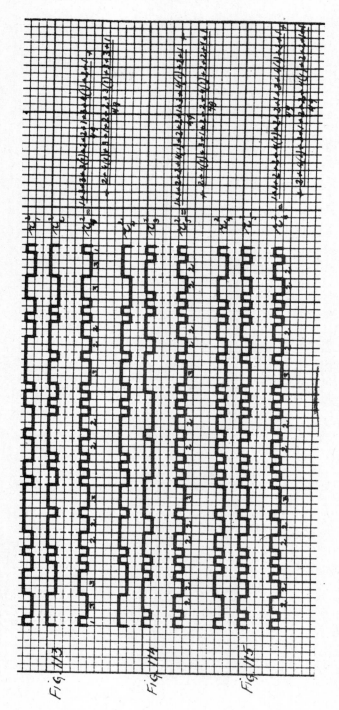

Squares of Trinomials

$$\frac{7}{7}$$

$$\frac{7}{7}\left(\frac{3}{7} + \frac{1}{7} + \frac{3}{7}\right) = \frac{21 + 7 + 21}{49}$$

$$\frac{7}{7}\left(\frac{1}{7} + \frac{3}{7} + \frac{3}{7}\right) = \frac{7 + 21 + 21}{49}$$

$$\frac{7}{7}\left(\frac{3}{7} + \frac{3}{7} + \frac{1}{7}\right) = \frac{21 + 21 + 7}{49}$$

$$r_1 = \frac{7 + 14 + 7 + 14 + 7}{49}$$

$$\frac{7}{7}\left(\frac{2}{7} + \frac{3}{7} + \frac{2}{7}\right) = \frac{14 + 21 + 14}{49}$$

$$\frac{7}{7}\left(\frac{3}{7} + \frac{2}{7} + \frac{2}{7}\right) = \frac{21 + 14 + 14}{49}$$

$$\frac{7}{7}\left(\frac{2}{7} + \frac{2}{7} + \frac{3}{7}\right) = \frac{14 + 14 + 21}{49}$$

$$r_2 = \frac{14 + 7 + 7 + 7 + 14}{49}$$

$$\frac{7}{7}\left(\frac{1}{7} + \frac{5}{7} + \frac{1}{7}\right) = \frac{7 + 35 + 7}{49}$$

$$\frac{7}{7}\left(\frac{5}{7} + \frac{1}{7} + \frac{1}{7}\right) = \frac{35 + 7 + 7}{49}$$

$$\frac{7}{7}\left(\frac{1}{7} + \frac{1}{7} + \frac{5}{7}\right) = \frac{7 + 7 + 35}{49}$$

$$r_3 = \frac{7 + 7 + 21 + 7 + 7}{49}$$

$$r_1 =$$

$$r_2 =$$

$$r_4 = \frac{7 + 7 + 7 + 7 + 7 + 7 + 7}{49}$$

$$r_2 =$$

$$r_3 =$$

$$r_5 = \frac{7 + 7 + 7 + 7 + 7 + 7 + 7}{49}$$

$$r_6 = r_5 = r_4$$

DISTRIBUTIVE INVOLUTION GROUPS

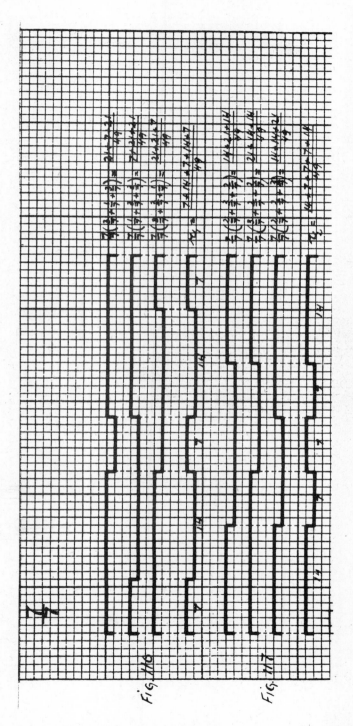

Squares of Trinomials

APPENDIX A

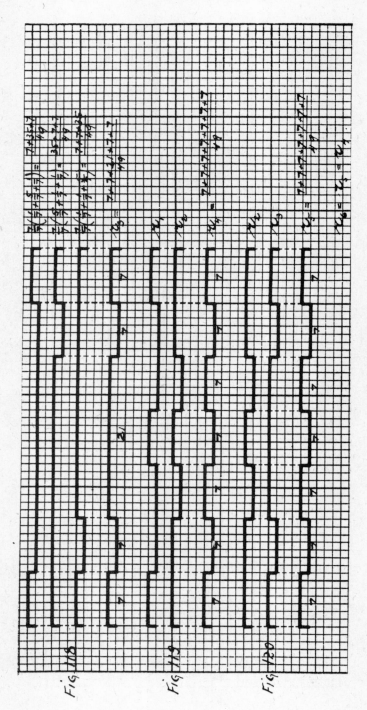

Squares of Trinomials

Fig. 118, Fig. 119, Fig. 120

$\frac{8}{8}$

$$\left(\frac{3}{8}+\frac{2}{8}+\frac{3}{8}\right)^2 = \frac{(9+6+9)+(6+4+6)+(9+6+9)}{64}$$

$$\left(\frac{2}{8}+\frac{3}{8}+\frac{3}{8}\right)^2 = \frac{(4+6+6)+(6+9+9)+(6+9+9)}{64}$$

$$\left(\frac{3}{8}+\frac{3}{8}+\frac{2}{8}\right)^2 = \frac{(9+9+6)+(9+9+6)+(6+6+4)}{64}$$

$$r_1^2 = \frac{4+5+1+5+1+2+4+2+6+1+2+1+}{64}$$

$$\frac{+6+2+4+2+1+5+1+5+4}{64}$$

$$\left(\frac{1}{8}+\frac{6}{8}+\frac{1}{8}\right)^2 = \frac{(1+6+1)+(6+36+6)+(1+6+1)}{64}$$

$$\left(\frac{6}{8}+\frac{1}{8}+\frac{1}{8}\right)^2 = \frac{(36+6+6)+(6+1+1)+(6+1+1)}{64}$$

$$\left(\frac{1}{8}+\frac{1}{8}+\frac{6}{8}\right)^2 = \frac{(1+1+6)+(1+1+6)+(6+6+36)}{64}$$

$$r_2^2 = \frac{1+1+5+1+1+1+4+2+6+6+8+}{64}$$

$$\frac{+6+6+2+4+1+1+1+5+1+1}{64}$$

$r_1^2 =$

$r_2^2 =$

$$r_3^2 = \frac{1+1+2+3+1+1+1+4+1+1+2+4+2+}{64}$$

$$\frac{+4+2+1+2+1+2+4+2+4+2+1+1+}{64}$$

$$\frac{+4+1+1+1+3+2+1+1}{64}$$

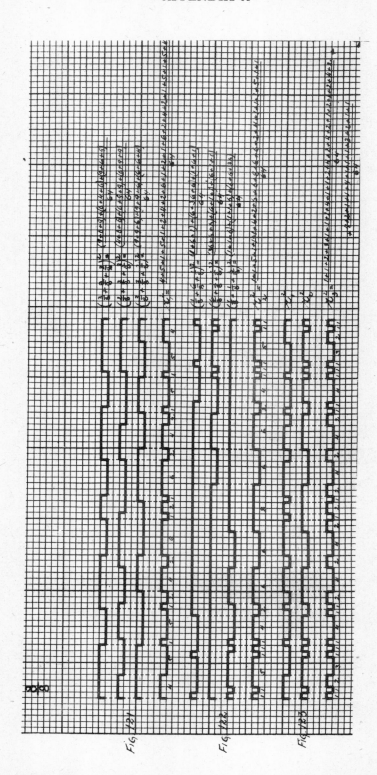

Squares of Trinomials

$\frac{8}{8}$

$$\frac{8}{8}\left(\frac{3}{8}+\frac{2}{8}+\frac{3}{8}\right) = \frac{24+16+24}{64}$$

$$\frac{8}{8}\left(\frac{2}{8}+\frac{3}{8}+\frac{3}{8}\right) = \frac{16+24+24}{64}$$

$$\frac{8}{8}\left(\frac{3}{8}+\frac{3}{8}+\frac{2}{8}\right) = \frac{24+24+16}{64}$$

$$r_1 = \frac{16+8+16+8+16}{64}$$

$$\frac{8}{8}\left(\frac{1}{8}+\frac{6}{8}+\frac{1}{8}\right) = \frac{8+48+8}{64}$$

$$\frac{8}{8}\left(\frac{6}{8}+\frac{1}{8}+\frac{1}{8}\right) = \frac{48+8+8}{64}$$

$$\frac{8}{8}\left(\frac{1}{8}+\frac{1}{8}+\frac{6}{8}\right) = \frac{8+8+48}{64}$$

$$r_2 = \frac{8+8+32+8+8}{64}$$

$$r_1 =$$

$$r_2 =$$

$$r_3 = \frac{8+8+8+16+8+8+8}{64}$$

APPENDIX A

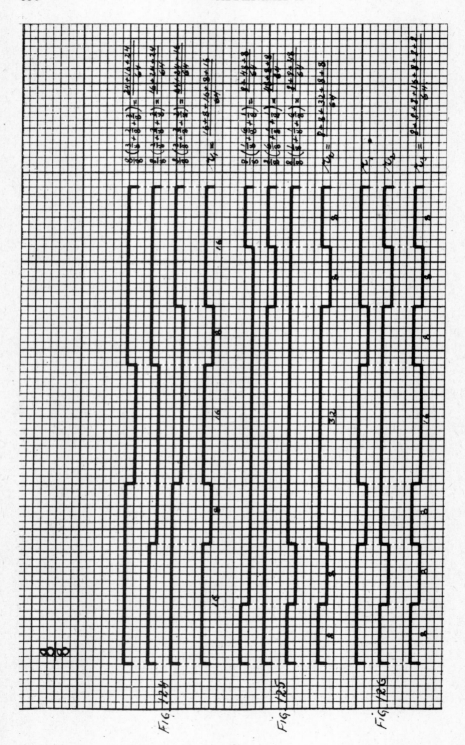

Squares of Trinomials

$\frac{9}{9}$

$$\left(\frac{4}{9}+\frac{1}{9}+\frac{4}{9}\right)^2 = \frac{(16+4+16)+(4+1+4)+(16+4+16)}{81}$$

$$\left(\frac{1}{9}+\frac{4}{9}+\frac{4}{9}\right)^2 = \frac{(1+4+4)+(4+16+16)+(4+16+16)}{81}$$

$$\left(\frac{4}{9}+\frac{4}{9}+\frac{1}{9}\right)^2 = \frac{(16+16+4)+(16+16+4)+(4+4+1)}{81}$$

$$r_1^2 = \frac{1+4+4+4+3+4+9+3+4+4+1+4+4+}{81}$$

$$\frac{+3+9+4+3+4+4+1}{81}$$

$$\left(\frac{2}{9}+\frac{5}{9}+\frac{2}{9}\right)^2 = \frac{(4+10+4)+(10+25+10)+(4+10+4)}{81}$$

$$\left(\frac{5}{9}+\frac{2}{9}+\frac{2}{9}\right)^2 = \frac{(25+10+10)+(10+4+4)+(10+4+4)}{81}$$

$$\left(\frac{2}{9}+\frac{2}{9}+\frac{5}{9}\right)^2 = \frac{(4+4+10)+(4+4+10)+(10+10+25)}{81}$$

$$r_2^2 = \frac{4+4+6+4+4+3+1+2+7+1+9+1+7+}{81}$$

$$\frac{+2+1+3+4+4+6+4+4}{81}$$

$$\left(\frac{1}{9}+\frac{7}{9}+\frac{1}{9}\right)^2 = \frac{(1+7+1)+(7+49+7)+(1+7+1)}{81}$$

$$\left(\frac{7}{9}+\frac{1}{9}+\frac{1}{9}\right)^2 = \frac{(49+7+7)+(7+1+1)+(7+1+1)}{81}$$

$$\left(\frac{1}{9}+\frac{1}{9}+\frac{7}{9}\right)^2 = \frac{(1+1+7)+(1+1+7)+(7+7+49)}{81}$$

$$r_3^2 = \frac{1+1+6+1+1+1+5+2+7+7+17+7+7+}{81}$$

$$\frac{+2+5+1+1+1+6+1+1}{81}$$

$r_1^2 =$

$r_2^2 =$

$$r_4^2 = \frac{1+3+1+3+1+4+1+2+2+2+2+3+1+}{81}$$

$$\frac{+2+1+3+3+1+4+1+4+1+3+3+1+2+}{81}$$

$$\frac{+1+3+2+2+2+2+1+4+1+3+1+3+1}{81}$$

$r_2^2 =$

$r_3^2 =$

$$r_5^2 = \frac{1+1+2+4+1+1+1+3+2+2+4+3+1+2+}{81}$$

$$\frac{+4+3+1+9+1+3+4+2+1+3+4+2+2+}{81}$$

$$\frac{+3+1+1+1+4+2+1+1}{81}$$

$r_4^2 =$

$r_5^2 =$

$$r_6^2 = \frac{1+1+2+1+3+1+1+1+2+1+2+2+2+2+}{81}$$

$$\frac{+3+1+2+1+3+3+1+4+1+4+1+3+}{81}$$

$$\frac{+3+1+2+1+3+2+2+2+2+1+2+1+}{81}$$

$$\frac{+1+1+3+1+2+1+1}{81}$$

DISTRIBUTIVE INVOLUTION GROUPS 539

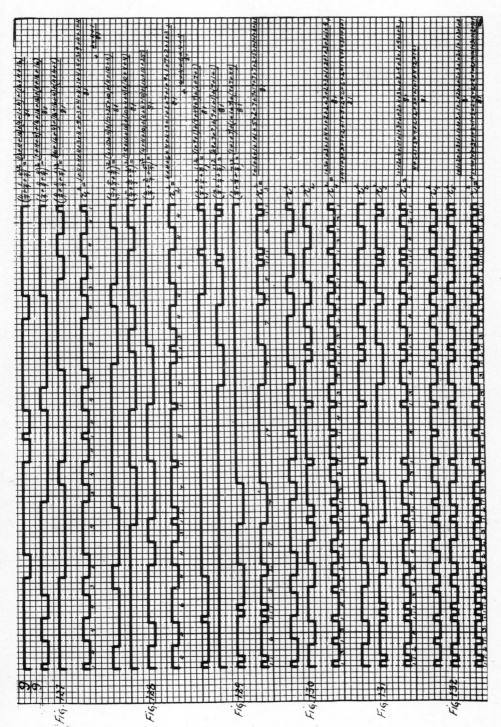

Squares of Trinomials

$$\frac{9}{9}\left(\frac{4}{9} + \frac{1}{9} + \frac{4}{9}\right) = \frac{36 + 9 + 36}{81}$$

$$\frac{9}{9}\left(\frac{1}{9} + \frac{4}{9} + \frac{4}{9}\right) = \frac{9 + 36 + 36}{81}$$

$$\frac{9}{9}\left(\frac{4}{9} + \frac{4}{9} + \frac{1}{9}\right) = \frac{36 + 36 + 9}{81}$$

$$r_1 = \frac{9 + 27 + 9 + 27 + 9}{81}$$

$$\frac{9}{9}\left(\frac{2}{9} + \frac{5}{9} + \frac{2}{9}\right) = \frac{18 + 45 + 18}{81}$$

$$\frac{9}{9}\left(\frac{5}{9} + \frac{2}{9} + \frac{2}{9}\right) = \frac{45 + 18 + 18}{81}$$

$$\frac{9}{9}\left(\frac{2}{9} + \frac{2}{9} + \frac{5}{9}\right) = \frac{18 + 18 + 45}{81}$$

$$r_2 = \frac{18 + 18 + 9 + 18 + 18}{81}$$

$$\frac{9}{9}\left(\frac{1}{9} + \frac{7}{9} + \frac{1}{9}\right) = \frac{9 + 63 + 9}{81}$$

$$\frac{9}{9}\left(\frac{7}{9} + \frac{1}{9} + \frac{1}{9}\right) = \frac{63 + 9 + 9}{81}$$

$$\frac{9}{9}\left(\frac{1}{9} + \frac{1}{9} + \frac{7}{9}\right) = \frac{9 + 9 + 63}{81}$$

$$r_3 = \frac{9 + 9 + 45 + 9 + 9}{81}$$

$$r_1 =$$

$$r_2 =$$

$$r_4 = \frac{9 + 9 + 18 + 9 + 18 + 9 + 9}{81}$$

$$r_2 =$$

$$r_3 =$$

$$r_5 = \frac{9 + 9 + 18 + 9 + 18 + 9 + 9}{81}$$

$$r_6 = r_5 = r_4$$

DISTRIBUTIVE INVOLUTION GROUPS

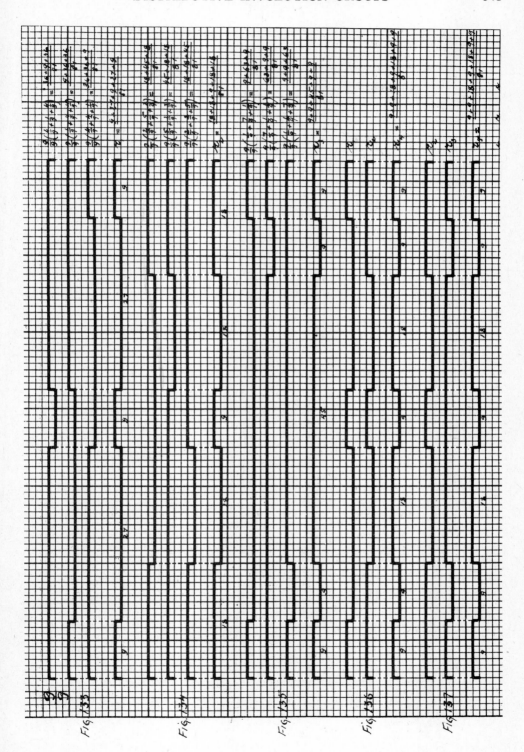

Squares of Trinomials

542   APPENDIX A

C. GENERALIZATION: DISTRIBUTIVE SQUARE OF POLYNOMIALS (N TERMS).

(1) Factorial: $(a + b + c + \ldots + m)^2 = (a^2 + ab + ac + \ldots + am) +$
$+ (ab + b^2 + bc + \ldots + bm) + (ac + bc + c^2 + \ldots + cm) + \ldots$
$\ldots + (am + bm + cm + \ldots + m^2);$

(2) Fractional: $\left(\dfrac{a}{a+b+c+\ldots+m} + \dfrac{b}{a+b+c+\ldots+m} + \right.$

$\left. + \dfrac{c}{a+b+c+\ldots+m} + \ldots + \dfrac{m}{a+b+c+\ldots+m} \right)^2 =$

$= \left[\dfrac{a^2}{(a+b+c+\ldots+m)^2} + \dfrac{ab}{(a+b+c+\ldots+m)^2} + \right.$

$\left. + \dfrac{ac}{(a+b+c+\ldots+m)^2} + \ldots + \dfrac{am}{(a+b+c+\ldots+m)^2} \right] +$

$+ \left[\dfrac{ab}{(a+b+c+\ldots+m)^2} + \dfrac{b^2}{(a+b+c+\ldots+m)^2} + \right.$

$\left. + \dfrac{bc}{(a+b+c+\ldots+m)^2} + \ldots + \dfrac{bm}{(a+b+c+\ldots+m)^2} \right] +$

$+ \left[\dfrac{ac}{(a+b+c+\ldots+m)^2} + \dfrac{bc}{(a+b+c+\ldots+m)^2} + \right.$

$\left. + \dfrac{c^2}{(a+b+c+\ldots+m)^2} + \ldots + \dfrac{cm}{(a+b+c+\ldots+m)^2} \right] + \ldots$

$\ldots + \left[\dfrac{am}{(a+b+c+\ldots+m)^2} + \dfrac{bm}{(a+b+c+\ldots+m)^2} + \right.$

$\left. + \dfrac{cm}{(a+b+c+\ldots+m)^2} + \ldots + \dfrac{m^2}{(a+b+c+\ldots+m)^2} \right].$

Synchronization of the first power group with the second power group.

(1) Factorial: $S = a(a+b+c+\ldots+m) + b(a+b+c+\ldots+m) +$
$+ c(a+b+c+\ldots+m) + \ldots + m(a+b+c+\ldots+m);$

(2) Fractional: $S = \dfrac{a}{a+b+c+\ldots+m} \cdot \left(\dfrac{a+b+c+\ldots+m}{a+b+c+\ldots+m} \right) +$

$+ \dfrac{b}{a+b+c+\ldots+m} \cdot \left(\dfrac{a+b+c+\ldots+m}{a+b+c+\ldots+m} \right) + \dfrac{c}{a+b+c+\ldots+m} \cdot$

$\cdot \left(\dfrac{a+b+c+\ldots+m}{a+b+c+\ldots+m} \right) + \ldots + \dfrac{m}{a+b+c+\ldots+m} \cdot \left(\dfrac{a+b+c+\ldots+m}{a+b+c+\ldots+m} \right).$

DISTRIBUTIVE INVOLUTION GROUPS

1. Squares of Quintinomials

$\frac{6}{6}$

$$\left(\frac{1}{6}+\frac{1}{6}+\frac{2}{6}+\frac{1}{6}+\frac{1}{6}\right)^2 = \frac{(1+1+2+1+1)+(1+1+2+1+1)+(2+2+4+2+2)+}{36}$$
$$\frac{+(1+1+2+1+1)+(1+1+2+1+1)}{36}$$

$$\left(\frac{1}{6}+\frac{2}{6}+\frac{1}{6}+\frac{1}{6}+\frac{1}{6}\right)^2 = \frac{(1+2+1+1+1)+(2+4+2+2+2)+(1+2+1+1+1)+}{36}$$
$$\frac{+(1+2+1+1+1)+(1+2+1+1+1)}{36}$$

$$\left(\frac{2}{6}+\frac{1}{6}+\frac{1}{6}+\frac{1}{6}+\frac{1}{6}\right)^2 = \frac{(4+2+2+2+2)+(2+1+1+1+1)+(2+1+1+1+1)+}{36}$$
$$\frac{+(2+1+1+1+1)+(2+1+1+1+1)}{36}$$

$$\left(\frac{1}{6}+\frac{1}{6}+\frac{1}{6}+\frac{1}{6}+\frac{2}{6}\right)^2 = \frac{(1+1+1+1+2)+(1+1+1+1+2)+(1+1+1+1+2)+}{36}$$
$$\frac{+(1+1+1+1+2)+(2+2+2+2+4)}{36}$$

$$\left(\frac{1}{6}+\frac{1}{6}+\frac{1}{6}+\frac{2}{6}+\frac{1}{6}\right)^2 = \frac{(1+1+1+2+1)+(1+1+1+2+1)+(1+1+1+2+1)+}{36}$$
$$\frac{+(2+2+2+4+2)+(1+1+1+2+1)}{36}$$

$$r^2 = \frac{36(1)}{36}$$

$$\frac{6}{6}\left(\frac{1}{6}+\frac{1}{6}+\frac{2}{6}+\frac{1}{6}+\frac{1}{6}\right) = \frac{6+6+12+6+6}{36}$$

$$\frac{6}{6}\left(\frac{1}{6}+\frac{2}{6}+\frac{1}{6}+\frac{1}{6}+\frac{1}{6}\right) = \frac{6+12+6+6+6}{36}$$

$$\frac{6}{6}\left(\frac{2}{6}+\frac{1}{6}+\frac{1}{6}+\frac{1}{6}+\frac{1}{6}\right) = \frac{12+6+6+6+6}{36}$$

$$\frac{6}{6}\left(\frac{1}{6}+\frac{1}{6}+\frac{1}{6}+\frac{1}{6}+\frac{2}{6}\right) = \frac{6+6+6+6+12}{36}$$

$$\frac{6}{6}\left(\frac{1}{6}+\frac{1}{6}+\frac{1}{6}+\frac{2}{6}+\frac{1}{6}\right) = \frac{6+6+6+12+6}{36}$$

$$r = \frac{6+6+6+6+6+6}{36}$$

DISTRIBUTIVE INVOLUTION GROUPS

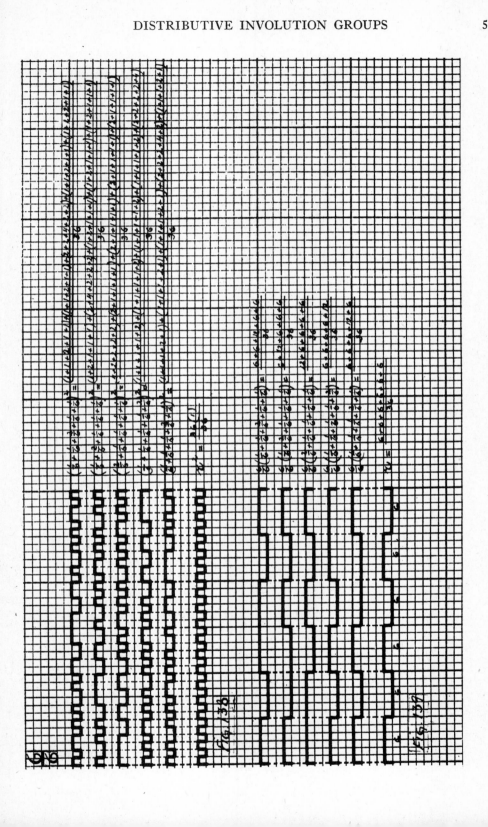

Squares of Quininomials

$\frac{7}{7}$

$$\left(\frac{1}{7}+\frac{2}{7}+\frac{1}{7}+\frac{2}{7}+\frac{1}{7}\right)^2 = \frac{(1+2+1+2+1)+(2+4+2+4+2)+(1+2+1+2+1)+}{49}$$
$$\frac{+\ (2+4+2+4+2)+(1+2+1+2+1)}{49}$$

$$\left(\frac{2}{7}+\frac{1}{7}+\frac{2}{7}+\frac{1}{7}+\frac{1}{7}\right)^2 = \frac{(4+2+4+2+2)+(2+1+2+1+1)+(4+2+4+2+2)+}{49}$$
$$\frac{+\ (2+1+2+1+1)+(2+1+2+1+1)}{49}$$

$$\left(\frac{1}{7}+\frac{2}{7}+\frac{1}{7}+\frac{1}{7}+\frac{2}{7}\right)^2 = \frac{(1+2+1+1+2)+(2+4+2+2+4)+(1+2+1+1+2)+}{49}$$
$$\frac{+\ (1+2+1+1+2)+(2+4+2+2+4)}{49}$$

$$\left(\frac{2}{7}+\frac{1}{7}+\frac{1}{7}+\frac{2}{7}+\frac{1}{7}\right)^2 = \frac{(4+2+2+4+2)+(2+1+1+2+1)+(2+1+1+2+1)+}{49}$$
$$\frac{+\ (4+2+2+4+2)+(2+1+1+2+1)}{49}$$

$$r_1^2 = \frac{49(1)}{49}$$

$$\left(\frac{2}{7}+\frac{1}{7}+\frac{1}{7}+\frac{1}{7}+\frac{2}{7}\right)^2 = \frac{(4+2+2+2+4)+(2+1+1+1+2)+(2+1+1+1+2)+}{49}$$
$$\frac{+\ (2+1+1+1+2)+(4+2+2+2+4)}{49}$$

$$\left(\frac{1}{7}+\frac{1}{7}+\frac{1}{7}+\frac{2}{7}+\frac{2}{7}\right)^2 = \frac{(1+1+1+2+2)+(1+1+1+2+2)+(1+1+1+2+2)+}{49}$$
$$\frac{+\ (2+2+2+4+4)+(2+2+2+4+4)}{49}$$

$$\left(\frac{1}{7}+\frac{1}{7}+\frac{2}{7}+\frac{2}{7}+\frac{1}{7}\right)^2 = \frac{(1+1+2+2+1)+(1+1+2+2+1)+(2+2+4+4+2)+}{49}$$
$$\frac{+\ (2+2+4+4+2)+(1+1+2+2+1)}{49}$$

$$\left(\frac{1}{7}+\frac{2}{7}+\frac{2}{7}+\frac{1}{7}+\frac{1}{7}\right)^2 = \frac{(1+2+2+1+1)+(2+4+4+2+2)+(2+4+4+2+2)+}{49}$$
$$\frac{+\ (1+2+2+1+1)+(1+2+2+1+1)}{49}$$

DISTRIBUTIVE INVOLUTION GROUPS

$$\left(\frac{2}{7}+\frac{2}{7}+\frac{1}{7}+\frac{1}{7}+\frac{1}{7}\right)^2 = \frac{(4+4+2+2+2)+(4+4+2+2+2)+(2+2+1+1+1)+}{49}$$
$$\frac{+(2+2+1+1+1)+(2+2+1+1+1)}{49}$$

$$r_2^2 = \frac{19(1)+2+3(1)+1+3(1)+2+19(1)}{49}$$

$$\left(\frac{1}{7}+\frac{1}{7}+\frac{3}{7}+\frac{1}{7}+\frac{1}{7}\right)^2 = \frac{(1+1+3+1+1)+(1+1+3+1+1)+(3+3+9+3+3)+}{49}$$
$$\frac{+(1+1+3+1+1)+(1+1+3+1+1)}{49}$$

$$\left(\frac{1}{7}+\frac{3}{7}+\frac{1}{7}+\frac{1}{7}+\frac{1}{7}\right)^2 = \frac{(1+3+1+1+1)+(3+9+3+3+3)+(1+3+1+1+1)+}{49}$$
$$\frac{+(1+3+1+1+1)+(1+3+1+1+1)}{49}$$

$$\left(\frac{3}{7}+\frac{1}{7}+\frac{1}{7}+\frac{1}{7}+\frac{1}{7}\right)^2 = \frac{(9+3+3+3+3)+(3+1+1+1+1)+(3+1+1+1+1)+}{49}$$
$$\frac{+(3+1+1+1+1)+(3+1+1+1+1)}{49}$$

$$\left(\frac{1}{7}+\frac{1}{7}+\frac{1}{7}+\frac{1}{7}+\frac{3}{7}\right)^2 = \frac{(1+1+1+1+3)+(1+1+1+1+3)+(1+1+1+1+3)+}{49}$$
$$\frac{+(1+1+1+1+3)+(3+3+3+3+9)}{49}$$

$$\left(\frac{1}{7}+\frac{1}{7}+\frac{1}{7}+\frac{3}{7}+\frac{1}{7}\right)^2 = \frac{(1+1+1+3+1)+(1+1+1+3+1)+(1+1+1+3+1)+}{49}$$
$$\frac{+(3+3+3+9+3)+(1+1+1+3+1)}{49}$$

$$r_3^2 = \frac{49(1)}{49}$$

$$r_1^2 =$$

$$r_2^2 =$$

$$r_4^2 = \frac{49(1)}{49}$$

$$r_6^2 = r_5^2 = r_4^2 = r_3^2 = r_1^2$$

Squares of Quintinomials

$$\frac{7}{7}\left(\frac{1}{7}+\frac{2}{7}+\frac{1}{7}+\frac{2}{7}+\frac{1}{7}\right) = \frac{7+14+7+14+7}{49}$$

$$\frac{7}{7}\left(\frac{2}{7}+\frac{1}{7}+\frac{2}{7}+\frac{1}{7}+\frac{1}{7}\right) = \frac{14+7+14+7+7}{49}$$

$$\frac{7}{7}\left(\frac{1}{7}+\frac{2}{7}+\frac{1}{7}+\frac{1}{7}+\frac{2}{7}\right) = \frac{7+14+7+7+14}{49}$$

$$\frac{7}{7}\left(\frac{2}{7}+\frac{1}{7}+\frac{1}{7}+\frac{2}{7}+\frac{1}{7}\right) = \frac{14+7+7+14+7}{49}$$

$$\frac{7}{7}\left(\frac{1}{7}+\frac{1}{7}+\frac{2}{7}+\frac{1}{7}+\frac{2}{7}\right) = \frac{7+7+14+7+14}{49}$$

$$r_1 = \frac{7+7+7+7+7+7+7}{49}$$

$$\frac{7}{7}\left(\frac{2}{7}+\frac{1}{7}+\frac{1}{7}+\frac{1}{7}+\frac{2}{7}\right) = \frac{14+7+7+7+14}{49}$$

$$\frac{7}{7}\left(\frac{1}{7}+\frac{1}{7}+\frac{1}{7}+\frac{2}{7}+\frac{2}{7}\right) = \frac{7+7+7+14+14}{49}$$

$$\frac{7}{7}\left(\frac{1}{7}+\frac{1}{7}+\frac{2}{7}+\frac{2}{7}+\frac{1}{7}\right) = \frac{7+7+14+14+7}{49}$$

$$\frac{7}{7}\left(\frac{1}{7}+\frac{2}{7}+\frac{2}{7}+\frac{1}{7}+\frac{1}{7}\right) = \frac{7+14+14+7+7}{49}$$

$$\frac{7}{7}\left(\frac{2}{7}+\frac{2}{7}+\frac{1}{7}+\frac{1}{7}+\frac{1}{7}\right) = \frac{14+14+7+7+7}{49}$$

$$r_2 = \frac{7+7+7+7+7+7+7}{49}$$

$$\frac{7}{7}\left(\frac{1}{7}+\frac{1}{7}+\frac{3}{7}+\frac{1}{7}+\frac{1}{7}\right) = \frac{7+7+21+7+7}{49}$$

$$\frac{7}{7}\left(\frac{1}{7}+\frac{3}{7}+\frac{1}{7}+\frac{1}{7}+\frac{1}{7}\right) = \frac{7+21+7+7+7}{49}$$

$$\frac{7}{7}\left(\frac{3}{7}+\frac{1}{7}+\frac{1}{7}+\frac{1}{7}+\frac{1}{7}\right) = \frac{21+7+7+7+7}{49}$$

$$\frac{7}{7}\left(\frac{1}{7}+\frac{1}{7}+\frac{1}{7}+\frac{1}{7}+\frac{3}{7}\right) = \frac{7+7+7+7+21}{49}$$

$$\frac{7}{7}\left(\frac{1}{7}+\frac{1}{7}+\frac{1}{7}+\frac{3}{7}+\frac{1}{7}\right) = \frac{7+7+7+21+7}{49}$$

$$r_3 = \frac{7+7+7+7+7+7+7}{49}$$

$$r_4 = r_3 = r_2 = r_1$$

550 APPENDIX A

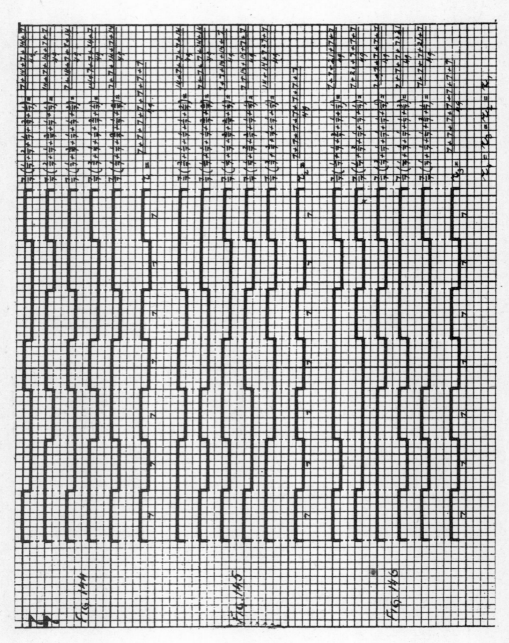

Fig. 144, Fig. 145, Fig. 146. Squares of Quintinomials

$\frac{8}{8}$

$$\left(\frac{2}{8}+\frac{1}{8}+\frac{2}{8}+\frac{1}{8}+\frac{2}{8}\right)^2 = \frac{(4+2+4+2+4)+(2+1+2+1+2)+(4+2+4+2+4)+}{64}$$

$$\frac{+(2+1+2+1+2)+(4+2+4+2+4)}{64}$$

$$\left(\frac{1}{8}+\frac{2}{8}+\frac{1}{8}+\frac{2}{8}+\frac{2}{8}\right)^2 = \frac{(1+2+1+2+2)+(2+4+2+4+4)+(1+2+1+2+2)+}{64}$$

$$\frac{+(2+4+2+4+4)+(2+4+2+4+4)}{64}$$

$$\left(\frac{2}{8}+\frac{1}{8}+\frac{2}{8}+\frac{2}{8}+\frac{1}{8}\right)^2 = \frac{(4+2+4+4+2)+(2+1+2+2+1)+(4+2+4+4+2)+}{64}$$

$$\frac{+(4+2+4+4+2)+(2+1+2+2+1)}{64}$$

$$\left(\frac{1}{8}+\frac{2}{8}+\frac{2}{8}+\frac{1}{8}+\frac{2}{8}\right)^2 = \frac{(1+2+2+1+2)+(2+4+4+2+4)+(2+4+4+2+4)+}{64}$$

$$\frac{+(1+2+2+1+2)+(2+4+4+2+4)}{64}$$

$$\left(\frac{2}{8}+\frac{2}{8}+\frac{1}{8}+\frac{2}{8}+\frac{1}{8}\right)^2 = \frac{(4+4+2+4+2)+(4+4+2+4+2)+(2+2+1+2+1)+}{64}$$

$$\frac{+(4+4+2+4+2)+(2+2+1+2+1)}{64}$$

$$r_1^2 = \frac{1+2+3(1)+6(2)+10(1)+2+2+2+2+10(1)+6(2)+3(1)+2+1}{64}$$

$$\left(\frac{1}{8}+\frac{1}{8}+\frac{4}{8}+\frac{1}{8}+\frac{1}{8}\right)^2 = \frac{(1+1+4+1+1)+(1+1+4+1+1)+(4+4+16+4+4)+}{64}$$

$$\frac{+(1+1+4+1+1)+(1+1+4+1+1)}{64}$$

$$\left(\frac{1}{8}+\frac{4}{8}+\frac{1}{8}+\frac{1}{8}+\frac{1}{8}\right)^2 = \frac{(1+4+1+1+1)+(4+16+4+4+4)+(1+4+1+1+1)+}{64}$$

$$\frac{+(1+4+1+1+1)+(1+4+1+1+1)}{64}$$

$$\left(\frac{4}{8}+\frac{1}{8}+\frac{1}{8}+\frac{1}{8}+\frac{1}{8}\right)^2 = \frac{(16+4+4+4+4)+(4+1+1+1+1)+(4+1+1+1+1)+}{64}$$

$$\frac{+(4+1+1+1+1)+(4+1+1+1+1)}{64}$$

$$\left(\frac{1}{8}+\frac{1}{8}+\frac{1}{8}+\frac{1}{8}+\frac{4}{8}\right)^2 = \frac{(1+1+1+1+4)+(1+1+1+1+4)+(1+1+1+1+4)+}{64}$$

$$\frac{+(1+1+1+1+4)+(4+4+4+4+16)}{64}$$

$$\left(\frac{1}{8}+\frac{1}{8}+\frac{1}{8}+\frac{4}{8}+\frac{1}{8}\right)^2 = \frac{(1+1+1+4+1)+(1+1+1+4+1)+(1+1+1+4+1)+}{64}$$

$$\frac{+(4+4+4+16+4)+(1+1+1+4+1)}{64}$$

$$r_2^2 = \frac{12(1)+2+6(1)+3+5(1)+4+4+5(1)+3+6(1)+2+12(1)}{64}$$

$$r_1^2 =$$

$$r_2^2 =$$

$$r_3^2 = \frac{12(1)+2+14(1)+2+2+2+2+14(1)+2+12(1)}{64}$$

$$\frac{8}{8}\left(\frac{2}{8}+\frac{1}{8}+\frac{2}{8}+\frac{1}{8}+\frac{2}{8}\right) = \frac{16+8+16+8+16}{64}$$

$$\frac{8}{8}\left(\frac{1}{8}+\frac{2}{8}+\frac{1}{8}+\frac{2}{8}+\frac{2}{8}\right) = \frac{8+16+8+16+16}{64}$$

$$\frac{8}{8}\left(\frac{2}{8}+\frac{1}{8}+\frac{2}{8}+\frac{2}{8}+\frac{1}{8}\right) = \frac{16+8+16+16+8}{64}$$

$$\frac{8}{8}\left(\frac{1}{8}+\frac{2}{8}+\frac{2}{8}+\frac{1}{8}+\frac{2}{8}\right) = \frac{8+16+16+8+16}{64}$$

$$\frac{8}{8}\left(\frac{2}{8}+\frac{2}{8}+\frac{1}{8}+\frac{2}{8}+\frac{1}{8}\right) = \frac{16+16+8+16+8}{64}$$

$$r_1 = \frac{8+8+8+8+8+8+8+8}{64}$$

$$\frac{8}{8}\left(\frac{1}{8}+\frac{1}{8}+\frac{4}{8}+\frac{1}{8}+\frac{1}{8}\right) = \frac{8+8+32+8+8}{64}$$

$$\frac{8}{8}\left(\frac{1}{8}+\frac{4}{8}+\frac{1}{8}+\frac{1}{8}+\frac{1}{8}\right) = \frac{8+32+8+8+8}{64}$$

$$\frac{8}{8}\left(\frac{4}{8}+\frac{1}{8}+\frac{1}{8}+\frac{1}{8}+\frac{1}{8}\right) = \frac{32+8+8+8+8}{64}$$

$$\frac{8}{8}\left(\frac{1}{8}+\frac{1}{8}+\frac{1}{8}+\frac{1}{8}+\frac{4}{8}\right) = \frac{8+8+8+8+32}{64}$$

$$\frac{8}{8}\left(\frac{1}{8}+\frac{1}{8}+\frac{1}{8}+\frac{4}{8}+\frac{1}{8}\right) = \frac{8+8+8+32+8}{64}$$

$$r_2 = \frac{8+8+8+8+8+8+8+8}{64}$$

$$r_3 = r_2 = r_1$$

APPENDIX A

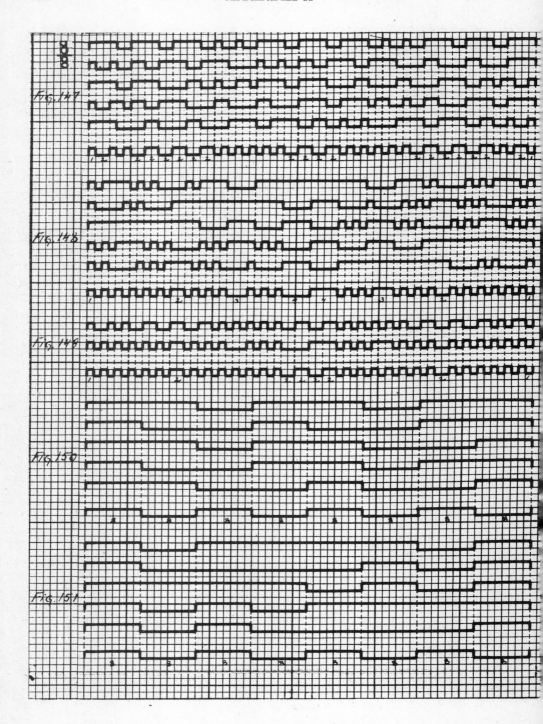

Fig. 147
Fig. 148
Fig. 149
Fig. 150
Fig. 151

Squares of Quintinomials

APPENDIX A

$\frac{9}{9}$

$$\left(\frac{1}{9}+\frac{3}{9}+\frac{1}{9}+\frac{3}{9}+\frac{1}{9}\right)^2 = \frac{(1+3+1+3+1)+(3+9+3+9+3)+(1+3+1+3+1)+}{81}$$
$$\frac{+\,(3+9+3+9+3)+(1+3+1+3+1)}{81}$$

$$\left(\frac{3}{9}+\frac{1}{9}+\frac{3}{9}+\frac{1}{9}+\frac{1}{9}\right)^2 = \frac{(9+3+9+3+3)+(3+1+3+1+1)+(9+3+9+3+3)+}{81}$$
$$\frac{+\,(3+1+3+1+1)+(3+1+3+1+1)}{81}$$

$$\left(\frac{1}{9}+\frac{3}{9}+\frac{1}{9}+\frac{1}{9}+\frac{3}{9}\right)^2 = \frac{(1+3+1+1+3)+(3+9+3+3+9)+(1+3+1+1+3)+}{81}$$
$$\frac{+\,(1+3+1+1+3)+(3+9+3+3+9)}{81}$$

$$\left(\frac{3}{9}+\frac{1}{9}+\frac{1}{9}+\frac{3}{9}+\frac{1}{9}\right)^2 = \frac{(9+3+3+9+3)+(3+1+1+3+1)+(3+1+1+3+1)+}{81}$$
$$\frac{+\,(9+3+3+9+3)+(3+1+1+3+1)}{81}$$

$$\left(\frac{1}{9}+\frac{1}{9}+\frac{3}{9}+\frac{1}{9}+\frac{3}{9}\right)^2 = \frac{(1+1+3+1+3)+(1+1+3+1+3)+(3+3+9+3+9)+}{81}$$
$$\frac{+\,(1+1+3+1+3)+(3+3+9+3+9)}{81}$$

$$r_1^2 = \frac{1+2+9(1)+2+1+1+2+5(1)+2+9(1)+2+1+1+2+1+2+1+1+2+}{81}$$
$$\frac{+9(1)+2+5(1)+2+1+1+9(1)+2+1}{81}$$

$$\left(\frac{2}{9}+\frac{2}{9}+\frac{1}{9}+\frac{2}{9}+\frac{2}{9}\right)^2 = \frac{(4+4+2+4+4)+(4+4+2+4+4)+(2+2+1+2+2)+}{81}$$
$$\frac{+\,(4+4+2+4+4)+(4+4+2+4+4)}{81}$$

$$\left(\frac{2}{9}+\frac{1}{9}+\frac{2}{9}+\frac{2}{9}+\frac{2}{9}\right)^2 = \frac{(4+2+4+4+4)+(2+1+2+2+2)+(4+2+4+4+4)+}{81}$$
$$\frac{+\,(4+2+4+4+4)+(4+2+4+4+4)}{81}$$

DISTRIBUTIVE INVOLUTION GROUPS

$$\left(\frac{1}{9}+\frac{2}{9}+\frac{2}{9}+\frac{2}{9}+\frac{2}{9}\right)^2 = \frac{(1+2+2+2+2)+(2+4+4+4+4)+(2+4+4+4+4)+}{81}$$
$$\frac{+(2+4+4+4+4)+(2+4+4+4+4)}{81}$$

$$\left(\frac{2}{9}+\frac{2}{9}+\frac{2}{9}+\frac{2}{9}+\frac{1}{9}\right)^2 = \frac{(4+4+4+4+2)+(4+4+4+4+2)+(4+4+4+4+2)+}{81}$$
$$\frac{+(4+4+4+4+2)+(2+2+2+2+1)}{81}$$

$$\left(\frac{2}{9}+\frac{2}{9}+\frac{2}{9}+\frac{1}{9}+\frac{2}{9}\right)^2 = \frac{(4+4+4+2+4)+(4+4+4+2+4)+(4+4+4+2+4)+}{81}$$
$$\frac{+(2+2+2+1+2)+(4+4+4+2+4)}{81}$$

$$r_2^2 = \frac{1+2+9(1)+2+1+1+2+5(1)+2+9(1)+2+1+1+2+1+2+1+1+2+}{81}$$
$$\frac{+9(1)+2+5(1)+2+1+1+9(1)+2+1}{81}$$

$$\left(\frac{1}{9}+\frac{1}{9}+\frac{5}{9}+\frac{1}{9}+\frac{1}{9}\right)^2 = \frac{(1+1+5+1+1)+(1+1+5+1+1)+(5+5+25+5+5)+}{81}$$
$$\frac{+(1+1+5+1+1)+(1+1+5+1+1)}{81}$$

$$\left(\frac{1}{9}+\frac{5}{9}+\frac{1}{9}+\frac{1}{9}+\frac{1}{9}\right)^2 = \frac{(1+5+1+1+1)+(5+25+5+5+5)+(1+5+1+1+1)+}{81}$$
$$\frac{+(1+5+1+1+1)+(1+5+1+1+1)}{81}$$

$$\left(\frac{5}{9}+\frac{1}{9}+\frac{1}{9}+\frac{1}{9}+\frac{1}{9}\right)^2 = \frac{(25+5+5+5+5)+(5+1+1+1+1)+(5+1+1+1+1)+}{81}$$
$$\frac{+(5+1+1+1+1)+(5+1+1+1+1)}{81}$$

$$\left(\frac{1}{9}+\frac{1}{9}+\frac{1}{9}+\frac{1}{9}+\frac{5}{9}\right)^2 = \frac{(1+1+1+1+5)+(1+1+1+1+5)+(1+1+1+1+5)+}{81}$$
$$\frac{+(1+1+1+1+5)+(5+5+5+5+25)}{81}$$

$$\left(\frac{1}{9}+\frac{1}{9}+\frac{1}{9}+\frac{5}{9}+\frac{1}{9}\right)^2 = \frac{(1+1+1+5+1)+(1+1+1+5+1)+(1+1+1+5+1)+}{81}$$
$$\frac{+(5+5+5+25+5)+(1+1+1+5+1)}{81}$$

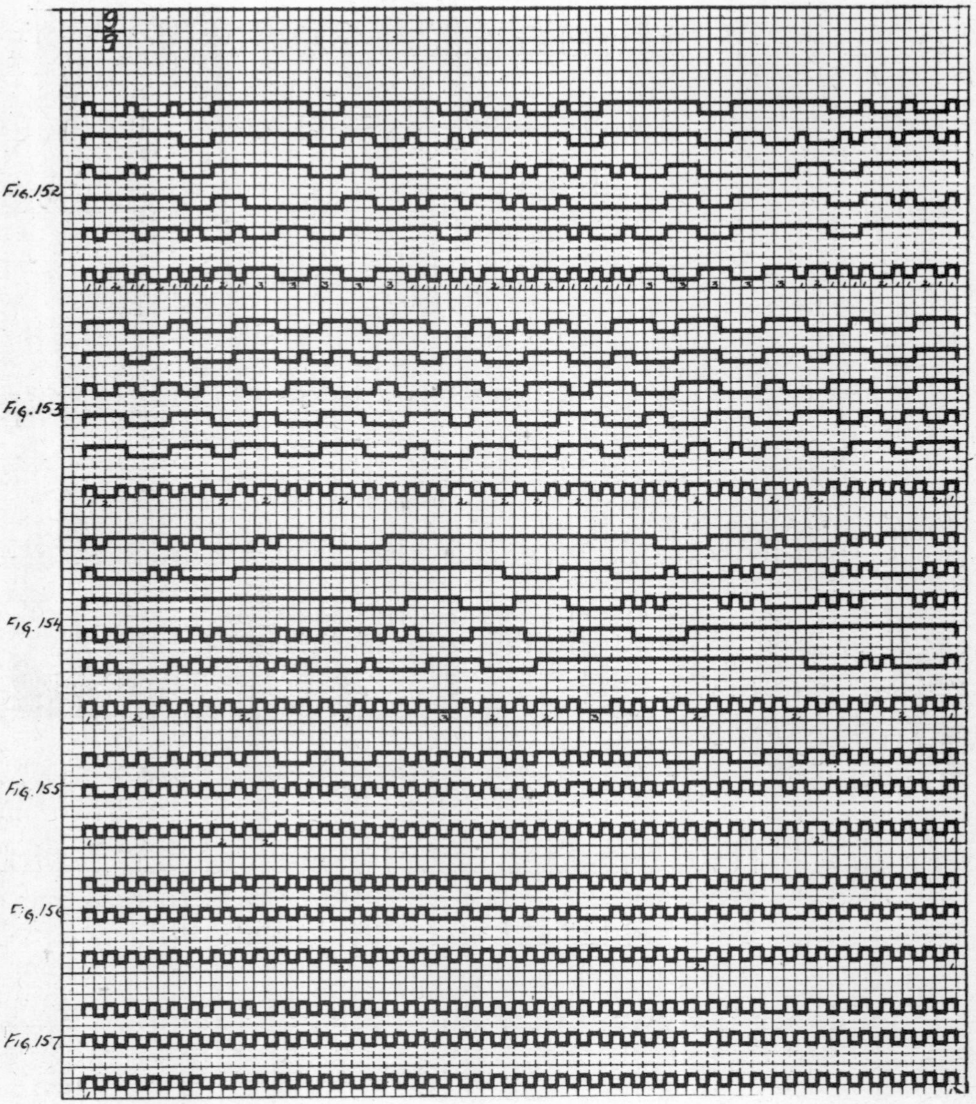
9/9
Fig. 152
Fig. 153
Fig. 154
Fig. 155
Fig. 156
Fig. 157

Squares of Quintinomials

$$r_3^2 = \frac{4(1)+2+8(1)+2+7(1)+2+7(1)+3+1+1+2+1+1+1+2+1+1+3}{81}$$
$$\frac{+\ 7(1)+2+7(1)+2+8(1)+2+4(1)}{81}$$

$r_1^2 =$

$r_2^2 =$

$$r_4^2 = \frac{12(1)+2+1+1+2+22(1)+1+22(1)+2+1+1+2+12(1)}{81}$$

$r_2^2 =$

$r_3^2 =$

$$r_5^2 = \frac{23(1)+2+15(1)+1+15(1)+2+23(1)}{81}$$

$r_4^2 =$

$r_5^2 =$

$$r_6^2 = \frac{81(1)}{81}$$

(See preceding page for Graph)

$$\frac{9}{9}\left(\frac{1}{9}+\frac{3}{9}+\frac{1}{9}+\frac{3}{9}+\frac{1}{9}\right) = \frac{9+27+9+27+9}{81}$$

$$\frac{9}{9}\left(\frac{3}{9}+\frac{1}{9}+\frac{3}{9}+\frac{1}{9}+\frac{1}{9}\right) = \frac{27+9+27+9+9}{81}$$

$$\frac{9}{9}\left(\frac{1}{9}+\frac{3}{9}+\frac{1}{9}+\frac{1}{9}+\frac{3}{9}\right) = \frac{9+27+9+9+27}{81}$$

$$\frac{9}{9}\left(\frac{3}{9}+\frac{1}{9}+\frac{1}{9}+\frac{3}{9}+\frac{1}{9}\right) = \frac{27+9+9+27+9}{81}$$

$$\frac{9}{9}\left(\frac{1}{9}+\frac{1}{9}+\frac{3}{9}+\frac{1}{9}+\frac{3}{9}\right) = \frac{9+9+27+9+27}{81}$$

$$r_1 = \frac{9+9+9+9+9+9+9+9+9}{81}$$

$$\frac{9}{9}\left(\frac{2}{9}+\frac{2}{9}+\frac{1}{9}+\frac{2}{9}+\frac{2}{9}\right) = \frac{18+18+9+18+18}{81}$$

$$\frac{9}{9}\left(\frac{2}{9}+\frac{1}{9}+\frac{2}{9}+\frac{2}{9}+\frac{2}{9}\right) = \frac{18+9+18+18+18}{81}$$

$$\frac{9}{9}\left(\frac{1}{9}+\frac{2}{9}+\frac{2}{9}+\frac{2}{9}+\frac{2}{9}\right) = \frac{9+18+18+18+18}{81}$$

$$\frac{9}{9}\left(\frac{2}{9}+\frac{2}{9}+\frac{2}{9}+\frac{2}{9}+\frac{1}{9}\right) = \frac{18+18+18+18+9}{81}$$

$$\frac{9}{9}\left(\frac{2}{9}+\frac{2}{9}+\frac{2}{9}+\frac{1}{9}+\frac{2}{9}\right) = \frac{18+18+18+9+18}{81}$$

$$r_2 = \frac{9+9+9+9+9+9+9+9+9}{81}$$

$$\frac{9}{9}\left(\frac{1}{9}+\frac{1}{9}+\frac{5}{9}+\frac{1}{9}+\frac{1}{9}\right) = \frac{9+9+45+9+9}{81}$$

$$\frac{9}{9}\left(\frac{1}{9}+\frac{5}{9}+\frac{1}{9}+\frac{1}{9}+\frac{1}{9}\right) = \frac{9+45+9+9+9}{81}$$

$$\frac{9}{9}\left(\frac{5}{9}+\frac{1}{9}+\frac{1}{9}+\frac{1}{9}+\frac{1}{9}\right) = \frac{45+9+9+9+9}{81}$$

$$\frac{9}{9}\left(\frac{1}{9}+\frac{1}{9}+\frac{1}{9}+\frac{1}{9}+\frac{5}{9}\right) = \frac{9+9+9+9+45}{81}$$

$$\frac{9}{9}\left(\frac{1}{9}+\frac{1}{9}+\frac{1}{9}+\frac{5}{9}+\frac{1}{9}\right) = \frac{9+9+9+45+9}{81}$$

$$r_3 = \frac{9+9+9+9+9+9+9+9+9}{81}$$

$$r_1 = r_2 = r_3$$

APPENDIX A

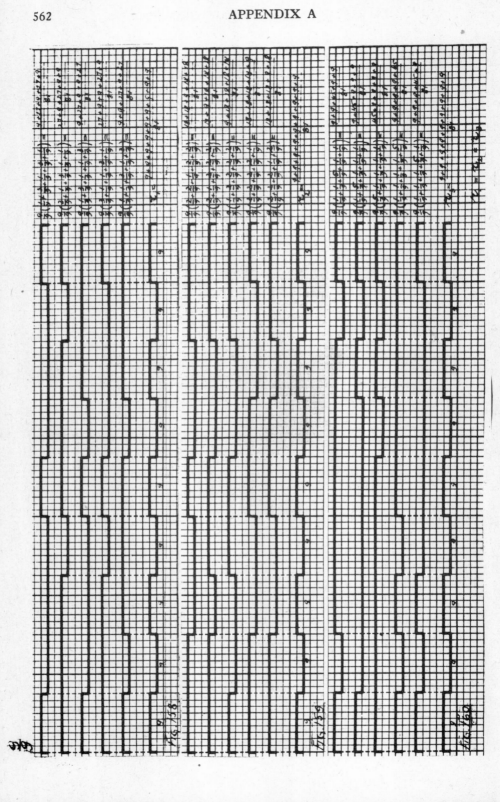

Squares of Quintinomials

D. Distributive Cube of Binomials.

(1) Factorial: $(a+b)^3 = a^3+a^2b+a^2b+ab^2+a^2b+ab^2+ab^2+b^3$;

(2) Fractional: $\left(\dfrac{a}{a+b} + \dfrac{b}{a+b}\right)^3 = \dfrac{a^3}{(a+b)^3} + \dfrac{a^2b}{(a+b)^3} + \dfrac{a^2b}{(a+b)^3} + \dfrac{ab^2}{(a+b)^3} +$

$+ \dfrac{a^2b}{(a+b)^3} + \dfrac{ab^2}{(a+b)^3} + \dfrac{ab^2}{(a+b)^3} + \dfrac{b^3}{(a+b)^3}$

Synchronization of the second power group with the third power group.

(1) Factorial: $S = a^2(a+b) + ab(a+b) + ab(a+b) + b^2(a+b)$;

(2) Fractional: $S = \dfrac{a^2}{(a+b)^2} \cdot \left(\dfrac{a+b}{a+b}\right) + \dfrac{ab}{(a+b)^2} \cdot \left(\dfrac{a+b}{a+b}\right) + \dfrac{ab}{(a+b)^2} \cdot \left(\dfrac{a+b}{a+b}\right) +$

$+ \dfrac{b^2}{(a+b)^2} \cdot \left(\dfrac{a+b}{a+b}\right).$

Synchronization of the first power group with the third power group.

(1) Factorial: $S = a(a+b)^2 + b(a+b)^2$;

(2) Fractional: $S = \dfrac{a}{a+b} \cdot \left(\dfrac{a+b}{a+b}\right)^2 + \dfrac{b}{a+b} \cdot \left(\dfrac{a+b}{a+b}\right)^2.$

1. Cube of Binomials

$\dfrac{3}{3}$

$\left(\dfrac{2}{3}+\dfrac{1}{3}\right)^3 = \dfrac{8+4+4+2+4+2+2+1}{27}$

$\left(\dfrac{1}{3}+\dfrac{2}{3}\right)^3 = \dfrac{1+2+2+4+2+4+4+8}{27}$

$r^3 = \dfrac{1+2+2+3+1+2+1+3+1+2+1+3+2+2+1}{27}$

$\dfrac{3}{3}\left(\dfrac{2}{3}+\dfrac{1}{3}\right)^2 = \dfrac{12+6+6+3}{27}$

$\dfrac{3}{3}\left(\dfrac{1}{3}+\dfrac{2}{3}\right)^2 = \dfrac{3+6+6+12}{27}$

$r^2 = \dfrac{3+6+3+3+3+6+3}{27}$

$\frac{3}{3}$ (concluded)

$$\frac{9}{9}\left(\frac{2}{3}+\frac{1}{3}\right) = \frac{18+9}{27}$$

$$\frac{9}{9}\left(\frac{1}{3}+\frac{2}{3}\right) = \frac{9+18}{27}$$

$$r = \frac{9+9+9}{27}$$

$\frac{4}{4}$

$$\left(\frac{3}{4}+\frac{1}{4}\right)^3 = \frac{27+9+9+3+9+3+3+1}{64}$$

$$\left(\frac{1}{4}+\frac{3}{4}\right)^3 = \frac{1+3+3+9+3+9+9+27}{64}$$

$$r^3 = \frac{1+3+3+9+3+8+1+8+1+8+3+9+3+3+1}{64}$$

$$\frac{4}{4}\left(\frac{3}{4}+\frac{1}{4}\right)^2 = \frac{36+12+12+4}{64}$$

$$\frac{4}{4}\left(\frac{1}{4}+\frac{3}{4}\right)^2 = \frac{4+12+12+36}{64}$$

$$r^2 = \frac{4+12+12+8+12+12+4}{64}$$

$$\frac{16}{16}\left(\frac{3}{4}+\frac{1}{4}\right) = \frac{48+16}{64}$$

$$\frac{16}{16}\left(\frac{1}{4}+\frac{3}{4}\right) = \frac{16+48}{64}$$

$$r = \frac{16+32+16}{64}$$

$\frac{5}{5}$

$$\left(\frac{3}{5}+\frac{2}{5}\right)^3 = \frac{27+18+18+12+18+12+12+8}{125}$$

$$\left(\frac{2}{5}+\frac{3}{5}\right)^3 = \frac{8+12+12+18+12+18+18+27}{125}$$

$\frac{5}{5}$ (concluded)

$$r^3 = \frac{8+12+7+5+13+5+12+1+12+5+13+5+7+12+8}{125}$$

$$\left(\frac{4}{5}+\frac{1}{5}\right)^3 = \frac{64+16+16+4+16+4+4+1}{125}$$

$$\left(\frac{1}{5}+\frac{4}{5}\right)^3 = \frac{1+4+4+16+4+16+16+64}{125}$$

$$r^3 = \frac{1+4+4+16+4+16+16+3+16+16+4+16+4+4+1}{125}$$

$$\frac{5}{5}\left(\frac{3}{5}+\frac{2}{5}\right)^2 = \frac{45+30+30+20}{125}$$

$$\frac{5}{5}\left(\frac{2}{5}+\frac{3}{5}\right)^2 = \frac{20+30+30+45}{125}$$

$$r^2 = \frac{20+25+5+25+5+25+20}{125}$$

$$\frac{5}{5}\left(\frac{4}{5}+\frac{1}{5}\right)^2 = \frac{80+20+20+5}{125}$$

$$\frac{5}{5}\left(\frac{1}{5}+\frac{4}{5}\right)^2 = \frac{5+20+20+80}{125}$$

$$r^2 = \frac{5+20+20+35+20+20+5}{125}$$

$$\frac{25}{25}\left(\frac{3}{5}+\frac{2}{5}\right) = \frac{75+50}{125}$$

$$\frac{25}{25}\left(\frac{2}{5}+\frac{3}{5}\right) = \frac{50+75}{125}$$

$$r = \frac{50+25+50}{125}$$

$$\frac{25}{25}\left(\frac{4}{5}+\frac{1}{5}\right) = \frac{100+25}{125}$$

$$\frac{25}{25}\left(\frac{1}{5}+\frac{4}{5}\right) = \frac{25+100}{125}$$

$$r = \frac{25+75+25}{125}$$

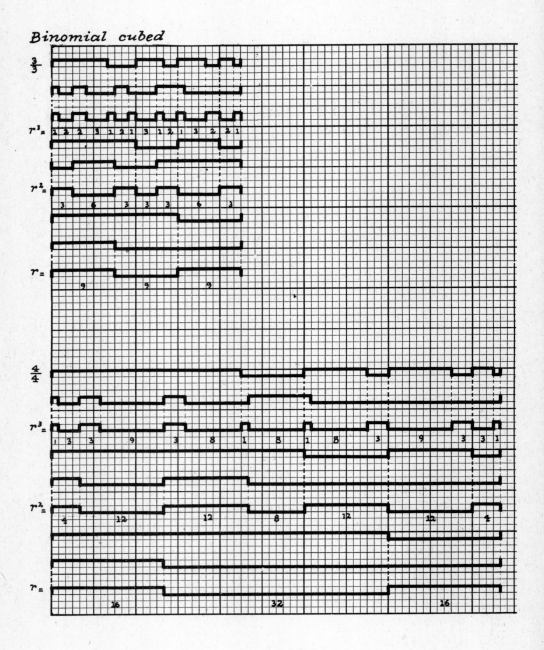

DISTRIBUTIVE INVOLUTION GROUPS

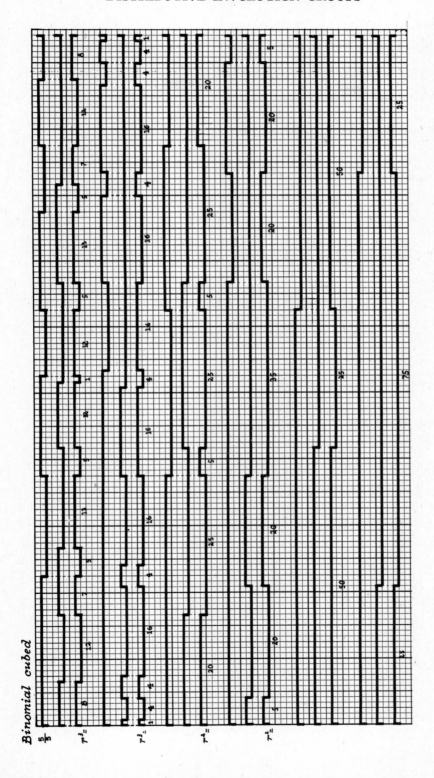

E. DISTRIBUTIVE CUBE OF TRINOMIALS.

(1) Factorial: $(a+b+c)^3 = (a^3+a^2b+a^2c+a^2b+ab^2+abc+a^2c+abc+ac^2) +$
$+ (a^2b+ab^2+abc+ab^2+b^3+b^2c+abc+b^2c+bc^2) + (a^2c+abc+ac^2+$
$+abc+b^2c+bc^2+ac^2+bc^2+c^3);$

(2) Fractional: $\left(\dfrac{a}{a+b+c} + \dfrac{b}{a+b+c} + \dfrac{c}{a+b+c}\right)^3 = \left[\dfrac{a^3}{(a+b+c)^3} + \right.$

$+ \dfrac{a^2b}{(a+b+c)^3} + \dfrac{a^2c}{(a+b+c)^3} + \dfrac{a^2b}{(a+b+c)^3} + \dfrac{ab^2}{(a+b+c)^3} + \dfrac{abc}{(a+b+c)^3} +$

$+ \dfrac{a^2c}{(a+b+c)^3} + \dfrac{abc}{(a+b+c)^3} + \dfrac{ac^2}{(a+b+c)^3}\biggr] + \biggl[\dfrac{a^2b}{(a+b+c)^3} + \dfrac{ab^2}{(a+b+c)^3} +$

$+ \dfrac{abc}{(a+b+c)^3} + \dfrac{ab^2}{(a+b+c)^3} + \dfrac{b^3}{(a+b+c)^3} + \dfrac{b^2c}{(a+b+c)^3} + \dfrac{abc}{(a+b+c)^3} +$

$+ \dfrac{b^2c}{(a+b+c)^3} + \dfrac{bc^2}{(a+b+c)^3}\biggr] + \biggl[\dfrac{a^2c}{(a+b+c)^3} + \dfrac{abc}{(a+b+c)^3} + \dfrac{ac^2}{(a+b+c)^3} +$

$+ \dfrac{abc}{(a+b+c)^3} + \dfrac{b^2c}{(a+b+c)^3} + \dfrac{bc^2}{(a+b+c)^3} + \dfrac{ac^2}{(a+b+c)^3} + \dfrac{bc^2}{(a+b+c)^3} +$

$+ \dfrac{c^3}{(a+b+c)^3}\biggr].$

Synchronization of the second power group with the third power group.

(1) Factorial: $S = [a^2(a+b+c)+ab\,(a+b+c)+ac\,(a+b+c)] +$
$+ [ab\,(a+b+c)+b^2(a+b+c)+bc\,(a+b+c)] + [ac\,(a+b+c)+$
$+bc\,(a+b+c)+c^2(a+b+c)];$

1. Cube of Trinomials

$\dfrac{4}{4}$

$$\left(\frac{1}{4}+\frac{2}{4}+\frac{1}{4}\right)^3 = \frac{(1+2+1)+(2+4+2)+(1+2+1)+(2+4+2)+(4+8+4)+}{64}$$
$$\frac{+(2+4+2)+(1+2+1)+(2+4+2)+(1+2+1)}{64}$$

$$\left(\frac{2}{4}+\frac{1}{4}+\frac{1}{4}\right)^3 = \frac{(8+4+4)+(4+2+2)+(4+2+2)+(4+2+2)+(2+1+1)+}{64}$$
$$\frac{+(2+1+1)+(4+2+2)+(2+1+1)+(2+1+1)}{64}$$

$$\left(\frac{1}{4}+\frac{1}{4}+\frac{2}{4}\right)^3 = \frac{(1+1+2)+(1+1+2)+(2+2+4)+(1+1+2)+(1+1+2)+}{64}$$
$$\frac{+(2+2+4)+(2+2+4)+(2+2+4)+(4+4+8)}{64}$$

$$r^3 = \frac{6(1)+2+2+2+1+2+3(1)+2+1+1+10(2)+1+1+2+3(1)+2+1+}{64}$$
$$\frac{+2+2+2+6(1)}{64}$$

$$\frac{4}{4}\left(\frac{1}{4}+\frac{2}{4}+\frac{1}{4}\right)^2 = \frac{(4+8+4)+(8+16+8)+(4+8+4)}{64}$$

$$\frac{4}{4}\left(\frac{2}{4}+\frac{1}{4}+\frac{1}{4}\right)^2 = \frac{(16+8+8)+(8+4+4)+(8+4+4)}{64}$$

$$\frac{4}{4}\left(\frac{1}{4}+\frac{1}{4}+\frac{2}{4}\right)^2 = \frac{(4+4+8)+(4+4+8)+(8+8+16)}{64}$$

$$r^2 = \frac{6(4)+8+8+6(4)}{64}$$

$$\frac{16}{16}\left(\frac{1}{4}+\frac{2}{4}+\frac{1}{4}\right) = \frac{16+32+16}{64}$$

$$\frac{16}{16}\left(\frac{2}{4}+\frac{1}{4}+\frac{1}{4}\right) = \frac{32+16+16}{64}$$

$$\frac{16}{16}\left(\frac{1}{4}+\frac{1}{4}+\frac{2}{4}\right) = \frac{16+16+32}{64}$$

$$r = \frac{16+16+16+16}{64}$$

$$\left(\frac{1}{5}+\frac{3}{5}+\frac{1}{5}\right)^3 = \frac{(1+3+1)+(3+9+3)+(1+3+1)+(3+9+3)+(9+27+9)+}{125}$$

$$\frac{+(3+9+3)+(1+3+1)+(3+9+3)+(1+3+1)}{125}$$

$$\left(\frac{3}{5}+\frac{1}{5}+\frac{1}{5}\right)^3 = \frac{(27+9+9)+(9+3+3)+(9+3+3)+(9+3+3)+(3+1+1)+}{125}$$

$$\frac{+(3+1+1)+(9+3+3)+(3+1+1)+(3+1+1)}{125}$$

$$\left(\frac{1}{5}+\frac{1}{5}+\frac{3}{5}\right)^3 = \frac{(1+1+3)+(1+1+3)+(3+3+9)+(1+1+3)+(1+1+3)+}{125}$$

$$\frac{+(3+3+9)+(3+3+9)+(3+3+9)+(9+9+27)}{125}$$

$$r^3 = \frac{1+1+2+4(1)+2+3+3+1+3+1+3+4(1)+2+1+1+3+1+1+1+}{125}$$

$$\frac{+2+1+4+4+1+3+1+2+1+3+5+3+1+2+1+3+1+4+4+1+2+}{125}$$

$$\frac{+1+1+1+3+1+1+2+4(1)+3+1+3+1+3+3+2+4(1)+3+1+1}{125}$$

$$\left(\frac{2}{5}+\frac{1}{5}+\frac{2}{5}\right)^3 = \frac{(8+4+8)+(4+2+4)+(8+4+8)+(4+2+4)+(2+1+2)+}{125}$$

$$\frac{+(4+2+4)+(8+4+8)+(4+2+4)+(8+4+8)}{125}$$

$$\left(\frac{1}{5}+\frac{2}{5}+\frac{2}{5}\right)^3 = \frac{(1+2+2)+(2+4+4)+(2+4+4)+(2+4+4)+(4+8+8)+}{125}$$

$$\frac{+(4+8+8)+(2+4+4)+(4+8+8)+(4+8+8)}{125}$$

$$\left(\frac{2}{5}+\frac{2}{5}+\frac{1}{5}\right)^3 = \frac{(8+8+4)+(8+8+4)+(4+4+2)+(8+8+4)+(8+8+4)+}{125}$$

$$\frac{+(4+4+2)+(4+4+2)+(4+4+2)+(2+2+1)}{125}$$

$\frac{5}{5}$ (continued)

$$r^3 = \frac{1+2+2+2+1+3+1+3+1+1+3+1+3+4(1)+2+1+4+1+2+1+}{125}$$

$$\frac{+1+2+2+3+1+2+4+1+1+2+1+1+2+1+2+1+1+2+1+1+4+}{125}$$

$$\frac{+2+1+3+2+2+1+1+2+1+4+1+2+4(1)+3+1+3+1+1+3+1+}{125}$$

$$\frac{+3+1+2+2+2+1}{125}$$

$$\frac{5}{5}\left(\frac{1}{5}+\frac{3}{5}+\frac{1}{5}\right)^2 = \frac{(5+15+5)+(15+45+15)+(5+15+5)}{125}$$

$$\frac{5}{5}\left(\frac{3}{5}+\frac{1}{5}+\frac{1}{5}\right)^2 = \frac{(45+15+15)+(15+5+5)+(15+5+5)}{125}$$

$$\frac{5}{5}\left(\frac{1}{5}+\frac{1}{5}+\frac{3}{5}\right)^2 = \frac{(5+5+15)+(5+5+15)+(15+15+45)}{125}$$

$$r^2 = \frac{5+5+10+6(5)+10+5+10+6(5)+10+5+5}{125}$$

$$\frac{5}{5}\left(\frac{2}{5}+\frac{1}{5}+\frac{2}{5}\right)^2 = \frac{(20+10+20)+(10+5+10)+(20+10+20)}{125}$$

$$\frac{5}{5}\left(\frac{1}{5}+\frac{2}{5}+\frac{2}{5}\right)^2 = \frac{(5+10+10)+(10+20+20)+(10+20+20)}{125}$$

$$\frac{5}{5}\left(\frac{2}{5}+\frac{2}{5}+\frac{1}{5}\right)^2 = \frac{(20+20+10)+(20+20+10)+(10+10+5)}{125}$$

$$r^2 = \frac{5+10+5(5)+10+5(5)+10+5(5)+10+5}{125}$$

$$\frac{25}{25}\left(\frac{1}{5}+\frac{3}{5}+\frac{1}{5}\right) = \frac{25+75+25}{125}$$

$$\frac{25}{25}\left(\frac{3}{5}+\frac{1}{5}+\frac{1}{5}\right) = \frac{75+25+25}{125}$$

$$\frac{25}{25}\left(\frac{1}{5}+\frac{1}{5}+\frac{2}{5}\right) = \frac{25+25+75}{125}$$

$$r = \frac{25+25+25+25+25}{125}$$

$\frac{5}{5}$ (continued)

$$\frac{25}{25}\left(\frac{2}{5}+\frac{1}{5}+\frac{2}{5}\right) = \frac{50+25+50}{125}$$

$$\frac{25}{25}\left(\frac{1}{5}+\frac{2}{5}+\frac{2}{5}\right) = \frac{25+50+50}{125}$$

$$\frac{25}{25}\left(\frac{2}{5}+\frac{2}{5}+\frac{1}{5}\right) = \frac{50+50+25}{125}$$

$$r = \frac{25+25+25+25+25}{125}$$

$\frac{7}{7}$

$$\left(\frac{3}{7}+\frac{1}{7}+\frac{3}{7}\right)^3 = \frac{(27+9+27)+(9+3+9)+(27+9+27)+(9+3+9)+}{343}$$

$$\frac{+(3+1+3)+(9+3+9)+(27+9+27)+(9+3+9)+(27+9+27)}{343}$$

$$\left(\frac{1}{7}+\frac{3}{7}+\frac{3}{7}\right)^3 = \frac{(1+3+3)+(3+9+9)+(3+9+9)+(3+9+9)+(9+27+27)+}{343}$$

$$\frac{+(9+27+27)+(3+9+9)+(9+27+27)+(9+27+27)}{343}$$

$$\left(\frac{3}{7}+\frac{3}{7}+\frac{1}{7}\right)^3 = \frac{(27+27+9)+(27+27+9)+(9+9+3)+(27+27+9)+}{343}$$

$$\frac{+(27+27+9)+(9+9+3)+(9+9+3)+(9+9+3)+(3+3+1)}{343}$$

$$r^3 = \frac{1+3+3+3+9+8+1+3+5+4+9+3+2+7+2+7+2+3+4+5+}{343}$$

$$\frac{+6+16+5+6+3+6+7+2+7+2+3+9+3+9+1+2+1+2+1+9+}{343}$$

$$\frac{+3+9+3+2+7+2+7+6+3+6+5+16+6+5+4+3+2+7+2+}{343}$$

$$\frac{+7+2+3+9+4+5+3+1+8+9+3+3+3+1}{343}$$

$\frac{7}{7}$ (continued)

$$\left(\frac{2}{7}+\frac{3}{7}+\frac{2}{7}\right)^3 = \frac{(8+12+8)+(12+18+12)+(8+12+8)+(12+18+12)+}{343}$$

$$\frac{+(18+27+18)+(12+18+12)+(8+12+8)+(12+18+12)+}{343}$$

$$\frac{+(8+12+8)}{343}$$

$$\left(\frac{3}{7}+\frac{2}{7}+\frac{2}{7}\right)^3 = \frac{(27+18+18)+(18+12+12)+(18+12+12)+(18+12+12)+}{343}$$

$$\frac{+(12+8+8)+(12+8+8)+(18+12+12)+(12+8+8)+}{343}$$

$$\frac{+(12+8+8)}{343}$$

$$\left(\frac{2}{7}+\frac{2}{7}+\frac{3}{7}\right)^3 = \frac{(8+8+12)+(8+8+12)+(12+12+18)+(8+8+12)+}{343}$$

$$\frac{+(8+8+12)+(12+12+18)+(12+12+18)+(12+12+18)+}{343}$$

$$\frac{+(18+18+27)}{343}$$

$$r^3 = \frac{8+8+4+7+1+8+4+4+1+11+2+5+5+2+8+2+1+9+3+5+7+}{343}$$

$$\frac{+1+4+4+9+3+2+6+1+5+2+5+7+4+7+1+11+1+7+4+7+5+}{343}$$

$$\frac{+2+5+1+6+2+3+9+4+4+1+7+5+3+9+1+2+8+2+5+5+2+}{343}$$

$$\frac{+11+1+4+4+8+1+7+4+8+8}{343}$$

$$\frac{7}{7}\left(\frac{3}{7}+\frac{1}{7}+\frac{3}{7}\right)^2 = \frac{63+21+63+21+7+21+63+21+63}{343}$$

$$\frac{7}{7}\left(\frac{1}{7}+\frac{3}{7}+\frac{3}{7}\right)^2 = \frac{7+21+21+21+63+63+21+63+63}{343}$$

$$\frac{7}{7}\left(\frac{3}{7}+\frac{3}{7}+\frac{1}{7}\right)^2 = \frac{63+63+21+63+63+21+21+21+7}{343}$$

APPENDIX A

$\frac{7}{7}$ (continued)

$$r^2 = \frac{7+21+21+14+7+14+42+7+14+21+7+21+14+7+42+14+7+}{343}$$

$$\frac{+14+21+21+7}{343}$$

$$\frac{7}{7}\left(\frac{2}{7}+\frac{3}{7}+\frac{2}{7}\right)^2 = \frac{28+42+28+42+63+42+28+42+28}{343}$$

$$\frac{7}{7}\left(\frac{3}{7}+\frac{2}{7}+\frac{2}{7}\right)^2 = \frac{63+42+42+42+28+28+42+28+28}{343}$$

$$\frac{7}{7}\left(\frac{2}{7}+\frac{2}{7}+\frac{3}{7}\right)^2 = \frac{28+28+42+28+28+42+42+42+63}{343}$$

$$r^2 = \frac{28+28+7+7+28+7+21+14+7+7+35+7+7+14+21+7+28+7+}{343}$$

$$\frac{+7+28+28}{343}$$

$$\frac{49}{49}\left(\frac{3}{7}+\frac{1}{7}+\frac{3}{7}\right) = \frac{147+49+147}{343}$$

$$\frac{49}{49}\left(\frac{1}{7}+\frac{3}{7}+\frac{3}{7}\right) = \frac{49+147+147}{343}$$

$$\frac{49}{49}\left(\frac{3}{7}+\frac{3}{7}+\frac{1}{7}\right) = \frac{147+147+49}{343}$$

$$r = \frac{49+98+49+98+49}{343}$$

$$\frac{49}{49}\left(\frac{2}{7}+\frac{3}{7}+\frac{2}{7}\right) = \frac{98+147+98}{343}$$

$$\frac{49}{49}\left(\frac{3}{7}+\frac{2}{7}+\frac{2}{7}\right) = \frac{147+98+98}{343}$$

$$\frac{49}{49}\left(\frac{2}{7}+\frac{2}{7}+\frac{3}{7}\right) = \frac{98+98+147}{343}$$

$$r = \frac{98+49+49+49+98}{343}$$

$$\frac{8}{8}$$

$$\left(\frac{3}{8}+\frac{2}{8}+\frac{3}{8}\right)^3 = \frac{(27+18+27)+(18+12+18)+(27+18+27)+(18+12+18)+}{512}$$
$$\frac{+(12+8+12)+(18+12+18)+(27+18+27)+(18+12+18)+}{512}$$
$$\frac{+(27+18+27)}{512}$$

$$\left(\frac{2}{8}+\frac{3}{8}+\frac{3}{8}\right)^3 = \frac{(8+12+12)+(12+18+18)+(12+18+18)+(12+18+18)+}{512}$$
$$\frac{+(18+27+27)+(18+27+27)+(12+18+18)+(18+27+27)+}{512}$$
$$\frac{+(18+27+27)}{512}$$

$$\left(\frac{3}{8}+\frac{3}{8}+\frac{2}{8}\right)^3 = \frac{(27+27+18)+(27+27+18)+(18+18+12)+(27+27+18)+}{512}$$
$$\frac{+(27+27+18)+(18+18+12)+(18+18+12)+(18+18+12)+}{512}$$
$$\frac{+(12+12+8)}{512}$$

$$r^3 = \frac{8+12+7+5+12+1+9+8+10+8+10+2+7+3+8+10+6+2+12+}{512}$$
$$\frac{4+3+11+4+3+11+4+12+2+16+9+2+1+18+6+2+4+8+4+2+}{512}$$
$$\frac{+6+18+1+2+9+16+2+12+4+11+3+4+11+3+4+12+2+6+}{512}$$
$$\frac{+10+8+3+7+2+10+8+10+8+9+1+12+5+7+12+8}{512}$$

$$\frac{8}{8}\left(\frac{3}{8}+\frac{2}{8}+\frac{3}{8}\right)^2 = \frac{72+48+72+48+32+48+72+48+72}{512}$$

$$\frac{8}{8}\left(\frac{2}{8}+\frac{3}{8}+\frac{3}{8}\right)^2 = \frac{32+48+48+48+72+72+48+72+72}{512}$$

$$\frac{8}{8}\left(\frac{3}{8}+\frac{3}{8}+\frac{2}{8}\right)^2 = \frac{72+72+48+72+72+48+48+48+32}{512}$$

$\frac{8}{8}$ (continued)

$$r^2 = \frac{32+40+8+40+8+16+32+16+48+8+16+8+48+16+32+16+8+}{512}$$

$$\frac{+40+8+40+32}{512}$$

$$\frac{64}{64}\left(\frac{3}{8}+\frac{2}{8}+\frac{3}{8}\right) = \frac{192+128+192}{512}$$

$$\frac{64}{64}\left(\frac{2}{8}+\frac{3}{8}+\frac{3}{8}\right) = \frac{128+192+192}{512}$$

$$\frac{64}{64}\left(\frac{3}{8}+\frac{3}{8}+\frac{2}{8}\right) = \frac{192+192+128}{512}$$

$$r = \frac{128+64+128+64+128}{512}$$

DISTRIBUTIVE INVOLUTION GROUPS

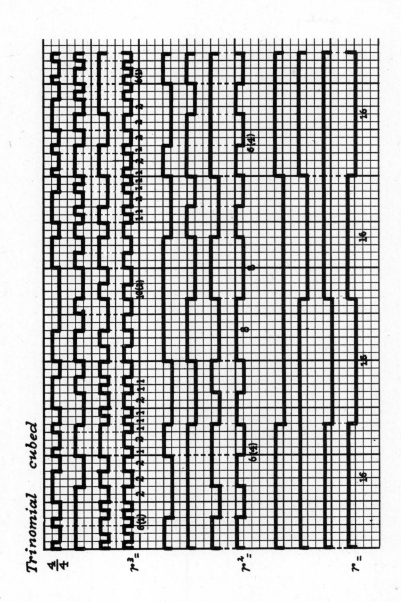

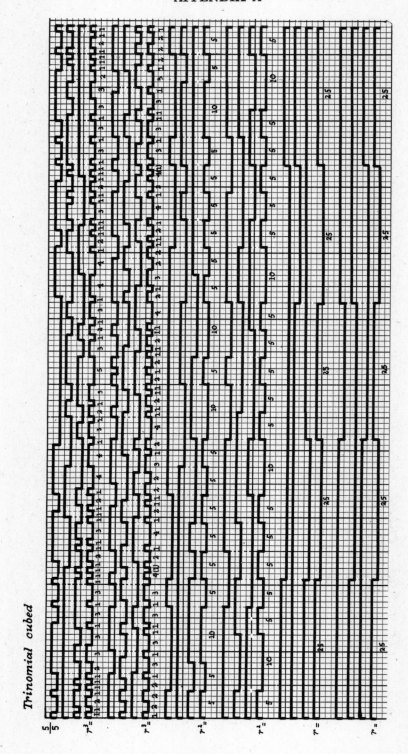

DISTRIBUTIVE INVOLUTION GROUPS

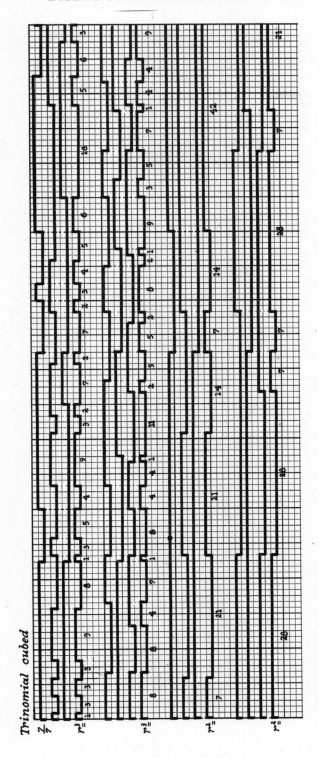

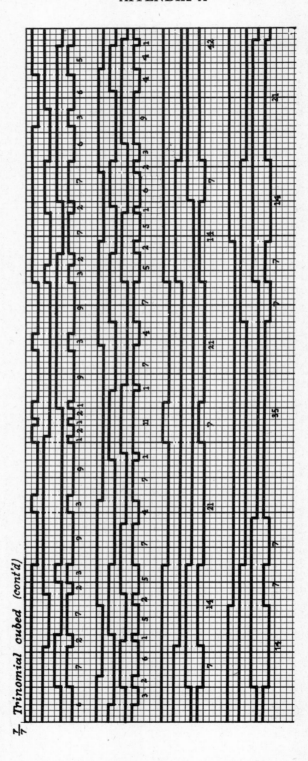

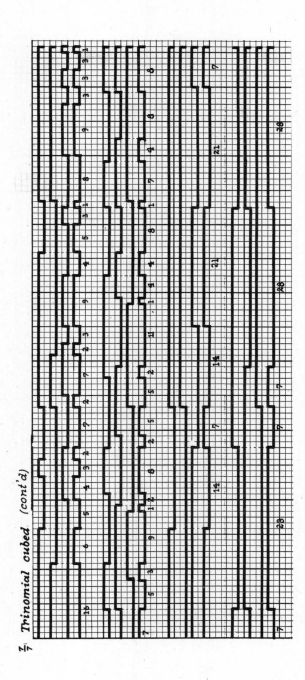

APPENDIX A

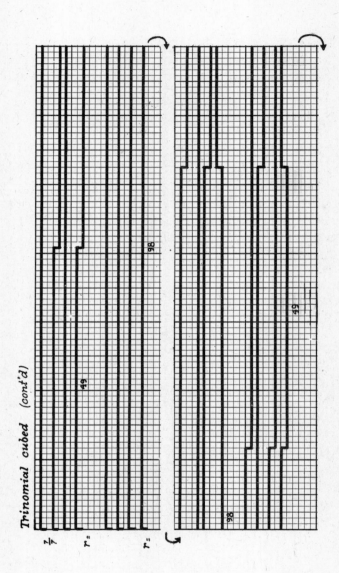

Trinomial cubed (cont'd)

DISTRIBUTIVE INVOLUTION GROUPS 583

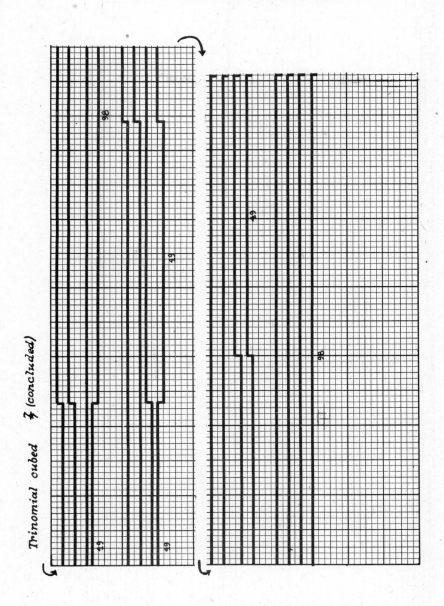

Trinomial cubed 7 (concluded)

APPENDIX A

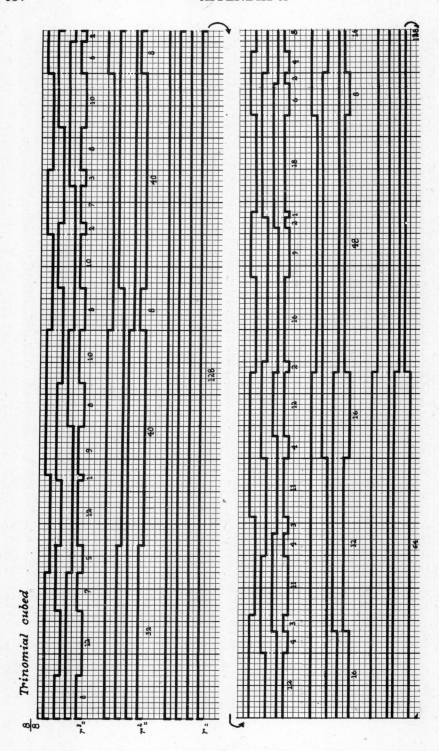

Trinomial cubed

DISTRIBUTIVE INVOLUTION GROUPS

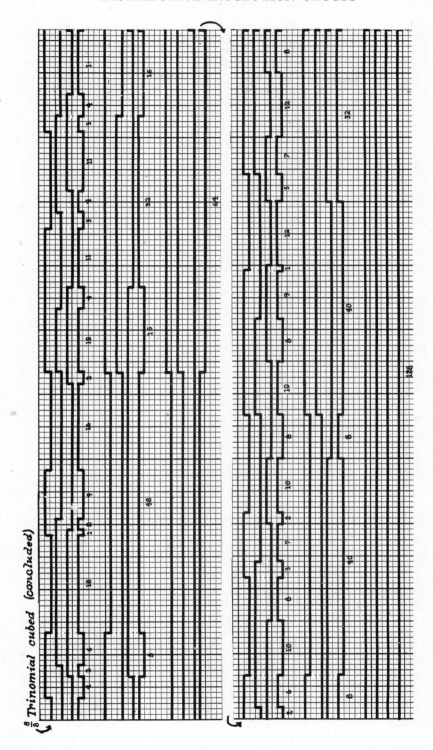

Trinomial cubed (concluded)

586   APPENDIX A

F. Generalization: Distributive Cube of Polynomials (n Terms).

(1) Factorial: $(a+b+c+\ldots+m)^3 = [(a^3+a^2b+a^2c+\ldots+a^2m) + (a^2b+ab^2+abc+\ldots+abm) + (a^2c+abc+ac^2+\ldots+acm) +\ldots+ (a^2m+abm+acm+\ldots+am^2)] + [(a^2b+ab^2+abc+\ldots+abm)+(ab^2+b^3+b^2c+\ldots+b^2m) + (abc+b^2c+bc^2+\ldots+bcm)+\ldots+(abm+b^2m+bcm+\ldots+bm^2)] + [(a^2c+abc+ac^2+\ldots+acm) + (abc+b^2c+bc^2+\ldots+bcm) + (ac^2+bc^2+c^3+\ldots+c^2m) +\ldots+ (acm+bcm+c^2m+\ldots+cm^2)] +\ldots+ [(a^2m+abm+acm+\ldots+am^2) + (abm+b^2m+bcm+\ldots+bm^2) + (acm+bcm+c^2m+\ldots+cm^2) +\ldots+ (am^2+bm^2+cm^2+\ldots+m^3)]$;

(2) Fractional: $\left(\dfrac{a}{a+b+c+\ldots+m}+\dfrac{b}{a+b+c+\ldots+m}+\dfrac{c}{a+b+c+\ldots+m}+\ldots+\dfrac{m}{a+b+c+\ldots+m}\right)^3 = \left\{\left[\dfrac{a^3}{(a+b+c+\ldots+m)^3}+\dfrac{a^2b}{(a+b+c+\ldots+m)^3}+\dfrac{a^2c}{(a+b+c+\ldots+m)^3}+\ldots+\dfrac{a^2m}{(a+b+c+\ldots+m)^3}\right]+\left[\dfrac{a^2b}{(a+b+c+\ldots+m)^3}+\dfrac{ab^2}{(a+b+c+\ldots+m)^3}+\dfrac{abc}{(a+b+c+\ldots+m)^3}+\ldots+\dfrac{abm}{(a+b+c+\ldots+m)^3}\right]+\left[\dfrac{a^2c}{(a+b+c+\ldots+m)^3}+\dfrac{abc}{(a+b+c+\ldots+m)^3}+\dfrac{ac^2}{(a+b+c+\ldots+m)^3}+\ldots+\dfrac{acm}{(a+b+c+\ldots+m)^3}\right]+\ldots+\left[\dfrac{a^2m}{(a+b+c+\ldots+m)^3}+\dfrac{abm}{(a+b+c+\ldots+m)^3}+\dfrac{acm}{(a+b+c+\ldots+m)^3}+\ldots+\dfrac{am^2}{(a+b+c+\ldots+m)^3}\right]\right\}+\left\{\left[\dfrac{a^2b}{(a+b+c+\ldots+m)^3}+\dfrac{ab^2}{(a+b+c+\ldots+m)^3}+\dfrac{abc}{(a+b+c+\ldots+m)^3}+\ldots+\dfrac{abm}{(a+b+c+\ldots+m)^3}\right]+\left[\dfrac{ab^2}{(a+b+c+\ldots+m)^3}+\dfrac{b^3}{(a+b+c+\ldots+m)^3}+\dfrac{b^2c}{(a+b+c+\ldots+m)^3}+\ldots+\dfrac{b^2m}{(a+b+c+\ldots+m)^3}\right]+\left[\dfrac{abc}{(a+b+c+\ldots+m)^3}+\dfrac{b^2c}{(a+b+c+\ldots+m)^3}+\dfrac{bc^2}{(a+b+c+\ldots+m)^3}+\ldots+\dfrac{bcm}{(a+b+c+\ldots+m)^3}\right]+\ldots+\left[\dfrac{abm}{(a+b+c+\ldots+m)^3}+\dfrac{b^2m}{(a+b+c+\ldots+m)^3}+\dfrac{bcm}{(a+b+c+\ldots+m)^3}+\ldots+\dfrac{bm^2}{(a+b+c+\ldots+m)^3}\right]\right\}+$

$$+ \left\{ \left[\frac{a^2c}{(a+b+c+\ldots+m)^3} + \frac{abc}{(a+b+c+\ldots+m)^3} + \frac{ac^2}{(a+b+c+\ldots+m)^3} + \ldots \right.\right.$$
$$\left.\ldots + \frac{acm}{(a+b+c+\ldots+m)^3} \right] + \left[\frac{abc}{(a+b+c+\ldots+m)^3} + \frac{b^2c}{(a+b+c+\ldots+m)^3} + \right.$$
$$+ \frac{bc^2}{(a+b+c+\ldots+m)^3} + \ldots + \frac{bcm}{(a+b+c+\ldots+m)^3} \bigg] + \bigg[\frac{ac^2}{(a+b+c+\ldots+m)^3} +$$
$$+ \frac{bc^2}{(a+b+c+\ldots+m)^3} + \frac{c^3}{(a+b+c+\ldots+m)^3} + \ldots + \frac{c^2m}{(a+b+c+\ldots+m)^3} \bigg] + \ldots$$
$$\ldots + \bigg[\frac{acm}{(a+b+c+\ldots+m)^3} + \frac{bcm}{(a+b+c+\ldots+m)^3} + \frac{c^2m}{(a+b+c+\ldots+m)^3} + \ldots$$
$$\ldots + \frac{cm^2}{(a+b+c+\ldots+m)^3} \bigg] \bigg\} + \ldots + \bigg\{ \bigg[\frac{a^2m}{(a+b+c+\ldots+m)^3} + \frac{abm}{(a+b+c+\ldots+m)^3} +$$
$$+ \frac{acm}{(a+b+c+\ldots+m)^3} + \ldots + \frac{am^2}{(a+b+c+\ldots+m)^3} \bigg] + \bigg[\frac{abm}{(a+b+c+\ldots+m)^3} +$$
$$+ \frac{b^2m}{(a+b+c+\ldots+m)^3} + \frac{bcm}{(a+b+c+\ldots+m)^3} + \ldots + \frac{bm^2}{(a+b+c+\ldots+m)^3} \bigg] +$$
$$+ \bigg[\frac{acm}{(a+b+c+\ldots+m)^3} + \frac{bcm}{(a+b+c+\ldots+m)^3} + \frac{c^2m}{(a+b+c+\ldots+m)^3} + \ldots$$
$$\ldots + \frac{cm^2}{(a+b+c+\ldots+m)^3} \bigg] + \ldots + \bigg[\frac{am^2}{(a+b+c+\ldots+m)^3} + \frac{bm^2}{(a+b+c+\ldots+m)^3} +$$
$$+ \frac{cm^2}{(a+b+c+\ldots+m)^3} + \ldots + \frac{m^3}{(a+b+c+\ldots+m)^3} \bigg] \bigg\}.$$

Synchronization of the first power group with the third power group.

(1) Factorial: $S = a(a+b+c+\ldots+m)^2 + b(a+b+c+\ldots+m)^2 +$
$$+ c(a+b+c+\ldots+m)^2 + \ldots + m(a+b+c+\ldots+m)^2;$$

(2) Fractional: $S = \dfrac{a}{a+b+c+\ldots+m} \cdot \left(\dfrac{a+b+c+\ldots+m}{a+b+c+\ldots+m}\right)^2 +$
$$+ \frac{b}{a+b+c+\ldots+m} \cdot \left(\frac{a+b+c+\ldots+m}{a+b+c+\ldots+m}\right)^2 + \frac{c}{a+b+c+\ldots+m} \cdot$$
$$\cdot \left(\frac{a+b+c+\ldots+m}{a+b+c+\ldots+m}\right)^2 + \ldots + \frac{m}{a+b+c+\ldots+m} \cdot \left(\frac{a+b+c+\ldots+m}{a+b+c+\ldots+m}\right)^2.$$

APPENDIX A

Synchronization of the second power group with the third power group.

(1) Factorial: $S = [a^2(a+b+c+\ldots+m)+ab(a+b+c+\ldots+m)+$
$+ac(a+b+c+\ldots+m)+\ldots+am(a+b+c+\ldots+m)]+$
$+[ab(a+b+c+\ldots+m)+b^2(a+b+c+\ldots+m)+bc(a+b+c+\ldots+m)+\ldots$
$\ldots+bm(a+b+c+\ldots+m)]+[ac(a+b+c+\ldots+m)+$
$+bc(a+b+c+\ldots+m)+c^2(a+b+c+\ldots+m)+\ldots$
$\ldots+cm(a+b+c+\ldots+m)]+\ldots+[am(a+b+c+\ldots+m)+$
$+bm(a+b+c+\ldots+m)+cm(a+b+c+\ldots+m)+\ldots$
$\ldots+m^2(a+b+c+\ldots+m)];$

(2) Fractional: $S = \left[\dfrac{a^2}{(a+b+c+\ldots+m)^2}\cdot\left(\dfrac{a+b+c+\ldots+m}{a+b+c+\ldots+m}\right)+\right.$
$+\dfrac{ab}{(a+b+c+\ldots+m)^2}\cdot\left(\dfrac{a+b+c+\ldots+m}{a+b+c+\ldots+m}\right)+\dfrac{ac}{(a+b+c+\ldots+m)^2}\cdot$
$\left.\left(\dfrac{a+b+c+\ldots+m}{a+b+c+\ldots+m}\right)+\ldots+\dfrac{am}{(a+b+c+\ldots+m)^2}\cdot\left(\dfrac{a+b+c+\ldots+m}{a+b+c+\ldots+m}\right)\right]+$
$+\left[\dfrac{ab}{(a+b+c+\ldots+m)^2}\cdot\left(\dfrac{a+b+c+\ldots+m}{a+b+c+\ldots+m}\right)+\dfrac{b^2}{(a+b+c+\ldots+m)^2}\cdot\right.$
$\left(\dfrac{a+b+c+\ldots+m}{a+b+c+\ldots+m}\right)+\dfrac{bc}{(a+b+c+\ldots+m)^2}\cdot\left(\dfrac{a+b+c+\ldots+m}{a+b+c+\ldots+m}\right)+\ldots$
$\left.\ldots+\dfrac{bm}{(a+b+c+\ldots+m)^2}\cdot\left(\dfrac{a+b+c+\ldots+m}{a+b+c+\ldots+m}\right)\right]+\left[\dfrac{ac}{(a+b+c+\ldots+m)^2}\cdot\right.$
$\left(\dfrac{a+b+c+\ldots+m}{a+b+c+\ldots+m}\right)+\dfrac{bc}{(a+b+c+\ldots+m)^2}\cdot\left(\dfrac{a+b+c+\ldots+m}{a+b+c+\ldots+m}\right)+$
$+\dfrac{c^2}{(a+b+c+\ldots+m)^2}\cdot\left(\dfrac{a+b+c+\ldots+m}{a+b+c+\ldots+m}\right)+\ldots+\dfrac{cm}{(a+b+c+\ldots+m)^2}\cdot$
$\left.\left(\dfrac{a+b+c+\ldots+m}{a+b+c+\ldots+m}\right)\right]+\ldots+\left[\dfrac{am}{(a+b+c+\ldots+m)^2}\cdot\left(\dfrac{a+b+c+\ldots+m}{a+b+c+\ldots+m}\right)+\right.$
$+\dfrac{bm}{(a+b+c+\ldots+m)^2}\cdot\left(\dfrac{a+b+c+\ldots+m}{a+b+c+\ldots+m}\right)+\dfrac{cm}{(a+b+c+\ldots+m)^2}\cdot$
$\left.\left(\dfrac{a+b+c+\ldots+m}{a+b+c+\ldots+m}\right)+\ldots+\dfrac{m^2}{(a+b+c+\ldots+m)^2}\cdot\left(\dfrac{a+b+c+\ldots+m}{a+b+c+\ldots+m}\right)\right].$

DISTRIBUTIVE INVOLUTION GROUPS

(2) Fractional: $S = \left[\dfrac{a^2}{(a+b+c)^2} \cdot \left(\dfrac{a+b+c}{a+b+c}\right) + \dfrac{ab}{(a+b+c)^2} \cdot \left(\dfrac{a+b+c}{a+b+c}\right) + \dfrac{ac}{(a+b+c)^2} \cdot \left(\dfrac{a+b+c}{a+b+c}\right)\right] + \left[\dfrac{ab}{(a+b+c)^2} \cdot \left(\dfrac{a+b+c}{a+b+c}\right) + \dfrac{b^2}{(a+b+c)^2} \cdot \left(\dfrac{a+b+c}{a+b+c}\right) + \dfrac{bc}{(a+b+c)^2} \cdot \left(\dfrac{a+b+c}{a+b+c}\right)\right] + \left[\dfrac{ac}{(a+b+c)^2} \cdot \left(\dfrac{a+b+c}{a+b+c}\right) + \dfrac{bc}{(a+b+c)^2} \cdot \left(\dfrac{a+b+c}{a+b+c}\right) + \dfrac{c^2}{(a+b+c)^2} \cdot \left(\dfrac{a+b+c}{a+b+c}\right)\right].$

Synchronization of the first power group with the third power group.

(1) Factorial: $S = a(a+b+c)^2 + b(a+b+c)^2 + c(a+b+c)^2;$

(2) Fractional: $S = \dfrac{a}{a+b+c} \cdot \left(\dfrac{a+b+c}{a+b+c}\right)^2 + \dfrac{b}{a+b+c} \cdot \left(\dfrac{a+b+c}{a+b+c}\right)^2 + \dfrac{c}{a+b+c} \cdot \left(\dfrac{a+b+c}{a+b+c}\right)^2.$

1. Cube of Quintinomials

$\dfrac{6}{6}$

$\left(\dfrac{1}{6}+\dfrac{1}{6}+\dfrac{2}{6}+\dfrac{1}{6}+\dfrac{1}{6}\right)^3 = \dfrac{(1+1+2+1+1)+(1+1+2+1+1)+(2+2+4+2+2)+}{216}$

$+\dfrac{(1+1+2+1+1)+(1+1+2+1+1)+(1+1+2+1+1)+}{216}$

$+\dfrac{(1+1+2+1+1)+(2+2+4+2+2)+(1+1+2+1+1)+}{216}$

$+\dfrac{(1+1+2+1+1)+(2+2+4+2+2)+(2+2+4+2+2)+}{216}$

$+\dfrac{(4+4+8+4+4)+(2+2+4+2+2)+(2+2+4+2+2)+}{216}$

$+\dfrac{(1+1+2+1+1)+(1+1+2+1+1)+(2+2+4+2+2)+}{216}$

$+\dfrac{(1+1+2+1+1)+(1+1+2+1+1)+(1+1+2+1+1)+}{216}$

$+\dfrac{(1+1+2+1+1)+(2+2+4+2+2)+(1+1+2+1+1)+}{216}$

$+\dfrac{(1+1+2+1+1)}{216}$

$\frac{6}{6}$ (continued)

$$\left(\frac{1}{6}+\frac{2}{6}+\frac{1}{6}+\frac{1}{6}+\frac{1}{6}\right)^3 = \frac{(1+2+1+1+1)+(2+4+2+2+2)+(1+2+1+1+1)+}{216}$$

$$\frac{+(1+2+1+1+1)+(1+2+1+1+1)+(2+4+2+2+2)+}{216}$$

$$\frac{+(4+8+4+4+4)+(2+4+2+2+2)+(2+4+2+2+2)+}{216}$$

$$\frac{+(2+4+2+2+2)+(1+2+1+1+1)+(2+4+2+2+2)+}{216}$$

$$\frac{+(1+2+1+1+1)+(1+2+1+1+1)+(1+2+1+1+1)+}{216}$$

$$\frac{+(1+2+1+1+1)+(2+4+2+2+2)+(1+2+1+1+1)+}{216}$$

$$\frac{+(1+2+1+1+1)+(1+2+1+1+1)+(1+2+1+1+1)+}{216}$$

$$\frac{+(2+4+2+2+2)+(1+2+1+1+1)+(1+2+1+1+1)+}{216}$$

$$\frac{+(1+2+1+1+1)}{216}$$

$$\left(\frac{2}{6}+\frac{1}{6}+\frac{1}{6}+\frac{1}{6}+\frac{1}{6}\right)^3 = \frac{(8+4+4+4+4)+(4+2+2+2+2)+(4+2+2+2+2)+}{216}$$

$$\frac{+(4+2+2+2+2)+(4+2+2+2+2)+(4+2+2+2+2)+}{216}$$

$$\frac{+(2+1+1+1+1)+(2+1+1+1+1)+(2+1+1+1+1)+}{216}$$

$$\frac{+(2+1+1+1+1)+(4+2+2+2+2)+(2+1+1+1+1)+}{216}$$

$$\frac{+(2+1+1+1+1)+(2+1+1+1+1)+(2+1+1+1+1)+}{216}$$

$$\frac{+(4+2+2+2+2)+(2+1+1+1+1)+(2+1+1+1+1)+}{216}$$

$$\frac{+(2+1+1+1+1)+(2+1+1+1+1)+(4+2+2+2+2)+}{216}$$

$$\frac{+(2+1+1+1+1)+(2+1+1+1+1)+(2+1+1+1+1)+}{216}$$

$$\frac{+(2+1+1+1+1)}{216}$$

$\dfrac{6}{6}$ (continued)

$$\left(\frac{1}{6}+\frac{1}{6}+\frac{1}{6}+\frac{1}{6}+\frac{2}{6}\right)^3 = \frac{(1+1+1+1+2)+(1+1+1+1+2)+(1+1+1+1+2)+}{216}$$
$$\frac{+(1+1+1+1+2)+(2+2+2+2+4)+(1+1+1+1+2)+}{216}$$
$$\frac{+(1+1+1+1+2)+(1+1+1+1+2)+(1+1+1+1+2)+}{216}$$
$$\frac{+(2+2+2+2+4)+(1+1+1+1+2)+(1+1+1+1+2)+}{216}$$
$$\frac{+(1+1+1+1+2)+(1+1+1+1+2)+(2+2+2+2+4)+}{216}$$
$$\frac{+(1+1+1+1+2)+(1+1+1+1+2)+(1+1+1+1+2)+}{216}$$
$$\frac{+(1+1+1+1+2)+(2+2+2+2+4)+(2+2+2+2+4)+}{216}$$
$$\frac{+(2+2+2+2+4)+(2+2+2+2+4)+(2+2+2+2+4)+}{216}$$
$$\frac{+(4+4+4+4+8)}{216}$$

$$\left(\frac{1}{6}+\frac{1}{6}+\frac{1}{6}+\frac{2}{6}+\frac{1}{6}\right)^3 = \frac{(1+1+1+2+1)+(1+1+1+2+1)+(1+1+1+2+1)+}{216}$$
$$\frac{+(2+2+2+4+2)+(1+1+1+2+1)+(1+1+1+2+1)+}{216}$$
$$\frac{+(1+1+1+2+1)+(1+1+1+2+1)+(2+2+2+4+2)+}{216}$$
$$\frac{+(1+1+1+2+1)+(1+1+1+2+1)+(1+1+1+2+1)+}{216}$$
$$\frac{+(1+1+1+2+1)+(2+2+2+4+2)+(1+1+1+2+1)+}{216}$$
$$\frac{+(2+2+2+4+2)+(2+2+2+4+2)+(2+2+2+4+2)+}{216}$$
$$\frac{+(4+4+4+8+4)+(2+2+2+4+2)+(1+1+1+2+1)+}{216}$$
$$\frac{+(1+1+1+2+1)+(1+1+1+2+1)+(2+2+2+4+2)+}{216}$$
$$\frac{+(1+1+1+2+1)}{216}$$

APPENDIX A

$\frac{6}{6}$ (continued)

$$r^3 = \frac{58(1)+2+1+1+2+32(1)+2+20(1)+2+32(1)+2+1+1+2+58(1)}{216}$$

$$\frac{6}{6}\left(\frac{1}{6}+\frac{1}{6}+\frac{2}{6}+\frac{1}{6}+\frac{1}{6}\right)^2 = \frac{(6+6+12+6+6)+(6+6+12+6+6)+}{216}$$
$$\frac{+(12+12+24+12+12)+(6+6+12+6+6)+(6+6+12+6+6)}{216}$$

$$\frac{6}{6}\left(\frac{1}{6}+\frac{2}{6}+\frac{1}{6}+\frac{1}{6}+\frac{1}{6}\right)^2 = \frac{(6+12+6+6+6)+(12+24+12+12+12)+}{216}$$
$$\frac{+(6+12+6+6+6)+(6+12+6+6+6)+(6+12+6+6+6)}{216}$$

$$\frac{6}{6}\left(\frac{2}{6}+\frac{1}{6}+\frac{1}{6}+\frac{1}{6}+\frac{1}{6}\right)^2 = \frac{(24+12+12+12+12)+(12+6+6+6+6)+}{216}$$
$$\frac{+(12+6+6+6+6)+(12+6+6+6+6)+(12+6+6+6+6)}{216}$$

$$\frac{6}{6}\left(\frac{1}{6}+\frac{1}{6}+\frac{1}{6}+\frac{1}{6}+\frac{2}{6}\right)^2 = \frac{(6+6+6+6+12)+(6+6+6+6+12)+}{216}$$
$$\frac{+(6+6+6+6+12)+(6+6+6+6+12)+(12+12+12+12+24)}{216}$$

$$\frac{6}{6}\left(\frac{1}{6}+\frac{1}{6}+\frac{1}{6}+\frac{2}{6}+\frac{1}{6}\right)^2 = \frac{(6+6+6+12+6)+(6+6+6+12+6)+}{216}$$
$$\frac{+(6+6+6+12+6)+(12+12+12+24+12)+(6+6+6+12+6)}{216}$$

$$r^2 = \frac{36(6)}{216}$$

$$\frac{36}{36}\left(\frac{1}{6}+\frac{1}{6}+\frac{2}{6}+\frac{1}{6}+\frac{1}{6}\right) = \frac{36+36+72+36+36}{216}$$

$$\frac{36}{36}\left(\frac{1}{6}+\frac{2}{6}+\frac{1}{6}+\frac{1}{6}+\frac{1}{6}\right) = \frac{36+72+36+36+36}{216}$$

$$\frac{36}{36}\left(\frac{2}{6}+\frac{1}{6}+\frac{1}{6}+\frac{1}{6}+\frac{1}{6}\right) = \frac{72+36+36+36+36}{216}$$

$$\frac{36}{36}\left(\frac{1}{6}+\frac{1}{6}+\frac{1}{6}+\frac{1}{6}+\frac{2}{6}\right) = \frac{36+36+36+36+72}{216}$$

DISTRIBUTIVE INVOLUTION GROUPS

$\frac{6}{6}$ (concluded)

$$\frac{36}{36}\left(\frac{1}{6}+\frac{1}{6}+\frac{1}{6}+\frac{2}{6}+\frac{1}{6}\right) = \frac{36+36+36+72+36}{216}$$

$$r = \frac{6(36)}{216}$$

$\frac{7}{7}$

$$\left(\frac{1}{7}+\frac{2}{7}+\frac{1}{7}+\frac{2}{7}+\frac{1}{7}\right)^3 = \frac{(1+2+1+2+1)+(2+4+2+4+2)+(1+2+1+2+1)+}{343}$$

$$\frac{+(2+4+2+4+2)+(1+2+1+2+1)+(2+4+2+4+2)+}{343}$$

$$\frac{+(4+8+4+8+4)+(2+4+2+4+2)+(4+8+4+8+4)+}{343}$$

$$\frac{+(2+4+2+4+2)+(1+2+1+2+1)+(2+4+2+4+2)+}{343}$$

$$\frac{+(1+2+1+2+1)+(2+4+2+4+2)+(1+2+1+2+1)+}{343}$$

$$\frac{+(2+4+2+4+2)+(4+8+4+8+4)+(2+4+2+4+2)+}{343}$$

$$\frac{+(4+8+4+8+4)+(2+4+2+4+2)+(1+2+1+2+1)+}{343}$$

$$\frac{+(2+4+2+4+2)+(1+2+1+2+1)+(2+4+2+4+2)+}{343}$$

$$\frac{+(1+2+1+2+1)}{343}$$

$$\left(\frac{2}{7}+\frac{1}{7}+\frac{2}{7}+\frac{1}{7}+\frac{1}{7}\right)^3 = \frac{(8+4+8+4+4)+(4+2+4+2+2)+(8+4+8+4+4)+}{343}$$

$$\frac{+(4+2+4+2+2)+(4+2+4+2+2)+(4+2+4+2+2)+}{343}$$

$$\frac{+(2+1+2+1+1)+(4+2+4+2+2)+(2+1+2+1+1)+}{343}$$

$$\frac{+(2+1+2+1+1)+(8+4+8+4+4)+(4+2+4+2+2)+}{343}$$

$$\frac{+(8+4+8+4+4)+(4+2+4+2+2)+(4+2+4+2+2)+}{343}$$

$\frac{7}{7}$ (continued)

$$\frac{+(4+2+4+2+2)+(2+1+2+1+1)+(4+2+4+2+2)+}{343}$$
$$\frac{+(2+1+2+1+1)+(2+1+2+1+1)+(4+2+4+2+2)+}{343}$$
$$\frac{+(2+1+2+1+1)+(4+2+4+2+2)+(2+1+2+1+1)+}{343}$$
$$\frac{+(2+1+2+1+1)}{343}$$

$$\left(\frac{1}{7}+\frac{2}{7}+\frac{1}{7}+\frac{1}{7}+\frac{2}{7}\right)^3 = \frac{(1+2+1+1+2)+(2+4+2+2+4)+(1+2+1+1+2)+}{343}$$
$$\frac{+(1+2+1+1+2)+(2+4+2+2+4)+(2+4+2+2+4)+}{343}$$
$$\frac{+(4+8+4+4+8)+(2+4+2+2+4)+(2+4+2+2+4)+}{343}$$
$$\frac{+(4+8+4+4+8)+(1+2+1+1+2)+(2+4+2+2+4)+}{343}$$
$$\frac{+(1+2+1+1+2)+(1+2+1+1+2)+(2+4+2+2+4)+}{343}$$
$$\frac{+(1+2+1+1+2)+(2+4+2+2+4)+(1+2+1+1+2)+}{343}$$
$$\frac{+(1+2+1+1+2)+(2+4+2+2+4)+(2+4+2+2+4)+}{343}$$
$$\frac{+(4+8+4+4+8)+(2+4+2+2+4)+(2+4+2+2+4)+}{343}$$
$$\frac{+(4+8+4+4+8)}{343}$$

$$\left(\frac{2}{7}+\frac{1}{7}+\frac{1}{7}+\frac{2}{7}+\frac{1}{7}\right)^3 = \frac{(8+4+4+8+4)+(4+2+2+4+2)+(4+2+2+4+2)+}{343}$$
$$\frac{+(8+4+4+8+4)+(4+2+2+4+2)+(4+2+2+4+2)+}{343}$$
$$\frac{+(2+1+1+2+1)+(2+1+1+2+1)+(4+2+2+4+2)+}{343}$$
$$\frac{+(2+1+1+2+1)+(4+2+2+4+2)+(2+1+1+2+1)+}{343}$$

$\frac{7}{7}$ (continued)

$$\frac{+(2+1+1+2+1)+(4+2+2+4+2)+(2+1+1+2+1)+}{343}$$

$$\frac{+(8+4+4+8+4)+(4+2+2+4+2)+(4+2+2+4+2)+}{343}$$

$$\frac{+(8+4+4+8+4)+(4+2+2+4+2)+(4+2+2+4+2)+}{343}$$

$$\frac{+(2+1+1+2+1)+(2+1+1+2+1)+(4+2+2+4+2)+}{343}$$

$$\frac{+(2+1+1+2+1)}{343}$$

$$\left(\frac{1}{7}+\frac{1}{7}+\frac{2}{7}+\frac{1}{7}+\frac{2}{7}\right)^3 = \frac{(1+1+2+1+2)+(1+1+2+1+2)+(2+2+4+2+4)+}{343}$$

$$\frac{+(1+1+2+1+2)+(2+2+4+2+4)+(1+1+2+1+2)+}{343}$$

$$\frac{+(1+1+2+1+2)+(2+2+4+2+4)+(1+1+2+1+2)+}{343}$$

$$\frac{+(2+2+4+2+4)+(2+2+4+2+4)+(2+2+4+2+4)+}{343}$$

$$\frac{+(4+4+8+4+8)+(2+2+4+2+4)+(4+4+8+4+8)+}{343}$$

$$\frac{+(1+1+2+1+2)+(1+1+2+1+2)+(2+2+4+2+4)+}{343}$$

$$\frac{+(1+1+2+1+2)+(2+2+4+2+4)+(2+2+4+2+4)+}{343}$$

$$\frac{+(2+2+4+2+4)+(4+4+8+4+8)+(2+2+4+2+4)+}{343}$$

$$\frac{+(4+4+8+4+8)}{343}$$

$$r^3 = \frac{9(1)+2+11(1)+2+19(1)+2+1+2+1+1+1+2+15(1)+2+14(1)+}{343}$$

$$\frac{+2+1+1+2+4(1)+2+27(1)+2+1+3+7(1)+2+10(1)+2+6(1)+2+}{343}$$

$$\frac{+27(1)+2+6(1)+2+10(1)+2+7(1)+3+1+2+27(1)+2+4(1)+2+1+}{343}$$

$\frac{7}{7}$ (continued)

$$\frac{+1+2+14(1)+2+15(1)+2+1+1+1+2+1+2+19(1)+2+11(1)+2+}{343}$$

$$\frac{+9(1)}{343}$$

$$\left(\frac{1}{7}+\frac{1}{7}+\frac{3}{7}+\frac{1}{7}+\frac{1}{7}\right)^3 = \frac{(1+1+3+1+1)+(1+1+3+1+1)+(3+3+9+3+3)+}{343}$$

$$\frac{+(1+1+3+1+1)+(1+1+3+1+1)+(1+1+3+1+1)+}{343}$$

$$\frac{+(1+1+3+1+1)+(3+3+9+3+3)+(1+1+3+1+1)+}{343}$$

$$\frac{+(1+1+3+1+1)+(3+3+9+3+3)+(3+3+9+3+3)+}{343}$$

$$\frac{+(9+9+27+9+9)+(3+3+9+3+3)+(3+3+9+3+3)+}{343}$$

$$\frac{+(1+1+3+1+1)+(1+1+3+1+1)+(3+3+9+3+3)+}{343}$$

$$\frac{+(1+1+3+1+1)+(1+1+3+1+1)+(1+1+3+1+1)+}{343}$$

$$\frac{+(1+1+3+1+1)+(3+3+9+3+3)+(1+1+3+1+1)+}{343}$$

$$\frac{+(1+1+3+1+1)}{343}$$

$$\left(\frac{1}{7}+\frac{3}{7}+\frac{1}{7}+\frac{1}{7}+\frac{1}{7}\right)^3 = \frac{(1+3+1+1+1)+(3+9+3+3+3)+(1+3+1+1+1)+}{343}$$

$$\frac{+(1+3+1+1+1)+(1+3+1+1+1)+(3+9+3+3+3)+}{343}$$

$$\frac{+(9+27+9+9+9)+(3+9+3+3+3)+(3+9+3+3+3)+}{343}$$

$$\frac{+(3+9+3+3+3)+(1+3+1+1+1)+(3+9+3+3+3)+}{343}$$

$$\frac{+(1+3+1+1+1)+(1+3+1+1+1)+(1+3+1+1+1)+}{343}$$

$\frac{7}{7}$ (continued)

$$\frac{+(1+3+1+1+1)+(3+9+3+3+3)+(1+3+1+1+1)+}{343}$$

$$\frac{+(1+3+1+1+1)+(1+3+1+1+1)+(1+3+1+1+1)+}{343}$$

$$\frac{+(3+9+3+3+3)+(1+3+1+1+1)+(1+3+1+1+1)+}{343}$$

$$\frac{+(1+3+1+1+1)}{343}$$

$$\left(\frac{3}{7}+\frac{1}{7}+\frac{1}{7}+\frac{1}{7}+\frac{1}{7}\right)^3 = \frac{(27+9+9+9+9)+(9+3+3+3+3)+}{343}$$

$$\frac{+(9+3+3+3+3)+(9+3+3+3+3)+(9+3+3+3+3)+}{343}$$

$$\frac{+(9+3+3+3+3)+(3+1+1+1+1)+(3+1+1+1+1)+}{343}$$

$$\frac{+(3+1+1+1+1)+(3+1+1+1+1)+(9+3+3+3+3)+}{343}$$

$$\frac{+(3+1+1+1+1)+(3+1+1+1+1)+(3+1+1+1+1)+}{343}$$

$$\frac{+(3+1+1+1+1)+(9+3+3+3+3)+(3+1+1+1+1)+}{343}$$

$$\frac{+(3+1+1+1+1)+(3+1+1+1+1)+(3+1+1+1+1)+}{343}$$

$$\frac{+(9+3+3+3+3)+(3+1+1+1+1)+(3+1+1+1+1)+}{343}$$

$$\frac{+(3+1+1+1+1)+(3+1+1+1+1)}{343}$$

$$\left(\frac{1}{7}+\frac{1}{7}+\frac{1}{7}+\frac{1}{7}+\frac{3}{7}\right)^3 = \frac{(1+1+1+1+3)+(1+1+1+1+3)+(1+1+1+1+3)+}{343}$$

$$\frac{+(1+1+1+1+3)+(3+3+3+3+9)+(1+1+1+1+3)+}{343}$$

$$\frac{+(1+1+1+1+3)+(1+1+1+1+3)+(1+1+1+1+3)+}{343}$$

$$\frac{+(3+3+3+3+9)+(1+1+1+1+3)+(1+1+1+1+3)+}{343}$$

598 APPENDIX A

$\frac{7}{7}$ (continued)

$$\frac{+(1+1+1+1+3)+(1+1+1+1+3)+(3+3+3+3+9)+}{343}$$

$$\frac{+(1+1+1+1+3)+(1+1+1+1+3)+(1+1+1+1+3)+}{343}$$

$$\frac{+(1+1+1+1+3)+(3+3+3+3+9)+(3+3+3+3+9)+}{343}$$

$$\frac{+(3+3+3+3+9)+(3+3+3+3+9)+(3+3+3+3+9)+}{343}$$

$$\frac{+(9+9+9+9+27)}{343}$$

$$\left(\frac{1}{7}+\frac{1}{7}+\frac{1}{7}+\frac{3}{7}+\frac{1}{7}\right)^3 = \frac{(1+1+1+3+1)+(1+1+1+3+1)+(1+1+1+3+1)+}{343}$$

$$\frac{+(3+3+3+9+3)+(1+1+1+3+1)+(1+1+1+3+1)+}{343}$$

$$\frac{+(1+1+1+3+1)+(1+1+1+3+1)+(3+3+3+9+3)+}{343}$$

$$\frac{+(1+1+1+3+1)+(1+1+1+3+1)+(1+1+1+3+1)+}{343}$$

$$\frac{+(1+1+1+3+1)+(3+3+3+9+3)+(1+1+1+3+1)+}{343}$$

$$\frac{+(3+3+3+9+3)+(3+3+3+9+3)+(3+3+3+9+3)+}{343}$$

$$\frac{+(9+9+9+27+9)+(3+3+3+9+3)+(1+1+1+3+1)+}{343}$$

$$\frac{+(1+1+1+3+1)+(1+1+1+3+1)+(3+3+3+9+3)+}{343}$$

$$\frac{+(1+1+1+3+1)}{343}$$

$$r^3 = \frac{25(1)+2+10(1)+2+28(1)+2+12(1)+2+1+1+1+2+6(1)+2+6(1)+}{343}$$

$$\frac{+2+5(1)+2+15(1)+2+3+6(1)+2+11(1)+2+6(1)+2+5(1)+2+7(1)+}{343}$$

$$\frac{+2+5(1)+2+6(1)+2+11(1)+2+6(1)+3+1+2+15(1)+2+5(1)+2+}{343}$$

$$\frac{+6(1)+2+6(1)+2+1+1+1+2+12(1)+2+28(1)+2+10(1)+2+25(1)}{343}$$

$\frac{7}{7}$ (continued)

$$\frac{7}{7}\left(\frac{1}{7}+\frac{2}{7}+\frac{1}{7}+\frac{2}{7}+\frac{1}{7}\right)^2 = \frac{(7+14+7+14+7)+(14+28+14+28+14)+}{343}$$
$$\frac{+(7+14+7+14+7)+(14+28+14+28+14)+(7+14+7+14+7)}{343}$$

$$\frac{7}{7}\left(\frac{2}{7}+\frac{1}{7}+\frac{2}{7}+\frac{1}{7}+\frac{1}{7}\right)^2 = \frac{(28+14+28+14+14)+(14+7+14+7+7)+}{343}$$
$$\frac{+(28+14+28+14+14)+(14+7+14+7+7)+(14+7+14+7+7)}{343}$$

$$\frac{7}{7}\left(\frac{1}{7}+\frac{2}{7}+\frac{1}{7}+\frac{1}{7}+\frac{2}{7}\right)^2 = \frac{(7+14+7+7+14)+(14+28+14+14+28)+}{343}$$
$$\frac{+(7+14+7+7+14)+(7+14+7+7+14)+(14+28+14+14+28)}{343}$$

$$\frac{7}{7}\left(\frac{2}{7}+\frac{1}{7}+\frac{1}{7}+\frac{2}{7}+\frac{1}{7}\right)^2 = \frac{(28+14+14+28+14)+(14+7+7+14+7)+}{343}$$
$$\frac{+(14+7+7+14+7)+(28+14+14+28+14)+(14+7+7+14+7)}{343}$$

$$\frac{7}{7}\left(\frac{1}{7}+\frac{1}{7}+\frac{2}{7}+\frac{1}{7}+\frac{2}{7}\right)^2 = \frac{(7+7+14+7+14)+(7+7+14+7+14)+}{343}$$
$$\frac{+(14+14+28+14+28)+(7+7+14+7+14)+(14+14+28+14+28)}{343}$$

$r^2 = \frac{49(7)}{343}$

$$\frac{7}{7}\left(\frac{1}{7}+\frac{1}{7}+\frac{3}{7}+\frac{1}{7}+\frac{1}{7}\right)^2 = \frac{(7+7+21+7+7)+(7+7+21+7+7)+}{343}$$
$$\frac{+(21+21+63+21+21)+(7+7+21+7+7)+(7+7+21+7+7)}{343}$$

$$\frac{7}{7}\left(\frac{1}{7}+\frac{3}{7}+\frac{1}{7}+\frac{1}{7}+\frac{1}{7}\right)^2 = \frac{(7+21+7+7+7)+(21+63+21+21+21)+}{343}$$
$$\frac{+(7+21+7+7+7)+(7+21+7+7+7)+(7+21+7+7+7)}{343}$$

$$\frac{7}{7}\left(\frac{3}{7}+\frac{1}{7}+\frac{1}{7}+\frac{1}{7}+\frac{1}{7}\right)^2 = \frac{(63+21+21+21+21)+(21+7+7+7+7)+}{343}$$
$$\frac{+(21+7+7+7+7)+(21+7+7+7+7)+(21+7+7+7+7)}{343}$$

APPENDIX A

$\frac{7}{7}$ (concluded)

$$\frac{7}{7}\left(\frac{1}{7}+\frac{1}{7}+\frac{1}{7}+\frac{1}{7}+\frac{3}{7}\right)^2 = \frac{(7+7+7+7+21)+(7+7+7+7+21)+}{343}$$
$$\frac{+(7+7+7+7+21)+(7+7+7+7+21)+(21+21+21+21+63)}{343}$$

$$\frac{7}{7}\left(\frac{1}{7}+\frac{1}{7}+\frac{1}{7}+\frac{3}{7}+\frac{1}{7}\right)^2 = \frac{(7+7+7+21+7)+(7+7+7+21+7)+}{343}$$
$$\frac{+(7+7+7+21+7)+(21+21+21+63+21)+(7+7+7+21+7)}{343}$$

$$r^2 = \frac{49(7)}{343}$$

$$\frac{49}{49}\left(\frac{1}{7}+\frac{2}{7}+\frac{1}{7}+\frac{2}{7}+\frac{1}{7}\right) = \frac{49+98+49+98+49}{343}$$

$$\frac{49}{49}\left(\frac{2}{7}+\frac{1}{7}+\frac{2}{7}+\frac{1}{7}+\frac{1}{7}\right) = \frac{98+49+98+49+49}{343}$$

$$\frac{49}{49}\left(\frac{1}{7}+\frac{2}{7}+\frac{1}{7}+\frac{1}{7}+\frac{2}{7}\right) = \frac{49+98+49+49+98}{343}$$

$$\frac{49}{49}\left(\frac{2}{7}+\frac{1}{7}+\frac{1}{7}+\frac{2}{7}+\frac{1}{7}\right) = \frac{98+49+49+98+49}{343}$$

$$\frac{49}{49}\left(\frac{1}{7}+\frac{1}{7}+\frac{2}{7}+\frac{1}{7}+\frac{2}{7}\right) = \frac{49+49+98+49+98}{343}$$

$$r = \frac{7(49)}{343}$$

$$\frac{49}{49}\left(\frac{1}{7}+\frac{1}{7}+\frac{3}{7}+\frac{1}{7}+\frac{1}{7}\right) = \frac{49+49+147+49+49}{343}$$

$$\frac{49}{49}\left(\frac{1}{7}+\frac{3}{7}+\frac{1}{7}+\frac{1}{7}+\frac{1}{7}\right) = \frac{49+147+49+49+49}{343}$$

$$\frac{49}{49}\left(\frac{3}{7}+\frac{1}{7}+\frac{1}{7}+\frac{1}{7}+\frac{1}{7}\right) = \frac{147+49+49+49+49}{343}$$

$$\frac{49}{49}\left(\frac{1}{7}+\frac{1}{7}+\frac{1}{7}+\frac{1}{7}+\frac{3}{7}\right) = \frac{49+49+49+49+147}{343}$$

$$\frac{49}{49}\left(\frac{1}{7}+\frac{1}{7}+\frac{1}{7}+\frac{3}{7}+\frac{1}{7}\right) = \frac{49+49+49+147+49}{343}$$

$$r = \frac{7(49)}{343}$$

$\frac{8}{8}$

$$\left(\frac{2}{8}+\frac{1}{8}+\frac{2}{8}+\frac{1}{8}+\frac{2}{8}\right)^3 = \frac{(8+4+8+4+8)+(4+2+4+2+4)+(8+4+8+4+8)+}{512}$$

$$\frac{+(4+2+4+2+4)+(8+4+8+4+8)+(4+2+4+2+4)+}{512}$$

$$\frac{+(2+1+2+1+2)+(4+2+4+2+4)+(2+1+2+1+2)+}{512}$$

$$\frac{+(4+2+4+2+4)+(8+4+8+4+8)+(4+2+4+2+4)+}{512}$$

$$\frac{+(8+4+8+4+8)+(4+2+4+2+4)+(8+4+8+4+8)+}{512}$$

$$\frac{+(4+2+4+2+4)+(2+1+2+1+2)+(4+2+4+2+4)+}{512}$$

$$\frac{+(2+1+2+1+2)+(4+2+4+2+4)+(8+4+8+4+8)+}{512}$$

$$\frac{+(4+2+4+2+4)+(8+4+8+4+8)+(4+2+4+2+4)+}{512}$$

$$\frac{+(8+4+8+4+8)}{512}$$

$$\left(\frac{1}{8}+\frac{2}{8}+\frac{1}{8}+\frac{2}{8}+\frac{2}{8}\right)^3 = \frac{(1+2+1+2+2)+(2+4+2+4+4)+(1+2+1+2+2)+}{512}$$

$$\frac{+(2+4+2+4+4)+(2+4+2+4+4)+(2+4+2+4+4)+}{512}$$

$$\frac{+(4+8+4+8+8)+(2+4+2+4+4)+(4+8+4+8+8)+}{512}$$

$$\frac{+(4+8+4+8+8)+(1+2+1+2+2)+(2+4+2+4+4)+}{512}$$

$$\frac{+(1+2+1+2+2)+(2+4+2+4+4)+(2+4+2+4+4)+}{512}$$

$$\frac{+(2+4+2+4+4)+(4+8+4+8+8)+(2+4+2+4+4)+}{512}$$

$$\frac{+(4+8+4+8+8)+(4+8+4+8+8)+(2+4+2+4+4)+}{512}$$

$$\frac{+(4+8+4+8+8)+(2+4+2+4+4)+(4+8+4+8+8)+}{512}$$

$$\frac{+(4+8+4+8+8)}{512}$$

$\frac{8}{8}$ (continued)

$$\left(\frac{2}{8}+\frac{1}{8}+\frac{2}{8}+\frac{2}{8}+\frac{1}{8}\right)^3 = \frac{(8+4+8+8+4)+(4+2+4+4+2)+(8+4+8+8+4)+}{512}$$

$$\frac{+(8+4+8+8+4)+(4+2+4+4+2)+(4+2+4+4+2)+}{512}$$

$$\frac{+(2+1+2+2+1)+(4+2+4+4+2)+(4+2+4+4+2)+}{512}$$

$$\frac{+(2+1+2+2+1)+(8+4+8+8+4)+(4+2+4+4+2)+}{512}$$

$$\frac{+(8+4+8+8+4)+(8+4+8+8+4)+(4+2+4+4+2)+}{512}$$

$$\frac{+(8+4+8+8+4)+(4+2+4+4+2)+(8+4+8+8+4)+}{512}$$

$$\frac{+(8+4+8+8+4)+(4+2+4+4+2)+(4+2+4+4+2)+}{512}$$

$$\frac{+(2+1+2+2+1)+(4+2+4+4+2)+(4+2+4+4+2)+}{512}$$

$$\frac{+(2+1+2+2+1)}{512}$$

$$\left(\frac{1}{8}+\frac{2}{8}+\frac{2}{8}+\frac{1}{8}+\frac{2}{8}\right)^3 = \frac{(1+2+2+1+2)+(2+4+4+2+4)+(2+4+4+2+4)+}{512}$$

$$\frac{+(1+2+2+1+2)+(2+4+4+2+4)+(2+4+4+2+4)+}{512}$$

$$\frac{+(4+8+8+4+8)+(4+8+8+4+8)+(2+4+4+2+4)+}{512}$$

$$\frac{+(4+8+8+4+8)+(2+4+4+2+4)+(4+8+8+4+8)+}{512}$$

$$\frac{+(4+8+8+4+8)+(2+4+4+2+4)+(4+8+8+4+8)+}{512}$$

$$\frac{+(1+2+2+1+2)+(2+4+4+2+4)+(2+4+4+2+4)+}{512}$$

$$\frac{+(1+2+2+1+2)+(2+4+4+2+4)+(2+4+4+2+4)+}{512}$$

$$\frac{+(4+8+8+4+8)+(4+8+8+4+8)+(2+4+4+2+4)+}{512}$$

$$\frac{+(4+8+8+4+8)}{512}$$

$\frac{8}{8}$ (continued)

$$\left(\frac{2}{8}+\frac{2}{8}+\frac{1}{8}+\frac{2}{8}+\frac{1}{8}\right)^3 = \frac{(8+8+4+8+4)+(8+8+4+8+4)+(4+4+2+4+2)+}{512}$$

$$\frac{+(8+8+4+8+4)+(4+4+2+4+2)+(8+8+4+8+4)+}{512}$$

$$\frac{+(8+8+4+8+4)+(4+4+2+4+2)+(8+8+4+8+4)+}{512}$$

$$\frac{+(4+4+2+4+2)+(4+4+2+4+2)+(4+4+2+4+2)+}{512}$$

$$\frac{+(2+2+1+2+1)+(4+4+2+4+2)+(2+2+1+2+1)+}{512}$$

$$\frac{+(8+8+4+8+4)+(8+8+4+8+4)+(4+4+2+4+2)+}{512}$$

$$\frac{+(8+8+4+8+4)+(4+4+2+4+2)+(4+4+2+4+2)+}{512}$$

$$\frac{+(4+4+2+4+2)+(2+2+1+2+1)+(4+4+2+4+2)+}{512}$$

$$\frac{+(2+2+1+2+1)}{512}$$

$$r^3 = \frac{1+2+1+1+1+7(2)+4+4(1)+6(2)+6(1)+7(2)+4+8(2)+4+4(2)+}{512}$$

$$\frac{+5(4)+8(2)+4+7(2)+6(1)+9(2)+4(1)+6(2)+3(1)+2+5(1)+6(2)+}{512}$$

$$\frac{+4+4+1+2+1+36(2)+1+2+1+4+4+6(2)+5(1)+2+3(1)+6(2)+}{512}$$

$$\frac{+4(1)+9(2)+6(1)+7(2)+4+8(2)+5(4)+4(2)+4+8(2)+4+7(2)+}{512}$$

$$\frac{+6(1)+6(2)+4(1)+4+7(2)+1+1+1+2+1}{512}$$

$$\frac{8}{8}\left(\frac{2}{8}+\frac{1}{8}+\frac{2}{8}+\frac{1}{8}+\frac{2}{8}\right)^2 = \frac{(32+16+32+16+32)+(16+8+16+8+16)+}{512}$$

$$\frac{+(32+16+32+16+32)+(16+8+16+8+16)+(32+16+32+16+32)}{512}$$

$$\frac{8}{8}\left(\frac{1}{8}+\frac{2}{8}+\frac{1}{8}+\frac{2}{8}+\frac{2}{8}\right)^2 = \frac{(8+16+8+16+16)+(16+32+16+32+32)+}{512}$$

$$\frac{+(8+16+8+16+16)+(16+32+16+32+32)+(16+32+16+32+32)}{512}$$

$\frac{8}{8}$ (concluded)

$$\frac{8}{8}\left(\frac{2}{8}+\frac{1}{8}+\frac{2}{8}+\frac{2}{8}+\frac{1}{8}\right)^2 = \frac{(32+16+32+32+16)+(16+8+16+16+8)+}{512}$$

$$\frac{+(32+16+32+32+16)+(32+16+32+32+16)+(16+8+16+16+8)}{512}$$

$$\frac{8}{8}\left(\frac{1}{8}+\frac{2}{8}+\frac{2}{8}+\frac{1}{8}+\frac{2}{8}\right)^2 = \frac{(8+16+16+8+16)+(16+32+32+16+32)+}{512}$$

$$\frac{+(16+32+32+16+32)+(8+16+16+8+16)+(16+32+32+16+32)}{512}$$

$$\frac{8}{8}\left(\frac{2}{8}+\frac{2}{8}+\frac{1}{8}+\frac{2}{8}+\frac{1}{8}\right)^2 = \frac{(32+32+16+32+16)+(32+32+16+32+16)+}{512}$$

$$\frac{+(16+16+8+16+8)+(32+32+16+32+16)+(16+16+8+16+8)}{512}$$

$$r^2 = \frac{8+16+8+8+8+6(16)+10(8)+4(16)+10(8)+6(16)+8+8+8+16+8}{512}$$

$$\frac{64}{64}\left(\frac{2}{8}+\frac{1}{8}+\frac{2}{8}+\frac{1}{8}+\frac{2}{8}\right) = \frac{128+64+128+64+128}{512}$$

$$\frac{64}{64}\left(\frac{1}{8}+\frac{2}{8}+\frac{1}{8}+\frac{2}{8}+\frac{2}{8}\right) = \frac{64+128+64+128+128}{512}$$

$$\frac{64}{64}\left(\frac{2}{8}+\frac{1}{8}+\frac{2}{8}+\frac{2}{8}+\frac{1}{8}\right) = \frac{128+64+128+128+64}{512}$$

$$\frac{64}{64}\left(\frac{1}{8}+\frac{2}{8}+\frac{2}{8}+\frac{1}{3}+\frac{2}{8}\right) = \frac{64+128+128+64+128}{512}$$

$$\frac{64}{64}\left(\frac{2}{8}+\frac{2}{8}+\frac{1}{8}+\frac{2}{8}+\frac{1}{8}\right) = \frac{128+128+64+128+64}{512}$$

$$r = \frac{8(64)}{512}$$

$\frac{9}{9}$

$$\left(\frac{1}{9}+\frac{3}{9}+\frac{1}{9}+\frac{3}{9}+\frac{1}{9}\right)^3 = \frac{(1+3+1+3+1)+(3+9+3+9+3)+(1+3+1+3+1)+}{729}$$

$$\frac{+(3+9+3+9+3)+(1+3+1+3+1)+(3+9+3+9+3)+}{729}$$

$$\frac{+(9+27+9+27+9)+(3+9+3+9+3)+(9+27+9+27+9)+}{729}$$

$$\frac{+(3+9+3+9+3)+(1+3+1+3+1)+(3+9+3+9+3)+}{729}$$

$$\frac{+(1+3+1+3+1)+(3+9+3+9+3)+(1+3+1+3+1)+}{729}$$

$$\frac{+(3+9+3+9+3)+(9+27+9+27+9)+(3+9+3+9+3)+}{729}$$

$$\frac{+(9+27+9+27+9)+(3+9+3+9+3)+(1+3+1+3+1)+}{729}$$

$$\frac{+(3+9+3+9+3)+(1+3+1+3+1)+(3+9+3+9+3)+}{729}$$

$$\frac{+(1+3+1+3+1)}{729}$$

$$\left(\frac{3}{9}+\frac{1}{9}+\frac{3}{9}+\frac{1}{9}+\frac{1}{9}\right)^3 = \frac{(27+9+27+9+9)+(9+3+9+3+3)+}{729}$$

$$\frac{+(27+9+27+9+9)+(9+3+9+3+3)+(9+3+9+3+3)+}{729}$$

$$\frac{+(9+3+9+3+3)+(3+1+3+1+1)+(9+3+9+3+3)+}{729}$$

$$\frac{+(3+1+3+1+1)+(3+1+3+1+1)+(27+9+27+9+9)+}{729}$$

$$\frac{+(9+3+9+3+3)+(27+9+27+9+9)+(9+3+9+3+3)+}{729}$$

$$\frac{+(9+3+9+3+3)+(9+3+9+3+3)+(3+1+3+1+1)+}{729}$$

$$\frac{+(9+3+9+3+3)+(3+1+3+1+1)+(3+1+3+1+1)+}{729}$$

$$\frac{+(9+3+9+3+3)+(3+1+3+1+1)+(9+3+9+3+3)+}{729}$$

$$\frac{+(3+1+3+1+1)+(3+1+3+1+1)}{729}$$

$\frac{9}{9}$ (continued)

$$\left(\frac{1}{9}+\frac{3}{9}+\frac{1}{9}+\frac{1}{9}+\frac{3}{9}\right)^3 = \frac{(1+3+1+1+3)+(3+9+3+3+9)+(1+3+1+1+3)+}{729}$$

$$\frac{+(1+3+1+1+3)+(3+9+3+3+9)+(3+9+3+3+9)+}{729}$$

$$\frac{+(9+27+9+9+27)+(3+9+3+3+9)+(3+9+3+3+9)+}{729}$$

$$\frac{+(9+27+9+9+27)+(1+3+1+1+3)+(3+9+3+3+9)+}{729}$$

$$\frac{+(1+3+1+1+3)+(1+3+1+1+3)+(3+9+3+3+9)+}{729}$$

$$\frac{+(1+3+1+1+3)+(3+9+3+3+9)+(1+3+1+1+3)+}{729}$$

$$\frac{+(1+3+1+1+3)+(3+9+3+3+9)+(3+9+3+3+9)+}{729}$$

$$\frac{+(9+27+9+9+27)+(3+9+3+3+9)+(3+9+3+3+9)+}{729}$$

$$\frac{+(9+27+9+9+27)}{729}$$

$$\left(\frac{3}{9}+\frac{1}{9}+\frac{1}{9}+\frac{3}{9}+\frac{1}{9}\right)^3 = \frac{(27+9+9+27+9)+(9+3+3+9+3)+}{729}$$

$$\frac{+(9+3+3+9+3)+(27+9+9+27+9)+(9+3+3+9+3)+}{729}$$

$$\frac{+(9+3+3+9+3)+(3+1+1+3+1)+(3+1+1+3+1)+}{729}$$

$$\frac{+(9+3+3+9+3)+(3+1+1+3+1)+(9+3+3+9+3)+}{729}$$

$$\frac{+(3+1+1+3+1)+(3+1+1+3+1)+(9+3+3+9+3)+}{729}$$

$$\frac{+(3+1+1+3+1)+(27+9+9+27+9)+(9+3+3+9+3)+}{729}$$

$$\frac{+(9+3+3+9+3)+(27+9+9+27+9)+(9+3+3+9+3)+}{729}$$

$$\frac{+(9+3+3+9+3)+(3+1+1+3+1)+(3+1+1+3+1)+}{729}$$

$$\frac{+(9+3+3+9+3)+(3+1+1+3+1)}{729}$$

$\frac{9}{9}$ *(continued)*

$$\left(\frac{1}{9}+\frac{1}{9}+\frac{3}{9}+\frac{1}{9}+\frac{3}{9}\right)^3 = \frac{(1+1+3+1+3)+(1+1+3+1+3)+}{729}$$

$$\frac{+(3+3+9+3+9)+(1+1+3+1+3)+(3+3+9+3+9)+}{729}$$

$$\frac{+(1+1+3+1+3)+(1+1+3+1+3)+(3+3+9+3+9)+}{729}$$

$$\frac{+(1+1+3+1+3)+(3+3+9+3+9)+(3+3+9+3+9)+}{729}$$

$$\frac{+(3+3+9+3+9)+(9+9+27+9+27)+(3+3+9+3+9)+}{729}$$

$$\frac{+(9+9+27+9+27)+(1+1+3+1+3)+(1+1+3+1+3)+}{729}$$

$$\frac{+(3+3+9+3+9)+(1+1+3+1+3)+(3+3+9+3+9)+}{729}$$

$$\frac{+(3+3+9+3+9)+(3+3+9+3+9)+(9+9+27+9+27)+}{729}$$

$$\frac{+3+3+9+3+9)+(9+9+27+9+27)}{729}$$

$$r^3 = \frac{1+1+2+1+1+2+1+1+1+1+2+1+3+3+3+3+6+3+1+3+1+}{729}$$

$$\frac{+1+3+7(1)+7(3)+1+3+1+3+1+1+1+1+2+1+3+1+1+1+2+1+}{729}$$

$$\frac{+4(3)+6+4(3)+1+1+3+1+4(3)+6+3+9+3(3)+6+3+9+10(3)+6+}{729}$$

$$\frac{+6(3)+9+7(3)+1+1+2+1+1+3+1+1+3+1+3+6+4(3)+1+2+1+}{729}$$

$$\frac{+1+1+3+7(1)+3+1+1+2+1+7(3)+1+1+1+2+1+1+2+1+1+1+}{729}$$

$$\frac{+2+1+1+2+1+1+1+7(3)+1+2+1+1+3+7(1)+3+1+1+1+2+}{729}$$

$$\frac{+1+4(3)+6+3+1+3+1+1+3+1+1+2+1+1+7(3)+9+6(3)+6+}{729}$$

$$\frac{+10(3)+9+3+6+3+3+3+9+3+6+4(3)+1+3+1+1+4(3)+6+4(3)+}{729}$$

$$\frac{+1+2+1+1+1+3+1+2+1+1+1+1+3+1+3+1+7(3)+7(1)+}{729}$$

$$\frac{+2+1+1+3+1+3+6+4(3)+1+2+4(1)+2+1+1+2+1+1}{729}$$

$\frac{9}{9}$ (continued)

$$\left(\frac{2}{9}+\frac{2}{9}+\frac{1}{9}+\frac{2}{9}+\frac{2}{9}\right)^3 = \frac{(8+8+4+8+8)+(8+8+4+8+8)+}{729}$$

$$\frac{+(4+4+2+4+4)+(8+8+4+8+8)+(8+8+4+8+8)+}{729}$$

$$\frac{+(8+8+4+8+8)+(8+8+4+8+8)+(4+4+2+4+4)+}{729}$$

$$\frac{+(8+8+4+8+8)+(8+8+4+8+8)+(4+4+2+4+4)+}{729}$$

$$\frac{+(4+4+2+4+4)+(2+2+1+2+2)+(4+4+2+4+4)+}{729}$$

$$\frac{+(4+4+2+4+4)+(8+8+4+8+8)+(8+8+4+8+8)+}{729}$$

$$\frac{+(4+4+2+4+4)+(8+8+4+8+8)+(8+8+4+8+8)+}{729}$$

$$\frac{+(8+8+4+8+8)+(8+8+4+8+8)+(4+4+2+4+4)+}{729}$$

$$\frac{+(8+8+4+8+8)+(8+8+4+8+8)}{729}$$

$$\left(\frac{2}{9}+\frac{1}{9}+\frac{2}{9}+\frac{2}{9}+\frac{2}{9}\right)^3 = \frac{(8+4+8+8+8)+(4+2+4+4+4)+}{729}$$

$$\frac{+(8+4+8+8+8)+(8+4+8+8+8)+(8+4+8+8+8)+}{729}$$

$$\frac{+(4+2+4+4+4)+(2+1+2+2+2)+(4+2+4+4+4)+}{729}$$

$$\frac{+(4+2+4+4+4)+(4+2+4+4+4)+(8+4+8+8+8)+}{729}$$

$$\frac{+(4+2+4+4+4)+(8+4+8+8+8)+(8+4+8+8+8)+}{729}$$

$$\frac{+(8+4+8+8+8)+(8+4+8+8+8)+(4+2+4+4+4)+}{729}$$

$$\frac{+(8+4+8+8+8)+(8+4+8+8+8)+(8+4+8+8+8)+}{729}$$

$$\frac{+(8+4+8+8+8)+(4+2+4+4+4)+(8+4+8+8+8)+}{729}$$

$$\frac{+(8+4+8+8+8)+(8+4+8+8+8)}{729}$$

$\frac{9}{9}$ (continued)

$$\left(\frac{1}{9}+\frac{2}{9}+\frac{2}{9}+\frac{2}{9}+\frac{2}{9}\right)^3 = \frac{(1+2+2+2+2)+(2+4+4+4+4)+}{729}$$

$$\frac{+(2+4+4+4+4)+(2+4+4+4+4)+(2+4+4+4+4)+}{729}$$

$$\frac{+(2+4+4+4+4)+(4+8+8+8+8)+(4+8+8+8+8)+}{729}$$

$$\frac{+(4+8+8+8+8)+(4+8+8+8+8)+(2+4+4+4+4)+}{729}$$

$$\frac{+(4+8+8+8+8)+(4+8+8+8+8)+(4+8+8+8+8)+}{729}$$

$$\frac{+(4+8+8+8+8)+(2+4+4+4+4)+(4+8+8+8+8)+}{729}$$

$$\frac{+(4+8+8+8+8)+(4+8+8+8+8)+(4+8+8+8+8)+}{729}$$

$$\frac{+(2+4+4+4+4)+(4+8+8+8+8)+(4+8+8+8+8)+}{729}$$

$$\frac{+(4+8+8+8+8)+(4+8+8+8+8)}{729}$$

$$\left(\frac{2}{9}+\frac{2}{9}+\frac{2}{9}+\frac{2}{9}+\frac{1}{9}\right)^3 = \frac{(8+8+8+8+4)+(8+8+8+8+4)+}{729}$$

$$\frac{+(8+8+8+8+4)+(8+8+8+8+4)+(4+4+4+4+2)+}{729}$$

$$\frac{+(8+8+8+8+4)+(8+8+8+8+4)+(8+8+8+8+4)+}{729}$$

$$\frac{+(8+8+8+8+4)+(4+4+4+4+2)+(8+8+8+8+4)+}{729}$$

$$\frac{+(8+8+8+8+4)+(8+8+8+8+4)+(8+8+8+8+4)+}{729}$$

$$\frac{+(4+4+4+4+2)+(8+8+8+8+4)+(8+8+8+8+4)+}{729}$$

$$\frac{+(8+8+8+8+4)+(8+8+8+8+4)+(4+4+4+4+2)+}{729}$$

$$\frac{+(4+4+4+4+2)+(4+4+4+4+2)+(4+4+4+4+2)+}{729}$$

$$\frac{+(4+4+4+4+2)+(2+2+2+2+1)}{729}$$

$\frac{9}{9}$ (continued)

$$\left(\frac{2}{9}+\frac{2}{9}+\frac{2}{9}+\frac{1}{9}+\frac{2}{9}\right)^3 = \frac{(8+8+8+4+8)+(8+8+8+4+8)+}{729}$$

$$\frac{+(8+8+8+4+8)+(4+4+4+2+4)+(8+8+8+4+8)+}{729}$$

$$\frac{+(8+8+8+4+8)+(8+8+8+4+8)+(8+8+8+4+8)+}{729}$$

$$\frac{+(4+4+4+2+4)+(8+8+8+4+8)+(8+8+8+4+8)+}{729}$$

$$\frac{+(8+8+8+4+8)+(8+8+8+4+8)+(4+4+4+2+4)+}{729}$$

$$\frac{+(8+8+8+4+8)+(4+4+4+2+4)+(4+4+4+2+4)+}{729}$$

$$\frac{+(4+4+4+2+4)+(2+2+2+1+2)+(4+4+4+2+4)+}{729}$$

$$\frac{+(8+8+8+4+8)+(8+8+8+4+8)+(8+8+8+4+8)+}{729}$$

$$\frac{+(4+4+4+2+4)+(8+8+8+4+8)}{729}$$

$$r^3 = \frac{1+2+2+2+1+1+2+1+3+1+3+1+3+1+3+1+1+3+1+3+1+}{729}$$

$$\frac{+3+1+1+2+1+1+1+3+1+1+2+1+1+3+1+2+4(1)+2+1+3+}{729}$$

$$\frac{+1+1+2+1+3+1+1+1+3+1+1+2+1+4+1+2+1+1+2+1+1+}{729}$$

$$\frac{+2+2+2+1+1+4+2+1+1+2+2+2+1+5+2+1+3+1+1+2+2+}{729}$$

$$\frac{+2+1+1+2+2+2+1+1+4+2+1+3+2+2+1+1+3+1+2+2+}{729}$$

$$\frac{+2+1+2+1+1+2+1+1+2+1+1+3+1+4+3+1+4+3+1+4+2+}{729}$$

$$\frac{+1+3+1+1+2+1+3+1+1+2+1+1+3+1+3+1+3+1+1+1+3+}{729}$$

$$\frac{+5(1)+2+2+1+2+1+1+1+2+1+2+2+4(1)+4+8(1)+2+1+4+}{729}$$

$$\frac{+1+2+1+4+1+2+1+1+2+1+1+2+2+2+1+1+4+2+1+3+4+}{729}$$

DISTRIBUTIVE INVOLUTION GROUPS

$\frac{9}{9}$ (continued)

$$\frac{+1+1+3+1+2+1+1+3+1+2+2+2+1+3+4+1+1+2+1+2+1+}{729}$$

$$\frac{+1+4+3+1+2+2+2+1+3+1+1+2+1+3+1+1+4+3+1+2+4+}{729}$$

$$\frac{+1+1+2+2+2+1+1+2+1+1+2+1+4+1+2+1+4+1+2+8(1)+}{729}$$

$$\frac{+4+4(1)+2+2+1+2+1+1+1+2+1+2+2+5(1)+3+1+1+1+3+}{729}$$

$$\frac{+1+3+1+3+1+1+2+1+1+3+1+2+1+1+3+1+2+4+1+3+4+}{729}$$

$$\frac{+1+3+1+1+2+1+1+2+1+1+2+1+2+2+2+1+3+1+1+2+2+}{729}$$

$$\frac{+3+1+2+4+1+1+2+2+2+1+1+2+2+2+1+1+3+1+2+5+1+}{729}$$

$$\frac{+2+2+2+1+1+2+4+1+1+2+2+2+1+1+2+1+1+2+1+4+1+}{729}$$

$$\frac{+2+1+1+3+1+1+1+3+1+2+1+1+3+1+2+1+1+1+1+2+1+}{729}$$

$$\frac{+3+1+1+2+1+1+3+1+1+1+2+1+1+3+1+3+1+3+1+1+3+}{729}$$

$$\frac{+1+3+1+3+1+3+1+2+1+1+2+2+2+1}{729}$$

$$\frac{9}{9}\left(\frac{1}{9}+\frac{3}{9}+\frac{1}{9}+\frac{3}{9}+\frac{1}{9}\right)^2 = \frac{(9+27+9+27+9)+(27+81+27+81+27)+}{729}$$

$$\frac{+(9+27+9+27+9)+(27+81+27+81+27)+(9+27+9+27+9)}{729}$$

$$\frac{9}{9}\left(\frac{3}{9}+\frac{1}{9}+\frac{3}{9}+\frac{1}{9}+\frac{1}{9}\right)^2 = \frac{(81+27+81+27+27)+(27+9+27+9+9)+}{729}$$

$$\frac{+(81+27+81+27+27)+(27+9+27+9+9)+(27+9+27+9+9)}{729}$$

$$\frac{9}{9}\left(\frac{1}{9}+\frac{3}{9}+\frac{1}{9}+\frac{1}{9}+\frac{3}{9}\right)^2 = \frac{(9+27+9+9+27)+(27+81+27+27+81)+}{729}$$

$$\frac{+(9+27+9+9+27)+(9+27+9+9+27)+(27+81+27+27+81)}{729}$$

$\frac{9}{9}$ (continued)

$$\frac{9}{9}\left(\frac{3}{9}+\frac{1}{9}+\frac{1}{9}+\frac{3}{9}+\frac{1}{9}\right)^2 = \frac{(81+27+27+81+27)+(27+9+9+27+9)+}{729}$$
$$\frac{+(27+9+9+27+9)+(81+27+27+81+27)+(27+9+9+27+9)}{729}$$

$$\frac{9}{9}\left(\frac{1}{9}+\frac{1}{9}+\frac{3}{9}+\frac{1}{9}+\frac{3}{9}\right)^2 = \frac{(9+9+27+9+27)+(9+9+27+9+27)+}{729}$$
$$\frac{+(27+27+81+27+81)+(9+9+27+9+27)+(27+27+81+27+81)}{729}$$

$$r^2 = \frac{9+9+18+9+9+18+4(9)+18+9+27+27+27+27+27+7(9)+}{729}$$
$$\frac{+18+9+9+9+18+7(9)+27+27+27+27+27+9+18+4(9)+18+9+}{729}$$
$$\frac{+9+18+9+9}{729}$$

$$\frac{9}{9}\left(\frac{2}{9}+\frac{2}{9}+\frac{1}{9}+\frac{2}{9}+\frac{2}{9}\right)^2 = \frac{(36+36+18+36+36)+(36+36+18+36+36)+}{729}$$
$$\frac{+(18+18+9+18+18)+(36+36+18+36+36)+(36+36+18+36+36)}{729}$$

$$\frac{9}{9}\left(\frac{2}{9}+\frac{1}{9}+\frac{2}{9}+\frac{2}{9}+\frac{2}{9}\right)^2 = \frac{(36+18+36+36+36)+(18+9+18+18+18)+}{729}$$
$$\frac{+(36+18+36+36+36)+(36+18+36+36+36)+(36+18+36+36+36)}{729}$$

$$\frac{9}{9}\left(\frac{1}{9}+\frac{2}{9}+\frac{2}{9}+\frac{2}{9}+\frac{2}{9}\right)^2 = \frac{(9+18+18+18+18)+(18+36+36+36+36)+}{729}$$
$$\frac{+(18+36+36+36+36)+(18+36+36+36+36)+(18+36+36+36+36)}{729}$$

$$\frac{9}{9}\left(\frac{2}{9}+\frac{2}{9}+\frac{2}{9}+\frac{2}{9}+\frac{1}{9}\right)^2 = \frac{(36+36+36+36+18)+(36+36+36+36+18)+}{729}$$
$$\frac{+(36+36+36+36+18)+(36+36+36+36+18)+(18+18+18+18+9)}{729}$$

$\frac{9}{9}$ (concluded)

$$\frac{9}{9}\left(\frac{2}{9}+\frac{2}{9}+\frac{2}{9}+\frac{1}{9}+\frac{2}{9}\right)^2 = \frac{(36+36+36+18+36)+(36+36+36+18+36)+}{729}$$

$$\frac{+(36+36+36+18+36)+(18+18+18+9+18)+(36+36+36+18+36)}{729}$$

$$r^2 = \frac{9+18+9(9)+18+9+9+18+5(9)+18+9(9)+18+9+9+18+9+18+}{729}$$

$$\frac{+9+9+18+9(9)+18+5(9)+18+9+9+18+9(9)+18+9}{729}$$

$$\frac{81}{81}\left(\frac{1}{9}+\frac{3}{9}+\frac{1}{9}+\frac{3}{9}+\frac{1}{9}\right) = \frac{81+243+81+243+81}{729}$$

$$\frac{81}{81}\left(\frac{3}{9}+\frac{1}{9}+\frac{3}{9}+\frac{1}{9}+\frac{1}{9}\right) = \frac{243+81+243+81+81}{729}$$

$$\frac{81}{81}\left(\frac{1}{9}+\frac{3}{9}+\frac{1}{9}+\frac{1}{9}+\frac{3}{9}\right) = \frac{81+243+81+81+243}{729}$$

$$\frac{81}{81}\left(\frac{3}{9}+\frac{1}{9}+\frac{1}{9}+\frac{3}{9}+\frac{1}{9}\right) = \frac{243+81+81+243+81}{729}$$

$$\frac{81}{81}\left(\frac{1}{9}+\frac{1}{9}+\frac{3}{9}+\frac{1}{9}+\frac{3}{9}\right) = \frac{81+81+243+81+243}{729}$$

$$r = \frac{9(81)}{729}$$

$$\frac{81}{81}\left(\frac{2}{9}+\frac{2}{9}+\frac{1}{9}+\frac{2}{9}+\frac{2}{9}\right) = \frac{162+162+81+162+162}{729}$$

$$\frac{81}{81}\left(\frac{2}{9}+\frac{1}{9}+\frac{2}{9}+\frac{2}{9}+\frac{2}{9}\right) = \frac{162+81+162+162+162}{729}$$

$$\frac{81}{81}\left(\frac{1}{9}+\frac{2}{9}+\frac{2}{9}+\frac{2}{9}+\frac{2}{9}\right) = \frac{81+162+162+162+162}{729}$$

$$\frac{81}{81}\left(\frac{2}{9}+\frac{2}{9}+\frac{2}{9}+\frac{2}{9}+\frac{1}{9}\right) = \frac{162+162+162+162+81}{729}$$

$$\frac{81}{81}\left(\frac{2}{9}+\frac{2}{9}+\frac{2}{9}+\frac{1}{9}+\frac{2}{9}\right) = \frac{162+162+162+81+162}{729}$$

$$r = \frac{9(81)}{729}$$

APPENDIX A

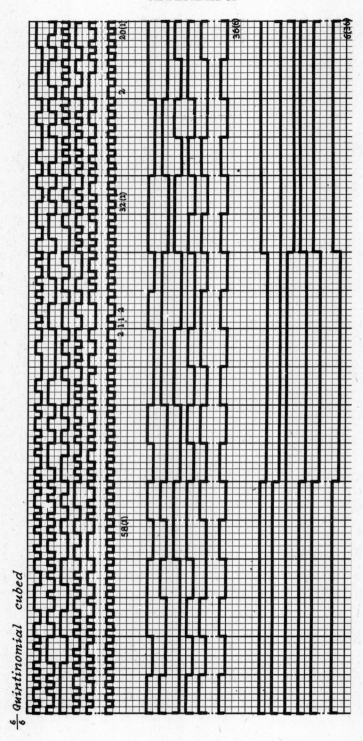

(continued)

DISTRIBUTIVE INVOLUTION GROUPS

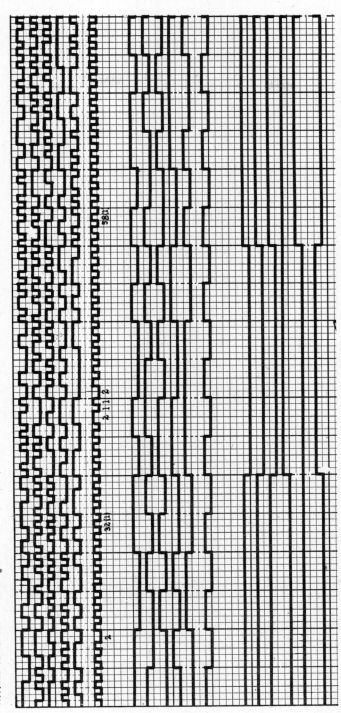

(concluded)

APPENDIX A

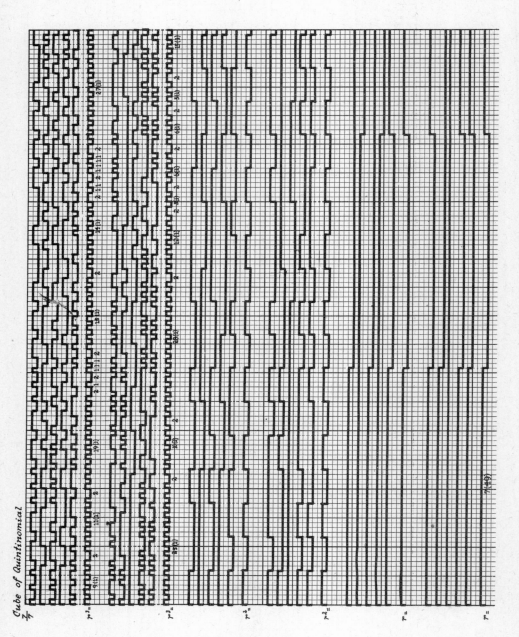

(continued)

DISTRIBUTIVE INVOLUTION GROUPS 617

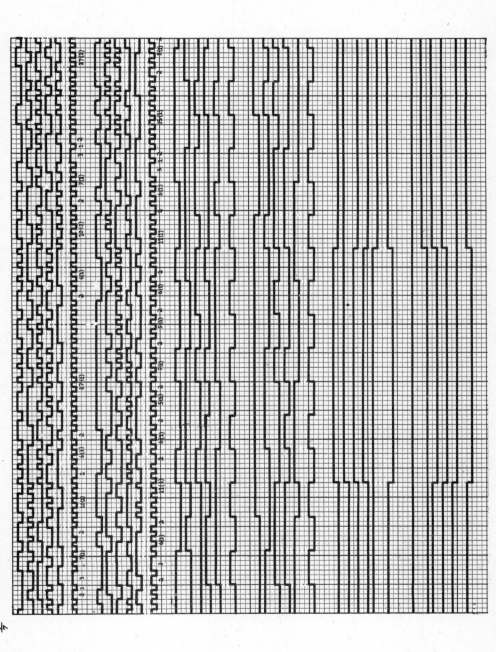

(continued)

618 APPENDIX A

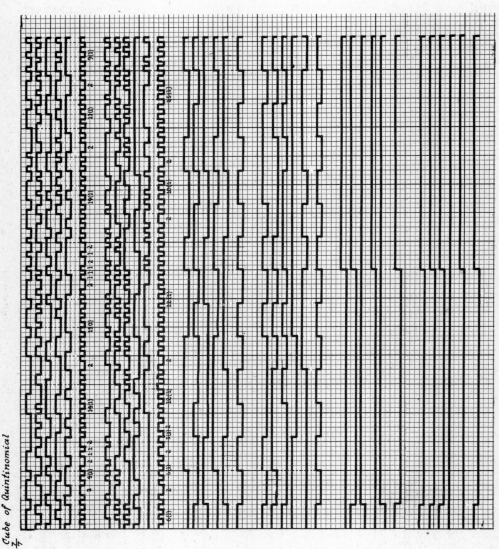

(concluded)

DISTRIBUTIVE INVOLUTION GROUPS 619

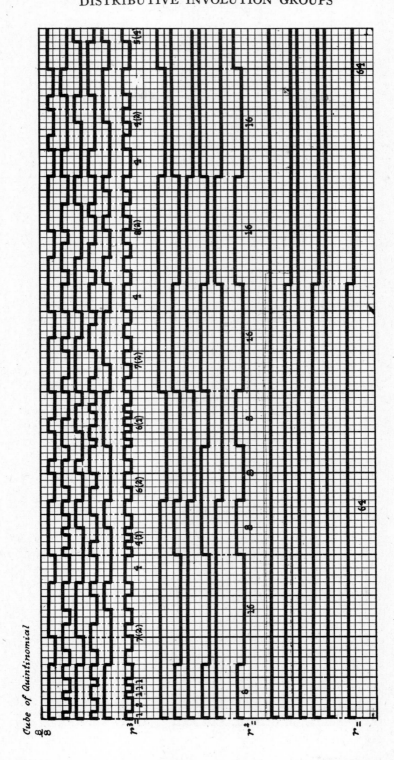

(continued)

APPENDIX A

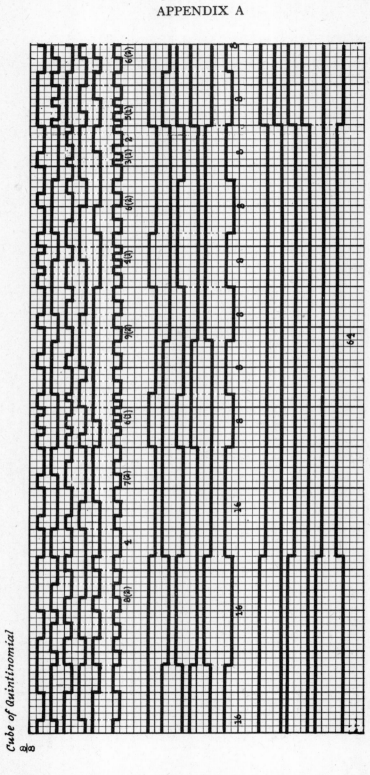

Cube of Quintinomial
8/8

(continued)

DISTRIBUTIVE INVOLUTION GROUPS

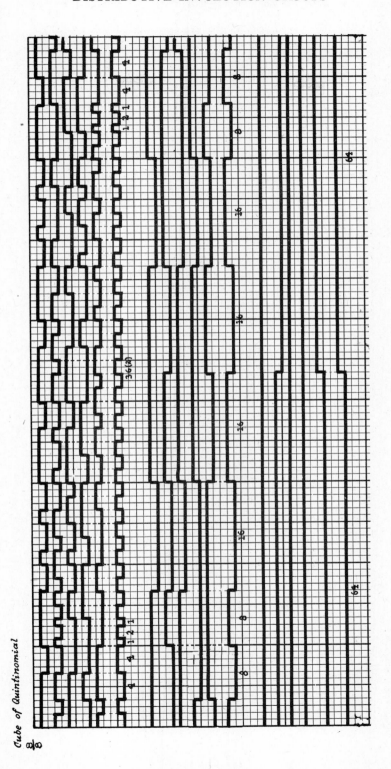

(continued)

APPENDIX A

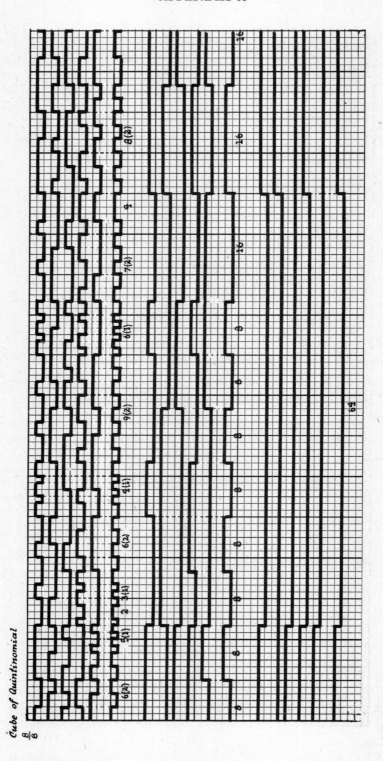

(continued)

DISTRIBUTIVE INVOLUTION GROUPS

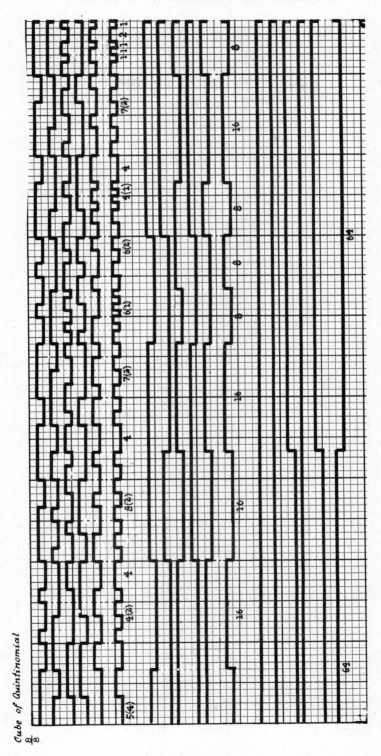

(concluded)

APPENDIX A

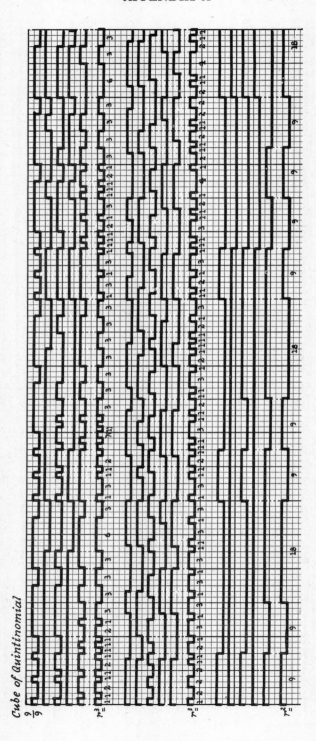

(continued)

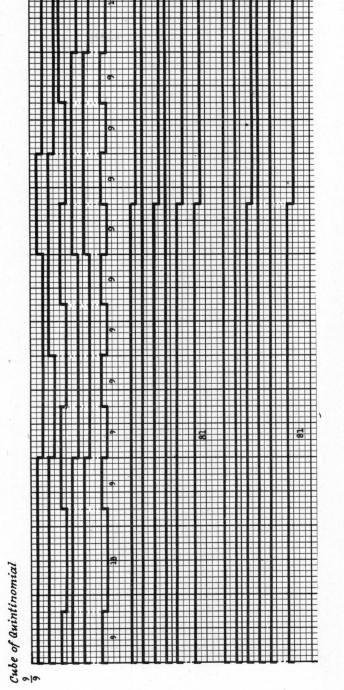

(continued)

APPENDIX A

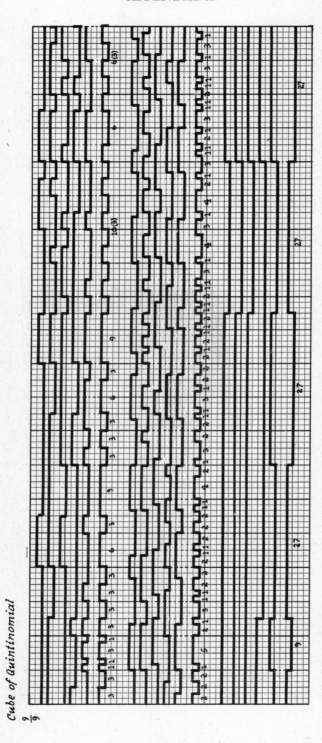

Cube of Quintinomial

(continued)

DISTRIBUTIVE INVOLUTION GROUPS

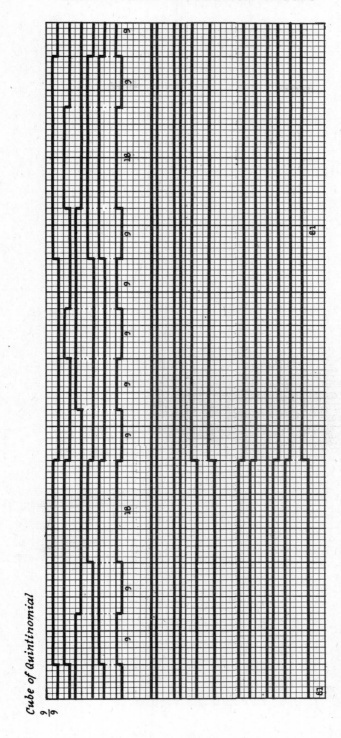

Cube of Quintinomial 9/9

(continued)

APPENDIX A

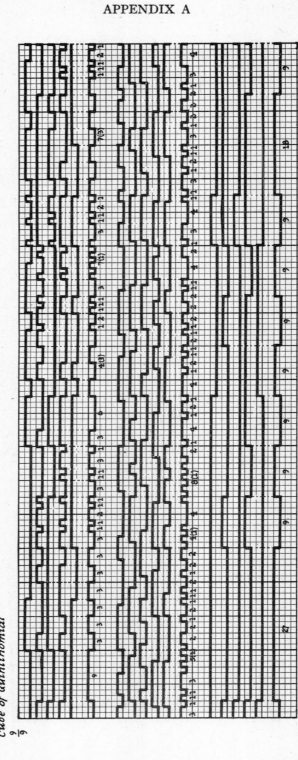

Cube of Quintinomial 9/9

(continued)

DISTRIBUTIVE INVOLUTION GROUPS 629

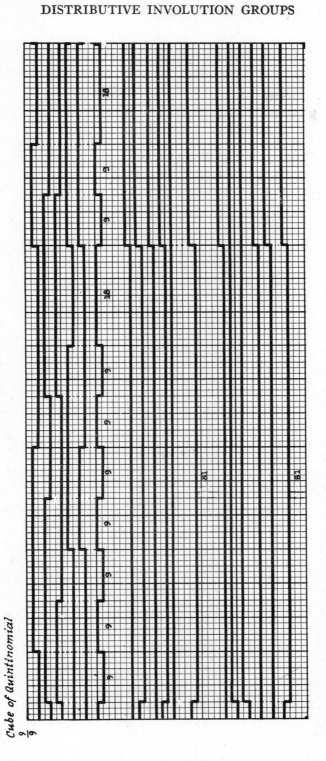

(continued)

630 APPENDIX A

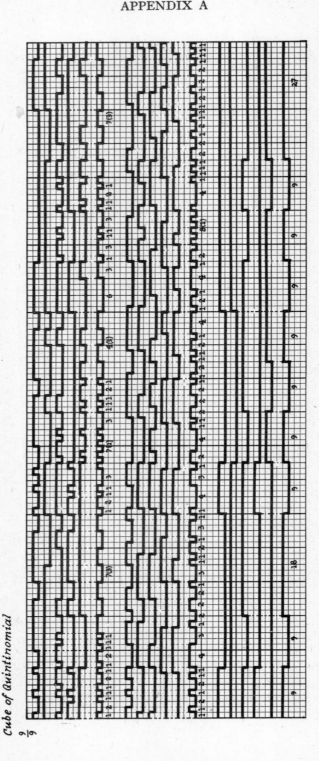

Cube of Quintinomial 9/9

(continued)

DISTRIBUTIVE INVOLUTION GROUPS

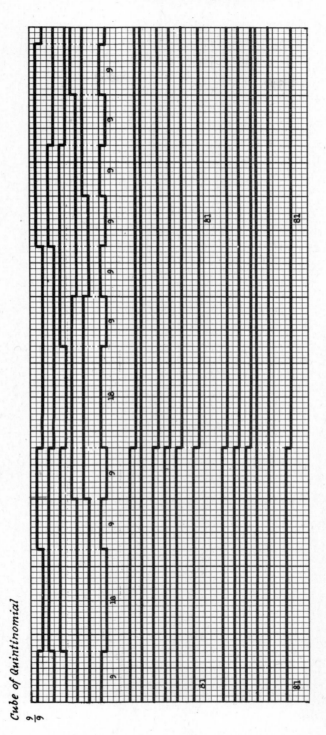

Cube of Quintinomial

(continued)

632 APPENDIX A

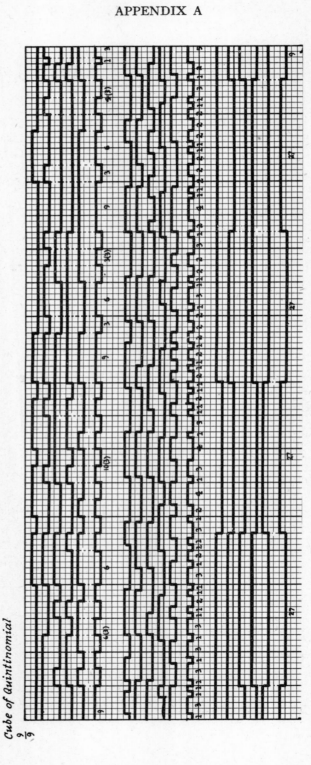

Cube of Quintinomial $\frac{9}{9}$

(continued)

DISTRIBUTIVE INVOLUTION GROUPS

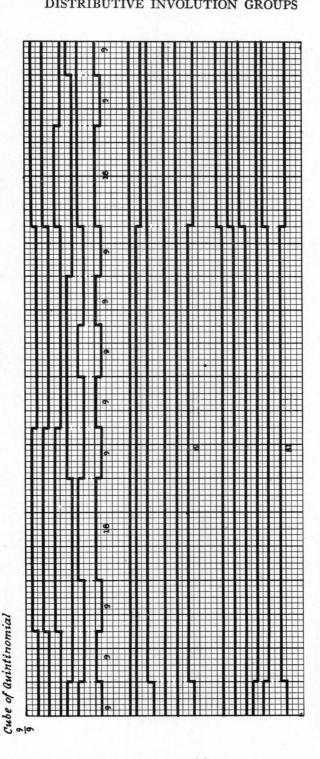

Cube of Quintinomial 9/9

(continued)

APPENDIX A

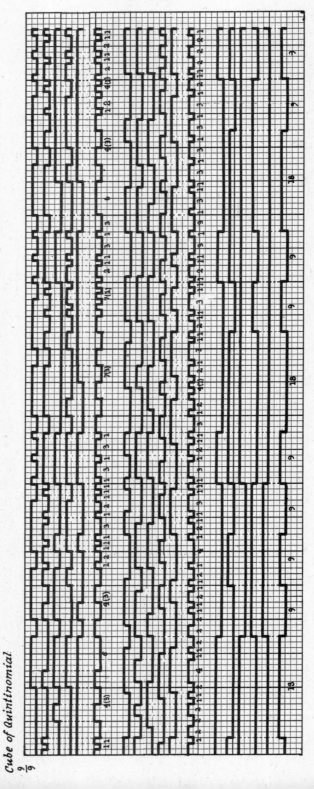

Cube of Quintinomial.

(continued)

DISTRIBUTIVE INVOLUTION GROUPS 635

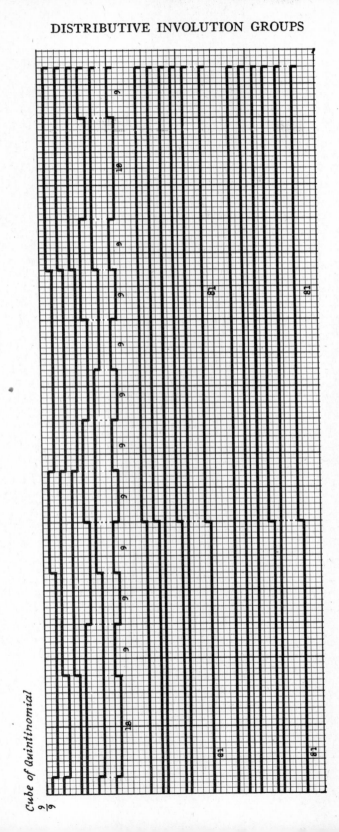

(concluded)

III. GROUPS OF VARIABLE VELOCITY.

These groups represent various forms of *acceleration* and *growth*.
Each group can be used in positive and in negative form (direction).
Both directions can be combined simultaneously and/or in sequence.
The resultant of interference of the same group in simultaneity of both directions (the resultant of acceleration) may be used as an additional component.
The resultant of acceleration has an axis of symmetry, and always consists of an odd number of terms, regardless of the origin of its generator.
The rate of acceleration in some of these groups is constant (as in the geometric progressions or the involution series) and in some—variable (as in the natural integers or the summation series).
The presentation of variable velocities in the form of series containing only integers makes it possible to solve many important problems in the practical execution of design and music.

A. Rhythms of Variable Velocities

1. Natural Harmonic Series

1, 2, 3, 4, 5, 6, 7, 8, 9
r = 1+2+3+3+1+5+2+4+3+4+2+5+1+3+3+2+1

2. Arithmetical Progressions

1, 3, 5, 7, 9, 11, 13, 15, 17, 19
r = 1+3+5+7+3+5+11+13+2+13+11+5+3+7+5+3+1

1, 4, 7, 10, 13, 16, 19, 22
r = 1+4+7+10+13+6+10+6+13+10+7+4+1

3. Geometrical Progressions

1, 2, 4, 8, 16, 32
r = 1+2+4+8+16+1+16+8+4+2+1

1, 3, 9, 27
r = 1+3+9+14+9+3+1

3, 6, 12, 24, 48
r = 3+6+12+24+3+24+12+6+3

2, 6, 18, 54
r = 2+6+18+28+18+6+2

GROUPS OF VARIABLE VELOCITY

4. Power Series

2, 4, 8, 16, 32
r = 2+4+8+16+2+16+8+4+2

3, 9, 27, 81
r = 3+9+27+42+27+9+3

5. Summation Series

1, 2, 3, 5, 8, 13, 21
r = 1+3+5+8+4+9+4+8+5+3+1

1, 3, 4, 7, 11, 18
r = 1+3+4+7+3+8+3+7+4+3+1

1, 4, 5, 9, 14, 23
r = 1+4+5+9+4+10+4+9+5+4+1

6. Arithmetical Progressions with Variable Differences

1, 2, 4, 7, 11, 16, 22, 29
r = 1+2+4+7+11+4+12+10+12+4+11+7+4+2+1

7. Prime Number Series

1, 2, 3, 5, 7, 11, 13, 17, 19, 23
r = 1+2+3+5+7+5+6+12+17+13+6+5+7+5+3+2+1

APPENDIX A

RHYTHMS OF VARIABLE VELOCITIES

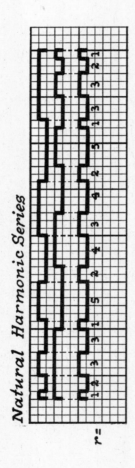

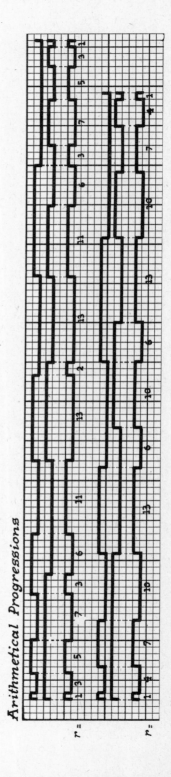

GROUPS OF VARIABLE VELOCITY

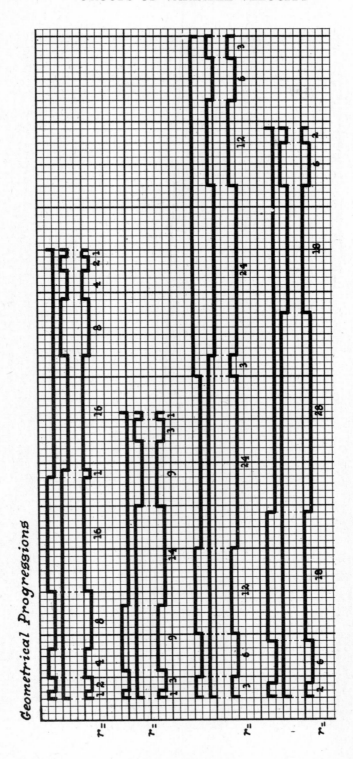

APPENDIX A

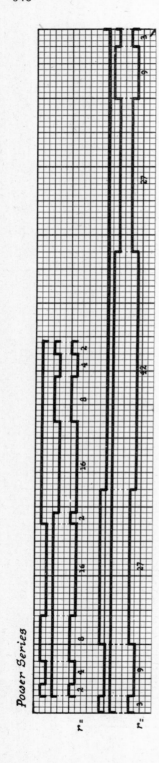

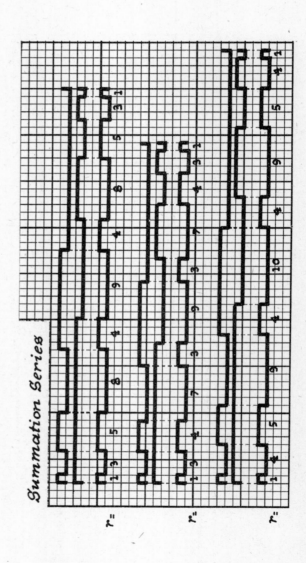

GROUPS OF VARIABLE VELOCITY

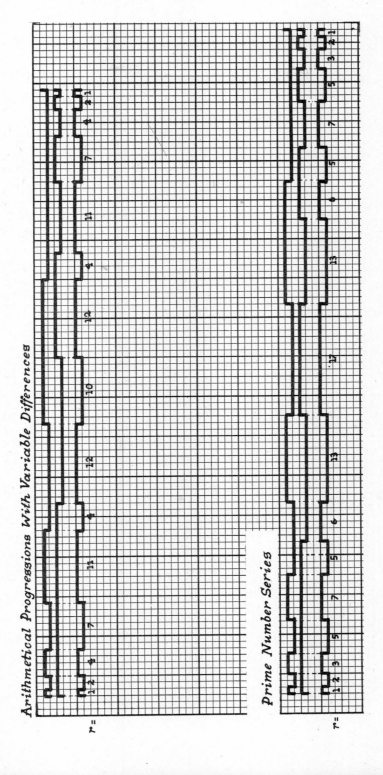

APPENDIX B.
RELATIVE DIMENSIONS [1]

---

[1] The reader is referred to Part II, Chapter 2, *Continuity*, Section E, "Ratios of the Rational Continuum," which offers a graphic presentation of the ratios producing these relative dimensions. (Ed.)

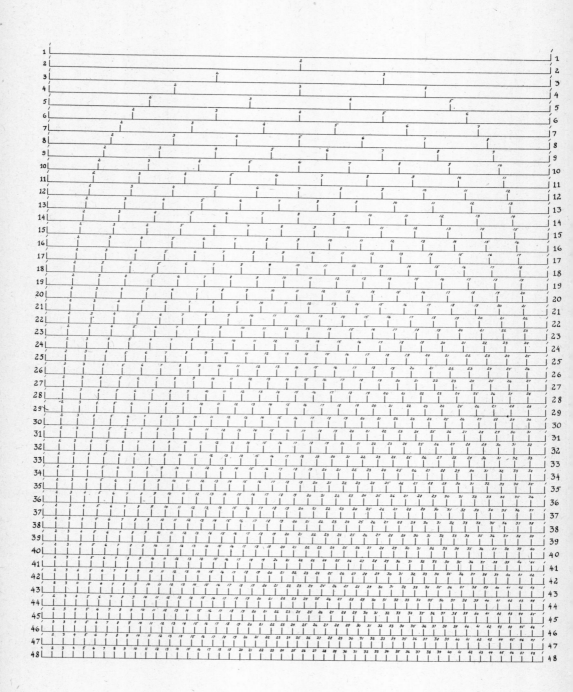

Ratios of the rational continuum.

I. Relative Dimensions

240,000		16,000	32,000
160,000		15,484	30,968
120,000	240,000	15,000	30,000
96,000	192,000	14,545	29,090
80,000	160,000	14,118	28,236
68,571	137,142	13,714	27,428
60,000	120,000	13,333	26,666
53,333	106,666	12,973	25,946
48,000	96,000	12,632	25,264
43,636	87,272	12,308	24,616
40,000	80,000	12,000	24,000
36,923	73,846	11,707	23,414
34,286	68,572	11,429	23,258
32,000	64,000	11,163	22,326
30,000	60,000	10,900	21,800
28,235	56,470	10,667	21,334
26,667	53,334	10,435	20,870
25,263	50,526	10,213	20,426
24,000	48,000	10,000	20,000
22,857	45,714		
21,818	43,636		
20,870	41,740		
20,000	40,000		
19,200	38,400		
18,462	36,924		
17,778	35,556		
17,143	34,286		

n	1/n	]	diff	n	1/n	]	diff
2.	1.00000					]	.00308
		]	.33334	26.	.07692		
3.	.66666					]	.00286
		]	.16666	27.	.07406		
4.	.50000					]	.00264
		]	.10000	28.	.07142		
5.	.40000					]	.00246
		]	.06667	29.	.06896		
6.	.33333					]	.00230
		]	.04761	30.	.06666		
7.	.28572					]	.00215
		]	.03572	31.	.06451		
8.	.25000					]	.00201
		]	.02778	32.	.06250		
9.	.22222					]	.00190
		]	.02222	33.	.06060		
10.	.20000					]	.00178
		]	.01819	34.	.05882		
11.	.18181					]	.00168
		]	.01515	35.	.05714		
12.	.16666					]	.159*
		]	.01282	36.	.05555		
13.	.15384					]	.150
		]	.01100	37.	.05405		
14.	.14284					]	.142
		]	.00951	38.	.05263		
15.	.13333					]	.135
		]	.00833	39.	.05128		
16.	.12500					]	.128
		]	.00736	40.	.05000		
17.	.11764					]	.122
		]	.00653	41.	.04878		
18.	.11111					]	.117
		]	.00585	42.	.04761		
19.	.10526					]	.110
		]	.00526	43.	.04651		
20.	.10000					]	.106
		]	.00477	44.	.04545		
21.	.09523					]	.101
		]	.00433	45.	.04444		
22.	.09090					]	.97
		]	.00395	46.	.04347		
23.	.08695					]	.92
		]	.00362	47.	.04255		
24.	.08333					]	.89
		]	.00333	48.	.04166		
25.	.08000						

* The figures in this column represent the difference between the bracketed numbers to the left. For the sake of visual simplicity, zeros are omitted after the decimal point. The full figure would be .00159 or .00097 or .01150, as the case may be. In some cases one zero is omitted, and in others, two or three zeros. By subtracting any bracketed figure from the one directly above it, the reader may quickly determine how many zeros have been omitted. (Ed.)

RELATIVE DIMENSIONS

#	Value		#	Value
4.	1.00000			
	] .20000		27.	.14812] .572
5.	.80000			] .528
	] .13334		28.	.14284
6.	.66666			] .492
	] .9522		29.	.13792
7.	.57144			] .460
	] .7144		30.	.13332
8.	.50000			] .430
	] .5556		31.	.12902
9.	.44444			] .402
	] .4444		32.	.12500
10.	.40000			] .380
	] .3638		33.	.12120
11.	.36362			] .356
	] .3030		34.	.11764
12.	.33332			] .336
	] .2564		35.	.11428
13.	.30768			] .318
	] .2200		36.	.11110
14.	.28568			] .300
	] .1902		37.	.10810
15.	.26666			] .284
	] .1666		38.	.10526
16.	.25000			] .270
	] .1472		39.	.10256
17.	.23528			] .256
	] .1306		40.	.10000
18.	.22222			] .244
	] .1170		41.	.09756
19.	.21052			] .234
	] .1052		42.	.09522
20.	.20000			] .220
	] .954		43.	.09302
21.	.19046			] .212
	] .866		44.	.09090
22.	.18180			] .202
	] .790		45.	.08888
23.	.17390			] .194
	] .724		46.	.08694
24.	.16666			] .184
	] .666		47.	.08510
25.	.16000			] .178
	] .616		48.	.08332
26.	.15384			

APPENDIX B

6.	1.00000				
	]	.14284		]	.792
7.	.85716		28.	.21426	
	]	.10716		]	.738
8.	.75000		29.	.20688	
	]	.8334		]	.690
9.	.66666		30.	.19998	
	]	.6666		]	.645
10.	.60000		31.	.19353	
	]	.5457		]	.603
11.	.54543		32.	.18750	
	]	.4545		]	.570
12.	.49998		33.	.18180	
	]	.3846		]	.534
13.	.46152		34.	.17646	
	]	.3300		]	.504
14.	.42852		35.	.17142	
	]	.2853		]	.477
15.	.39999		36.	.16665	
	]	.2499		]	.450
16.	.37500		37.	.16215	
	]	.2208		]	.426
17.	.35292		38.	.15789	
	]	.1959		]	.405
18.	.33333		39.	.15384	
	]	.1755		]	.384
19.	.31578		40.	.15000	
	]	.1578		]	.366
20.	.30000		41.	.14634	
	]	.1431		]	.351
21.	.28569		42.	.14283	
	]	.1299		]	.330
22.	.27270		43.	.13953	
	]	.1185		]	.318
23.	.26085		44.	.13635	
	]	.1086		]	.303
24.	.24999		45.	.13332	
	]	.999		]	.291
25.	.24000		46.	.13041	
	]	.924		]	.276
26.	.23076		47.	.12765	
	]	.858		]	.267
27.	.22218		48.	.12498	

See footnote on page 646.

8.	1.00000				
	]	.11112		]	.984
9.	.88888		29.	.27584	
	]	.8888		]	.920
10.	.80000		30.	.26664	
	]	.7276		]	.860
11.	.72724		31.	.25804	
	]	.6060		]	.804
12.	.66664		32.	.25000	
	]	.5128		]	.760
13.	.61536		33.	.24240	
	]	.4400		]	.702
14.	.57136		34.	.23528	
	]	.3804		]	.672
15.	.53332		35.	.22856	
	]	.3332		]	.636
16.	.50000		36.	.22220	
	]	.2944		]	.600
17.	.47056		37.	.21620	
	]	.2612		]	.568
18.	.44444		38.	.21052	
	]	.2340		]	.540
19.	.42104		39.	.20512	
	]	.2104		]	.512
20.	.40000		40.	.20000	
	]	.1908		]	.488
21.	.38092		41.	.19512	
	]	.1732		]	.468
22.	.36360		42.	.19044	
	]	.1580		]	.440
23.	.34780		43.	.18604	
	]	.1448		]	.424
24.	.33332		44.	.18180	
	]	.1332		]	.404
25.	.32000		45.	.17776	
	]	.1232		]	.388
26.	.30768		46.	.17020	
	]	.1144		]	.368
27.	.29624		47.	.17020	
	]	.1056		]	.356
28.	.28568		48.	.16664	

APPENDIX B

10.	1.00000			
	]	.9095	]	.1150
11.	.90905		30. .33330	
	]	.7575	]	.1075
12.	.83330		31. .32255	
	]	.6410	]	.1005
13.	.76920		32. .31250	
	]	.5500	]	.950
14.	.71420		33. .30300	
	]	.4755	]	.890
15.	.66665		34. .29410	
	]	.4165	]	.840
16.	.62500		35. .28570	
	]	.3680	]	.795
17.	.58820		36. .27775	
	]	.3265	]	.750
18.	.55555		37. .27025	
	]	.2925	]	.710
19.	.52630		38. .26315	
	]	.2630	]	.675
20.	.50000		39. .25640	
	]	.2385	]	.640
21.	.47615		40. .25000	
	]	.2165	]	.610
22.	.45450		41. .24390	
	]	.1975	]	.585
23.	.43475		42. .23805	
	]	.1810	]	.550
24.	.41665		43. .23255	
	]	.1665	]	.530
25.	.40000		44. .22725	
	]	.1540	]	.505
26.	.38460		45. .22220	
	]	.1430	]	.485
27.	.37030		46. .21735	
	]	.1320	]	.460
28.	.35710		47. .21275	
	]	.1230	]	.445
29.	.34480		48. .20830	

See footnote on page 646.

RELATIVE DIMENSIONS

12.	1.00000			
	]	.7696		
13.	.92304	31.	.38706	] .1290
	]	.6600		
14.	.85704	32.	.37500	] .1206
	]	.5706		
15.	.79998	33.	.36360	] .1140
	]	.4998		
16.	.75000	34.	.35333	] .1068
	]	.4416		
17.	.70584	35.	.34284	] .1008
	]	.3918		
18.	.66666	36.	.33330	] .954
	]	.3510		
19.	.63156	37.	.32430	] .900
	]	.3156		
20.	.60000	38.	.31578	] .852
	]	.2862		
21.	.57138	39.	.30768	] .810
	]	.2598		
22.	.54540	40.	.30000	] .768
	]	.2510		
23.	.52170	41.	.29268	] .732
	]	.2172		
24.	.49998	42.	.28566	] .702
	]	.1998		
25.	.48000	43.	.27906	] .660
	]	.1848		
26.	.46152	44.	.27270	] .636
	]	.1716		
27.	.44436	45.	.26664	] .606
	]	.1584		
28.	.42852	46.	.26082	] .582
	]	.1476		
29.	.41376	47.	.25530	] .552
	]	.1380		
30.	.39996	48.	.24996	] .534

APPENDIX B

14.	1.00000				]	.1407
	]	.6669		32.	.43750	
15.	.93331				]	.1330
	]	.5831		33.	.42420	
16.	.87500				]	.1246
	]	.5152		34.	.41174	
17.	.82348				]	.1176
	]	.4571		35.	.39998	
18.	.77777				]	.1113
	]	.4095		36.	.38885	
19.	.73682				]	.1050
	]	.3682		37.	.37835	
20.	.70000				]	.994
	]	.3339		38.	.36841	
21.	.66661				]	.945
	]	.3031		39.	.35896	
22.	.63630				]	.896
	]	.2765		40.	.35000	
23.	.60865				]	.854
	]	.2554		41.	.34146	
24.	.58331				]	.819
	]	.2331		42.	.33327	
25.	.56000				]	.770
	]	.2156		43.	.32527	
26.	.53844				]	.742
	]	.2002		44.	.31815	
27.	.51842				]	.707
	]	.1848		45.	.31108	
28.	.49994				]	.679
	]	.1722		46.	.30429	
29.	.48272				]	.644
	]	.1610		47.	.29785	
30.	.46662				]	.623
	]	.1505		48.	.29162	
31.	.45157					

See footnote on page 646.

16.	1.00000	
	]	.5888
17.	.94112	
	]	.5224
18.	.88888	
	]	.4680
19.	.84208	
	]	.4208
20.	.80000	
	]	.3816
21.	.76184	
	]	.3464
22.	.72720	
	]	.3160
23.	.69560	
	]	.2896
24.	.66664	
	]	.2664
25.	.64000	
	]	.2464
26.	.61536	
	]	.2288
27.	.59248	
	]	.2112
28.	.57136	
	]	.1968
29.	.55168	
	]	.1840
30.	.53328	
	]	.1720
31.	.51608	
	]	.1608
32.	.50000	
	]	.1520
33.	.48480	
	]	.1424
34.	.47056	
	]	.1344
35.	.45712	
	]	.1272
36.	.44440	
	]	.1200
37.	.43240	
	]	.1136
38.	.42104	
	]	.1080
39.	.41024	
	]	.1024
40.	.40000	
	]	.976
41.	.39024	
	]	.936
42.	.38088	
	]	.880
43.	.37208	
	]	.848
44.	.36360	
	]	.808
45.	.35552	
	]	.776
46.	.34776	
	]	.736
47.	.34040	
	]	.712
48.	.33328	

18.	1.00000			33.	.54540	
		]	.5266			] .1602
19.	.94734			34.	.52938	
		]	.4734			] .1512
20.	.90000			35.	.51426	
		]	.4293			] .1431
21.	.85707			36.	.49995	
		]	.3897			] .1350
22.	.81810			37.	.48645	
		]	.3555			] .1278
23.	.78255			38.	.47367	
		]	.3258			] .1215
24.	.74997			39.	.46152	
		]	.2997			] .1152
25.	.72000			40.	.45000	
		]	.2772			] .1098
26.	.69228			41.	.43902	
		]	.2574			] .1053
27.	.66654			42.	.42849	
		]	.2376			] .990
28.	.64278			43.	.41859	
		]	.2214			] .954
29.	.62064			44.	.40905	
		]	.2070			] .909
30.	.59994			45.	.39996	
		]	.1935			] .873
31.	.58059			46.	.39123	
		]	.1809			] .828
32.	.56250			47.	.38295	
		]	.1710			] .801
				48.	.37494	

See footnote on page 646.

RELATIVE DIMENSIONS

20.	1.00000				
		] .4770			] .1680
21.	.95230		35.	.57140	
		] .4330			] .1590
22.	.90900		36.	.55550	
		] .3950			] .1500
23.	.86950		37.	.54050	
		] .3620			] .1420
24.	.83330		38.	.52630	
		] .3330			] .1350
25.	.80000		39.	.51280	
		] .3080			] .1280
26.	.76920		40.	.50000	
		] .2860			] .1220
27.	.74060		41.	.48780	
		] .2640			] .1170
28.	.71420		42.	.47610	
		] .2460			] .1100
29.	.68960		43.	.46510	
		] .2300			] .1060
30.	.66660		44.	.45450	
		] .2150			] .1010
31.	.64510		45.	.44440	
		] .2010			] .970
32.	.62500		46.	.43470	
		] .1900			] .920
33.	.60600		47.	.42550	
		] .1780			] .890
34.	.58820		48.	.41660	

22.	1.00000				
		] .4355		]	.1749
23.	.95645		36.	.61105	
		] .3982		]	.1650
24.	.91663		37.	.59455	
		] .3663		]	.1562
25.	.88000		38.	.57893	
		] .3388		]	.1485
26.	.84612		39.	.56408	
		] .3146		]	.1408
27.	.81466		40.	.55000	
		] .2904		]	.1342
28.	.78562		41.	.53658	
		] .2706		]	.1287
29.	.75856		42.	.52371	
		] .2530		]	.1210
30.	.73326		43.	.51161	
		] .2365		]	.1166
31.	.70961		44.	.49995	
		] .2211		]	.1111
32.	.68755		45.	.48884	
		] .2090		]	.1067
33.	.66660		46.	.47817	
		] .1958		]	.1012
34.	.64702		47.	.46805	
		] .1848		]	.979
35.	.62854		48.	.45826	

See footnote on page 646.

24.	1.00000					
		]	.4000		]	.1800
25.	.96000			37.	.64860	
		]	.3696		]	.1704
26.	.92304			38.	.63156	
		]	.3432		]	.1620
27.	.88872			39.	.61536	
		]	.3168		]	.1536
28.	.85704			40.	.60000	
		]	.2952		]	.1464
29.	.82752			41.	.58536	
		]	.2760		]	.1404
30.	.79992			42.	.57132	
		]	.2580		]	.1320
31.	.77412			43.	.55812	
		]	.2412		]	.1272
32.	.75000			44.	.54540	
		]	.2280		]	.1212
33.	.72720			45.	.53328	
		]	.2136		]	.1164
34.	.70584			46.	.52164	
		]	.2016		]	.1104
35.	.68568			47.	.51060	
		]	.1908		]	.1068
36.	.66660			48.	.49992	

26.	1.00000				
	]	.3722		]	.1872
27.	.96278		38.	.68419	
	]	.3432		]	.1846
28.	.92846		39.	.66664	
	]	.3198		]	.1755
29.	.89648		40.	.65000	
	]	.2990		]	.1586
30.	.86658		41.	.63414	
	]	.2795		]	.1521
31.	.83863		42.	.61893	
	]	.2613		]	.1430
32.	.81250		43.	.60463	
	]	.2470		]	.1378
33.	.78780		44.	.59085	
	]	.2314		]	.1313
34.	.76466		45.	.57772	
	]	.2184		]	.1261
35.	.74282		46.	.56511	
	]	.2067		]	.1196
36.	.72215		47.	.55315	
	]	.1950		]	.1157
37.	.70265		48.	.54158	

28.	1.00000				
	]	.3456		]	.1890
29.	.96544		39.	.71792	
	]	.3220		]	.1792
30.	.93324		40.	.70000	
	]	.3010		]	.1708
31.	.90314		41.	.68292	
	]	.2814		]	.1638
32.	.8750		42.	.66654	
	]	.2660		]	.1540
33.	.84840		43.	.65114	
	]	.2492		]	.1484
34.	.82348		44.	.63630	
	]	.2352		]	.1414
35.	.79996		45.	.62216	
	]	.2226		]	.1358
36.	.77770		46.	.60858	
	]	.2100		]	.1288
37.	.75670		47.	.59570	
	]	.1988		]	.1246
38.	.73682		48.	.58324	

See footnote on page 646.

RELATIVE DIMENSIONS

30.	1.00000			32.	1.00000	
		]	.3235		]	.3040
31.	.96765			33.	.96960	
		]	.3015		]	.2848
32.	.93750			34.	.94112	
		]	.2850		]	.2688
33.	.90900			35.	.91425	
		]	.2670		]	.2544
34.	.88230			36.	.88880	
		]	.2520		]	.2400
35.	.85710			37.	.86480	
		]	.2385		]	.2272
36.	.83325			38.	.84208	
		]	.2250		]	.2160
37.	.81075			39.	.82048	
		]	.2130		]	.2048
38.	.78945			40.	.80000	
		]	.2025		]	.1952
39.	.76920			41.	.78048	
		]	.1920		]	.1872
40.	.75000			42.	.76176	
		]	.1830		]	.1760
41.	.73170			43.	.74416	
		]	.1755		]	.1696
42.	.71415			44.	.72720	
		]	.1650		]	.1616
43.	.69765			45.	.71104	
		]	.1590		]	.1552
44.	.68175			46.	.69552	
		]	.1515		]	.1472
45.	.66660			47.	.68080	
		]	.1455		]	.1424
46.	.65205			48.	.66656	
		]	.1380			
47.	.63825					
		]	.1335			
48.	.62490					

#	Value			#	Value	
34.	1.00000			36.	1.00000	
		]	.2862			] .2710
35.	.97138			37.	.97290	
		]	.2703			] .2556
36.	.94435			38.	.94734	
		]	.2550			] .2430
37.	.91885			39.	.92304	
		]	.2414			] .2304
38.	.89471			40.	.90000	
		]	.2295			] .2196
39.	.87176			41.	.87804	
		]	.2176			] .2106
40.	.85000			42.	.86698	
		]	.2074			] .1980
41.	.82926			43.	.83718	
		]	.1989			] .1908
42.	.80937			44.	.81810	
		]	.1870			] .1818
43.	.79067			45.	.79992	
		]	.1870			] .1746
44.	.77265			46.	.78246	
		]	.1717			] .1656
45.	.74448			47.	.76590	
		]	.1649			] .1602
46.	.73899			48.	.74988	
		]	.1564				
47.	.72335					
		]	.1513				
48.	.70822					

See footnote on page 646.

RELATIVE DIMENSIONS

38.	1.00000			42.	1.00000	
		]	.2568			] .2329
39.	.97432			43.	.97671	
		]	.2432			] .2226
40.	.95000			44.	.95445	
		]	.2318			] .2121
41.	.92672			45.	.93324	
		]	.2223			] .2037
42.	.90459			46.	.91287	
		]	.2090			] .1932
43.	.88369			47.	.89355	
		]	.2014			] .1869
44.	.86355			48.	.87486	
		]	.1919			
45.	.84436					
		]	.1843			
46.	.82593			44.	1.00000	
		]	.1748			] .2232
47.	.80845			45.	.97768	
		]	.1691			] .2034
48.	.79154			46.	.95634	
						] .2024
				47.	.93610	
						] .1958
40.	1.00000			48.	.91652	
		]	.2440			
41.	.97560					
		]	.2340			
42.	.95220					
		]	.2200			
43.	.93020			46.	1.00000	
		]	.2120			] .2135
44.	.90900			47.	.97865	
		]	.2020			] .2047
45.	.88880			48.	.95818	
		]	.1940			
46.	.86940					
		]	.1840			
47.	.85100					
		]	.1780			
48.	.83320					

APPENDIX C.
NEW ART FORMS

I. DOUBLE EQUAL TEMPERAMENT
II. RHYTHMICON
III. SOLIDRAMA

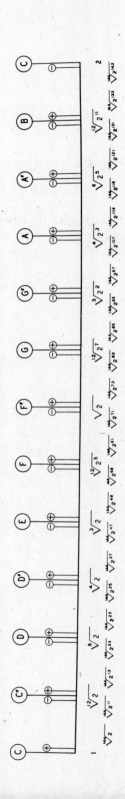

Double Equal Temperament

I. Double Equal Temperament[1]

Double Equal Temperament is a primary selective system of tuning, designed in accordance with this theory. Double Equal Temperament successfully unifies all systems of intonation used in the western world of today. It consists of the basic intervals: $\sqrt[12]{2}$, combined with micro-intervals: $\sqrt[144]{2}$, which serve as deviation units and are arranged bifoldedly in relation to each basic unit. The micro-units are best averages for all differences between the units of the $\sqrt[12]{2}$ and just intonation (natural scale). This tuning permits one to execute with a high degree of precision: twelve-unit equal temperament, mean temperament, just intonation, and the string and vocal inflections of special types of intonation (chamber, jazz, Gypsy music, etc.).

An electronic organ with micro-tuning and a specially designed keyboard was built in 1932 for the author by Leon Theremin for the performance of Double Equal Temperament.

II. Rhythmicon*

The Rhythmicon is an instrument constructed by Leon Theremin. It is the first modern instrument that composes music automatically. The present model is confined to the composition and automatic performance of rhythmic patterns in the acoustical scale of intonation. The forms of rhythmic groups produced by this instrument are the resultants of interference of generators from one to sixteen.

The author found a special use for the Rhythmicon: reproduction of the most intricate forms of aboriginal African drumming. Many phonograms have been made to illustrate this, by means of eliminating the middle and the low frequencies in both the recording and the performance.

The drum sounds obtained from the Rhythmicon are fully realistic, and the configurations reach the intricacy of completely saturated sets.

The total number of resultants is 65,535.

The Rhythmicon is in the possession of The Schillinger Estate.

---

[1] The reader is referred back to Chapter 1, "Selective Systems" of Part III, *Technology of Art Production*. Double Equal Temperament illustrates a *primary selective system* of tuning devised by Schillinger to accommodate intonations not possible in our present system of tuning. (Ed.)

*The Rhythmicon, in Schillinger's terminology, produces secondary selective systems. The reader is referred back to Chapter 1, "Systems" of Part III, *Technology of Art Production*. (Ed.)

A. The Number of Forms Produced by the Rhythmicon
(C = combinations)

$$_{16}C_1 = \frac{16!}{1!(16-1)!} = \frac{16!}{1!15!} = 16$$

$$_{16}C_2 = \frac{16!}{2!(16-2)!} = \frac{16!}{2!14!} = 120$$

$$_{16}C_3 = \frac{16!}{3!(16-3)!} = \frac{16!}{3!13!} = 560$$

$$_{16}C_4 = \frac{16!}{4!(16-4)!} = \frac{16!}{4!12!} = 1820$$

$$_{16}C_5 = \frac{16!}{5!(16-5)!} = \frac{16!}{5!11!} = 4368$$

$$_{16}C_6 = \frac{16!}{6!(16-6)!} = \frac{16!}{6!10!} = 8008$$

$$_{16}C_7 = \frac{16!}{7!(16-7)!} = \frac{16!}{7!9!} = 11440$$

$$_{16}C_8 = \frac{16!}{8!(16-8)!} = \frac{16!}{8!8!} = 12870$$

$$_{16}C_9 = \frac{16!}{9!(16-9)!} = \frac{16!}{9!7!} = 11440$$

$$_{16}C_{10} = \frac{16!}{10!(16-10)!} = \frac{16!}{10!6!} = 8008$$

$$_{16}C_{11} = \frac{16!}{11!(16-11)!} = \frac{16!}{11!5!} = 4368$$

$$_{16}C_{12} = \frac{16!}{12!(16-12)!} = \frac{16!}{12!4!} = 1820$$

$$_{16}C_{13} = \frac{16!}{13!(16-13)!} = \frac{16!}{13!3!} = 560$$

$$_{16}C_{14} = \frac{16!}{14!(16-14)!} = \frac{16!}{14!2!} = 120$$

$$_{16}C_{15} = \frac{16!}{15!(16-15)!} = \frac{16!}{15!1!} = 16$$

$$_{16}C_{16} = \frac{16!}{16!(16-16)!} = \phantom{\frac{16!}{16!0!}} = 1$$

TOTAL = 65535

NEW ART FORMS

It would take 10,922.5 hours to play all the combinations if we gave each combination the average duration of 10 seconds.

$$10,922.5 \text{ hours} = 455 \text{ days, 2 hours, 30 minutes.}$$

III. Solidrama

The eighteenth art form,[1] representing motion and transformation of solids. Its components are: time and the three spatial coordinates.

Motion of screens and solids is executed by means of a magnetic drive in the working model designed by the author. The name of the instrument is *Solidrive*.[2] This instrument gives a simultaneous performance of space-time configurations in an automatic form.

The Solidrive may be synchronized with light and sound. The present model has a 45-inch diameter and four symmetrically arranged drives. The Solidrive is in the possession of The Schillinger Estate.

A. Forms of Solidrama

(1) Motion of solids: through their *own trajectories* or through a *built-in-trajecto-form*.

(2) Motion of planes: *vertical, horizontal, curved*.

(3) Use of solids and planes as *platforms, stages*, etc.

(4) Use of solids and planes as *screens for illumination* (light projection [colors]).

(5) Use of solids and planes as *luminescent or partly luminescent forms*.

(6) Use of solids and planes as *shadow-casters*.

(7) Use of all the previous devices combined with *mirror reflexions* and *multi-reflexions*.

(8) All the previous forms combined with *diffusing screens*.

(9) Motion of solids and planes combined with *music*.

(10) Combined arts: *solids and planes, light* (colors) and *music* (recitation).

---

[1] See the table in Chapter 1 of Part II.
[2] The Solidrive is protected by patent.

APPENDIX D.

POETRY AND PROSE [1]

---

[1] In 1934 Schillinger lectured on the application of his theory to poetry and prose before the Faculty Club of Columbia University (Mathematics Dept.). The title of the lecture was "Poetry and Prose Mathematically Devised." Schillinger had planned to include a chapter on the subject in Part III of the present work. The material in this Appendix represents the beginning of such a chapter. (Ed.)

A. Elements of Poetic Structure

Different literary forms require different degrees of precision.
The structural unit of poetry may be:

      (1) a syllable;

      (2) a word;

      (3) a sentence;

      (4) a stanza; etc.

Poetic structures may be arranged into sonic and semantic scales. Sonic scales represent: syllabic configurations, with specified accents, assonance and alliteration. Semantic scales represent: direct and indirect (metaphoric) associational classifications, and are arranged through the degrees of connotational intensity (associative power).

This theory abolishes the duality of meter and rhythm, unifying both into temporal structures.

B. Rhythmic Composition of Sentences

Each plot is subdivided into a uniform scale of events (episodes). The importance of event defines temporal stress. Thus events are rhythmically arranged, each episode being expressed through a different number of sentences.

Application of the technique of expansion through growth of the determinant of a series makes it possible to extend a short story into a novel.

APPENDIX E.
PROJECTS [1]

I. BOOKS
II. INSTRUMENTS

---

[1] To present the application of his basic ideas to all the arts, Schillinger planned a series of books and instruments. His sudden death in 1943 prevented the completion of a project as fabulous as it is fraught with the most significant implications for the future of the arts. Here, in typical rhythmic form, is a graphic presentation of the project "Books" as Schillinger set it down. (Ed.)

APPENDIX E

I. Books[1]

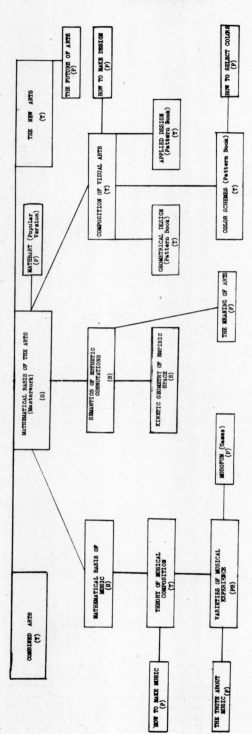

[1] S = scientific
T = technical
P = popular

II. Instruments[1]

Mechanical realization of this method is a natural consequence. Various instruments may be constructed for the automatic production, reproduction, and variation of works of art. Instruments of the analyzer type may also be constructed for the automatic testing of the esthetic quality of works of art. The following types of instruments are planned by the author in the form of engineering design and kinetic diagrams:

1. Instruments for production, variation and reproduction of industrial design. This group of instruments serves the purpose of automatic composition of design in the following forms:

   a. drawing
   b. projection
   c. printing
   d. weaving

2. Instruments for automatic variation of design, using design in the following forms:

   a. drawing
   b. films or slides
   c. fabrics

3. Instruments for automatic composition of music:

   a. limited to specified components, such as rhythm, melody, harmony, harmonization of melodies, counterpoint, etc.
   b. combining the above functions, and capable of composing an entire piece with variable tone qualities (choral, instrumental chamber music, symphonic and other orchestral music)

4. Instruments for automatic variation of music of the following types:

   a. quantitative reproductions and variations of existing music
   b. modernizing old music
   c. antiquating modern music

5. Instruments of groups 3 and 4, combined with sound production for the purpose of performance during the process of composition or variation.

6. Semi-automatic instruments for composing music. These instruments will be used as a hobby for everyone interested in musical composition, whether amateur or professional, and will not require any special training. The prospective name for instruments of this type will be "Musamaton."[2] These instruments

---

[1] Plans for the construction of various instruments are in the possession of Mrs. Joseph Schillinger. The right to construct these instruments is protected by patent, and no instrument may be constructed or used for private profit without the written consent of The Schillinger Estate.

[2] These words, coined by Schillinger to describe various instruments, have been registered, and may not be used in connection with these or other instruments without the written permission of The Schillinger Estate.

may be used for the purpose of entertainment and study, and will be suitable for schools, clubs. public amusement places, and homes.

7. Instruments of the type described in paragraph 6, but in the field of design. The prospective general name for such instruments will be "Artomaton."[2] The two fundamental types of Artomatons will be:

    a. "graphomaton"[2]—an instrument producing linear design
    b. "luminaton"[2]—an instrument producing design projected by light source

8. Projecting optical instruments with the mechanism for automatic composition of form by color, and capable of projecting the latter during the process of composing.

9. Instruments for kinetic displays, which may be used in exhibitions, permanent exhibits, stores, store windows.

10. Instruments for kinetic theatrical productions (as conceived through this theory) including kinetic stage, light, sound, scent, taste, and tactile effects.

GLOSSARY

Compiled by ARNOLD SHAW

(This glossary is limited to terms that have a special significance and that are basic in Schillinger's system of thought. Mathematical terms as such are not included. Terms that appear within a definition in bold face are explained in their alphabetical order. The index at the close of this volume indicates the pages in the text where the term occurs.)

A

a or **A** Denotes **Attack**

ⓐ Denotes the original position in **Quadrant Rotation**.

a ÷ b Signifies the **Resultant** of **Interference** between a and b. The division sign, as used by Schillinger, does not mean a divided by b, or a : b. It means **Interference**, not division, and may be algebraically represented as follows: $a \div b = b + (a-b) + (2a-2b) + (2a-2b) + (a-b) + a$. Arithmetically $4 \div 3 = 1\frac{1}{3}$. In interference $4 \div 3 = \frac{3+1+2+2+1+3}{12} = \frac{12}{12} = 1$. $R_{a \div b}$, $r_{a \div b}$, and $T_{a \div b}$ are other variant symbols of interference.

a ÷ b Symbol for the **Resultant** of another type of interference between a and b known as Interference with **Fractioning**.

ABSCISSA ROTATION. A technique for varying the pitch or spatial pattern of an art work. We begin with a graph on a plane. If we rotate the graph around its horizontal coordinate, or abscissa, we secure a cylinder in horizontal position on which the lowest and highest parts of the score or design meet.

ACCELERATION SERIES. Any series in which successive terms are the result of an increasing differential, e.g., 1①, 2②, 4③, 7④, 11⑤, 16.... When the differential decreases, the series may be known as a **Retardation Series**. **Arithmetical** and **Geometrical Progressions,** the various **Power** and **Summation Series, Natural Harmonic Series, Prime Number Series,** etc., are sometimes also known as Acceleration Series.

AMPLITUDE. The loudness or intensity of a sound is measured by the amplitude, or greatest displacement, of its air vibrations. In the graph of a sound wave, the amplitude is the distance between the time axis and the highest point of the wave.

ARITHMETICAL PROGRESSION. A series formed by the addition of a constant number to each successive term of the series: e.g., 1, 4, 7, 10, 13, 16.... in which the constant is 3. A slightly different form of this progression is encountered in the **Acceleration Series** 1, 2, 4, 7, 11, 16, 22, 29..... Here the number being added is 1, then 2, then 3; 4; 5; 6; etc. In this form of progression, the constant is the difference (1) between the successive terms in the additive group. This series is known as an Arithmetical Progression with Variable Difference.

ARTISTIC SCALE. See **Operand Group**.

ATTACK. In music, any tonal event. "Four attacks per measure" means four musical events in each measure. These events may be four single tones, four chords, or four string attacks without reference to rhythmic pattern.

ATTACK-GROUP. Simply a series of attacks considered as a unit.

AXIS OF SYMMETRY. See **Symmetry**.

B

B Symbol for balance.

ⓑ Denotes the backward position in **Quadrant Rotation**.

BI-COORDINATE ROTATION. Rotation through abscissa and ordinate. We begin with a graph, or bi-coordinate system, on a plane. By rotating the graph around the vertical coordinate, or ordinate, ($\phi\ \circlearrowright$ or $\phi\ \circlearrowleft$ in reference to time) we obtain a cylinder in vertical position. Rotating the graph around the abscissa, or horizontal coordinate, ($\phi\ \circlearrowright$ or $\phi\ \circlearrowleft$ in reference to pitch or density) produces a cylinder in horizontal position. In the first instance, beginning and terminal durations meet. In the second case, when the vertical coordinate represents pitch, the highest and the lowest pitches meet. In music, bi-coordinate rotation is a technique for varying the textural density of the composition.

BINOMIAL. An algebraic expression consisting of two elements, e. g., $a + b$. In general, any group composed of two elements.

C

ⓒ Denotes the backward and upside down position in **Quadrant Rotation**.

CIRCULAR PERMUTATION. A type of **Permutation** in which the original group or series is re-arranged one step at a time. Such re-arrangement may proceed in a clockwise $\circlearrowright$ or counter-clockwise $\circlearrowleft$ direction. 1, 2, 3, in clockwise circular permutation, becomes 2, 3, 1 and 3, 1, 2 before returning to 1, 2, 3. In counter-clockwise direction, the circular permutations are 1, 3, 2; 3,2, 1; and 2, 1, 3. Circular permutation is also known as **Displacement** or Circular Displacement.

COEFFICIENTS OF RECURRENCE. When the **Resultants** of **Interference** are used as a form of regularity, they are known as coefficients of recurrence. Such coefficients serve to control the periodicity of a given element in an art form—the number of times a color recurs in a rhythmic design or the number of times a given interval or duration recurs in a musical composition. Any rhythmic number series may serve as a coefficient of recurrence. Schillinger's procedure is to begin with a **Primary Selective System,** the color spectrum or the tuning system. When coefficients of recurrence are applied to this system, we obtain a **Color Scale** or a pitch scale. The further application of coefficients of recurrence to this **Secondary Selective System** produces a color scheme or a melody.

COLOR SCALE. A sequence of colors produced by applying some rhythmic pattern or form of regularity to the color spectrum. Such a sequence involves increasing or decreasing wavelengths, increasing or decreasing densities.

COMMON PRODUCT. In the process of deriving **Resultants**, a number obtained by multiplying the **Generators**.

COMPLEMENTARY FACTOR. The number of times a given **Generator** recurs in the process of deriving **Resultants of Interference**. This is calculated by dividing the given generator into the **Common Product** of all the generators involved.

CONFIGURATION. A pattern or rhythmic form. Schillinger uses *configuration scale* as another term for **Secondary Selective System.**

CONTINUITY. A continuity is a finite portion of a **Continuum**. It is apparent because the particular points either belong to an axiomatic class (e.g., the natural integers) or are part of a harmonic group (e.g., **Summation Series**). In terms of a graph, a continuity is formed by selecting rhythmic points along any coordinate or **Parameter**. In terms of an art form, a continuity is an ordered sequence of elements.

CONTINUUM. A system of unlimited **Parameters**, or measuring lines. In music, the total manifold of all possible frequencies. In design, the total manifold of points, lines, arcs, colors, etc.

CONTRACTION GROUP. Any complex rhythmic group in which a longer duration group is followed by a shorter duration group—both groups being derived from the same **Style**

GLOSSARY

Series. The longer group is generally the resultant of interference with fractioning ($r_{a \div b}$) while the shorter group is the resultant of simple interference ($r_{a \div b}$). An Expansion Group consists of the same groups, with the shorter preceding the longer.

COORDINATE CONTRACTION. See **Coordinate Expansion**.

COORDINATE EXPANSION. May be performed with reference either to the abscissa, thereby affecting the general component, time, or with reference to the ordinate, expressing some special art component. The process is one of multiplying the given component by a coefficient of 2, 3, 4 or more. In music, coordinate expansion is distinguished from tonal expansion in that the simple semitone units of the graph serve as the basis in the former, and some selected diatonic scale provides the basis in the latter. The reverse process of coordinate expansion is coordinate contraction, which cannot generally be achieved in music with our present tuning system. In the visual arts, coordinate expansion produces forms of optical aberration, elongation or distension such as we find in El Greco and other artists.

CORRELATION. Schillinger deals with three main types of correlation: *parallel* when two series (quantities, directions or phases) increase at the same rate; *contrary* when one series increases in value as the other decreases; *oblique* when one series remains constant while the other increases or decreases. These correlations apply to the motion of two voices in music, to the correspondence between pitch and time ratios, to the relation of density and time in the composition of density groups, and to the correlation of any two components in design.

COSINE CURVE. See **Sine Curve**.

D

④   The fourth position in **Quadrant Rotation**—forward and upside down.

Δ, δ The Greek letter Delta. Symbols for **Density**. Δ = compound density-group. $\Delta^{\rightarrow}$ = sequent compound density group.

D, d   Symbol used in composition of **Density**. d = density unit. D = simultaneous density-group and $D^{\rightarrow}$ = sequent density-group.

d_0 Denotes zero displacement in **Circular Permutation**. The zero displacement is the initial position and no displacement at all. d_1, d_2, d_3 denote successive displacements.

DENSE SET. If we take a straight line, itself finite, we can insert an infinite number of points or number values. All these points or numbers constitute a dense set. Viewed more generally, a dense set consists of all the number values, both rational and irrational, in the space-time **Continuum**. **Primary** and **Secondary Selective Systems**, which are developed on the basis of uniform symmetric ratios or non-uniform rhythmic forms, are derived from the continuum.

DENSITY. The quantity of sound per unit of time in music. In design, the number of lines, areas, arcs, colors, etc., per unit of space. In general, the criterion of judgment is the approach to a **Dense Set**. In music, Density Groups vary according to whether they involve the use of all available pitches and parts. In design, they vary as they involve all possible configurations and colors. Density groups are subject to **Phasic Rotation**, symbolized by the Greek letter ϕ. Such rotation may involve both coordinates and produce intercomposition of their phases; this process is symbolized by the Greek letter Θ.

DERIVATIVE SCALE. Any scale developed by a process of **General** or **Circular Permutation** from an original or **Parent Scale**.

DETERMINANT. The original value in a **Style Series**. $\frac{1}{t_n} \ldots \frac{1}{t_3} \ldots \frac{1}{t_2} \ldots \frac{1}{t} \ldots \frac{t}{t} \ldots t \ldots$ $\ldots t^2 \ldots t^3 \ldots t^n \ldots \frac{1}{27} \ldots \frac{1}{9} \ldots \frac{1}{3} \ldots \frac{3}{3} \ldots 3 \ldots 9 \ldots 27 \ldots \frac{t}{t}$, or $\frac{3}{3}$ in this instance, is the determinant.

GLOSSARY

DEVIATION SERIES. A series involving forms of expansion and contraction, and produced by adding a unit of growth (τ) to a constant unit (t). A term in such a series would be represented as $t = t' + \tau$. The simplest relation of t' to τ is $\tau = \dfrac{1}{t'}$. The numerator in this relationship may grow through any of the **Acceleration Series,** such as **Arithmetical and Geometric Progressions,** etc.

DEVIATION, STANDARD. A balanced binomial may be thrown out of balance through the use of a standard deviation unit. This unit is added to one term of the binomial and subtracted from the other. Formula for such a unit is τ (unit of growth) $= \dfrac{1}{t'}$ (constant unit).

DISPLACEMENT. A method of permutating the elements in a group by consecutively rearranging their order, generally in one direction. For example:

$$d_0 = a + b + c + \ldots + m$$
$$d_1 = b + c + \ldots + m + a$$
$$d_2 = c + \ldots + m + a + b$$

Such re-arrangement may be clockwise ↻ (forward) or counterclockwise ↺ (backward). Also known as **Circular Permutation** or mechanical permutation as distinguished from **General Permutation.** The initial group is known as d_0, denoting zero or no displacement. d_1 is the first displacement, d_2 the second, etc.

DISTRIBUTIVE CUBE. See **Distributive Involution.**

DISTRIBUTIVE INVOLUTION. A process of raising a binomial or polynomial to any power and arranging the product into its summary parts—in short, power differentiation. The binomial $a + b$ squared is $a^2 + 2ab + b^2$. The distributive square, however, is $aa + ab + ba + bb$. The non-distributive square of $4 + 3$ is 49. Distributively the square is $16 + 12 + 12 + 9$. Schillinger found that the distributive use of powers was extremely valuable in design and music, where they are occasionally used as **Coefficients of Recurrence.**

DISTRIBUTIVE POWERS. See **Distributive Involution.**

DISTRIBUTIVE SQUARE. See **Distributive Involution.**

DISTRIBUTIVE USE. See **Distributive Involution.**

DURATION. The time within which a sound lasts. Represented on a graph by extension along the abscissa or horizontal coordinate. Any duration may be used as a unit in developing a **Continuity** or as a phase in **Positional Rotation.**

DURATION GROUP. A group of one or more durations used as a pattern for rhythm in music.

E

E_0 Symbol for a **Pitch-Scale** in zero expansion, meaning that the scale cannot be contracted in the given tuning system. E_1, E_2, etc., indicate the first **Expansion,** second **Expansion,** etc.

EQUAL TEMPERAMENT. The present system of tuning, which serves as the basis of most Occidental music. This tuning system, like all **Primary Selective Systems,** is composed of a particular set of pitches selected from the manifold, or **Dense Set,** of all possible pitches. Equal temperament, developed by Andreas Werckmeister in 1691, involves the division of the octave into 12 pitches whose frequencies are related to each other as the logarithmic series based on the twelfth root of 2 ($\sqrt[12]{2}$). $C = 2^{\frac{0}{12}}$ or 1, $C\sharp = 2^{\frac{1}{12}}$, $D = 2^{\frac{2}{12}}$, $D\sharp = 2^{\frac{3}{12}}$, $E = 2^{\frac{4}{12}}$, $F = 2^{\frac{5}{12}}$, $F\sharp = 2^{\frac{6}{12}}$, $G = 2^{\frac{7}{12}}$, $G\sharp = 2^{\frac{8}{12}}$, $A = 2^{\frac{9}{12}}$, $A\sharp = 2^{\frac{10}{12}}$, $B = 2^{\frac{11}{12}}$, $C = 2^{\frac{12}{12}}$ or 2. The frequency of each octave is twice the original tone, and remaining pitches are derived by octave duplication. The frequency of concert A today is 440 vibrations per second.

GLOSSARY

EXPANSION. A process applied particularly to **Pitch-Scales** by which the successive pitches are increased by some constant factor. When the results are determined tonally (i.e., diatonically) we have a Tonal Expansion. When they are determined exactly, we have a **Geometrical Projection** or expansion. See **Coordinate Expansion.** Applicable also to durations in music, and to quantities, extensions, densities, etc., in design.

EXPANSION GROUP. See **Contraction Group.**

F

F denotes factorial continuity while f denotes fractional continuity.

FACTORIAL. Refers to the organization of an art form as a whole while fractional concerns individual units. The rhythmic structure of the whole is a factorial problem whereas the rhythmic organization within a unit, a bar in music or a unit-area in design, is fractional.

FACTORIAL-FRACTIONAL CONTINUITY. A progression developed by inserting terms between the existing terms in a **Normal Series.** The values at the left side of such a progression determine the fractional continuity (unit rhythms). The values on the right side control the **Factorial** continuity (work as a whole). With the determinant $\frac{2}{2} = \frac{t}{t}$, the following factorial-fractional continuity may be developed: ..... $\frac{1}{16}$ .... $\frac{1}{8}$ .... $\frac{1}{4}$ .... $\frac{1}{2}$ .... $\frac{2}{2}$ ..... 2 ..... 4 ..... 8 ..... 16 .....

FAMILY-SERIES. See **Style-Series.**

FIBONNACI SERIES. See **Summation Series.**

FRACTIONING. The process of dividing a rhythmic group into fragments, generally on the basis of polynomials in the **Style Series.** Also the specific process of producing resultants known as **Interference with Fractioning.** The **Resultant** of $3 \div 2 = 2+1+1+2$. The resultant of $3 \div 2$ with fractioning $= 2+1+1+1+1+1+2$.

FREQUENCY. The number of vibrations per second of a vibrating medium. The frequency of vibrations for middle C is 256 per second, while the frequency of C one octave higher is 512 vibrations per second.

FUNDAMENTAL TONE. The tone produced by the vibrations of the whole string or column of air, as distinguished from overtones or **Partials** produced by vibrations of portions of the string or air column.

G

GENERAL PERMUTATION. See **Permutation.**

GENERATOR. A series of numbers (generally composing a uniform group) used in combination with another uniform series to produce a new non-uniform group. The new group is known as the **Resultant,** and the process, as **Interference.** The numbers may be converted into durations, pitches, etc., in music and extensions, angles, colors, etc., in design.

GENETIC FACTOR. Schillinger regards phasic or periodic differences as the genetic factor in art. If we take two uniform groups of durations with such differences and synchronize them, **Interference** occurs. The **Resultant** is a non-uniform group, which may become the basis of general or specific art components.

GEOMETRICAL PROGRESSION. Various number series formed by multiplying each successive term by a constant number: e.g., 1, 3, 9, 27, 81...... The multiplier is 3. In 3, 6, 12, 24, 48, 96.... the multiplier is 2. **Power Series** frequently have the appearance of geometrical progressions, but they are formed by a process of raising the initial term to its different powers: e.g., 3, 9, 27, 81...., which is evolved through 3^1, 3^2, 3^3, 3^4 .....

GEOMETRICAL PROJECTION. The general technique for varying art forms. Includes 1) **Quadrant Rotation,** 2) **Coordinate Expansion** or Geometric Expansion, and 3) **Coordinate Contraction.**

H

HARMONIC GROUP. A group whose numbers, durations, extensions, etc., display **Harmonic Relations** or a perceptible rhythmic regularity.

HARMONICS. In acoustics, the subcomponents of a sound wave, accessory to the fundamental tone. Also known as **Partials** and produced by physical factors that transform a simple sound wave (**Sine wave**) into one of more complex form. See **Natural Harmonic Series**.

HARMONIC RELATIONS. Refers to the rhythm, design, or underlying regularity of a work of art.

HIGHER ORDER. An operation of any type performed on the *results* of a previous operation of the same type: e.g., squaring a square, cubing a cube, or grouping a group. Schillinger also employs a variation technique that he calls **Permutations of a Higher Order**. If we designate a and b as our original elements, a_1 and b_1 become elements of the first order. $a_2 = a_1 + b_1$ and $b_2 = b_1 + a_1$, are elements of a second or higher order. $a_3 = a_2 + b_2$ and $b_3 = b_2 + a_2$ is a further permutation of a higher order, etc., until $a_n = a_{n-1} + b_{n-1}$ and $b_n = b_{n-1} + a_{n-1}$.

HYBRID SERIES. A series that involves a mixture of types. The **Determinants** of a uniform series are associated with the natural integer sets — $\frac{2}{2}$ series, $\frac{3}{3}$ series, $\frac{4}{4}$ series ..... $\frac{n}{n}$ series. When the members of a family belong to one series, it is considered pure. When the members belong to several series, it is considered hybrid.

I

INTERFERENCE. One of the basic concepts of Schillinger's approach to the arts, interference is a phenomenon observed in all fields of wave motion. When two sound waves or two light waves of varying **Periodicity** cross, they interfere or combine to form a third wave that is the summation of the two. Schillinger regarded this phenomenon as a process of growth and evolution. For him, it became the fundamental procedure for combining two or more uniform periodicities to produce a new non-uniform group. This procedure, which he calls interference, is the foundation stone of the Theory of Regularity and Coordination in the present work, and of the Theory of Rhythm in the *Schillinger System of Musical Composition*.

INVARIANT OF INVERSION. Denotes the axis or element around which an inversion or **Geometrical Projection** is performed.

INVERSION. See **Geometrical Projection** and **Quadrant Rotation**.

INVOLUTION. See **Distributive Involution**.

K

KINETIC ARTS. The arts that evolve in time, music, poetry, etc., as contrasted with those that exist in space (Static Arts). The latter, which are generally perceived by sight, are crystallized in space and do not change in time. The kinetic arts, which do change in time after the process of composition has been completed, are perceived by such sense organs as touch, smell, taste, and, of course, hearing. Television and motion pictures are kinetic arts.

L

LOGARITHM. When a number is expressed as a power of ten, the exponent of that power is called the logarithm to the base ten.

LOGARITHMIC SCALE. A series of points within two rational limits, selected by a process of determining logarithmic ratios. Within the limits, b and a, intermediate points of uniform symmetry may be designated as follows: $\sqrt[n]{\frac{a}{b}} \ldots \sqrt[n]{\left(\frac{a}{b}\right)^2} \ldots \sqrt[n]{\left(\frac{a}{b}\right)^3} \ldots \sqrt[n]{\left(\frac{a}{b}\right)^n}$.

M

m denotes determinant of a series.

MAJOR GENERATOR. The larger of two numbers or durations in the process of producing **Resultants** by **Interference**.

MANIFOLD. Mathematically, a number of elements related under one system. In this system, a **Secondary Selective System**, or one that is the result of selection and serves as the limit of another selection. In music, e.g., a scale. In design, a color pattern, etc.

MECHANICAL PERMUTATION. See **Displacement**.

METHOD OF SERIES. A process of developing a related group of terms by inserting connecting terms between given limits. Two terms become three, three become five, five become nine, etc. Suppose we begin with Y and B. This becomes

$$Y\ldots\ldots G\ldots\ldots B$$
$$Y\ldots YG\ldots G\ldots BG\ldots B$$

A basic technique in Schillinger's approach.

MINOR GENERATOR. The smaller of two numbers or periodicities combined in the process of **Interference** or **Synchronization**.

MONOMIAL. An expression consisting of a single term.

MONOMIAL PERIODICITY. A group or series composed of one number repeated several times. Two monomial or uniform periodicities, in the process of **Interference**, produce a non-uniform periodicity.

N

NATURAL HARMONIC SERIES. The series of overtones produced by a vibrating string or air column. The original tone is known as the **Fundamental Tone,** and the overtones are sometimes called Partials, because they are produced by vibrations of parts of the string. When the C two octaves below middle C on the piano is struck, the fundamental tone is C with a vibration frequency of 64. The first overtone or partial is C an octave higher, with a frequency of 128. Next overtone is G, a fifth higher; then C, a fourth higher, with a frequency of 256; then E, a major third higher; G, a major third higher; B♭, a minor third higher; C, major second; D, E, F♯, G, A♭, B♭, B; and the 16th in the series, C, five octaves above the fundamental.

NATURAL SERIES. The series of natural integers: 1, 2, 3, 4, 5, 6, 7, 8, 9; and the natural fractional series, $\frac{1}{2}, \frac{1}{3}, \frac{1}{4}, \frac{1}{5}, \frac{1}{6} \ldots \ldots \frac{1}{n}$. Schillinger also refers to these two series as the Natural Harmonic Series, which should not be confused with the Overtone series or **Natural Harmonic Series**.

NOMOGRAPHY. The graph method of notation. In general, any scientific system for notating natural phenomena.

NORMAL SERIES. A series of numbers or terms evolved by the **Method of Series**.

O

OPERAND GROUP. Mathematical term for a melody, design pattern, color pattern, used as the basis of an art work. In music, we have a tuning system (or **Primary Selective System**) from which we select certain symmetric or asymmetric points known as scales (or **Secondary Selective System**) from which in turn we abstract a melodic pattern (or operand group).

ORDINATE ROTATION. See **Bi-Coordinate Rotation**.

OVERTONE SERIES. See **Natural Harmonic Series**.

P

ϕ The Greek letter Phi. Refers to process of phasic rotation. See **Density**.

p. r. Used to denote **Positional Rotation**.

PARAMETER. A measuring line. In the Cartesian sense, the bi-coordinate system of measuring lines (horizontal and vertical) that make up a graph. In the Einsteinian sense, a system of measuring lines involving four coordinates: x_1 (width), x_2 (depth), x_3 (height), and x_4 (time). Schillinger employs a special version of Einstein's correlated time-space parameters. Time and space are regarded as *general* parameters, while *special* parameters are established to measure each physical component of an art object appealing to a different organ of sensation.

PARENT SCALE. A scale selected from our tuning system, from which a series of related scales are derived through **General** and **Circular Permutation**.

PARTIALS. See **Natural Harmonic Series**.

PERIODICITY. The recurrence in time or space of some phenomenon — in time, of a note or sound; in space, of an area, angle, arc, color, etc. The simplest periodicity is **Monomial**: $a + a + a + \ldots + a$ where each consecutive term is equivalent in extension (space) or duration (time). In *uniform* periodicity, the repetitious factor may consist of one or *more* terms.

PERIODIC MOTION. Commonly known as rhythm, periodic motion is regarded by Schillinger as the basis of all art forms and of phenomena in the world of nature.

PERMUTATION. The process of modifying or varying the elements in a group by rearranging their order or sequence. General permutation (also known as logical permutation) yields all possible variations since it does not proceed in one direction as in the case of **Circular Permutation (or Displacement)**. A polynomial of 5 terms yields only 10 circular permutations (5 clockwise and 5 counter-clockwise), but 120 general permutations ($1 \times 2 \times 3 \times 4 \times 5$). Both figures include the original order as a permutation.

PERMUTATIONS OF A HIGHER ORDER. See **Higher Order**.

PHASIC ROTATION. A process for varying the density of a sound or space composition by rotation or displacement along the time-axis, density-axis, or both: ⊂ ⊃ ( ) ( ). Also known as phasic displacement.

PITCH. The highness or lowness of a tone as determined by the number of vibrations per second, or **Frequency**. In our tuning system, concert A has 440 vibrations per second. The brilliant tone of the Boston Symphony is sometimes attributed to the fact that it tunes its concert A at 444 vibrations per second.

PITCH-SCALE. A sequence of pitch-units selected from our tuning system according to some definite pattern of increasing or decreasing frequencies. In Schillinger's terminology — a **Secondary Selective System** as distinguished from our tuning system, which is a **Primary Selective System**. Schillinger does not follow the traditional system of classifying scales as major and minor, but devises an exhaustive system of classification under four headings:

> Group One: Scales with one tonic and consisting of any number of notes up to and including a range of one octave.
>
> Group Two: Scales with one tonic, a range of more than one octave, and evolved by **Expanding** the scales in the first group.
>
> Group Three: Scales of more than one tonic, a range of not more than one octave, and constructed symmetrically.
>
> Group Four: Scales of more than one tonic, a range of more than one octave, and constructed symmetrically.

POSITIONAL ROTATION. A general technique for varying the simultaneity phase (ordinate), the continuity phase (abscissa), or both, of an art form. Involves the application of **Circular Permutation** to structures and sequences. In design, p. r. may produce superimposed images. In music, p. r. produces variations in textural density. See **Bi-coordinate Rotation** and **Phasic Rotation**.

POLYNOMIAL. An algebraic expression containing two or more terms. In general, any group made up of more than one element.

POWER SERIES. An acceleration series formed by raising the initial term to its successive powers. 2, 4, 8, 16, 32.... is evolved by squaring 2, cubing 2, etc.

PRE-SET. The process of selecting the components of any art form in advance of the actual composition, the specific components being chosen according to the effects desired.

PRIMARY SELECTIVE SYSTEM. In music our tuning system is the result of a selection from the complete manifold (or dense set) of all possible frequencies. Schillinger designates it a primary system. Mathematically speaking, such a system is a series of fixed points selected symmetrically from the sound **Continuum**. A **Secondary Selective System** in turn involves a selection from the primary system. In music, certain pitch-units are chosen to form a scale. In design, certain colors or angles or arcs are selected to form color scales, etc.

PRIME NUMBER SERIES. An acceleration series composed of numbers that are divisible only by 1 or themselves: 1, 2, 3, 5, 7, 11, 13, 17....

PROGRESSIVE SYMMETRY. See **Symmetry**.

Q

QUADRANT ROTATION. One of the fundamental techniques for varying an art form. The original form denoted as ⓐ, is developed backwards in time ⓑ, then backwards in time and upside down ⓒ, and finally forward in time and upside down ⓓ. The relation of these four forms to each other is made clear by seeing them in relation to the four quadrants of a graph

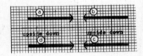

R

r Denotes **Resultant**.

$R_{a \div b}$. See a ÷ b.

RATIONAL SET. Refers to numbers, specifically the positive and negative integers 0, 1, 2, 3, 4, 5.... and positive and negative fractions $\frac{1}{2}, \frac{1}{3}, \frac{1}{4}, \frac{1}{5}$, etc. Contrasts with **Real Set**, which includes irrational numbers $\sqrt{2}, \sqrt{3}, \sqrt{5}, \sqrt[3]{2}$, as well as rational. A real set of numbers refers to all the number values necessary to describe a **Dense Set**.

RATIONALIZATION. In design, the process of inscribing a structure, originally evolved in an unbounded space, within a boundary. Also the process of subjecting a spatial form to the tendency of its own ratio.

REAL SET. See **Rational Set** and **Dense Set**.

REGULARITY. The simplest form of regularity is uniformity or **Monomial Periodicity**. In its more complex forms, regularity results from the combination of different uniformities. Regularity is another word for **Rhythm** and the process of producing non-uniform forms of regularity is **Interference**.

REGULARITY and COORDINATION. The foundation of Schillinger's approach to the arts is his Theory of Regularity and Coordination, also known as the Theory of Rhythm in the composition of music. This theory embraces the manipulation of duration, frequency, intensity and quality factors underlying the process of artistic creation. The principle of rhythm governs the periodicity (recurrence) of art components, from the most elementary, such as attacks in music and extensions in design, to the most complex questions of form.

RESULTANT. The product of the **Interference** of two or more uniform periodicities of different frequencies, brought into synchronization. Such resultants may be obtained either by graphs or direct computation, and are the parent-shapes of all rhythms. The interfering periodicities are known as **Generators**.

RETARDATION SERIES. A series of numbers in which successive terms are the result of a decreasing differential. See **Acceleration Series**.

RHYTHM. To Schillinger, rhythm is any form of periodic motion that may be discerned in natural, social and artistic phenomena. Such periodic motion is reducible to a mathematically conceivable **Regularity**, i.e., to numbers. Broadly speaking, rhythm is of two types: *fractional* when it refers to units of a composition or design, and **Factorial** when it refers to the structure of the whole.

RHYTHMIC CENTER. The center is generally regarded as a point that is equally distant from the circumference of a circle or the sides of a rectangle. In contrast, the *rhythmic* center is the result of ratios, not simple measurement.

RHYTHMICON. See Appendix.

RUBATO. In the performance of music, an alteration in the duration of given notes, literally by *stealing* time from other notes. Mathematically, rubato is accomplished by introducing a standard unit of deviation and throwing a balanced binomial or polynomial out of balance. $4 + 4$ thus becomes $5 + 3$, and $2 + 2 + 2 + 2$ becomes $1 + 3 + 2 + 2$, etc.

S

S Used to denote Structural group in **Positional Rotation**. **s** denotes structural unit. **S** also used to denote **Synchronization**, the coordination of durations, or **Symmetrization**, the coordination of extensions, as in $S(a:b)$.

Σ The Greek letter Sigma. Used to denote a large structure or something compounded of a group of smaller structures. In music, the sigma is the same as the **Expansion** of a scale, save that it is a chord, not a sequence. If we use numbers to denote the pitch-units (1, 2, 3, 4, 5, 6, 7, 8), the Σ (or E_1) would be 1, 3, 5, 7, 2(9), 4(11), 6(13). Σ is also used to denote a series.

SCALES OF LINEAR CONFIGURATION. In design, the equivalent of pitch-scales in music: a sequence of linear forms selected from the time-space **Continuum** according to some specific pattern of increasing or decreasing quantities, extensions, etc.

SECONDARY SELECTIVE SYSTEM. Secondary systems vary in density and become identical with the primary system when they reach a saturation point. Full saturation of the primary system in turn produces a **Dense Set**. In short, the creation of music or design involves a series of selections. From the **Continuum** or total manifold, we select a series of pitch-units which constitute our tuning system. From this primary selective system, we make a second selection which yields a series of scales or secondary selective systems. From these we make further selections, which result in melodies, color schemes, spatial designs, etc.

SELECTIVE SYSTEM. See **Primary Selective System** and **Secondary Selective System**.

SEMANTICS. For Schillinger the study, not of the evolution of meaning, but of meaning itself. More specifically, study of the relationship between form and sensation. Semantic requirements define the purpose or meaning of a work of art.

SERIES OF NATURAL DIFFERENCES. A series in which each succeeding term is based on an increasing differential or difference: e.g., 1, 2, 4, 7, 11, 16, 22, 29...... The difference between 1 and 2 is 1; between 2 and 4 = 2; between 4 and 7 = 3; etc. Also known as an **Acceleration Series**.

SINE CURVE. The simple curve described by a pendulum swinging from a fixed point and leaving a trace on a sheet of paper moving perpendicular to its line of motion. The curve, which is regular and symmetrical, represents the motion of a simple sound wave, sometimes referred to as a *sine wave*. Sine curves differ considerably in appearance depending upon the relation of **Amplitude** (height of a crest above the axis) and **Wave-length** (crest and trough). A curve of the same general form but differing in phase by a quarter period, or 90°, is known as a cosine curve.

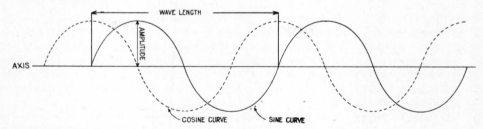

SOLIDRAMA. See Appendix C.

SPLIT UNITS. The result of dividing a single unit in a **Resultant** or other harmonic series by some divisor. The process of selecting the units to be split is itself controlled by **Coefficients of Recurrence** or **Permutation**.

STATIC ARTS. See **Kinetic Arts**.

STYLE SERIES. A series of numbers that serves as the source of all rhythmic patterns: $\frac{1}{n}$ .....
.... $\frac{1}{5}, \frac{1}{4}, \frac{1}{3}, \frac{1}{1}, \frac{2}{1}, \frac{3}{1}, \frac{4}{1}, \frac{5}{1}$ .... $\frac{n}{1}$. Each of these yields a series of **Determinants** known as the $\frac{2}{2}$ series, $\frac{3}{3}$ series, $\frac{4}{4}$ series, ... $\frac{n}{n}$ series. Rhythmic patterns derived from any one group of determinants are known as a family, which accounts for Schillinger's occasional use of *Family-Series* as an alternative for style series. If we take the $\frac{5}{5}$ series, for example, we may generate a full family of related **Duration Groups**. $\frac{5}{5}$ may be split into $4+1$ or $3+2$. The former binomial synchronized with itself produces the trinomial $1+3+1$. This trinomial synchronized with its various permutations produces $1+2+1+1$ and $1+1+2+1$. All of these groups bear an apparent relation to each other and the parent series. Translated into design or music components, such a series produces an identifiable style. **Hybrid** style is based on several series, or several determinants, instead of one, which produces a pure style.

SUMMATION SERIES. A series in which each number is the sum of the preceding two. From an esthetic point of view, the most useful summation series are $1+2=3, 5, 8, 13, 21, 34, 57......1+3=4, 7, 11, 18, 29, 47, 76......$ Also known as Fibonacci Series after Leonard of Pisa, who developed these additive series. Leonard was the son of Bonaccio (filius Bonacci), which gave the series its name.

SYMMETRIC SCALES. **Pitch-Scales** in music and **Color Scales** or **Scales of Linear Configurations** in design, marked by uniformity of sequence and reversibility of pattern. Schillinger classifies pitch-scales into four groups, two of which are symmetrical. Group III, confined to a range of less than one octave, contains, 2, 3, 4 and 6 tones between the units of each scale. Group IV, with a range of more than one octave, contains 8, 9, 10 and 11 semitones between units. Any of these scales is reversible: e.g., C — E — A♭ — C, with 4 semitones between each tone, contains these specific tones whether one begins at the bottom or top.

SYMMETRY. The quality of a sequence, series, or form whereby the pattern is reversible. The point around which the group balances or may be reversed is designated as the axis of symmetry. In progressive symmetry the successive terms or elements, instead of being identical in relation to a central point of reference, are marked by factors of growth: e.g., $A + (A+B) + (A+B+C) + (B+C) + C$.

SYNCHRONIZATION. The process of making two or more duration groups (or number series) coincide as to period of time. When the two series are not identical, **Interference** results.

T

T or t Symbol for time. Small t also used to designate deviation unit. Capital T is also used to denote a definite periodic portion of continuity along a line or extension, as $T = t_1 + t_2 + + t_3 + \ldots \ldots t_n$. In this formula, small t denotes duration units.

$T_{a \div b}$. See $a \div b$. Used interchangeably with $R_{a \div b}$.

τ The Greek letter Tau. Denotes a unit of growth or a unit of deviation in the development of asymmetrical patterns.

Θ The Greek letter Theta. Used to denote a compound rotation group in the control of **Density**.

TEMPORAL SCALE. Any sequence of durations abstracted from the uniform set of natural integers. Such a sequence may be developed through the process of **Interference**, by the application of **Coefficients of Recurrence**, through **Permutation**, or **Quadrant Rotation**.

TONIC. The root, starting point, or first tone of a Pitch-Scale. In Schillinger's classification of scales there are systems of multiple tonics, e.g., two-tonic system (C and F♯), three-tonic system (C, E, A♭), four-tonic system (C, E♭, G♭, A), six-tonic system (C, D, E, F♯, G♯, A♯), and the twelve-tonic system (employing all the semitones in our equal temperament tuning system).

TUNING SYSTEM. See **Equal Temperament**; also **Primary Selective System**.

U

UNIFORMITY. A concept associated with and evolved from various axiomatic proportions. In the field of reasoning: the system of count based on the so-called natural integers. In the field of sight: the system of measurement of space. And in the field of hearing: the clock system of time measurement. Uniformity may be regarded as a special case of **Periodicity**.

V

VARIABLE VELOCITY, GROUPS OF. A general term for groups embodying various forms of acceleration and growth, such as **Arithmetical and Geometrical Progressions, Power Series, Summation Series,** etc.

W

WAVE-LENGTH. The length of the crest and trough of a wave produced by a vibrating medium —sound, light or radio. While color in general depends upon wave-length, pitch is determined by **Frequency,** or the number of waves per second. The oscillograph and phonodeick are used to make photographic records of sound waves.

INDEX

abscissa, 44, 234, 255
   phases, 215
   rotation, 255
abstract cinema, 68
acceleration, positive and negative, 44
accent, 41
accentuation, 24
Aeolian mode, 224
African drumming, 665
albumen, 254
algebraic powers, 173–183
American Indians, 98
American Institute for the Study of Advanced Education, 46
American Institute of the City of New York, Mathematics Division, 46
American Library of Musicology, 98
American Musicological Society, 46
American jazz, 140
angle units, 329–330
angle-perspective, 232
angles, symmetric construction of, 312
Angstrom units, 272
Annixter, Paul, 173
"antiquation," 253
Archipenko, 68
archipentura,
   Archipenko, 62, 67, 68
architecture, Gothic, 57
arcs
   alternating sin movement of, 320, 321
   automatic continuity of, 341
   moving in constant alternating direction, 317, 318
   moving in constant direction, 314, 315
   variable direction of, 336
   variable lengths of, 324, 328
A Rebours (Huysmans), 80
Aristotle, 5, 13, 14, 27
Aristoxenes, 14
arithmetical progression, 636
   series, 87, 88
art
   ancient Egyptian, 89
   ancient Greek, 89
   creation and criteria of, 30–32
   development of, 11
   European, 43
   experimental, 33
   imitative, 35
   kinetic, 67
   static, 63
art and evolution, 7–10
art and nature, 3–6
art creation, intuitive period of, 5
   pre-esthetic period of, 4, 5
   rational period of, 5
art for art's sake, 17
art forms, 8, 10
   complex, chart for combination of, 79
   complex heterogeneous, 78
   complex homogeneous, 74, 75
   compound, 54
   correspondences between, 80–84
   diagram of, 60
   first group, 58–71
   individual table of, 59
   kinetic, 74
   measurable quantity of, 6
   new, 663–667
   technique of engineering in, 6
art, mathematics and, 38–47
art of audible sound, 64
art of audible word, 64
art of smellable odor, 65
art of tastable flavor, 66
art of texture of visible surface, 67
art of touchable mass, 65
art of visible light, 66
art of visible pigment, 66
art product, development of, by method of normal series, 55
   definition of, by method of series, 51–55
art production, the technology of, 3, 4
"artistic differential," 84
Artomaton, 674
arts, combined, production of, 429
arts, fine, 41
   history of, 17
   kinetic (musical), 61, 62, 63
   liberal, 41
   physical source of the, 18–22
   static (plastic), 61, 62
   technical, 41
Asia, music of, 221
attack group, synchronization of, 402, 403

auditory and visual forms, correlation of, 432–444
Aztec, 34

Bach, J. S., 16, 45, 55, 236, 237, 247, 400
balance, 184–192
barcarolles, 14
Barr, Professor, 10
beauty
  mathematical definition of, 84
  psychological definition of, 83
Beethoven, Ludwig van, 238, 239, 247, 400
Berg, 247
binomial, 95, 676
  distributive cube of, 563–565
  distributive square of, 502–519
  elements, 168, 169
  forms of standard deviation in, 153–156
  groups, balanced, 140
  groups, unbalanced, 140
  growth of trinomials, 169, 170
  series, 96
Biological Bases of Evolution of Music, The (Ivan Kryzhanovsky), 184
bithematic composition, 398
"blues," 26
B Minor Sonata (Liszt), 247
Brancusi, 70

calculus, differential, 365
Cartesian manner, 4
Caruso, Enrico, 20, 22
cathode-ray tube, 38
Catholic liturgical music, **186**, 219
Caucasian rug patterns, 221
Cenozoic age, 8
Cezanne, 7
Chansons Madecasses (Ravel), 223
Charleston, 140, 141
  music, 52
"Charles V, 1533" (painting), 233
Chinese pentatonic scales, 224
Chopin, 54, 247
chord structures, 12
chord progressions, 42
Chromatic Fantasy and Fugue (J. S. Bach), 247
cinema, 68, 215
cinematic design, time-space unit in, 429
"clavilux," 22
cochlea, 58
coefficient groups, 42
color
  audition, 80
  interpretation, **51**
  origin, 346
  scales, 194, 274, 346–362
  sectors, distribution of, 346
  sequence, 215
color-axis, 346
  angle of, 346
Columbia University, Faculty Club of (Mathematics Dept.), 669
combinatory analysis, 173
Compton, 365
Conditioned Reflexes (Ivan Pavlov), 58
configuration, 27, 39, 40, 45
  scales, 44
consistency, 43
contemplative music, 14
continuity, 72, 85–108, 146, 215, 253, 260
  composition of, 267
  coordinate components, 85
  definition of, 85
  factorial-fractional, 92–98, 173
  harmonic, 157
  parametral components, 85
  perfect form of, 265
  rhythm of, 400
  of rotary groups, 254
continuum, 51–84, 676
  class, 85
  definition of, 57
  psycho-physiological, 146
  space-time, 365
  spatial, 39
contraction patterns, 28
contrapuntalists, 240
coordinate expansion, 44, 244–254
cortis arch, 58
Cortis organ, 27
counterpoint, 161, **215**, **255**
  rhythmic, in graphic form, 161
cradle songs, 14
crescendo, 146
crystallization, 186
  of event, 184–192
crystals, harmonic structure of, 4

Dada: *Surrealism and Fantastic Art*, exhibit, 33
Dali, 36
dance, 71
  African, 73
  Asiatic, 73
  Balinese, 224
De Divina Proporzione (Luca Pacioli), 32
Debussy, 12, 45, 247

delta, 254, 255, 261
  phasic rotation of, 262–266
  variants of, 258
Democritus, 365
"dense set," 39, 40, 41, 43, 273
density, 45, 184
  composition of, 254–270
  in proportion to mobility, 266
  scheme of, 255
  variable, composition of from strata, 266–270
density-groups, 45
  binomial, 256, 257
  compound, 254, 258, 261
  monomial, 256
  permutation of, 260
  polynomial, 257
  simultaneous, 254
  sequent, 254
  smaller than Δ, 259
density-time relations, 256
density-unit, 254, 256, 266
derivative series, 113
Des Esseintes, 80
Descartes, 365
design analysis, 375–390
design,
  geometrical variation of, 373
  ornamental, African, 43
  ornamental, Asiatic, 43
  linoleum, 389, 390
  problems in, 397, 398
  production of, 363–398
  rug, 387
  stereometric, 74
  wall paper, 388
design, linear, 73, 74
  elements of, 367
  variation of, 369
design, rhythmic, 365–374
  definition of, 365
  evolution of, through elements of, 372
determinant, 98–107
  series with one, 101
deviation
  in binomial or polynomial, 153, 154, 156
  series of, 149–153
diffused polygon, 32
"diminished 7th" chord, 247
Dirac, Professor, 7
discontinuity, 218
discontinuous 180° arcs, 313
displacement, 158–162
  graphic representation of, 159, 160

displacement, 163
  arranged in simultaneity, 159, 160
distributive
  cubing, 415
  involution, 43, 173, 416–418
  involution groups, 502–635
  involution in linear design, 418–421
  properties, 173
  squaring, 414
"Djanger," 224
Double Equal Temperament, 273, 274, 665
Duerer, 7
duration group, 402–404, 678
  distribution of, 400–402
  synchronized, distribution of, 404–405
duration-units, 255
Dynamic Symmetry (Jay Hambidge), 32
dynamic symmetry, theory of, 89

ecclesiastic music, 14
Eddington, 365
Eiffel Tower, 31
Einstein, Albert, 56, 73, 365
Einsteinian manner, 4
El Greco, 45, 245
electro-magnetic behaviour, 56
electro-mechanical synthesis, 47
Empire State Building, 30, 250
equal temperament system, 273, 277
equilibrium, unstable, 184–192
esthetic experience
  varieties of, 14, 15, 16
esthetic expression, the semantics of, 3, 4
esthetic satisfaction, 84
esthetic symbols, nature of, 25
Euclidean geometry, 73
evolutionary series, method of interference, 43
expansion
  coordinate, 244–254
  geometrical, of harmonized melody, 253
  group, 248
  patterns, 28
  pitch, 248–251
  tonal, 248
expansion-contraction series, 148
exponent scale, 272

factorial continuity,
  determination of, 93
factorial-fractional continuity, 679
Fall of the House of Usher (Watson and Webber), 69
family series, 679

Fibonacci (Leonard of Pisa), 33
"Fibonacci series," 88, 679
folklore, 98
formalism, 14
Forms of Primitive Music (Helen Roberts), 98
forms of unity
  logical association, 38
  perceptive auditory association, 38
  perceptive visual association, 38
fractional series, natural, plotted, 87
fractional continuity, determination of, 93
Franciscan Order, 219
frequency, 19, 41, 147
fresco, 67
Fugues (Bach), 55
futurists, 67

gamelan, 3
General and Special Theory of Relativity (Albert Einstein), 56
general determinant,
  cubing of, 185
generator
  in music, 52
  in photography, 52
geometrical expansions, 244, 251, 252
geometrical inversion, 232, 236, 238, 240
geometrical progression, 636
  series, 88
geometry, 345
  fluent, 57
Gershwin, George, 238
Gibbon, 173
Golden Gate Bridge, 30
Goldschmidt, 4
graph, musical, 244
graph, unit of measurement of, 244
"graphomaton," 674
Greeks, ancient, 52
  music of, 223

Hambidge, Jay, 32, 89
Handel, 45, 247
harmonic accompaniment, 255
harmonic contrasts, 43
  coordination and continuity of, 173
harmonic motion, simple, 109
  graphed, 110
harmonic ratios, 442
harmonic relations, 34, 35, 146
  and harmonic coordination, 41–44
harmonic series, natural, 247
harmonic waves, 365

harmonic whole, 35
harmonics, 24, 680
harpsichord, 55
Haydn, Joseph, 238, 250
Hebrews, 186
Hellenic, 43
Helmholtz, Hermann Ludwig Ferdinand von, 24, 57
Herzog, George, 25
Hindemith, 45, 247
Hinton, 74
History of Music (Karl Stork), 25
honeycomb, 67, 221
  cells, 43
Hungary, folk dances of, 146
Huysmans, 80
hymns, revolutionary, 14

Indians, Oklahoma, music of, 220
Institute of Light, Grand Central Palace, New York, 69
instrumental group, synchronization of, 406, 407
instrumental interference, composition of resultant of, 408, 409
integer series, fractional, 86–87
  natural, 85, 86
  natural, plotted, 86
intensity, 19, 42
interference, 365, 447
  of periodicities, 366
interior decorating, 67
intonation, 25
International Exposition in Paris (1937), 16
inversions, geometric, 234, 235, 253
inverted symmetry, 140
involution
  distributive, 274
  groups, 43
  series, 44
Isolde's Love-Death (Wagner), 28
Ivan the Terrible Killing His Son (Repin), 33

Kandinsky, 33, 36
Kant, 58
Kasner, 4
kinetic art of light projected on 3-dimensional or 2-dimensional screen in motion, 70
kinetic art of visible mass, 71
kinetic art of visible pigment, 71
kinetic art of visible pigment transforming on moving surface, 68, 69

kinetic art of visible texture of surface or volume, 71
kinetic arts, 680
kinetic design, production of, 414–427
kinetic geometry, 50, 345
kinetic light of visible art projected on plane surface, 68
kinetic linear design, elementary illustration of, 442
kinetic sequence of image, 215
kinetic visual forms, 36
Klee, 33
Klein, 68
Krenek, 247
Kryzhanovsky, Ivan, 184

Leibnitz, 365
Leonardo da Vinci, 32
"Les Fontaines Lumineuses," Paris International Exposition, 1937, 70
Liebling, Leonard, 238
li-ki, 6
linear configurations, scales of saturation, 385, 386
linear design, 42
  compositions in, 392–396
  definition of, 363
  distributive involution in, 418–422
  elements of, 363, 364
  illustration of proportions, 245
  rhythmic groups in, 335, 336
Lipchitz, Jacques, collection (Paris), 233
Liszt, Franz, 52, 247
Lobachevsky, 73
logarithm, 680
logarithmetic contraction of time, 232
  dependence of ratios, 39, 273
  selection, 272
Lorentz, 365
love songs, 14
lumia, 16, 22
"luminaton," 674

Mach, 40
Madagascar, 223
magic, 11, 37
major genetic factor, 447
Marinetti, 61
master patterns, 44, 45
Mathematicians Faculty Club of Columbia University, 46
Mathematics Museum of Teachers College, Columbia University, 46
melodic curve, 23

melodic trajectory, 23
melody, 255
  analysis of, 23
  definition of, 24, 29
  dualism of problem of, 23
  function of, 25
  geometrical expansion of, 251
  insufficiency of, 25
  semantics of, 25–28
memorial rites of the ancient Chinese, 6
Mendel, 43
metronome, 274
Metropolitan Museum of Art, 246
Michael Angelo, 32
Michelson, 365
military music, 52
Miller, D. C., 64
Millikan, 365
mimicry, 11
Minkowsky, 56, 365
minor genetic factor, 447
mobility, density in proportion to, 266
"modernization," 253
Modigliani, 245
"Mogen Dovid," Hebrew, 186, 219
monodic musical culture, 399
monomial, 94
Moonlight Sonata (L. van Beethoven), 247
Moore, Prof. Douglas, 28
mosaics, mobile, 68
motion picture, 36
Mozart, 400
Musamaton, 673
Museum of Modern Art, 33, 246
music
  African, 43
  Arabian, 26
  Asiatic, 43
  analysis of dramatic qualities of, 57
  Chinese, 25, 41
  elements of, 432
  European, 140, 219
  expression, 232
  Greek, ancient, 223
  medicinal application of, 14
  modern school, 247
  natural sources of, 13
  of civilized world, averages in, 97
  of civilized world, series of averages in, 98
  production of,, 399–413
  real meaning of, 26, 27
  reflexological origin of, 26
musical instruments, development of, 11

natural fraction and integer series plotted together, 90
natural harmonic series, 636
natural integers, 38, 43
natural integer and fraction series plotted together, 90
natural series
  fractional continuity, 92
  fractional continuity, rational ratios of, 91
Navajo sand paintings, 36
Neanderthal man, 8
New School for Social Research, 46
Newton, 365
New York University, 46
New York World's Fair, 16
Niagara Falls, 67, 70
1922 (Piano Suite) (Hindemith), 247
nocturnes, 14
notation, 232
Novalis, 15

operand group, 272
optical illusion, 73
optical perspective, 146
optical projection, through extension of ordinate, 244
orchestral writing, 255
ordinate, 255
  phases, 215
  rotation, 255, 681
organ, electronic, 665
organic art, 32
  nature of, 32–34
Origin of Music, The (Karl Stumpf), 13
oscillations, 52, 109
  transverse, of a pendulum (illus.), 109
oscillogram, 23, 56
oscillograph, 38
overtones, 146

Pacioli, Luca, 32
paint, luminescent, 66
"painted sculpture," 69
painting, 67
  Navajo sand, 36
parameter, 56, 57
  correlated, 4
  frequency, 64
  general, 146
  intensity, 64
  quality, 64
  special, 146
parametral interpretation of a system, 56–58

Paris International Exposition, 70
"partials," 24, 682
pastorals, 14
Pavlov, Ivan, 27, 58
"pelog," 224
pendulum, 109
pentacle, 10
pentagonal form, sequence of angles in, 220
perception, 39
  esthetic and selective, 30
periodic motion, 37
periodic regularity, 41
periodic series
  binomial, binomial relations of, 129–134
  binomial, polynomial relations of, 136, 137
  polynomial, binomial relations of, 137, 138
  polynomial, polynomial relations of, 129–140
  practical application in art, 140–141
  synchronization of, 129
periodic waves, 41
periodicities
  monomial, polynomial relations of, 123–129
  monomial, synchronization of three, 124
  monomial, synchronization of several, 123, 125
  monomial, synchronization of two, 112, 113
  simultaneous monomial, 112–123
periodicity, 27, 109–157, 288, 682
  consecutive displacement of, 120
  derivative, 114, 120, 121
  factorial, 174
  fractional, 174, 175
  in general parameters, 146
  in phases, 109
  in special parameters, 146
  monomial, 110
  monomial, relation of two (illus.), 111
  monomial, of sector radii, 339
  of angles, 293–313
  of angles, binomial, 297
  of angles, monomial, 293
  of angles, trinomial, 301, 312
  of arcs and radii, 324–328
  of dimensions, 284–292
  of expansion and contraction, 146–156
  of radii and angle values, 314–323
  of rectilinear segments, binomial, 287, 288
  of rectilinear segments, trinomial, 288
  synchronization and derivative, 115
permutation, 42, 141, 158–172
  binomial, 168
  circular, 158, 215

INDEX

general or logical, 162
  in mechanical sequence, tables of, 164–168
  mechanical, 162, 163
  of a higher order, 168
  of four elements, mechanical scheme for, 172
  of sequent density-groups, 260–261
Persia, 221
Persian popular songs, 224
phase, 109
  arrangement, 340
phase-units, 40
phasic differences, 18
phasic rotation, 682
  of delta, 262–266
phonograms, 665
photography, 67
Phrygian scale, 240
Physical Optics (Robert W. Wood), 80
"Piano Sonata No. 8" (Ludwig van Beethoven), 238, 239
Picasso, Pablo, 33
Pissarro, 32
pitch, 24, 682
  axis, 252
  contraction of, 247, 248
  discrimination, 54
  expansion of, 244–248
  frequencies, 40
  geometrical expansion of, 45
  musical, 51
pitch-coordinate, 45
pitch-scales, 42, 682
  Arabian, 223
pitch-unit, 244, 247, 255
  scale of, and corresponding expansions, 253
planets, 223
planimetric clusters, 69
planimetric linear design, 73
Plato, 14, 27, 52
"poetic image," 29
"Poetry and Prose Mathematically Devised," 669
poetic structure, elements of, 670
poetry, structural unit of, 670
"pointillism," 35, 67
Politeia (Plato), 14, 52
polygons, closed, conceived as monomial periodicity of angles, dimensions and directions, 344–345
polynomial
  distributive cube of, 586
  distributive **property of,** 173

  distributive square of, 542
  forms of standard deviation in, 153
  groups, balanced, 140
  series, 96
Porgy and Bess (George Gershwin), 238
positional rotation, 45, 215–218, 423–428, 682
  definition of, 215
  dimensionality of, 215–218
  in design, 215
  in music, 215
positive and negative values, series, 97
power series, 637
powers, general treatment of, 179–183
primary selective series, 42
primary selective systems, 38, 39, 40
prime number, 10
  series, 90, 637
progressions
  arithmetic, 44, 636
  geometric, 44, 636
progressive series, 44
Prokofieff, 252
proportionate distribution within rectangular areas, 414–418
Ptolemy, 365
"pyramids," 400
Pythagoras, 15
Puerto Rico, dance, songs of, 141

quadrant rotation, 44, 232–243, 272
quintinomials
  cube of, 589–613
  squares of, 543–561

range-contraction, 252, 253
Raphael, 52
ratio and rationalization, 193–214
ratio of radii and scale of curvature of arcs, 328
ratio, uniform, series, 277
ratio-realization of space, 193
rational behaviour, 193
rational continuum, ratios of, 107–108
rational composition, 193
rational set, 272
rational values, factorial-fractional continuity of, 97
Ravel, 12, 223
real set, 272
rectangle, rationalization of, 195–214
rectilinear segments, 302, 304, 305
  infinite series, 288–290
  symmetric contruction of, 312
reflexological origin of music, 26

regularity
  and coordination, theory of, 34, 41, 49
  law of, 109
  pragmatic validity of theory of, 45–46
Repin, 33
Republic (Plato), 27
resultant, 44
revolutionary songs, 14
rhythm, 37, 39, 41, 147, 400, 684
  attacks of, 42
  coordinators of, 400
  definition of, 109
  intensities of, 42
  of durations, 42, 274
  of variable velocities, 636, 637
  qualities of, 42
  renaissance of, 400
  sequences of chord progressions of, 42
  spatial, 431
  temporal, 431
  the laws of, 4
  theory of, 265
rhythmic
  center, 336, 337, 338, 370, 371
  composition of sentences, 670
  durations, 141
  groups in linear design, 329–336
  patterns, 11, 12
  reality, expansion-contraction as a, 148
  resultants, 275
  resultants, with fractioning, 275–276
  series of deviation, 149
Rhythmicon, 274, 665
  -produced forms, 666, 667
Riemann, 56, 73, 365
Rimington, 68
Roberts, Helen, 98
Rome, 221
root-tone, 252
rotary phases, sequence of, 262
rotation
  bi-coordinate, 255
  of phases, 255
  scales of, 262
rotation-phase, compound, 254
rubato, 153
Russia, European, music of, 221
Russian folk music, 140, 223

"St. Anthony of Padua" (painting), 233
Saint-Martin, 15
Sappho, 7
scale of twelve hues, 351–362
scale units, 40

scales, 44
  Aeolian, 223
  assymetric, 278
  Balinese, 224
  color, 274, 346–362
  derivative, 42
  Dorian, 223
  expanded, 278, 280
  Hindu, 39
  Javanese, 39, 223, 224
  Locrian, 223
  Lydian, 223
  Mixolydian, 223
  of linear configuration, 274, 684
  of linear configuration and area, 284–345
  Phrygian, 223
  pitch, 274, 277, 278, 279, 280, 281, 282
  sonic, 670
  symmetric, 278, 281, 282, 283
  symmetrical, 278
  temporal, 274
Schillinger System of Musical Composition, The (Joseph Schillinger), 23, 232, 244, 254, 277, 674
Schlegel, Friedrich, 57
scholasticism, 14
Schopenhauer, 15
science and esthetics, 1–47
Science of Musical Sounds, The (D. C. Miller), 64
screen (illustration), 429
screen, distributive squaring of, 430, 431
sculpture, 70
sea urchin, 220
Seashore, Prof. Carl, 58
secondary selective series, 42
secondary selective systems, 46, 684
sector radii, 313
selective systems, 273–362, 684
  primary, 273, 274
  secondary, 273, 274
"selenders," 224
semantics, 23, 25, 26
semitone, 244
Sensations of Tone (Hermann von Helmholtz), 57
sensory perceptions, 58
sequence of simultaneity, 161
serenades, 14
serial determinant, 43
series of natural differences, 89–90
series of prime numbers, 90
Seurat, Georges, 32, 35
Shakespeare, 7

Shaw, Arnold, vii, 675
shiva, 34
sight, sense of, 68
Σ sigma, 267, 268
simple harmonic motion, 365
simultaneity, 215
    rhythm of, 400
sine curve, 110, 684
sine waves, 18, 24, 27, 109
"slendro," 224
sliding pitch, 146
"slow motion," 33, 146
snowflake formations, 222
solidrama, 71
    forms of, 667
Solidrive, 71, 667
sound, 164
    animal, 13
    biological factors of, 13
    cinema, 54
    description of, 58
    frequency, 54
    waves, 58, 109, 146
sounding texture, 254
sources of art of music, 5–6
space theory, 50
"space-time continuum," 39, 40, 41
space-time relations, 58
spatial design, 146
spatial kinetic configuration, 43
spectrum, 51, 146
    full, 346, 351
    full (illus.), 347
Spencer, 15
Sposalizio (Franz Liszt), 52
Sposalizio (Raphael), 52
star (illustration), 221
starfish, 43, 220
static art
    of 3-dimensional visible mass, 70
    of visible light placed inside 3-dimensional spatial form, 69
    of visible pigment covering surfaces of 3-dimensional form, 69
    of visible texture of 3-dimensional forms, 69
static optical forms, 63
Steiner, Ralph, 69
Stony Indians
    expressions in song, 25
Stork, Karl, 25
strata, 255
    composition of variable density from, 266–270

structure
    density of, 184
    in stable equilibrium, 184
Stumpf, Karl, 13
style determinants, 141
summation series, 34, 44, 88, 89, 637
superimposition, 267
supersonic waves, 37
Surf and Seaweed (Ralph Steiner), 69
surrealism, 7, 33
symmetrization, 447
symmetric parallelisms, 219–222
symmetric scales, 685
symmetry, 34, 219–231
    bifold, 223
    dynamic, 32, 34
    esthetic evaluation on the basis of, 223–224
    progressive, 157
    rectangular (extensions), 224–231
synchronization, 112, 447, 686
    binary, 274, 447–457
    binary and ternary, 445–501
    binary, with fractioning, 458–475
    double process of, 145
    of a motive, 134
    of a musical motive, 135
    of different rhythmic series, 144–146
    of rhythmic series with itself, 141–143
    of second order, 141–146
    ternary, 274, 476

Tactilism (Marinetti), 61
tangent trajectories, 4
Teachers College of Columbia University (Depts. of Mathematics, Fine Arts and Music), 46
telecasting, 47
television, 16, 47
temporal plasticity, 397
temporal structures, coordination of, 399–401
thematic
    components, 146
    entity, 140
    motif, 42
    textures, 255
Theory of Regularity, 42
Theory of Relativity, 56, 73
Theremin, Leon, 58, 665
theta, 254
time
    as general parameter, 71–74
    positions implied by conception of, 232
    psycho-physical, 72

Titania Palast, Berlin, 69
tonic, 686
trajectory, 73, 74
transformer, 52
transmitter, 53
triads, 241
trinomials
   distributive cube of, 568–576
   distributive square of, 520–541
tuning, 277
   Javanese, 221
tuning fork, 24, 38
tuning system, 273, 274
   contraction of, 45
Two-Part Inventions (J. S. Bach), 236, 237

ultra-violet rays, 66
uniformity, 38, 39
   and primary selective systems, 38–41
unstable equilibrium, 254

variable velocity groups of, 274, 636–641
variation and composition, techniques, 44–45

velocities, 44
   constant, 147
vertical coincidence, 340
Vinci, Leonardo da, 16
visual and auditory forms, correlation of, 432–444
visual art, 45, 232
visual kinetic composition, elements of, 432

Wagner, 28
Watson, 69
wave-length, color and sound, 21
wave motion, 37
Webber, 69
Webern, Anton von, 247, 250, 252
Whitlock, 4
Wilfred, Thomas, 22, 68
With the Greatest of Ease (Paul Annixter), 173
Wood, Robert W., 80
Woolworth, 65

"zer ef kend," 219, 220, 223

DATE DUE

#47-0108 Peel Off Pressure Sensitive